MANUAL OF PLASTICS ANALYSIS

MANUAL OF PLASTICS ANALYSIS

T. R. CROMPTON
Formerly of Shell Research Co.
Cheshire, United Kingdom

PLENUM PRESS • NEW YORK AND LONDON

Library of Congress Cataloging-in-Publication Data

On file

ISBN 0-306-45912-4

A Division of Plenum Publishing Corporation
233 Spring Street, New York, N.Y. 10013

http://www.plenum.com

10 9 8 7 6 5 4 3 2 1

PREFACE

The book aims to cover all analytical aspects using both chemical and physical methods. Its contents derive from the author's experience as previous Head of the Analytical Research Department concerned with plastics at a major plastic producer and are also based on a complete review of world literature on the subject.

The subject matter is broken down into six chapters, each dealing with a particular aspect of polymer analysis as shown below. The book offers the student and general reader a complete picture of all aspects of polymer analysis, including not only the polymer itself, but also deliberately added non-polymeric processing chemicals used during manufacture, chemicals added to improve the polymer properties during service life and impurities such as water and processing solvents, unreacted monomers etc.

Chapters 1 and 2 deal, respectively with the analysis of a wide range of polymers and copolymers, and under each polymer discusses the determination of volatiles, water and various types of polymer additives. Numerous detailed methods, many previously unpublished, are included for carrying out these analyses. Also included in Chapter 2 is a discussion of the available methods for determining total unsaturation, ester groups, amide and amino groups, sulphonyl groups, carbonyl groups, nitrile groups, alkyl groups, hydroxyl, alkoxyl, carboxyl, epoxy and anhydride groups. This is supported by examples of results obtained in this type of analysis. The chapter concludes with a section on the occurrence of end-groups in polymers and their determination.

In Chapter 3 are discussed the determination in polymers and copolymers of non-metallic and metallic elements whether these be major constituents of the polymer, catalyst residues, adventicious impurities or pigments. A wide variety of chemical and physical methods of analysis are discussed in detail. Chapter 4 is devoted entirely to copolymers and discusses chemical and physical methods available for the determination of comonomer ratios in those materials.

Methods discussed in Chapters 1 and 2 for the determination of additives are usually concerned with the determination of a single additive, e.g. an antioxidant. Many polymer formulations contain several additives which it may be required to determine. In these cases, as discussed in Chapter 5, in order to avoid interference effects it is usually necessary to apply a chromatographic procedure to a solvent extract of the polymer in order to obtain fractions containing the different additives which are then separately determined without interference by appropriate chemical or physical methods.

Methods are discussed utilising thin-layer chromatography, high performance liquid chromatography and gas chromatography, used in conjunction with an infrared or ultraviolet spectroscopic or mass spectroscopic finish.

Frequently, the identification of an unknown plastic can be made on the basis of simple physical tests such as solubility in solvents, odours produced on ignition, etc. Such tests will be discussed in as much as they provide a simple practical means of identification. In Chapter 6 the principles of infrared spectroscopy and pyrolysis - gas chromatography will be discussed and infrared and pyrograms presented for a wide range of polymers and copolymers showing how they may be used to identify unknown polymers by the fingerprint approach.

Detailed analytical methods are presented in Chapter 7.

The book aims to cover all analytical aspects using both chemical and physical methods and is intended as a manual for use in the laboratory.

The book offers the student and general reader a complete picture of all aspects of polymer analysis, including not only the polymer itself, but also deliberately added non-polymeric processing chemicals used during manufacture, chemicals added to improve polymer properties during service life, and impurities such as water and processing solvents, unreacted monomers etc.

ACKNOWLEDGEMENTS

The author wishes to express his gratitude to the publishers of various journals for permission to reproduce the following illustrations.

Figure 1.2	L, Ho.F.F. Analytical Chemistry 45 603 (1973). American Chemical Society.
Figures 5.1 - 5.3	Squirrell, D.C.M. Analyst (London) 106 1042 (1981). Royal Society of Chemistry, London.
Figure 6.4	Dennig, R. Kunststoffe 75 8 (1985). Carol Hauser Verlag. Munich.
Figure 6.5	May, R.W. Pearson, E.F. Porter, J. Scothern, M.D. Analyst (London) 98 364 (1973). Royal Society of Chemistry, London.
Figures 6.6 and 7.45	Folmer, O.F. Analytical Chemistry 43 1057 (1971). American Chemical Society.
Figure 7.4	Soncek, J. Jelinkova, E. Analyst (London) 107 623 (1982). Royal Society of Chemistry, London.
Figure 7.10	Allen, B.J. Elsea, G.M. Keller, K.P. Kinder, H.D. Analytical Chemistry 49 741 (1977). American Chemical Society.
Figure 7.11	Frankoski, S. P. Siggia, S. Analytical Chemistry 44 507 (1972) . American Chemical Society.
Figure 7.15	Anderson, D.G. Isakson, K.E. Snow, D.L. Tessari, D.J. Vandeberg, J.T. Analytical Chemistry 43 894 (1971). American Chemical Society.
Figure 7.16	Korenaga, T. Analyst (London) 106 40 (1981). Royal Society of Chemistry, London.
Figure 7.17	Tittareli, P. Zerlia, T. Calli, A. Ferrari, G. Analytical Chemistry 55 220 (1983). American Chemical Society.
Figure 7.27	Krishen, A. Analytical Chemistry 44 494 (1972). American Chemical Society.
Figure 7.28	Johnson, D.E. Lyeria, J.R. Horikawa, T.T. Pederson, L.A. Analytical Chemistry 49 77 (1977). American Chemical Society.
Figures 7.29 and 7.30	Blackwell, J.T. Analytical Chemistry 48 1883 (1976). American Chemical Society.

Figures 7.34 - 7.36	Majors, R.E. J. Chromatographic Science 8 338 (1970). Preston Publications Inc, New York.
Figures 7.38 - 7.40	Juvet, R.S. Smith, L.S. Li, K P. Analytical Chemistry 44 49 (1972). American Chemical Society.
Figure 7.6	Crompton, T.R. Myers, L.W. Plastics and Polymers p205 June (1968). Pergamon Press, Oxford.
Figures 7.7 - 7.9	Udris, J. Analyst (London) 96 130 (1971). Royal Society of Chemistry, London.
Figure 7.37	Rudewicz, P. Munson, B. Analytical Chemistry 58 358 (1986). American Chemical Society.
Figures 7.41 - 7.44	Anderson, D.M.W. Analyst (London) 84 50 (1959). Royal Society of Chemistry, London.

CONTENTS

PART 1, REVIEW OF METHODOLOGY

CHAPTER 1 POLYMER ANALYSIS

PART 2, ANALYTICAL METHODS

7.1 POLYMER ANALYSIS

CHAPTER 1

POLYMER ANALYSIS

1.1. POLYETHYLENE AND POLYPROPYLENE

1.1.1. Residual Volatiles and Water

The solid polymer head space analysis technique developed by Crompton and Myers[1] has been applied to the identification and determination of residual polymerization solvent residues in polyethylene and polypropylene.

Polyethylene is not appreciably soluble in organic solvents and the determination of existing volatile impurities cannot be conveniently carried out by the analysis of solutions of the polymer. Many organic solvents will extract such volatiles from polyethylene, but these are likely to be lost during the solvent removal stage prior to gas chromatographic analysis. A further problem is that extraction procedures cannot be used for investigating those volatiles that are not inherently present but are produced only when the polymer is heated. The following examples illustrate the use of this method for examining both types of volatiles released from polyethylene.

The technique was used to study the effect of temperature on the nature of the volatiles released. To investigate this, polyethylene was heated for 15 min in air at 125°, 150°, 175° and 200°C and the volatiles were examined by gas chromatography. Figure 1.1 shows the chromatograms of the volatiles released at 125° and 200°C and Table 1.1 gives the peak height ratios of the major components (1 - 7) released at four temperatures. Replicate runs indicated good reproducibility.

The results in Table 1.1 show that the only significant variation of peak height ratios with temperature occurs when component 1 is compared with components 2 - 7 and component 2 is compared with components 4 - 7. As the temperature increases to 200°C components 1 and 2 increase while components 3 - 7 decrease (Fig 1.1). Components 1 and 2 were eluted in the C_2 - C_4 hydrocarbon region and components 3 - 7 were eluted coincident with the major components of the polymerization solvent known to be used in this polyethylene manufacturing process. These observations suggest that between 125°C and 200°C there is some thermal degradation of the polymerization solvent to C_2 - C_4 hydrocarbons.

Food and drink containers extruded or moulded from polyethylene may occasionally possess unpleasant odours which are likely to taint the packaged product and are unacceptable to the consumer. In one such case it was found that by heating a sample of an odour-producing polyethylene for 15 min at 200°C under helium the chromatogram of the liberated volatiles contained certain peaks which were absent from the corresponding chromatogram from a polyethylene which produced non-odourous food containers. The temperature of 200°C was chosen to simulate the extrusion temperature.

Method 1 at the end of this chapter has been used[2] for determining water and volatiles in polyolefins in which the solid sample is placed in a gas chromatograph sample loop which is then heated under controlled conditions and the volatiles swept by helium carrier gas on to a gas chromatographic column prior to estimation.

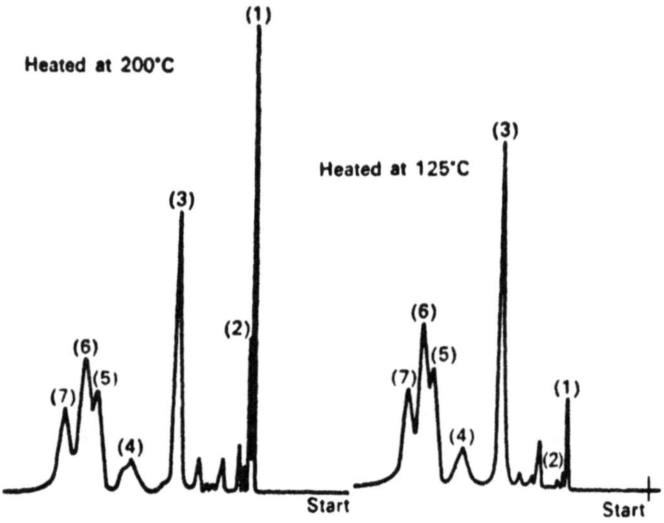

Figure 1.1. Chromatograms of volatiles liberated from polyethylene heated at different temperatures for 15 minutes in air. Chromatographed on 200 ft x 1/16 ID dibutyl-phthalate-coated copper column at 30°C and 100 ml/min helium flow, with flame ionization detection.

Table 1.1 - Peak Height Ratios of Volatiles Liberated by Polyethylene at Various Temperatures

Component Number Ratio	Peak Height Ratio			
	125°C	150°C	175°C	200°C
1:2	4.4	5.4	3.1	3.1
1:3	0.3	0.3	0.3	1.7
1:4	2.1	3.2	2.4	14.6
1:5	0.7	1.0	0.8	4.7
1:6	0.5	0.8	0.6	3.5
1:7	0.9	1.2	1.0	5.9
2:3	0.06	0.06	0.08	0.06
2:4	0.5	0.6	0.8	4.6
2:5	0.2	0.2	0.3	1.9
2:6	0.1	0.1	0.2	1.1
2:7	0.2	0.2	0.3	1.9
3:4	8.1	9.0	9.1	8.3
3:5	2.9	3.1	3.1	2.7
3:6	2.1	2.3	2.3	2.0
3:7	3.4	3.7	3.8	3.4
4:5	0.4	0.3	0.3	0.3
4:6	0.3	0.2	0.3	0.2
4:7	0.4	0.4	0.4	0.4
5:6	0.7	0.7	0.7	0.8
5:7	1.2	1.2	1.2	1.3
6:7	1.7	1.6	1.6	1.7

Conventional weight loss methods for determining water in polymers have several disadvantages not the least of which is that any other volatile constituents of the polymer such as solvents, dissolved gases, volatile substances produced by decomposition of the polymer or additives therein are included in the determination. The Karl Fischer method described in method 2 for the determination of water in polyalkenes has the advantage that it is absolutely specific for water. In addition, by controlling the temperature of the sample, for example, by temperature programming the sample, information can be obtained regarding the rate of release of water from the polymer and its dependence on sample temperature.

Evolved gas analysis [372 - 374] has been evolved to study the chemical nature of the volatiles produced by breakdown of a polymer when it is heated to various controlled temperatures. The polymer is heated in a thermogravimetric analyser and the volatiles lead into a gas chromatograph for identification and determination.

1.1.2. - Additives

1.1.2.1 General Discussion

In order to appreciate fully the techniques which have been developed for the analysis of additives in polymers, it is necessary to be familiar with the difficulties involved in such an undertaking and also with the chemical and physical properties of the additives themselves.

Most of the analytical problems arise from three factors; the situation of the additive in a more or less insoluble polymer matrix, the high reactivity and low stability of many types of additives, especially antioxidants, and the low concentrations of additives present in many instances in the polymer matrix. The first factor severely limits the choice of analytical techniques that can be applied to the sample without prior separation of the additive from polymer, a procedure which is itself hindered by the nature of the polymer matrix. In addition, any extract of the polymer is liable to contamination by low molecular weight polymer "wax" which may interfere with subsequent analysis and is difficult to remove.

The second and third factors mentioned above combine to make the handling of extracts an exacting job if quantitative information is required. Antioxidants, particularly, are labile unstable compounds, forming complex decomposition products; this considerably complicates interpretation of analytical data and any loss of material by decomposition is liable to be a significant since the quantities present are initially so low. It is recommended that polymer extracts are kept in actinic glassware and used for subsequent analysis without delay. If any storage of solutions is necessary, this should be done under nitrogen, in the dark and in a refrigerator to minimize the effects of oxygen, light and heat on any labile compounds present. Lorenz et al[3] have published data on sample changes during handling of antioxidant extracts, including losses during concentration by evaporation. Generally, however, this aspect of additive analysis does not seem to have received the consideration it deserves.

Apart from these factors which complicates the processing of the sample, there are others which complicate the interpretation of the data obtained, the principle ones being the wide range of additives used nowadays in polymer technology, which make positive identification difficult by all but the most sophisticated analytical techniques, the presence of several types of additives in a single polymer formulation, e.g. plasticizers, ultraviolet stabilizers, slip agents and possibly two antioxidants, one for processing and one for service, may all be present in a single formulation and, finally, depending on processing history and age of the polymer, the possibility that additive decomposition products may also be present to complicate the analytical problem in hand. The latter type of additive decomposition should be distinguished from that occurring during analytical processing. For example, a particular type of polymer additive may undergo partial thermal degradation during extrusion operations involved in its manufacture and then during analysis may degrade by another route under the influence of light.

To summarise then, the determination of additives in polymers presents the analyst with some difficult problems. Only small concentrations are present, complex mixtures may be involved and, moreover, frequently the mixture is of compounds of completely unknown type. Most methods for the determination of additives in plastics come under one of three categories:

i. Direct examination of polymer, ie. non-destructive testing
ii. Examination of solvent extracts of polymers
iii. Examination of volatiles released upon heating the polymer.

An example of direct examination is the examination of the polymer film by infrared or ultra-violet spectroscopy or of thicker sections of polymer by attenuated total reflectance (ATR) infrared spectroscopy.[375] Such techniques have severe limitations in that, because the additive is in effect heavily diluted with polymer, detection limits are usually well above the low concentration of additive present and this method is only applicable if the additive has distinct sharp absorption bands in regions where the polymer itself shows little or no absorption. In-situ spectroscopic techniques are not likely to be of value, then, in the analysis of samples of unknown composition. If known amounts of additive can be incorporated into additive-free polymer, however, these techniques are likely to be extremely useful in the study of solvent extraction procedures, and the study of additive ageing processes (ie. the effects of heat, light, sterilization, radiation, etc.), since the rate of disappearance of or decay can be measured directly by the decrease in absorbance of the sample at a suitable wave-length.

The direct analysis of additives in polyethylene film by means of ultraviolet spectroscopy is limited by excessive beam dispersion due to light scattering from the polymer crystalline regions. Additives at low concentrations (0.1%) require sample thickness such that analysis must be performed in the presence of a high level of light scattering which may change unpredictably with wavelength. At lower levels of concentration and corresponding greater sample thickness, unacceptable signal to noise ratios exist. Nevertheless, ultraviolet spectroscopy remains an attractive method for analysis for many additives; principal advantages over infrared spectroscopy include greater sensitivity arising from higher extinction coefficients and a lack of interfering absorptions from the polymer matrix. These advantages can be realised, however, only if background scattering from the polymer can be reduced.

Solvent extraction procedures are commonly used in additive analysis. In these procedures the polymer is refluxed with a solvent which either dissolves the polymer or softens it so that additives dissolve into the solvent phase. A non-solvent for the polymer is then added to reprecipitate the polymer, leaving the additives in solution. A good example of this is the extraction of additives from polyethylene in which the polymer is refluxed with toluene which partly dissolves the polymer. Upon the addition of absolute ethanol the polymer is reprecipitated leaving the additives in solution in the toluene - ethanol solvent mixtures which is then filtered off and carefully evaporated to dryness to provide the additive extract which is essentially free from polymer. Such extracts when prepared quantitatively are extremely useful for either direct analysis for known additives or for subsequent separation of their components by one of a number of chromatographic techniques prior to examination of the individual separated compounds by spectroscopy for identification and for quantitation.

A list of soluble solvent extraction procedures for various types of polymers is given in Table 1.2.

1.1.2.2 Phenolic Antioxidants

Generally, methods are based on solvent extraction of the additive followed by analysis for the extracted additive by a suitable physical technique such as visible spectrophotometry of the coupled antioxidant, redox spectrophotometric methods, ultra-violet spectroscopy, infrared spectroscopy, gas chromatography, thin-layer chromatography or column chromatography. In general, direct chemical methods of analysis have not found favour. These include potentiometric titration with standard sodium isopropoxide in pyridine medium[26] or reaction of the antioxidant with excess standard potassium bromide-potassium bromate (ie. free bromine) and estimation of the unused bromine by addition of potassium iodide and determination of the iodine produced by titration with sodium thiosulphate to the starch end-point.[27]

1.1.2.2a Visible Spectrophotometric Method. Metcalf and Tomlinson[8] have described a very useful general colorimetric procedure for the determination of phenolic and other types of antioxidant in polyethylenes (method 3). This procedure involves oxidation of the antioxidant (A) under controlled conditions with an absolute ethanol solution of ferric chloride:

$$\text{A reduced} + Fe^{3+} = \text{A oxidized} + Fe^{2+}$$

followed by reaction of the ferrous iron produced with 2,2'- dipyridyl to form a coloured complex, the intensity of which is proportional to the concentration of antioxidant present.

The procedure has been applied to various phenolic and amine type antioxidants, viz. Succanox 18, butylated hydroxy toluene, Ionol (2,6-di-tert-butyl-p-cresol) and Nonox CI (N-N-di-2-napthyl-p-phenylenediamine)

Coupling with diazotized sulphanilic acid has also been used as the basis of a method for determining phenolic antioxidants in polyethylene.[29]

Hilton[28] published a method (method 4) for the determination of phenolic antioxidants in polymers based on the preparation of a methanol or ethanol extract of the polymer, followed by coupling the extracted phenol with diazotized p-nitroaniline in strongly acidic medium. The solution is then made alkaline and the visible absorption spectrum determined.

1.1.2.2b Ultraviolet Spectroscopy. Ultraviolet spectroscopy is liable to be in error owing to interference by other highly absorbing impurities that may be present in the sample.[30 - 33] However, within prescribed limits, ultraviolet spectroscopy is of use and, as an example,[34 - 37] two procedures are discussed (methods 5 and 6) for the determination of Ionol (2,6-di-tert-butyl-p-cresol) and of Santonox R (4,4'-thio-bis-6-tert-butyl-m-cresol) in polyolefins.

In an attempt to overcome the difficulty of interference effects by other polymer additives in the ultraviolet spectroscopic determination of phenolic antioxidants, Wexler [38] makes use of the bathochromic shift exhibited by phenols on changing from a neutral or acidic medium to an alkaline one. This shift is due to the change of absorbing species

Table 1.2 - Solvent Extraction Methods of Additive Extraction from Polymers

Polymer Type	Substances Extracted	Extracting Solvent	Comments	References	Type of Extraction
Polyethylene	cresolic and phenolic antioxidants	chloroform	heat at 50°C for 3 h in a closed container	4	fractional extraction
Polyethylene	cresolic antioxidants	hexane	heat at 50°C for 24 h	5	fractional extraction
Polyethylene	cresolic antioxidants	ether	in the dark at 20°C	6	fractional extraction
Polyethylene	phenolic antioxidants	chloroform		7	fractional extraction
Polyethylene	antioxidants	toluene	reflux to dis-solve polymer in precipitate with methanol	8	fractional precipitation
Polyethylene	antioxidants	water	at 70°C under nitrogen	9	fractional extraction
Polyethylene	phenolic antioxidants	carbon disulphide iso-octane		10	
Polyolefins	2,6 di-t-butyl p-cresol	cydo-hexane	reflux 30 min	11	
PVC	diphenyl thio-urea 2-phenylindole dicyandiamide	methanol or ether			
PVC	Stabilizers, lubricants, plasticizers	ether		12	
Rubbers	amine and phenolic antioxidants	ethanol/ HC1	reflux, then steam-distil amines from extract	13,14	
Rubbers	phenyl salicylate, resorcinol benzoate	ether			
Rubbers	antioxidants	acetone		15,16	
Rubbers	ketone-amine condensates phenols 2-mercaptobenz-imidazole	acetone		17	
General	p-Phenylen-ediamine-derivatives	95% methanol or ethanol	reflux 16 h in an extraction cup	18	
Polypropylene	antioxidants	tetrahydrof uran		19	
Polyethylene	antioxidants	chloroform		20	
Polyolefins	antioxidants	methylene dichloride		21	
Polyethylene	antioxidants	ground polymer chloroform extraction on soxhlet		22	
Polypropylene (8 mesh)	antioxidants	tetrahydrof uran for 24		23	

Table 1.2 continued

Polyethylene and Polypropylene	antioxidants	dekalin at 110°for 30 mins	24,25

because of solute-solvent interaction. Using a double-beam recording spectrophotometer, he measured a difference spectrum by placing an alkaline solution of the polymer extract in the sample beam, and an identical concentration of sample in acid solution in the reference beam. The resulting difference spectrum is a characteristic and useful indication of the concentration and chemical identity of the phenolic substance. Possible interference due to non-ionizing, non-phenolic species are usually cancelled out in the difference spectrum which should make the technique of interest to the polymer analyst.

Soncek and Jelinkova[39] have also used this differential principle to determine in polypropylene two antioxidants (2,6-ditert-butyl-4-methylphenol and 4-substituted 2,6-xylenol) which have virtually identical ultraviolet absorption spectra in the absence of alkali (method 7). The antioxidants can be distinguished in alkaline medium, where 4 substituted 2,6-xylenol forms phenolate readily, thus allowing the utilisation of the bathochromic shift for its determination. The use of derivative spectroscopy reduces light scattering and matrix interferences when extracts from polypropylene samples are measured.

Ruddle and Wilson[34] used observed shifts in the ultraviolet following treatment of polymer extracts with alkali and nickel peroxide as the basis of a low interference method for determining Topanol OC, Binox M and Ionox 330 in polyethylene (Method 8).

1.1.2.2c Infrared Spectroscopy. Spell and Eddy [41] have described infrared spectroscopic procedures for the determination of up to 500 ppm of various additives in polyethylene pellets following solvent extraction of additives at room temperature. They showed that Ionol (2,6-di-tert-butyl-p-cresol) and Santonox R (4,4-thio-bis-(6-tert-butyl-m-cresol) are extracted quantitatively from polyethylene pellets by carbon disulphide in 2 - 3 h and by iso-octane in 50 - 75 h.

Miller and Willis[42] obtained infrared spectra of antioxidants from polymer films. They compensated with additive free polymer in the reference beam. Infrared spectroscopy is more specific than ultraviolet spectroscopy, but some workers [43] find that the antioxidant level in polymers is too low to give suitable spectra. In-situ spectroscopic techniques are not likely to be of value then, in the analysis of samples of unknown composition.

Luongo[44] and Drushel and Sommers[45] described methods for incorporating known amounts of additives into polymer.

1.1.2.2d Thin Layer Chromatography. This technique comes into its own when dealing with known mixtures of antioxidants in polymer extracts as discussed in more detail in Chapter 5.

In Method 9 is described a quantitative thin-layer procedure for determining a single additive, namely Santonox R (4,4'-thio-bis-6-tert-butyl-m-cresol) in polyethylene.

Dobies[46] has described detailed thin-layer procedures for determining 0.02 - 0.2% of various antioxidants in polyethylene and polypropylene films (method 10). Simpson and Currell[47] use thin-layer chromatography in the determination of additives such as antioxidants, ultraviolet absorbers and organotin stabilizers in polyolefins and other polymers (Method 11)

1.1.2.2e Direct Ultraviolet Spectroscopy Of Polyalkene Films. Albarino [48] stated that analysis of polyalkene additives by means of direct ultraviolet spectroscopy of the polymer film is limited by excessive beam dispersion due to light scattering from the polymer crystalline regions. Additives at low concentrations (0.1%) require sample thicknesses such that analysis must be performed in the presence of a high level of scattering which may change unpredictably with wave-length. At lower levels of concentration and correspondingly greater sample thicknesses, unacceptable signal-to-noise ratios exist. Nevertheless, ultraviolet spectroscopy remains an attractive method for analysis of many additives. Principal advantages over infrared analysis include greater sensitivity arising from higher extinction coefficients and a lack of interfering absorptions from the polyalkene matrix. These advantages can be realized however, only if background scattering from the polymer can be reduced.

Albarino[48] demonstrated the feasibility of quantitative ultraviolet analysis of Irganox 1010 antioxidant in polyethylene film at temperatures above the polymer melting point where the crystallites, which account for much of the scattering, are eliminated.

1.1.2.2f High Performance Liquid Chromatography. Schabon and Fenska[24] have described a rapid extraction procedure (method 12) followed by high performance liquid chromatography for the determination in polyethylene of down to 10 ppm of Irganox 1076 (octadecyl 3,5-di-tert-4-hydroxy hydrocinnamate), Irganox 1010 (tetrakis (methylene(3,5-di-tert-butyl-4-hydroxhydrocinnamate)) methane), and butylated hydroxytoluene (2,6-di-tert-butyl-4-methyl-phehol).

Vargo and Olsen [377] used absorbance and mass spectrometric detectors to detect antioxidant, and UV stabilizers leaving an hplc column in acetonitrile extracts of poly-propylene.

Selective mass spectrometry. Rudewicz and Munson [376] used selective mass spectrometry (SIMS) for the determination of additives in polypropylene without the need for prior solvent extraction of additives.

1.1.2.3 Amine Antioxidants

1.1.2.3a Spectrophotometric Method. In a procedure recommended by the British Standards Institution[50] for determining Nonox CI (N,N'-di-2 napthyl-p-phenylene-diamine) in low density and high density polyethylene, the polymer is refluxed with toluene, reprecipitated with ethanol and filtered to provide a clear extract. The antioxidant content of the filtrate is determined colorimetrically by oxidation with hydrogen peroxide, in the presence of sulphuric acid. This reagent produces a green colour with Nonox CI which gradually reaches a maximum intensity. The colour is evaluated at 430 nm when the maximum depth of colour is reached.

Crompton[51] has described a modification to this procedure (method 13) for the determination of Nonox CI in high density polyethylene which, compared to low density polymer, has only a small solubility in toluene. He has also extended the procedure to the determination of oxidized Nonox CI, and a further product which he calls degraded Nonox CI.

The procedure described under phenolic antioxidants involving oxidation of the antioxidant with ferric iron followed by spectrophotometric estimation of the ferrous iron produced using 2,2'-dipyridyl is also applicable to the determination of amine antioxidants. Other chromogenic reagents that have been used for the determination of amine antioxidants include p-diazobenzene sulphonic acid,[52] diphenyl-ß-picrylhydrazyl and benzothiazolin-2-onehydrazone.[54]

The Hilton [35,56] diazotization procedure has been used for the determination of amine antioxidants in polyolefins and other polymers (Method 14).

1.1.2.4 Diorganosulphide (and Tertiary Phosphite) Types of Antioxidants in Polyolefins

1.1.2.4a Selective Oxidation Procedure. Kellum[57] has described a method (Method 15) based on selective oxidation, for the determination of diorganosulphide and tertiary phosphite types of secondary antioxidants in polyolefins. These two classes of compounds were selectively determined in the presence of each other by oxidation using m-chloro-peroxy-benzoic acid to sulphones and phosphates. In this method, a heptane extract of the polyolefins containing the antioxidants is treated with a two-fold excess of the oxidant and allowed to react before the unreacted oxidant is decomposed with sodium iodide to produce iodine which is estimated by sodium thiosulphate titration. This method has the advantage of being free from interference by hindered phenolic antioxidants, benzophenones and substituted hydroxy phenyl benzotriazole type light stabilizers, triazoles, fatty acid amides and stearate salts, all of which might be present in the polymer extract.

1.1.2.4b Gas Chromatographic Procedure. Gas chromatography has been employed to determine diorganosulphide antioxidant, such as dilauryl ß,ß' thiodipropionic acid in polyolefins (method 16). This method has the advantage of being less subject to interference than the selective oxidation procedure described above.

1.1.2.4c Thin Layer Chromatography. Thin layer procedures have been described[252] for the determination of dilauryl ß,ß' - thiodipropionate antioxidant in polyalkenes. In this procedure the total additives are first extracted from the polymer by extraction under nitrogen with a 1:1:4 mixture of chloroform, ethanol and n-hexane. To separate dilauryl ß,ß' thiodipropionate from its own oxidation products and from other additives, the extract was subject to thin layer chromatography on a silica gel coated plate. The spot containing dilauryl ß,ß' thiodipropionate is refluxed for 30 minutes at 80^0C with 5N methanolic potassium hydroxide to hydrolyse the ester to lauryl alcohol which is then extracted from the aqueous alcoholic phase with chloroform containing n-octadecene internal standard. This extract is gas chromatographed on an isothermal column packed with 1.5% fluor-silicone oil FS 1265 on 60 to 100 mesh Chromosorb W operated at 300^0C and utilizing a flame ionization detector. Suitable reference solutions of dilauryl ß,ß' - thiodiproprionate run in parallel through the whole procedures gave a recovery of 97%.

1.1.2.5 Ultraviolet Absorbers and Optical Brighteners in Polyalkenes

Among the numerous additives commonly used in plastic materials, the ultraviolet absorbers are increasing in importance because they are often used in food packaging materials to protect the plastic material as well as the foodstuff packaged from the actinic action of ultraviolet radiation. Actinic effects may cause discolouration of both the plastic material and the foodstuff and, on occasion, also changes in taste and loss of vitamins in the food.

The ultraviolet absorbers can be divided in different groups (Table 1.3)

(a) Benzophenone derivatives
(b) Salicylic acid esters
(c) Resorcinol esters
(d) Benzotriazole compounds
(e) Coumarin derivatives

Methods 17 and 18, respectively, determine Cyasorb UV 531 (2-hydroxy-4-n-octoxybenzophenone) in polyolefins by direct infrared spectroscopy and thin-layer chromatography. Phenolic and organophosphorus type antioxidants do not interfere in these procedures. Method 19 describes an ultraviolet spectroscopic method for the determination of Tinuvin 326 ultraviolet absorber in polypropylene.

The fluorescence emitted by optical brighteners under ultraviolet light on a thin-layer plate has been utilized as a means of estimating these compounds[59] (method 19). To estimate Tinuvin 326 7(6-butoxy-5-methyl-benzotriazol-2-yl) 3-phenyl-coumarin) in polymer granules, the sample is extracted from the polymer with chloroform. The extract is applied to two Kieselgel G plates and chromatograms are developed with benzene-chloroform (2:3) and benzene, respectively; the spots being detected by their fluorescence in ultraviolet radiation.

1.1.2.5a Gas Chromatographic Procedure. Lappin and Zannucci[60] showed that a number of ultraviolet light absorbers as well as antioxidants could be quantitatively determined on an SE-30 column. SE-30 has a maximum operating temperature of about 350^{0}C. This high temperature permits elution of some high molecular weight additives. When the polymer was dissolved in p-xylene and reprecipitated with an equal volume of p-dioxane a relatively clean chromatogram was obtained from the filtrate. The position of decomposition peaks from unstable compounds, such as dilauryl ß,ß' -thiodipropionate and distearyl pentaerythritol diphosphite, are predictable and do not interfere with the determination of those additives studied by Lappin and Zannucci.[60]

Lappin and Zannucci[60] determined 4-(dodecyloxy)2-hydroxybenzophenone (DOBP) and 2,6-di-tert-butyl-p-cresol (BHT) in polypropylene by this technique (Table 1.4). The precision of the BHT determination is good but the quantity found (0.02%) was less than the amount added (0.05%) possibly due to some losses of volatile BHT during polymer compounding. DOBP determinations ranged from 0.20 to 0.35% for a sample originally containing 0.30% of the additive.

Denning and Marshall[61] devised a method in which toluene extracts of polyethylene are examined for ultraviolet absorbers (and antioxidants) at two column temperatures in order to overcome the problem that whereas some compounds have a relatively low retention time, others of higher molecular weight have very long retention time at 250^{0}C. Relative retention data obtained for ultraviolet absorbers, are given in Table 1.5.

1.1.3 Fractionation

Low molecular weight high density polyethylene has been fractionated by supercritical fluid chromatography using carbon dioxide, propane and propane modified carbon dioxide as eluting agents.[408]

Table 1.3 - Some Ultraviolet Absorbers Used in Plastic Materials

	Chemical Formula	Trade Name	Manufacturer
1	2-hydroxy-4-methoxy-benzophenone	Uvinal M40 Uvistat 24 Cyasorb UV 9	General Aniline Co Ward & Blenkinsop Cyanamid
2.	2,4-dihydroxy-benzophenone	Uvistat 12 Uninul 400	Ward & Blenkinsop General Aniline Co
3.	2-hydroxy-4-methoxy-4-methyl-benzophenone	Uvistat 2211	Ward & Blenkinsop
4.	2,4,5-trihydroxy butyrophenone	Inhibitor THBP	Eastman
5.	4-dodecyloxy-2-hydroxy benzophenone	Inhibitor DOBP	Eastman
6.	2-hydroxy-5-n-octoxy-benzophenone	Cyasorb UV 531	Cyanamid
7.	2,2'dihydroxy-4-methoxy benzophenone	Cyasorb UV 24	Cyanamid
8.	2,2'dihydroxy-4,4' dimethoxy benzophenone	Uvinul D49	General Aniline Co
9.	p-tert-butylphenylsalicilate		
10.	resorcinol mono benzoate	Inhibitor RMB	Eastman
11.	hydroxyphenylbenzo-triazole	Tinuvin P	Geigy
12.	7-diethylamino-4-methyl coumarin		Ward

Table 1.4 - Analyses of Polypropylene for BHT and DOBP

Sample Composition	Additive found (%) Run 1	Run 2	Run 3	Run 4
0.05% 2,6-di-tert-butyl-p-cresol	0.02	0.02	0.03	0.02
0.30% 4(dodecyloxy) 2-hydroxy-benzophenone	0.25	0.32	0.20	0.35

From Lapin and Zanncci[60] with permission. American Chemical Society.

1.2 POLYSTYRENE

1.2.1 Volatiles and Water

1.2.1.1 Monomers and Aromatic Volatiles

1.2.1.1a Gas Chromatography. The styrene monomer used in the manufacture of polystyrene generally contains low concentrations of aromatic impurities. These fall into two categories (a) saturated compounds such as benzene, toluene, ethyl benzene, xylenes, propyl benzene, butyl benzene, ethyl toluenes, diethyl benzenes, traces of which remain in the final polymer and (b) unsaturated impurities such as methyl styrenes which either wholly or partially copolymarize with the styrene. The presence of all these substances in addition to residual styrene monomer in the final polymer has implications from the points of view of the mechanical properties of the final polymer and toxicity in the case

Table 1.5 - Separation of Ultraviolet Absorbers

Ultra violet absorber	Class of compound	Relative retention time (relative to Santonox R)	Column Temperature (°C)
Tinuvin P	Benzotriazole	0.20	250
Tinuvin 326	Benzotriazole	0.64	250
Tinuvin 327	Benzotriazole	0.85	250
Cyasorb UV 531	Aromatic ketone	1.0	250
Tinuvin P	2-(2'-Hydroxy-5'-methylphenyl benzotriazole)		
Tinuvin 326	2-(2'Hydroxy-3'-t-butyl-5'methyl ethyl phenyl-3-)chlorobenzo triazole		
Cyasorb UV 531	2-Hydroxy-4-n-octoxybenzophenone		

From Denning and Marshall with permission.[61] American Chemical Society

of foodstuff packing grades of polystyrene. Method 20 is capable of determining all these aromatics in amounts down to 0.001% w/w in polystyrene.

A further apparatus for carrying out solid polymer head space analysis of polystyrene for styrene monomer and aromatic volatiles has been described by Crompton and Myers.[1] In this apparatus the polymer is heated up to 300°C in the absence of solvents, prior to their examination by gas chromatography. The technique (Method 21) avoids the disadvantages resulting from the use of extraction or solution procedures.

Davies and Bishop[62] have developed a further head space analysis technique (Method 22) in which the polystyrene film or sheet, together with internal standard, is placed in a 250 cm^3 sealed container and heated at 100°C for 90 min. This method has been used to determine many different solvents in several different substrates. The solvents include ethanol, ethyl acetate, ethyl methyl ketone, 2-ethoxyethanol, propan-l-ol and toluene. The substrates include polyethylene, polypropylene and cellophane, which occur individually, coated with saran or combined in laminates.

Roper[63] has discussed the problem of determining very low concentrations of volatiles in polymers. To achieve this Roper[63] employed a trap-tube. The capillary portion of the tube is packed with a gas chromatographic column packing consisting of 20 to 30% of a suitable liquid phase on a granular diatomaceous type support. The tube is fitted with a suitable hypodermic needle so that it can be connected to the gas chromatograph through the injection port. The polymer sample to be analysed is weighed into the trap-tube and the tube is connected to an inert-gas stream and heated to the appropriate temperature while the gas stream sweeps the volatile material into the packed section of the tube, which is usually cooled with dry ice. Melting the polymer is often necessary to rid it of volatiles. After the volatiles have been absorbed in its packed section, the trap-tube is disconnected and moved to the chromatograph. The heater, already at temperature, is then moved to the capillary portion of the tube-trap where the heat and carrier gas flow sweeps the volatiles from the packing through the hypodermic needle and into the gas chromatograph.

Various other workers have described procedures for the determination of volatiles in polyethylene[1,2,64] and polypropylene.[66]

1.2.1.1 Ultraviolet Spectroscopic Procedure. An alternate procedure for determining styrene monomer in polystyrene is based on ultraviolet spectroscopy. Styrene monomer has a strong absorption maximum of 292 nm. Unfortunately, this is subject to interference by aromatic additives, eg. antioxidants, which are often present in the polymer at a concentration similar to that of the monomer. Methods which handle this interference are discussed below.

The baseline correction technique can obviously be applied to the determination of styrene monomer in polystyrene only if any other ultraviolet absorbing constituents in the polymer extract (eg. lubricant, antioxidants) absorb linearly in the wavelength range 288 - 300 nm). If the polymer extract contains polymer constituents other than styrene with non-linear absorptions in this region, then incorrect styrene monomer contents will be obtained.

1.2.1.1c Distillation Ultraviolet Absorption Procedure. In the distillation technique[64] the polystyrene (0.5g) is dissolved in chloroform or ethylene dichloride (20 ml) in a stoppered flask and the solution is poured into an excess of methyl alcohol (110 ml) to reprecipitate dissolved polymer. The polymer is filtered off and washed with methanol (120 ml) and the combined filtrate and washings gently distilled to provide 200 ml of a distillate containing only styrene monomer and any other distillable component of the original polystyrene sample. Non-volatile polymer components (viz. stabilizers, lubricants and low molecular weight polymer) remain in the distillation residue.

The optical density of the distillate is measured at 292 nm or by the baseline method against the distillate obtained in a polymer-free blank distillation. Calibration is performed by applying the distillation procedure to solutions of known weights of pure styrene monomer in the appropriate quantities of methyl alcohol and the chlorinated solvent.

Tables 1.6 and 1.7 show results obtained for styrene monomer determinations carried out on samples of polystyrene by the direct ultraviolet method and by the distillation modification of this method. It is seen that the distillation method gives results that are consistently higher than those obtained by direct spectroscopy.

Similar effects to these were also observed with polystyrene containing other ultraviolet absorbing additives. Thus, the influence of a mixture of 0.4% w/w tris-(nonylated phenyl phosphite (Polygard) and 0.2% w/w 2,6-di-tert-butyl-p-cresol (Ionol CP) on the determination of styrene monomer is shown in Table 1.8.

1.2.1.1d Mass Spectrometry. Peltonen[378] applied mass spectrometry to the identification of volatile breakdown products of heated polystyrene. Evolved gas analysis involving the coupling of a gas chromatograph to a mass spectrometer has been used to identify volatile products formed on heating polystyrene.[379-382]

1.2.1.1e Thermogravimetric Analysis. Thermogravimetric analysis has been used to determine the concentration of solvents in polymers and in neat polymer additives and to determine the temperatures of which additives in polymers start to degrade.[383]

1.2.1.1f Supersonic Jet Spectrometry. Imasaki et al[409] carried out supersonic jet spectrometry of chemical species resulting from the thermal decomposition of polystyrene at 400^0C. Styrene monomer and its dimer and trimer, also toluene were identified in the decomposition products.

1.2.1.2 Aliphatic expanding agents

Expandable grades of polystyrene are manufactured by steeping volatile polystyrene granules in a low boiling hydrocarbon solvent until the polymer becomes saturated with the solvent. When the granules are put in a mould and are treated with steam, they undergo an expansion to many times their original volume and the expanded

Table 1.6 - Comparison of Direct Ultraviolet and Distillation/Ultraviolet Methods for the Determination of Styrene Monomer

Method	Solvent	Styrene Monomer (%w/w) in Polystyrene Sample	
		No. 1	No. 2
Direct UV Method	Chloroform/	<0.05	0.13
	carbon tetrachloride	<0.05	0.13 0.14 0.16 0.18
	Ethyl acetate	<0.05	0.14 0.12 0.14
			0.14 0.15
Distillation/UV	Ethylene di	0.16 0.18	0.27 0.29
	chloride/	0.16 0.18	0.26 0.29
	methanol	0.20	0.29

Table 1.7 - Influence of Phenolic Antioxidant* on the Determination of Styrene Monomer by Direct Ultraviolet and by Distillation/Ultraviolet Methods

	Styrene Found (% w/w)			
	Direct UV Method[a]		Distillation/UV Method	
Styrene added to polystyrene (% w/w)	No phenolic anti-oxidant* addition	0.5% phenolic anti-oxidant* added on polymer	No phenolic anti-oxidant* addition	0.5% phenolic anti-oxidant* added on polymer
0.11	0.11 0.11	0.04 0.08	0.12	0.12
0.22	0.22	0.11	-	-
0.27	0.26	0.18	0.29	0.26
0.41	0.41 0.40	0.30 0.27	0.40	0.40

* Wingstay T. [a] Chloroform used as a sample solvent

Table 1.8 - Influence of Polygard/Ionol CP Mixture on Determination of Styrene Monomer by Direct Ultraviolet and by Distillation/Ultraviolet Methods

	Styrene Found (% w/w)			
	Direct UV Method[a]		Distillation/UV Method[b]	
Styrene added to polystyrene (% w/w)	No additive	0.4% w/w Polygard and 0.2% w/w Ionol CP added on polymer	No additive	0.4% w/w Polygard and 0.2% w/w Ionol CP added on polymer
0.10	0.10 0.10	- [b]	0.11	0.11 0.09
0.26	0.25 0.26	0.16 0.16	0.26	0.27 0.25
0.41	0.42 0.41	0.29 0.29	0.41	0.40 0.42

[a] Chloroform used as a sample solvent. [b] Not measurable due to strong interference by additives

granules coalesce. This process is used for the manufacture of insulating polystyrene board and insulated hot beverage vending vessels. It is important to be able to determine the concentration of residual volatiles in the original and the expanded polymer and gas chromatographic methods for carrying out the analysis of several expanding agents are discussed in Methods 23 to 26.

1.2.1.2a Solid Polymer Head Space Analysis Techniques. A further method which does not involve the use of solvents is available for the determination of expanding agents in polystyrene (method 27). This method involves heating the polymer, without solvent, in a heated copper block and, using the carrier gas, sweeping the released volatiles from the expanded polymer into the gas chromatograph for quantiative analysis. An advantage of this procedure is that no sample solvents are present which could interfere in the interpretation of the gas chromatogram.

1.2.2. Oligomers

High performance liquid chromatography on silica gel columns and 13C NMR spectroscopy[387] have been employed to determine polystyrene oligomers in the molecular weight range up to 4800 in polystyrene. Typical elution gradient solvents used in hplc are n-hexane-tetrahydrofuran or n-hexane-methylene dichloride.

Particle beam liquid chromatography mass spectrometry has been used to determine up to n18 oligomers in polystyrene.[410]

Petro et al[431] used a molded monolithic rod of macroporous polystyrene divinyl benzene copolymer as a separation medium for the high performance liquid chromatography of styrene oligomers and copolymers eg. styrene - 2 naphthyl (methacrylate). "On column" precipitation redissolution chromatography is an alternative to size exclusion chromatography. The solvent gradient used comprises a poor solvent (water, methanol or acetonitrile) and increasing amounts of a good solvent such as tetrahydrofuran. Excellent separations of oligomers were achieved by this technique.

1.2.3. Impurities in styrene monomers

All the non-polymerisable volatile impurities which occur in styrene monomer used in polystyrene manufacture will also be present in the finished polymer. Analysis of the monomer is therefore important. Gas chromatographic methods, for determining up to 40 impurities in styrene monomer are discussed in methods 28 to 30.

In Method 28 the sample was first analysed on the squalene column in the presence of an added internal standard (eg. n-butyl benzene) to determine down to 5 ppm n-propyl benzene, butyl benzenes, diethyl benzenes, ethyl toluenes, methyl styrenes, ethyl styrenes and divinyl benzenes. The sample was then re-analysed on the Carbowax column to determine benzene, toluene, xylenes, ethyl benzene, cumene, n-propyl benzene and m-/p-ethyl toluenes. By temperature programming it is possible to reduce the analysis time to about one hour using a single gas chromatographic column (Carbowax 15 - 20 M-Celite) and to separate up to 40 hydrocarbons up to the divinyl benzenes and benzaldehyde. Experimental conditions are given in Method 29 (hydrocarbons) and Method 30 (benzaldehyde).

1.2.4. Additives

1.2.4.1 Phenolic Antioxidants

1.2.4.1a Coupling Visible Spectrophotometric Methods. In one such procedure[28] described in Method 31 the polymer is extracted with ethanol and the extracted phenolic antioxidant coupled with diazotized p-nitroaniline in strongly acidic medium. The solution is then made alkaline and the visible absorption spectrum determined.

1.2.4.2 Ultraviolet Absorbers

1.2.4.2a Ultraviolet Spectroscopic Procedure. By their nature, many of these types of compounds are amenable to analysis by fluorimetric analysis, thus Uvitex OB has an intense ultraviolet absorption at a wavelength of 378 nm which is high enough to be outside the region where many potentially interfering substances present in the polymer extract would be excited to fluoresce. This is illustrated in the fluorimetric procedure described in Method 32 for the determination of down to 10 ppm Uvitex OB in polystyrene.

1.2.4.3 Organic Peroxide Residues

Organic peroxides can occur in small amounts in some types of polymers such as polystyrenes as a result of the fact that a peroxide has been used as a polymerization catalyst in polymer manufacture. Also, stable organic peroxides such as dicumyl peroxide has been used as synergists, in conjunction with bromine and or phosphorus containing additives, to impart fire resistance to cellular expanded polystyrene and other types of plastics

1.2.4.3a Polarographic Procedure. A polarographic procedure for the determination of p-tert-butyl perbenzoate in polystyrene is described in method 33.

1.2.4.3b Thin Layer Chromatographic Procedure. Certain types of peroxides used in polymer formulations are extremely stable and unreactive. This applies to substances such as dicumyl peroxide used as an ingredient of some self-extinguishing grades of polymers.

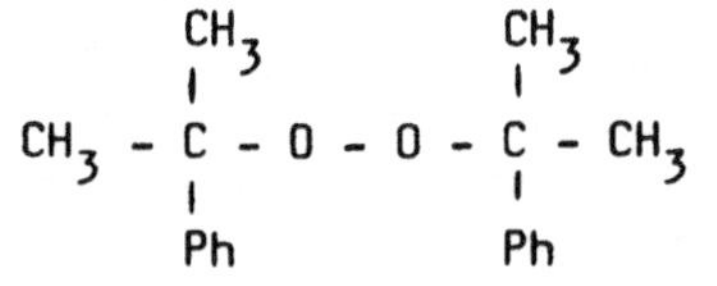

1.1

This substance cannot be determined by polarography and will not react with many of the reagent normally used for determining organic peroxides.

Brammer et al[68] have described a method for determining dicumyl peroxide in polystyrene, which is not subject to interference by other organic peroxides or additives that may be present in the polymer. The dicumyl peroxide is extracted from the polymer with acetone and then separated from any other additives present by thin-layer chromatography on silica gel. The gel in the area of the plate containing dicumyl peroxide is then isolated and digested with potassium iodide in glacial acetic acid followed by titration of the liberated iodine with dilute sodium thiosulphate solution. This procedure has a precision of $\pm$12% of the determined value with polymers containing 0.25 to 0.5% dicumyl peroxide. It is a rather time-consuming procedure but has the advantage of avoiding all risk of interference from other types of peroxides present in the sample.

Thin layer chromatography has also been applied to reactive organic peroxide residues such as tertiary butyl perbenzoate and benzyl peroxide in polystyrene (Method 34).

1.2.4.3c Gas Chromatographic Procedures. Gas chromatography has also been used to determine certain types of organic peroxides. Bukata et al[69] for example, describe procedures involving chromatography of heptane solutions of peroxide (Table 1.9) on phthalate/distomaceous earth or silicone/diatomaceous earth columns and using helium as the carrier gas. No doubt, this type of procedure could be easily adapted to the examination of solvent extracts of polymers.

Hyden[70] describes a gas chromatographic procedure for the determination of di-tert-butyl peroxide. This is based on the thermal decomposition of the peroxide in benzene solution into acetone and ethane when the solution is injected into the gas chromatographic column at 310^{0}C. The technique is calibrated against standard solutions of pure di-tert-butyl peroxide of known concentration.

1.2.4.4 Inhibitors in Styrene Monomer

Many monomers used in polymer manufacture must be stabilised with a low concentration of an inhibitor, such as hydroquinone in order to prevent oxidation of the monomer during storage and consequent effects on its polymerization qualities. As the inhibitor would, itself, interfere in the polymerization, it is usually stripped from the monomer immediately prior to use by processes such as flash distillation or washing with aqueous alkali solution.

General methods are described in Methods 35 - 38 for the determination of inhibitors in monomers.

1.2.5. Fractionation and Molecular Weight

Flow field - flow fractionation has been used[405] to fractionate polystyrene and Kirkland and Rementer[411] used thermal field flow fractionation using Mark Houweq constants to determine the molecular weight distribution of polystyrene and poly methylstyrene.

Desorption chemical ionization mass spectrometry has been used to determine the molecular weight distribution of polystyrene.[412] .The liquid chromatographic behaviour has been studied of polystyrene homopolymers on a C4 pore diameter reversed phase column. Separation was achieved on a molecular weight basis using a tetrahydrofuran-acrylonitrile eluting solvent system.[413]

Hancock and Synovec[414] have carried out rapid characterization of linear and star branched polystyrene by gradient detection using methylene dichloride solutions of the polymers. This technique measures weight average molecular weights and is more specific than results obtained using a refractive index detector.

Time of flight secondary ion mass spectrometry has been used to determine the molecular weight distribution of polystyrene.[419]

Table 1.9 - Gas Chromatography Retention Times for Organic Peroxides

Compound	Column[a]	Temp °C	Helium[b] pressure psi	Retention time min
tert-Butyl hydroperoxide	2 m-A	80	20	22.7
tert-Pentyl hydroperoxide	1 m-A	80	15	19.9
tert-Butyl peracetate	1 m-A	100	20	6.5
tert-Butyl peroxyisobutrate	1 m-A	100	20	15.3
Di-tert-butyl peroxide	2 m-A	80	20	8.1
Di-tert-pentyl peroxide	1 m-A	80	15	15.4
2,5-Dimethyl-2, 5-di (tert-butyl peroxy)-3-hexyne	1 m-0	138	15	4.9
2,5-Dimethyl-2, 5-di(tert-butyl peroxy) hexane	1 m-0	138	15	6.9
n-Heptane	2 m-A	80	20	6.1
n-Dodecane	1 m-0	138	15	3.1
n-Pentane	1 m-A	100	20	0.4
n-Nonane	1 m-A	100	20	4.9

[a] Length (metres) and type of column used.
[b] Inlet pressure.
A, didecyl phthalate on diatomaceous earth.
O, silicon grease on diatomaceous earth.

From Buketa et al.[69] American Chemical Society

Schriemer and Li[432] used matrix assisted laser desorbtion/ionization (MALDI) mass spectrometry to detect high molecular weight narrow polydisperse polystyrene with molecular weights up to 1.5 x 10^6 Daltons. The method is rapid and agrees well with results obtained by chemical methods. Retinoic acid was used as the organic matrix for the laser absorption procedure and the samples were analysed as their silver cation adducts.

1.3. POLYVINYL CHLORIDE

1.3.1. Water and Volatiles (Monomers, Chlorohydrocarbons) in Vinyls and Acrylics

1.3.1.1 Gas Chromatography Procedures.

The difference between Method 1 for the determination of water and volatiles in polyalkenes and that described below for the determination of water and volatiles in vinyls and acrylics, is that in the latter method, the polymer is heated in the presence of the residual air in the sample tube, and in the first method (polyalkenes) the residual air in the sample tube is replaced with inert carrier gas, helium, before the sample is heated. With the non-olefinic polymers the same results for water content are obtained by both methods, ie. it is not necessary to replace the air with inert gas. Polyalkenes, however, are oxidized if heated in the presence of air, thus producing additional amounts of water. Jeffs[2] recommends that before carrying out any quantitative work on the volatile constituents obtained from a polymer powder, a preliminary gas-chromatographic investigation should be carried out, of their complexity. For example, one polymer showed nine components other than air, water and the original monomer. These components are chlorinated hydrocarbons, such as 1,1- and 1,2- dichloromethane and cis- and trans-dichloroethylene, that are present as impurities in the original monomer.

Having chosen a suitable column, the temperature at which a given polymer liberates all of the moisture has to be found. This is determined by keeping the gas chromatographic conditions constant and heating known amounts of polymer at various split-heater temperatures. The peak height (or peak area) per gram of sample is then plotted against the sample temperature.

Jeffs[2] showed that a plot of water liberated against temperature for PVC indicated a constant amount of water was evolved at temperatures between 115^0 and 160^0C. A heater temperature of 125^0C and the following gas chromatographic conditions are, therefore, recommended - column, 10 foot (1/4 inch o.d.) stainless steel tubing, packed with 10 per cent w/w poly(ethylene glycol) 400 on "Fluon" CD4; oven temperature, 90^0C; and helium inlet pressure, 10 p.s.i. A blank determination is carried out in a similar manner with an "empty" glass tube containing only two quartz-wool plugs, and any blank subtracted from the water in the sample.

The major volatile constituents in vinyl chloride - vinyl acetate copolymers is unpolymerised vinyl acetate, and this monomer is difficult to eliminate from the polymer by ordinary drying methods. The plot of volatile constituents (water, vinyl chloride monomer and vinyl acetate) against the sample temperature shows that the maximum amount of vinyl acetate is eliminated from the sample under the test conditions, only in the temperature range 130^0 to 145^0C. Above this temperature the amount of vinyl acetate detected starts to decrease. A similar effect is found for water. At the same time, another peak begins to appear in the chromatogram. This peak corresponds, in retention time, to acetaldehyde. This is consistent with the hydrolysis of the vinyl acetate monomer by water at the elevated temperatures, to give the unstable vinyl alcohol and, hence, acetaldehyde. A heater temperature of 135^0C is recommended for these copolymers.

In the case of acrylic mouldings powders based on methyl methacrylate, the method described above for vinyl and acrylic polymers is again used. The chromatographic conditions used were similar to those described above for poly(vinyl chloride) polymer but, if the chromatogram is less complex, a shorter column (4 feet long) of the same packing can be used. A plot of moisture liberated against sample temperature indicates that 160^0C is the most useful temperature to operate the split heater. A peak due to unpolymerised methyl methacrylate is also obtained and, at first sight, this also appears to be a method for determining residual monomer as well as moisture. Unlike vinyl chloride and vinyl acetate, however, the unchanged methyl methacrylate in the polymer appears to undergo further polymerisation during the 5 min heating period, as the results obtained by the gas chromatographic method are low.

A method involving head space analysis gas chromatography has been described for the determination of monomers and volatiles in PVC.[90,91]

Residual vinylchloride monomer from 0.02 to 0.1 ppm in PVC has been determined by gas chromatographic - mass spectrometric monitoring.[92,93]

The volatiles produced during PVC degradation has been monitored using Raman microline focus spectrometry. Modal polyene chains containing 11-12 or 13-14 double bonds occur in the polymer backbone.[415]

1.3.2 Additives

1.3.2.1 Phenolic and Amine Antioxidants

Vasil'eva et al[71] described a method for the determination of Ionol and 4,4'-isopropylidenediphenol in PVC by anodic voltammetry. The sample is dissolved in dimethylformamide and methanol added to precipitate PVC. It is then diluted with 1.5M aqueous sodium acetate. A polarogram of an aliquot of the clear solution is recorded (graphite rod indicator - electrode and 0.5 M cadmium sulphate - cadmium reference-electrode) to include the steps for 2,6-di-tert-butyl-p-cresol at 0.21 V vs the S.C.E. and for 4,4′-isopropylidene diphenol at 0.53 V vs the S.C.E. The application of voltammetry to the determination of antioxidants has also been discussed by Barendrecht.[72]

Other procedures which have been described include the conversion of an antioxidant into a polarographically reducible form[73] and a general method for antioxidants which involved measuring a decrease in the height of the wave, due to the reduction of dissolved oxygen by antioxidants.[75] Ward[74] has discussed the determination of phenolic and amine types of antioxidants and antiozonants in PVC by the chronopotentiometric technique, using a paraffin wax impregnated graphite indicating electrode[76] and solutions of lithium chloride and lithium perchlorate and acetonitrile in 95% ethanol as supporting electrolytes. Precision obtainable for repeated chronopotentiometric runs in acetonitrile was found to be better than + 1.0% in cases in which electrode fouling did not occur.

Phenolic (and amine) antioxidants in extracts of PVC have been titrated electrometrically with lithium aluminium hydride, using platinum or silver electrodes.[78] Small amounts of water in the sample or analysis solvent have an influence on the results obtained by these procedures.

Electrophoresis is a technique worthy of further consideration for the analysis of antioxidants[79] in PVC. Sawada et al[80] report successful separations by coupling the antioxidants with p-diazobenzene sulphonic acid before electrophoresis. Amine antioxidants are coupled in acetic acid and phenolic antioxidants in sodium hydroxide-ethanol. Electrophoresis was carried out in 1% w/v methanolic sodium borate.

Procedures involving solvent extraction followed by thin-layer chromatography have been described for the determination of phenolic and amine types of antioxidants[47] and optical whiteners in PVC.

1.3.2.2 Plasticizers

Due to their volatility the esters used as plasticizers in polymers such as flexible PVC can be determined by gas chromatographic analysis of a solvent extract of the polymer. Most published methods of analysis are based on this technique and this is the favoured technique for determining such substances (see Table 1.10)

The identification and determination of plasticizers is discussed further in Chapter 4.

Wideline nuclear magnetic resonance spectroscopy has been used (Mansfield[94]) for the determination of the di-iso-octyl phthalate content of PVC. The principle of the method is that the narrow line liquid-type NMR signal of the plasticizer is easily separated from the very broad signal due to the resin; integration of the narrow line signal permits determination of the plasticizer. A Newport Quantity Analyser Mk I low-resolution NMR instrument, equipped with a 40 cm^3 sample assembly and digital read-out, has been used to determine 20 to 50% of plasticizer in poly(vinyl chloride) A curvilinear relationship exists between the signal per g and the percentage by weight of the plasticizer. For highest precision, it is necessary to know the type of plasticizer present; use of the appropriate calibration graph gives a precision of ±0.5%.

Table 1.10 - Gas Chromatographic Methods for the Determination of Plasticizers in PVC

Polymer Extraction	Analysis of Extract	Reference
Extraction of polymer with carbon tetra-chloride-methanol azeotrope for 15 h	Weighing and gas chromatography	82
Extraction of polymer with diethyl ether for 12 h	Gas chromatography	83 - 86, 88
Extraction of polymer with methanol	Gas chromatography	87
Solution of polymer in tetrahydrofuran, precipitation of polymer with methanol, filtration of clear phase	Gas chromatography of THF - methanol extract	89

1.3.2.3 Organotin Heat Stabilizers

Organotin compounds are widely used in the plastics industry as stabilizers for poly(vinyl chloride) compositions. The most important compounds used for this purpose are based on dialkyltin groups R_2 Sn<, especially where R = butyl or octyl.

For analytical considerations it is convenient to divide tin stabilizers into those containing sulphur and those without sulphur. To the first group belong compounds like dialkyltin mercaptides, mercaptoesters and mercapto carboxylates (see method 39) to the second, dialkyltin carboxylates and their esters (see method 40). The identification of acids and alcohols present in tin stabilizers containing no sulphur and the identification of alcohols from stabilizers containing mercaptoesters presents little or no difficulty. However, the identification of the thioacid and of the alkyl groups attached directly to tin can prove more difficult, especially if long-chain acids form part of the compound or if the stabilizer is not pure, eg. if it contains plasticisers. The thioacid may also tend to decompose during hydrolysis procedures.[95]

Methods based on titration and thin-layer chromatography have also been used to determine organotin compounds in solvent extracts of PVC (Table 1.11).

1.3.2 4 Other Types of Heat Stabilizers

Other types of non-tin organometallic stabilizers have been used in the formulation of PVC. Mal'kova et al[99] described an alternating current polarographic method for the determination of cadmium, zinc and barium stearates or laurates in PVC. The samples are prepared for analysis by being ashed in a muffle-furnace at 500^{0}C, a solution of the ash in hydrochloric acid being made molar in lithium chloride and adjusted to pH 4.0 $\pm$ 0.2. The solution obtained is de-aerated by passage of argon and the polarogram is recorded. Cadmium, zinc and barium give sharp peaks at - 0.65, -1.01 and -1.90V respectively, vs the mercury-pool anode.

Other techniques that have been used for the examination of organometallic stabilizers extracted from PVC include column chromatography (barium, cadmium and zinc salts of fatty acids[100]), paper chromatography (cadmium, lead and zinc salts of fatty acids)[101] and polarography (cadmium, lead and zinc salts)[102].

Table 1.11 - Other Techniques for the Determination of Organotin Compounds in PVC Extracts

Compounds Determined	Method	Reference
Organotin heat stabilizers	Potentrometric titration of organotin in polymer extract with standard sodium methoxide in pyridine using antimony and calomel electrodes	95, 96
Dialkyltin thiol compounds	Titration in benzene methanol extract with standard silver nitrate (in IPA water medium) using sulphide coated wire electrode	95
Organotin thioacids and thiols	TLC on kielselgel G, elution with hexane - glacial acetic acid and detection with catechol violet	97
Organotin stabilizers	TLC on kieselgel GF 254, elution with butanol-glacial acetic acid, detection with catechol violet and UV irradiation	47

1.3.3. Fractionation and Molecular Weight

Kirkland and Rementer[411] determined the molecular weight distribution of polyvinyl chloride by thermal field flow fractionation using Mark Houweg constants.

1.4. VULCANIZED AND UNVULCANIZED RUBBERS

1.4.1. Volatiles

Thermal degradation and volatilization of chloroprene rubbers and blends thereof have been studied using thermogravimetric analysis[416].

1.4.2. Additives

1.4.2.1 Accelerators

Millingen[103] has described a thin-layer chromatographic method for the identification of accelerators and antioxidants in nonvulcanized (natural) rubber compounds (Method 41). This procedure discusses interference from other compounding ingredients and describes, in detail, means for overcoming this.

1.4.2.2 Antioxidants and Antiozonants

Lattimer et al[104] have very successfully applied mass spectrometry to the determination and identification of organic additives (antioxidants and antiozonants) in rubber vulcanizates (Method 42). A wide variety of components are involved - polymers, fillers, solvents and organic and inorganic additives. Field desorption/ionization (FD/FI) is the most efficient for identifying typical organic additives in rubber vulcanizates.

1.4.3. Fractionation and Molecular Weight

Kirkland and Rementer[411] studied the molecular weight distribution of polyisoprene using thermal field flow fractionation using Mark Houweg constants.

1.5. POLYESTERS AND POLYAMIDES

1.5.1. Functional Groups

1.5.1.1 Hydroxyl Groups

Houwelingen[105,106] have described a procedure for the determination of hydroxyl groups in the following types of polyesters:

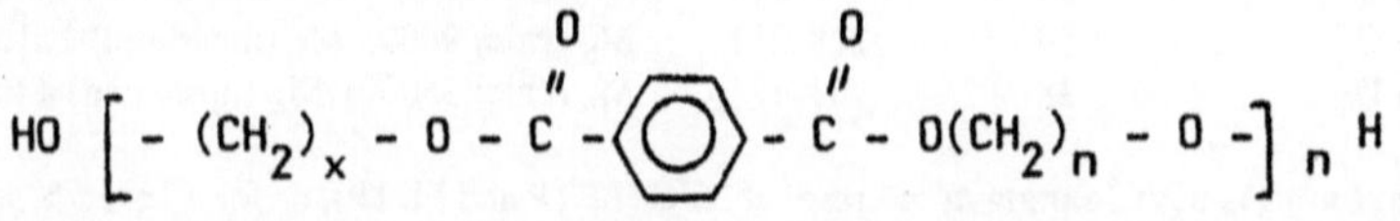

1.2

with n = 1-100 and x = 2 (polyethylene terephthalate) or 4 (polybutylene terephthalate) and ester-interchange elastomers of 4-polybutylene terephthalate and polypropylene glycol.

The hydroxyl groups in these products are determined by acetylation with an excess of dichloroacetic anhydride in dichloroacetic acid at 60^{0}C and measurement of the amount of acetylation by a chlorine determination either by potentiometric titration with silver ions after combustion or by X-ray fluorescence spectroscopy of a compressed disc of the polymer. The suitability of this method for a number of samples is demonstrated in Table 1.12. The standard deviation of the method for high relative molecular weight polymers is 0.7 mmol kg^{-1} whereas for the low relative molecular weight materials, it is about 10 mmol kg^{-1}.

In an alternate method for hydroxy groups described by these workers, the polyester is reacted with a dimethyl formamide - monochlorobenzene solution of vanadium 8-hydroxyquinolinate (V8HQ). According to Tanaka and Kojima[107] a coloured complex with the following structure is formed:

1.3

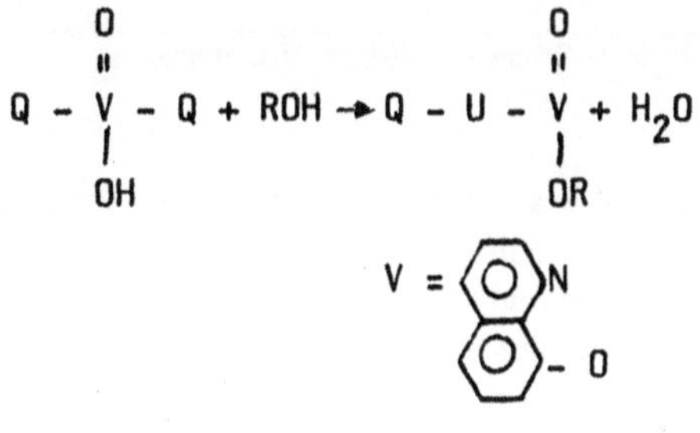

Table 1.12 - Hydroxyl End Group Content of Several PETP, PBTP and PBTP-PPG Samples

Sample	Viscosity ratio*	Hydroxyl End Groups/mmol kg^{-1}	Remarks
PETP A	1.80	30.9-29.8 30.6-30.4	Total end groups** 76.6 mmol kg^{-1}
PETP B	1.82	14.4-14.8	Total end groups 79.1 mmol kg^{-1}
PBTP A	2.07	70.7-70.6	Total end groups 95.2 mmol kg^{-1}
PBTP B	2.08	50.6-49.3	Total end groups 92.6 mmol kg^{-1}
PBTP C	1.24	664-653	
PBTP-PPG A	1.25	742-729	M_N (calc)*** 2700; M_N (measured) 2400****
PBTP-PPG B	1.34	329-345	
PBTP-PPG C	1.50	229-212	M_N (calc) 9000; M_N (measured) 10100
PBTP-PPG D	2.38	51-47	M_N (calc) 35000; M_N (measured) 44800

* Measured for a 1% m/m solution in m-cresol at 25°C (PETP and PBTP), or for a 1% m/V solution in chlorophenol at 25°C (PBTP-PPG)
** Total end groups in sum of OH + COOH + methyl ester end groups
*** M_N measured by gel-permeation chromatography in m-cresol as a solvent
**** Calculated from OH + COOH content

From Van Houwelingen.[105] Royal Society of Chemistry, London

Table 1.13 - Comparison of the V8HQ Method with the Acetylation Method

	OH content/mmol kg^{-1}	
Sample	V8HQ	Acetylation
I	47.9-47.6 43.1-46.8	50-70
II	35.5-34.9	35-50
III	130.1-126.7	170-120
IV	73.8-74.9	50-70

From Tanaka and Kojima.[107] Elsevier Science Publishers, Netherlands

After removal of the excess of reagent by extraction, the complex is acidified with dichloroacetic acid and the blue colour formed is measured at 620 nm. Calibration is carried out with an alcohol as internal standard.

The application of this method to some esters of new types of acids showed that a precise determination is possible (Table 1.13). The standard deviation of this method is 1.5 mmol kg^{-1} which is about ten times more precise than that of the classical acetylation procedure.

Groom and co-workers[108] employed trichloroacetyl isocyanate and trifluoroacetic anhydride acetylations to determine hydroxy end-groups in polyester polyols. The isocyanate reagent was best suited for samples having molecular weights of less than 4500; this anhydride method was more sensitive and applicable to higher molecular weight polymers than is a ^{19}F NMR method.

1.5.1.2 Alkene Groups And Vinyl Esters

Chemical methods for the determination of alkene groups are often based on halogenation or hydrogenation.

A survey of applications has been given by Hirozawa et al[109]. Von Houwelingen[105] have used coulometric bromination to determine vinyl ester end-groups in poly(ethylene terephthalate) formed by thermal chain scission.

1.4

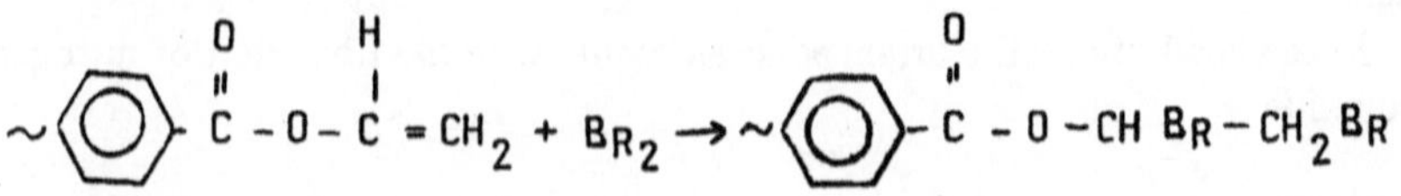

Polyethylene terephthalate

The constant-current generation of bromine is carried out in a medium of dichloroacetic acid, water, potassium bromide and mercury II chloride. To this medium an amount of the polymer, previously dissolved in hexafluoroisopropanol and diluted with anhydrous dichloroacetic acid, is added and bromine is generated. The end of the reaction is detected biamperometrically. The suitability of this method was tested against methyl vinyl terephthalate.

1.5

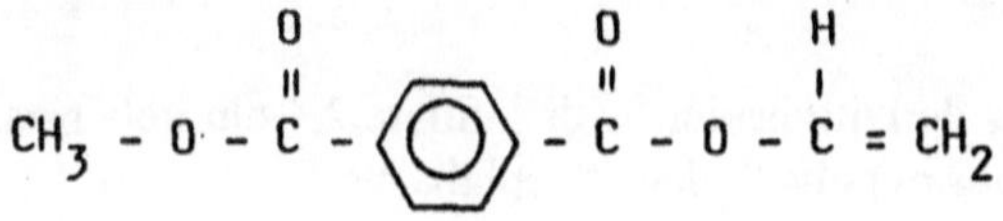

Methyl vinyl terephthalate

Additions of 14.2 and 1.0 umol of methyl vinyl terephthalate (corresponding to 30 and 2 mmol of vinyl ester end-groups per kilogram of polymer) were recovered quantitatively (recoveries of 99.8 and 98.5% respectively).

1.5.1.3 Carboxyl Groups

The determination of these groups in high relative molecular weight polyesters and polyamides gives, in combination with other data, important information about the degree of polymerisation, chain branching, degradation and thermal stability. The choice of the method depends on the solubility of the polymer. Most methods are based on titration in non-aqueous media with visual, potentiometric or photometric indication of the end-point.

A very convenient technique for the determination of carboxyl groups in polyethylene and nylon-6 is photometric titration. The sample is dissolved in o-cresol at 125^0C, cooled, diluted with chloroform and, after addition of bromophenolblue or bromocresol green, the titration is carried out in a spectrophotometer with tetrabutyl ammonium hydroxide. The change in absorbance is recorded continuously and the end-point is determined graphically. A detailed description has been given by Van Lingen.[111]

Table 1.14 compares applications of the photometric titration method together with a comparison with a potentiometric titration with tetrabutyl ammonium hydroxide in aniline of low molecular weight poly(ethylene terephthalate) and nylon 6. Generally the agreement is very satisfactory.

An attractive method for determining carboxyl groups is based on the Schmidt reaction, in which the carboxyl group is subjected to reaction with sodium azide in sulphuric acid.[112]

The acylazide formed rearranges to an amine with the liberation of nitrogen and carbon dioxide

$$2RCOOH + 2NaN_3 + H_2SO_4 = 2RNH_2 + 2CO_2 + 2N_2 + Na_2SO_4 \qquad 1.6$$

The carbon dioxide evolved is measured quantitatively by an automatic non-aqueous titration. This method is applicable to poly(m-phenylene isophthalamide) which is soluble in dimethylformamide, but is not applicable to poly (p-phenylene terephthalamide) for which no organic solvent could be found.

1.5.1.4 C-O-C Groups In Polyesters

Simak[388] investigated absorptions of the C-O-C group in polyethylene terephthalate in the infrared at 632, 640 and 380 cm^{-1}.

1.5.1.5 Amino Groups

Heterogeneous derivitivisation with 1-fluoro,2,4-dinitrobenzene has been used to determine amino groups in polyethylene terephthalate.[105]

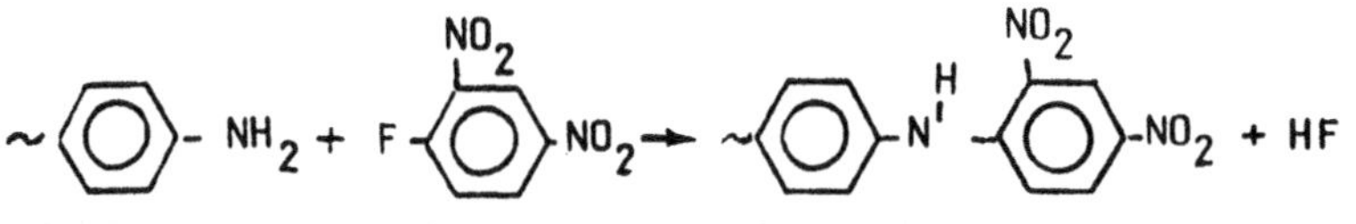

1.7

Polyethylene terephthalate

The polymer is treated for 4 h at 80^{0}C with 1 fluoro,2,4 dinitrobenzene in an ethanol hydrogen carbonate medium. After washing out the excess of reagent, the dinitrophenyl group introduced is measured spectrophotometrically at 430 nm after dissolution of the derivatised polymer in methane sulphonic acid.

The spectrophotometric method was calibrated with the 1-fluoro,2,4-dinitrobenzene derivative of the model compound NN'-bis(p-amino-phenyl) terephthalamide:

Table 1.14 - Application of the Photometric Titration Method to Various Samples

Sample		Content of Carboxyl Groups/mmol kg^{-1}			
		Photometric Titration		Potentiometric Titration	
		x [a]	standard deviation	x [a]	standard deviation
PETP (M_N low	C	4.3	0.25	4.0[b]	0.10
	D	12.0	0.34	12.0[b]	0.30
	E	17.4	0.56	17.5[b]	0.49
PETP (M_N high	F	36.8	0.32	35.2	0.9
	G	68.3	0.29	68.2	1.2
	H	113.4	0.52	112.8	1.8
Nylon 6	I	73.3	0.31	72.4[c]	1.0
	J	53.2	0.22	52.5[c]	0.37
	K	35.3	0.45	36.2[c]	0.32

[a] Mean of four independent determinations.
[b] Titration in aniline at 40°C.
[c] Titration in benzyl alcohol-methanol-water.
From Van Lingen.[111] Springer Verlag Heidelberg.

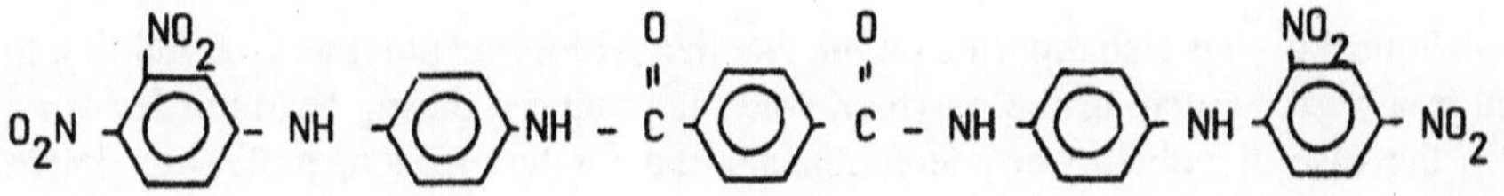

1.8

1-fluoro,2,4 dinitro benzene derivative of N,N′-bis (p-amino-phenyl) terephthalamide

1.5.1.6 Acidic And Glycol Repeat Units In Terylene (Dacron)

Terylene is manufactured from terephthalic acid and ethylene glycol as follows:

1.9

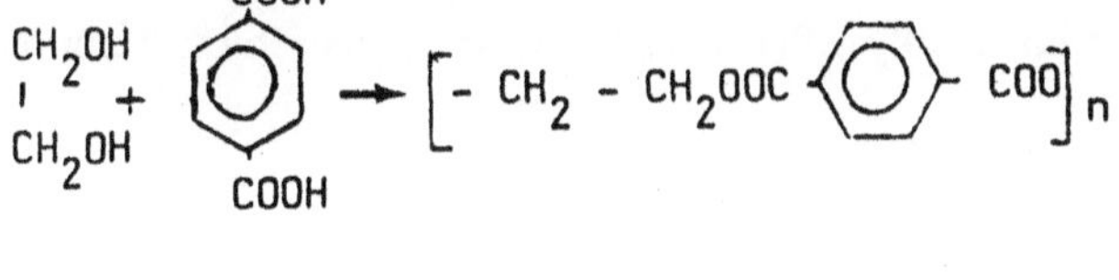

Synthesis of Terylene

Such polymers contain repeat units based on ethylene glycol and indeed other glycols such as 1,4-butane diol, and terephthalic acid and proportions of isophthalic acid. Allen et al[113] developed a precise quantitative method (method 43) for determining such units in terylene. The sample is subject to an alkaline hydrolysis and glycol and acidic products after conversion to the trimethylsilyl derivatives are analysed by gas chromatography.

It has been observed that in the preparation of polyester for compositional analysis by NMR spectroscopy much higher rates of hydrolysis are achieved by refluxing polyethylene terephthalate with a mixture of sodium hydroxide, an alcohol (methanol) and a protic solvent (dimethyl sulphoxide) than is achieved with the usual alcoholic alkali reagents.[417]

1.5.2. Volatiles and Water

1.5.2.1 Water In Polyesters.

Hoffman[114] has described a procedure (Method 44) for the determination of water in polyesters. In this method, water is allowed to react quickly with a mixture of hexamethyldisiloxane and trimethylchlorosilane (2:1) in the presence of pyridine to form hexamethyldisiloxane.

$$2(Me_3Si)_2NH + 3H_2O + 2Me_3SiCI = 3Me_3SiOSiMe_3 + 2NH_4CI \qquad 1.10$$

The hexamethyldisiloxane is separated from other silanized active hydrogen compounds and determined by gas chromatography.

1.5.2.2 Water In Poly(Ethylene Terephthalate).

Squirrel[115] has also reported on the fact that with some polymers, in addition to the initial free water content of the polymer, water is produced during heating. He observed this in the case of polyethylene terephthalate and for this reason, prefers to determine water by Karl Fischer titration.

1.5.2.3 Volatiles.

Gas chromatographic methods have been described for the determination of volatile constituents in polyester coated film[116] and in polyesters.[116]

Welsch et al[116] used gas chromatography combined with accumulated dosage to determine volatiles. Volatiles from a heated sample are transferred in a carrier gas stream to a cooled initial part of the GC column and are subsequently discharged onto the column by heating. The products were identified by mass spectrometry.

Urbanski[117] measured small amounts of propylene oxide and epichlorohydrin in polyesters colorimetrically after reaction with a chloroform solution of 2,4-dinitrosulphonic acid.

1.5.3. Oligomers

Supercritical fluid extraction and chromatography[418] and comparative complimentary phasing desorption mass spectrometry secondary ion mass spectrometry have been used to determine oligomers in polyethylene terephthalate.[420]

1.5.4. Additives

1.5.4.1 Flame Retardants.

Surface tris (2,3-dibromopropyl) phosphate has been determined on the surface of flame retardant polyester fabrics[118]. The technique used to determine the surface TDP levels involved extraction of the fabric with an organic solvent followed by analysis of the solvent by x-ray fluorescence for surface bromine and by high pressure liquid chromatography for molecular tris (2,3-dibromopropyl) phosphate.

Cope[121] has described a pyrolysis - gas chromatographic procedure for the determination of tetrakis(hydroxymethyl)phosphonium hydroxide and tris (2,3-dibromopropyl)phosphate flame retardants on polyesters and cotton.

1.5.4.2 Residual Organic Peroxides.

Iodometric methods have been described[119] for the determination of organic peroxides in polyesters.

1.5.4.3 Unreacted Starting Materials.

Tengler and Von Falkai[120] determined diols, adipic acid and volatiles cyclomonodiol adipates qualitatively and quantitatively in aliphatic polyesters based on adipic acid. The identification of the cyclic mon-esters is accomplished with the help of coupling a gas chromatograph with a mass spectrometer, the diols were classified by gas chromatographic retention volume.

1.6. POLYACRYLATES

1.6.1. Functional Groups

1.6.1.1 Hydroxy Groups.

Dickie et al[389] derivitivized surface hydroxy groups on acrylic copolymers with ammonia then characterized them by x-ray photoelectron microscopy.

Kim et al[390] examined the strong infrared band at 3450cm^{-1} of tetrahydrofuran associated hydroxy groups in dinitropropyl acrylate prepolymer in order to determine hydroxy groups.

1.6.2 Monomers and Water

Polyacrylates are used extensively as dispersants for emulsion paints and as boiler-scale inhibitors and copolymers of acrylic acid and acrylamide are used to aid water clarification at water treatment works and for the treatment of coal tailings. The manufacture of such polyelectrolytes could leave a small residue of unpolymerised acrylic acid monomer in the polymer.

Brown[122] has described a high performance liquid chromatographic method for the determination of acrylic acid monomer in polyacrylates (method 45).

A method for the determination of water in polyacrylate is discussed in section 1.1 under polyethylene.

1.6.3. Oligomers

Comparative complimentary plasma desorption mass spectrometry/secondary ion mass spectrometry has been used to determine oligomers in polyacrylates.[420]

1.6.4. Additives

1.6.4.1 Phenolic Antioxidants

Budylina et al[123] have described methods based on anodic voltammetry for the determination of Ionol (2,6-di-tert-butyl-p-cresol) and quinol in polyester acrylates. To determine Ionol the sample is dissolved in acetone:methanol (1:2 v/v) and an aliquot diluted with a solution of lithium chloride (0.1M) and sodium tetraborate (0.02M). A polarogram is recorded with a graphite-rod indicator-electrode. To determine quinol, the sample is dissolved in methanol or methanol:acetone (1:1 v/v) and diluted to 100 ml with the lithium chloride-sodium tetraborate solution. A polarogram is recorded under the same conditions. The $E_{½}$ values (vs. the S.C.E.) are 0.25V for Ionol and 0.16V for quinol.

1.6.5. Fractionation and Molecular Weight

Kirkland and Rementer[411] determined the molecular weight distribution of polyacrylates using thermal field flow fractionation using Mark Houweg constants.

Jackson et al[433] compared the most probable peak values as measured for polymer distributions by matrix assisted laser desorption/ionization (MALDI) mass spectrometry and by size exclusion chromatography (SEC). MALDI provided the most probable peak (Mp) values for polymethyl methacrylate polymers that are low reletive to Mp values supplied by the polymer manufacturers which were obtained by SEC. The reasons for this discrepancy are discussed.

1.7. POLYETHYLENE GLYCOL AND POLYPROPYLENE GLYCOL

1.7.1. Oligomers

Comparative complimentary plasma absorption mass spectrometry has been used to determine oligomers in polyethylene glycol.[420]

Thirty polyethylene glycol oligomers have been separated on K^+ - form cation exchange resin.[423]

Time of flight secondary ion mass spectrometry has been used to determine the molecular weight distribution of polyethylene glycol oligomers.[421]

1.7.2. Functional Groups

1.7.2.1 Hydroxyl Groups.

Earlier methods for the determination of hydroxyl groups in polyethylene and polypropylene glycols were based on reaction with p-toluene sulphonic acid catalysed acetyl chloride[124-126], phthalic anhydride[126], perchloric acid catalysed acetylation[126-127,129], ß-nitro phthalic anhydride[129], pyromelliticanhydride[131], phenyl isocyanate[129,132] and potassium periodate.[125]

Stetzler and Smullin[126] found that for poly(propylene glycols) the classical phthalation procedure gave consistently low results as did the perchloric acid catalysed acetylation procedure, developed by Fritz and Shlenk[127]. In addition to its greater intrinsic accuracy, the advantages claimed for the Stelzer and Smullin[126] toluene p-sulphonic acid catalyzed procedure over phthalation include shorter reaction time and lower reaction temperature also good reproducibility, indicating no reaction with the ether groups. In Table 1.15 are compared hydroxy values obtained by three methods. It is seen that in every case the acid catalyzed acetylation method gives a higher result than that obtained by phthalation, the average difference between the two methods being 2.5% of the determined value. The agreement between the p-toluenesulphonic acid method and the phenyl isocyanate method is good.

Kaduji and Rees[134] have described a direct injection enthalpimetric method to measure the hydroxy value of glycerol alkylene oxide condensates and butane-1,4-diol-adipic acid polyesters.

Smith and Dawson[135] determined traces of poly(ethylene glycol) by fission with hydrogen bromide to produce dibromomethane and dibromo-propane followed by gas chromatographic determination of these two substances.

A method (Method 46) for the determination of hydroxy group in polyethylene glycols has been described by Fritz et al[133]. The method consists of substituting the hydroxyl group with a chromophoric siloxy group, purification of the silylated polymer, and a photometric determination of the chromophor concentration. The silanization of the primary hydroxyls with dimethylaminosilanes proceeds quantitatively under very mild conditions and elimination of the excess reagent by precipitation is easy. The precision of the method is $\pm$ 4.5% (95% confidence level) down to 5×10^{-4} mol kg^{-1}. The method has about the same precision as acylation methods but yields a 1000 fold gain in sensitivity.

Most of the chemical methods for determining hydroxy groups in polymers do not distinguish between primary, secondary or tertiary hydroxy groups. An exception to this is a kinetic approach to the determination of primary and secondary hydroxy groups as will be discussed under copolymers (Chapter 2.4) under ethylene oxide polyglycol condensates.

A further method for distinguishing between different types of hydroxy groups in polymers is the NMR method described by L'Ho[136]. This procedure is based on the reaction of the hydroxy compound with hexafluoroacetone to form an adduct which is amenable to F-19 NMR spectroscopy.

$$(CF_3)_2\ C = O + ROH = C(CF_3)_2\ OHOR \qquad 1.11$$

A fluorine resonance spectrum of a mixture of several hexalfuoroacetone adducts is shown in Figure 1.2. There are two interesting features in this spectrum. First, it illustrates clearly the high resolution and the information on structural aspects of the molecules one can obtain by this method. The hydroxyl adduct from each alcohol gives a sharp resonance, with the tertiary adducts at high field, followed by secondary and then primary. The chemical shift of the hexafluoroacetone alcohol adduct is determined by the structural environment of the hydroxyl. For example, the gradual upfield shift observed in the series methanol, ethanol, isopropanol and tert-butanol, results from the increase in shielding caused by replacing a hydrogen atom with a methyl group. The high degree of resolution under fig. 1.2 was also observed for polymeric materials. Therefore, not only can the total hydroxyl concentration be determined but also the type or types present in the polymer.

Table 1.15 - Comparison of Results Obtained on Polypropylene-glycol by Catalyzed Acetylation, Phthalation and Phenyl Isocyanate Reaction

MOl wt	Stetzler and Smullin [126] PTSA Method Average OH No. mg/KOH/g	Phthalation Method Average OH No. mg/KOH/g	Reaction with phenyl isocyanate Average OH No. mg/KOH/g
760	353	346	
260	276	265	
2000	34.9	34.0	34.9
3000	34.3	33.8	35.1
3390	58.3	57.0	

From Stelzer and Smullin.[126] American Chemical Society

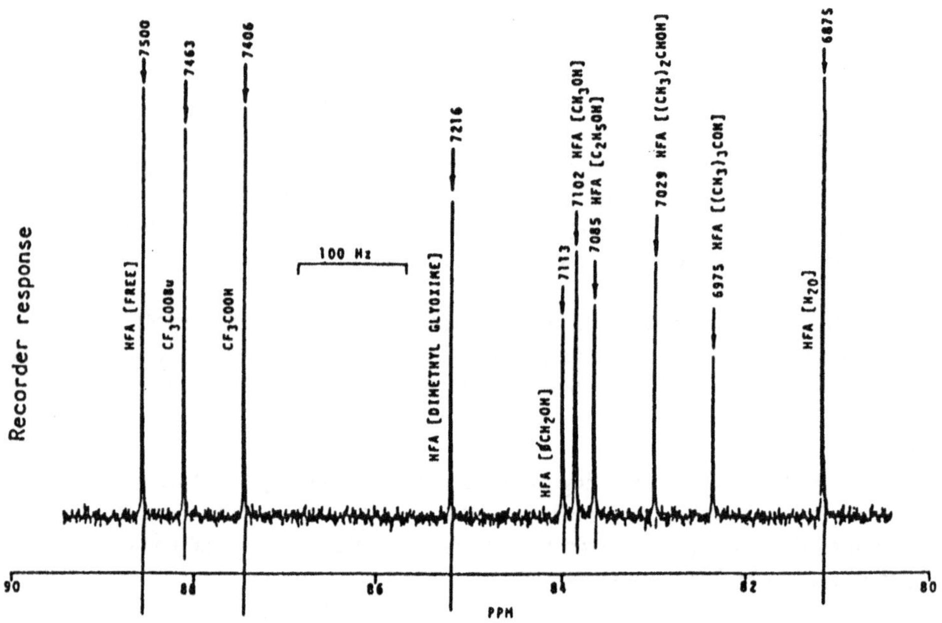

Figure 1.2. Fluorine-19 NMR spectra of several HFA adducts. From left: HFA (free); Butyl trifluoroacetate; trifluoroacetic acid; HFA dimethyl glyoxime); HFA (benzyl alcohol); HFA (methanol); HFA (ethanol); HFA (isopropanol); HFA (tert-butanol) and HFA H_2O). The numbers on top are Hz downfield from C_8F_8 lock. From L'Ho with permission.[136] American Chemical Society.

The quantitative aspect of hydroxyl determination is illustrated in Table 1.16 for the analysis of hydroxyl in some polymeric materials.

1.7.2.2 Ether Linkages.

Smith and Dawson[135] determined traces of polyethylene glycol and polypropylene glycol by reacting the sample with hydrobromic acid then determining the dibromoethane and dibromopropane fission products by gas chromatography.

1.7.3. Fractionation and Molecular Weight

Various workers have discussed the fractionation of polyethylene glycols using laser desorption / Fourier transform ion cyclotron resonance mass spectrometry[403,404] and flow field - flow fractionation[405] techniques.

Kirkland et al[420] determined the molecular weight distribution of polyethylene oxide in the 10^4 to 2×10^6 molecular weight range by field flow fractionation.

Vincenti et al[412] used chemical ionization mass spectrometry to determine the molecular weight distribution of polyethylene glycol.

Polyethylene glycols have been detected using an indirect conductiometric detector following chromatographic separation.[422]

Dey et al[434] used matrix assisted laser desorption/ionization (MALDI) Fourier Transform mass spectrometry (FTMS) to determine the molecular weight distribution of polyethylene glycol samples containing oligomers. This combination of techniques create a powerful tool for the characterisation of polymers. MALDI permits the creation of intact, singly charged, high mass (m/z 10^3-10^4) ions of polar and non-polar analytes, whilst FTMS gives high resolution and mass accuracy. The technique was applied to polyethylene glycols containing oligomers with masses covering the IOk Dalton mass range.

1.8 POLYBUTADIENE

1.8.1 Additives

1.8.1.1 Amine Antioxidants And Antiozonants.

Protivova and Pospisil[137] have reported on the behaviour of some amine antioxidants and antiozonants and some model substances (phenols, aromatic hydrocarbons and amines) during gel permeation chromatography and have applied this technique (method 47) to the analysis of rubber extracts.

The diazotized p-nitroaniline procedure has been applied to the determination of amine antioxidants in ethanol extracts of thin films of rubber.[56]

Gaeta et al[146] have described a gas chromatographic method for determining a number of commercial antioxidants such as N-phenyl-2-naphthylamine and N,N′-sec-heptyl phenyl phenylene diamine in oil-extended synthetic polymers such as poly-butadiene or styrene-butadiene rubber. This involves extracting the antioxidant from the polymer with ethanol, taking up the concentrated extract with the appropriate solvent and analysing the resulting solution by gas chromatography. They found by the use of standard solutions of the antioxidants in carbon tetrachloride, acetone or carbon disulfide and by the careful choice of chromatographic conditions that the elution of these materials had little or no interference from the extending -oil present. In addition, the oil was completely soluble in the solvent and all but the heaviest fractions eluted prior to that of the antioxidant.

High temperature gas chromatography has been used[144,145] for the analysis of mixtures of amine type antidegradants in rubber. These workers used a separation column constructed of aluminium packed with 20% Apiezon L on 30 - 60 mesh Chromasorb W. Analysis was carried out on an acetone extract of the rubber sample, employing diphenylamine as an internal standard. Using column temperatures up to 310^0C they were able to separate a range of antidegradants including 1,2-dihydroxy-2-

Table 1.16 - Quantitative Determination of Hydroxyls in Several Commercial Polymers

Polymers	Hydroxyls	
	Found	Expected
Aliphatic polyethers		
Polyethylene glycol		
Carbowax 200	17.40	16.2-17.9[a]
Carbowax 1000	3.54	3.24-3.58[b]
Polypropylene glycol		
Varonol P1010	3.46	3.37[d]
Voranol P4000	0.85	0.85[e]
Polyepichlorohydrin[f]	0.84	0.85
Aromatic polyethers		
Bakelite phenoxy resin PKHC	5.8	6.0[g]
Monosanto R, J-100 resin	5.4	5.4-6.0
Polyesters		
Atlac 382E	0.92	0.93
Aromatic ester resin No. 1[h]	8.20	8.30
Aromatic ester resin No. 2[h]	12.20	12.35

[a] Based on literature molecular weight of 190 to 210.
[b] Based on literature molecular weight of 950 to 1050
[c] Determined after addition of 4.7 molar equivalent of trifluroacetic acid and 7.6 molar equivalent of water.
[d] Based on an average molecular weight of 1010.
[e] Based on an average molecular weight of 4000.
[f] Polyepichlorihydrin 2000 from Shell Chemical Co.
[g] Calculated from the idealized molecular structure with a unit weight of 284.
[h] Experimental resin from Hercules Incorporated.

From L'Ho.[136] American Chemical Society

2,4-tri-methyl-6-ethoxy-quinaline, N-isopropyl-N^1-phenyl-p-phenylenediamine, N-phenyl, -2-2naphthalamine,N,N'-di-2-octyl-p-phenylene-diamine and N,N'-diphenyl-p-phenylenediamine. Near quantitative determinations were obtained for all these substances. Apiezon L was found to be distinctly superior to other substances tried as column packings.

Solvent extraction gas chromatographic methods have been described[146] for determining alkylated cresols (2,6-di-tert-butyl-p-cresol) and amine antioxidants (N-phenyl-2-naphthylamine, p-phenylene diamine type) and Santoflex (NN'sec-heptyl phenyl-p-phenylene diamine) in polybutadiene.

1.9 CELLULOSE

1.9.1 Water

Details of a Karl Fischer titration procedure which is suitable for polymers such as cellulose are given in Method 48.

1.10 POLYAMIDES AND POLYIMIDES AND POLY(AMIDES-IMIDES) (SEE ALSO SECTION 1.5)

1.10.1 Functional groups

1.10.1.1 Amino Groups In Aromatic Polyamides, Polyimides And Poly(Amides-Imides)

Schleuter and Siggia[147-148] and Frankoski and Siggia[150] used the technique of alkali fusion reaction gas chromatography for the analysis of imide monomers and aromatic polyimides, polyamides and poly(amide-imides) (Method 49).

Two indirect titration procedures have been described for the determination of amino end-groups in aromatic polyamides. One method involves the reaction of the amine with salicyladehyde to form the Schiff base. After precipitation of the polymer, the excess, unreacted aldehyde is titrated with potassium hydroxide.[151]

Amino groups have also been acetylated with acetic anhydride in dimethylacetamide. Diethylamine was added, and the excess amine was titrated potentiometrically[152]. The sequence distribution of an aromatic polyamide terpolymer prepared under various reaction conditions was determined by nuclear magnetic resonance spectrometry[153]. Infrared spectroscopy[154,155] and mass spectrometry[156] have been used to estimate the degree of conversion of polyimides, ie. the extent of polyamic acid ring closure.

Gomoryova[157] hydrolyzed amide and imide linkages with hydrochloric acid and identified the diamines using paper or thin-layer chromatography. The diamine portion of polyamides has also been determined by fusing the sample with an alkali reagent and separating the products by thin-layer chromatography[158] Acidification of the melt allowed the separation and identification of the diacid components. Polyimides have been decomposed with hydrazine hydrate and the diamine products identified by gas chromatography[159]. Polyamides and poly(amide-imides) required a prehydrolysis step. The di, tri- or tetracarboxylic acid segments were determined by reaction of the polymer with a 10% aqueous solution of tetramethylammonium hydroxide, pyrolysis of the resulting salt and identification of the volatile methyl ester by gas chromatography.[159]

Kalinina and Doroshina[160] have reviewed qualitative and quantitative methods for the analysis of polyamides and polyimides.

1.10.1.2 Carboxyl Groups.

A method for the determination of carboxyl groups is discussed in section 1.5.1.

Carboxyl end-groups in poly(m-phenyleneisophthalamide) have been titrated potentiometrically in dimethylformamide using a non-aqueous potassium hydroxide solution.[151]

1.11 POLYACRYLAMIDES

1.11.1 Monomers

Due to its toxicity, very sensitive methods are required to determine acrylamide monomer in polyacrylamide.

Betso and McLean[161] have described a differential pulse polarographic method for carrying out this determination (method 50). A measurement of the acrylamide electrochemical reduction peak current at -2V v, SCE is used to quantitate the acrylamide concentration.

High performance liquid chromatography, using the reverse phase mode has been used[162] to determine acrylonitrile monomer and related compounds, including meth-acrylonitrile in polyacrylamide (Method 51).

1.12 POLYURETHANES

1.12.1 Volatiles

Guthrie and McKinney[164] have described a thin-layer chromatography - fluorimetric method (Method 52) for the determination of 2,4 and 2,6-diaminotoluene in flexible polyurethane foams.

High performance liquid chromatography has been used[391] to determine methylene bis aniline volatiles in polyurethanes.

Joseph and Browner[392] studied the types of volatiles produced upon the combustion of polyurethane foams.

1.12.2 Characterization

Time of flight secondary ion mass spectrometry (SIMS) has been used[406] to carry out structure characterizations on polyurethanes based on diols (ethylene glycol, 1,4 butane diol, 1.6-hexanediol) and diisocyanates, (methylene disocyanate, toluene diisocyanate, dicyclohexylmethane diisocyanate). Fragments and oligomers produced in the m/z range 500 - 3200 enabled characterizations to be carried out.

1.13 POLYVINYL ALCOHOL

1.13.1 Functional groups

1.13.1.1 Hydroxy.

Baumgartner[371] has shown that the solution absorbance of a PVA-iodine-boric acid complex depends on the concentration of cadmium in ions in solution.

1.14 SILICONES AND SILOXANES

1.14.1 Functional groups

1.14.1.1 Alkyl and Aryl Groups.

The technique of alkali fusion reaction gas chromatography has been applied to the determination of alkyl and aryl groups in polysiloxanes. The method involves the quantitative cleavage of all organic substituents bonded to silicon, producing the corresponding hydrocarbons. Reactions are driven to completion, with no apparent decomposition, in less than 10 minutes by fusing the sample with potassium hydroxide in an inert atmosphere. After concentration of the volatile products, they are separated and determined by gas chromatography. The structure of the silicone can be elucidiated from the nature and concentrations of the decomposition products. Fluids, gum rubbers and resins are handled with equal ease. The percent relative standard deviation of the method is 1.0%; the average deviation between experimental and theoretical or check method results is 0.5% absolute. This technique for carrying out the potassium hydroxide fusion procedure is the same as the analysis described in method 49, in the previous section,[147] for the analysis of polyamides and polyimides. Various other reagents have been used for the determination of organic substituents bonded to silicon in organosilicon polymers (Table 1.17).

1.14.1.2 Si OH and Si H Groups.

Dubiel et al[178] described methods for determining these reactive components in room temperature vulcanized silicone foams. Total SiOH and SiH are determined by Fourier transform infrared (FTIR) spectrometry, the SiOH peak at 3687 cm^{-1} and the SiH peak at 2168 cm^{-1} were used for quantitation. The tetrapropoxysilane content was determined by gas chromatography using a solid capillary open tubular (SCOT) column and linear programmed temperature control. The diphenylmethylsilanol content was determined by gel permeation chromatography using the tetrahydrofuran solvent.

1.14.2 Oligomers

Comparative complimentary plasma desorption mass spectrometry/secondary ion mass spectrometry has been used to determine oligomers in siloxanes.[420]

Supercritical fluid chromatography combined with time of flight mass spectrometry has been used to determine the molecular weight distribution of poly(dimethylsiloxane) oligomers in the mass range 10^3 to 10^4.

1.14.3 Fractionation and Molecular Weight

Vincenti et al[412] determined the molecular weight distribution of polysiloxanes by desorption chemical ionization mass spectrometry.

Supercritical fluid chromatography coupled with time of flight mass spectrometry has been used to determine the molecular weight distribution of poly(dimethyl siloxane) oligomers in the mass range 10^3 to 10^4.

1.15 PHENOL FORMALDEHYDE RESINS

1.15.1 Oligomers

Ludwig[393] used high performance liquid chromatography to determine oligomers in p-alkyl phenol-formaldehyde resins.

1.16 POLYOXYMETHYLENE GLYCOLS

1.16 Oligomers

To determine formaldehyde oligomers in polyoxymethylene glycols and the corresponding polyoxymethylene ethers Utterback et al[394] first derivitivized them with ammonia then determined the oligomers by capillary column gas chromatography.

Polyoxymethylene oligomers have been separated by temperature programmed gas chromatography.[425]

Table 1.17 - Reaction Methods for the Quantitative Determination of Organic Subtituents Bonded to Silicon

Group Determined	Reagent	Reaction Conditions	Product	Analysis	Ref.
Phenyl	60% aqueous KOH in DMSO	2 h at 120°C	Benzene	GC	166
Phenyl	Bromine in glacial acetic acid	Boiling solution	Bromo-benzene	Titration of excess bromine	167
Ethyl and Phenyl	Phosphorus pentoxide and water	30-580°C over 45 min	Ethane and benzene	GC-FID	168
Methyl and Ethyl	Powdered potassium hydroxide	2 h at 250-270°C	Methane and ethane	Gas buret	169
Methyl	Sulphuric acid	20 min at 280-300°C	Methane	Gas buret	170,171
Phenyl	Ethyl bromide in the presence of aluminium chloride		Hexaethyl benzene	Gravimetric	172
Vinyl	Phosphorus pentoxide and water	80-600°C over 40 min	Ethylene	GC-FID	173
Vinyl	Phosphorus pentoxide	Ambient to 500°C	Ethylene	GC-FID	174
Vinyl	90% sulfuric acid	75-250°C at 10°C/min and 1 h at 250°C	Ethylene	GC-TC	175
Vinyl	Sodium hydroxide	300°C for 15 min	Ethylene	Colorimetric	176
Vinyl	Potassium hydroxide pellets	Heat with Meker burner	Ethylene	GC-FID	177

1.17 POLY(STYRENE SULPHONATES)

1.17.1 Fractionation and Molecular Weight

Poli and Schure[407] used capillery electrophoresis to fractionate poly(styrene sulphonates) over a wide molecular weight range. The technique was superior to size exclusion chromatography in terms of resolution, efficiency and fractionating power.

Field flow fractionation has been used to fractionate and determine the molecular weight of sodium polystyrene sulphonate[426,427] and polystyrene sulphonate (molecular weight range 6.6×10^3 to 6.9×10^5).

Thielking and Kuliche[435] used a combination of flow field flow fractionation and multiple laser light scattering for the characterization of a sulphonated polystyrene solutions. These workers examined seven sulphonated polystyrene standards ($1800 - 3 \times 10^6$ g $mole^{-1}$), which were taken as model substances for macromolecular poly-electrolytes. The technique was applied to 0.1M sodium nitrate solutions of the polyelectrolytes.

1.18 FLUOROPOLYMERS

1.18.1 Oligomers

Comparative complimentary plasma desorption mass spectrometry/secondary ion mass spectrometry has been used to determine oligomers in polytetrafluoroethylene.[420]

Davidson et al[436] characterized perflurotetradecahydrophenanthroline oligomers using a combination of nuclear magnetic resonance spectroscopy, electron spin resonance spectroscopy for chemical analysis and time of flight secondary ion mass spectrometry, infrared spectroscopy and ultraviolet-visible spectroscopy.

They determined the following distribution of groups in this polymer;

Olefin carbon 8%, CF_3 8%, CF_2 51%, CF 33%

A structure $C_{14}F_{23}(C_{14}F_{22})_n C_{14}H_{23}$ is proposed where n = 0, 1 or 2 corresponding to dimer, trimer and tetramer.

1.18.2 Fractionation and Molecular Weight

The molecular weight and composition of a perfluorinated polyether has been determined from fragment intensities using time of flight secondary ion mass spectrometry[428]. Continuose flow fast atom bombardment and field desorption mass spectrometry of 3,3,3 trifluorophenyl propyne oligomers determined chain length, molecular weight range and identities of end-groups[429].

1.19 POLYCARBONATE

1.19.1 Composition

Supersonic jet spectrometry of the chemical species resulting from the thermal decomposition of polycarbonate at temperatures up to 400^{0}C showed that p-cresol is the major decomposition product.[409]

1.19.2 Oligomers

Comparative complimentary plasma desorption mass spectrometry/secondary ion mass spectrometry has been used to determine oligomers in polycarbonate.[420]

1.20 POLYVINYLPYRIDINE

1.20.1 Fractionation and Molecular Weight

Field flow fractionation has been used to fractionate and determine the number average molecular weight distribution of polyvinylpyridine.[430]

1.21 POLYALKYLKETENES

1.21.1 Dimers

Zimmermann et al[437] used time of flight secondary ion mass spectrometry to determine alkylketene dimer in kraft papers.

Further methods for the determination of functional groups, volatiles and additives in polymers are reviewed in Tables 1.18 to 1.21.

Table 1.18 - Functional Groups in Polymers

Group	Polymer	Principle of Method	Ref.
Primary amino	Amino containing polymers	Fluorescence	179
Primary amino	Amino containing polymers	Spectrofluorimetric, (p-dimethyl-aminobenzaldehyde)	180
6-caprolactam	Poly(ethylene terephthalate)	Spectrophotometry (ninhydrin)	181
Carbonyl	Vinyl chloride-vinyl-acetate copolymer	Spectrophotometric, (2:4 dinitrophenyl hydrazine)	182
Amino	Amino containing polymers	Spectrophotometry of 4,4' dimethoxytrityl chloride derivative at 498 nm	395
Carboxyl	Polyacrylic acid	Titration with standard organic base in non-aqueous medium	183,184
Carboxyl	Carboxy terminated polybutadiene	Infrared spectroscopy	185
Carboxyl	Carboxyl methyl cellulose	Precipitation as copper or uranyl salt	186,187
Chloro groups	Polystyrene	13 C NMR	188
Epoxy groups	Epoxy resins	Titration with halogen acid spectrophotometric	189
Epoxy groups	Epoxy resins	(2,4 dinitrobenzene sulphonic acid)	190
Ester	Methacrylate copolymers	Pyrolysis to alkyl iodides-gas chromatography	191,192
Ester	Alkyd resins, polyesters	Saponification in alcoholic potassium hydroxide	193,195
Ester	Cellulose, polyvinyl esters, polyacrylates polymethacrylates	Saponification in ethanolic or glycol solution of potassium hydroxide	196-198
Ether	Cellulose, polyvinyl ethers	Zeisel method	199
Ethylene	Polyethylene oxides	NMR	200
Hydroxyl	Epoxy resins	Reaction with lithium aluminium hydride	201
Hydroxyl	Polycarbonates	Spectrophotometric (ceric ammonium nitrate)	202
Hydroxyl	Hydroxypolybutadienes		185
Nitric ester groups	Nitrocellulose	Saponification, reduction with devades alloy determination of ammonia	204
Nitrile	Styrene-acrylonitrile copolymer	Infrared spectroscopy	205,206
Oxirane rings	Epoxy resins	Ringcleavage with pyriding-HCl, glacial acetic acid HBr, dioxan-HCl	207,208
Unsaturation	Di and polyolefins	Halogen addition	210,211
Unsaturation	Poly(methylmethacrylate)	Bromination	212
Unsaturation	Butyl rubber	Chlorination with radio chlorine	213,214
Unsaturation	1,2-polybutadiene	NMR	215

Table 1.19 - Residual Volatiles in Polymers

Volatile	Polymer	Method	Reference
Styrene o-xylene	Polystyrene	Head-space analysis, glc	216
Styrene	Polystyrene	Head-space analysis, glc	217,218
Dichloromethane	Polycarbonate	Head-space analysis	216,
Styrene ethyl benzene	Polystyrene	Solution, glc	219-224

Table 1.19 continued

Styrene	Polystyrene	Solution in o-dichlorobenzene or methylene di-chloride, glc	225
Styrene	Polystyrene	Solution in DMF, glc	226
Styrene, o-methyl styrene	Polystyrene	Solution in propylene oxide, glc	64,228
Benzene, toluene, ethyl benzene, xylene, cumene, propyl benzenes, ethyl toluenes, butyl benzenes, styrene	Polystyrene	Solution in THF, glc, methanol precipitation of polymer, ultraviolet spectroscopy	224-231 237
Styrene	Polystyrene	Polarography	232,233
Alpha o-methylstyrene	Polystyrene	Polarography	232
Methylmethacrylate	Poly(methyl-methacrylates)	Infrared spectroscopy	235
Methylmethacrylate and methacrylic acid	Poly(methyl-methacrylate)	Gas chromatography	236
Plasticizers	Polyacrylates	Ziesel reaction - gas chromatography	192
Hexane, tridecane, butylated hydroxy toluene	Polypropylene	Head-space analysis	396
Ethyl acetate, ethanol	Ziegler-Natta catalyst support	FTIR attenuated total reflectance spectroscopy	397

Table 1.20 - Determination of Water in Polymers

Polymer	Procedure	Reference
PVC	Devolatilization - gas chromatography	238
Polypropylene	Heating under nitrogen - Karl Fischer titration	66
Anionic resins	Oven-drying - Karl Fischer titration	240,243
Polyamide and polyethane diol terephthalate		245
Polypropylene	Removal of water in vacuo or a stream of nitrogen then Karl Fischer titration	244
Several polymers		241,242

Table 1.21 - Determination of Additives in Polymers

(a) Antioxidants in Polyethylene Gas Chromatography		Reference
Antioxidants, uv absorbers, lubricants, antistatic agents, optical brighteners	solvent extraction, glc	246,247
2,6 di-t-butyl-4-methylphenol	solvent extraction, glc	248
butylated hydroxy anisole	solvent extraction, glc	249
butylated hydroxy toluene	solvent extraction, glc	249
2,6-butyl-4-methylphenol 2,6 di-tert-butyl-4-methylphenol, p-tert-butylphenol	solvent extraction, glc	7
antioxidants	solvent extraction glc	60,61
antioxidants	solvent extraction, glc	61
Antioxidants, amine and phenolic	solvent extraction, gas chromatography	254
2,6 di-tert-butyl p-cresol	solvent extraction, glc	255,256
hydroxy-5-methyl phenyl benzotriazole	solvent extraction, glc	255,254

Table 1.21 continued

2,6 di-tert-butyl phenol, diphenylamine	solvent extraction, glc	256
phenyl-2 napthylamine	solvent extraction, glc	258
Halogenated bisphenols	solvent extraction, glc	254
Antioxidants	solvent extraction, glc	360
Ionox 330	solvent extraction, glc	261
dilaurylthiodipropionate secondary antioxidant	hydrolysis to lauryl alcohol, extraction with chloroform-ethanol-hexane and thin-layer chromatography, then glc	262
2,6 ditert butyl-4-methyl-phenol	thermal analysis/gas chromatography	263
Topanol O C, 2,6 di-tert-butyl-p-cresol	toluene extraction, glc	61
Topanol CA (1,1 3 tris(2-methyl-4 hydroxy-5-tert butylphenyl) butane)		
Polygard, (tris(nonyl phenyl) phosphite)		
Irganox 1010 (pentaerythritol tetra-3(3,5-di-tert-butyl-4-hydroxyphenyl propionate)		
Irganox 1076 (Octadecyl 3(3,5 di-t-butyl-4-hydroxyphenyl)propionate)		
Santonox R (4,4 thiobis (6-t-butyl-m-cresol))		
Annulex PBA15 (1,1 di (3-tert-butyl-4-hydroxy-6-methylphenyl)butane)		
Nonox DCP (2,2 bis (3-methyl-4-hydroxy-phenyl) - (propane)		
Nonox-WSP (bis (2-hydroxy-5-methyl-3-(methyl-cyclohexyl)phenyl methane)		
Ionox 330 (1,3,5 trimethyl-2,4,6-tris (3,5 di-tert-butyl-4 hydroxybenzyl) benzene)		
Visible Spectrophotometric Methods		
butylated hydroxytoluene	solvent extraction, spectrophotometric finish using 2,6 di-chloro-p-benzoquinone-4-chlorimine	265
Antioxidants	solvent extraction, spectrophotometric finish using diphenyl-beta-picryl-hydrazyl	53,267
bis-(3,5-di-tert-butyl-4-hydroxy phenol) methane	solvent extraction, oxidation of alkaline solution and spectrophotometric evaluation	268
Ionol (2,6 di-tert-butyl-o-cresol)	cyclohexane extraction, visible spectrophotometric finish (low levels)	10
Phenolic antioxidants including dicresylol propane and Santonox R ((4-4, -thio-bis-(3-napthyl-6-tert butyl phenol	toluene extraction, coupling with sulphanilic acid spectrophotometric finish (low levels)	29
Phenolic antioxidants including Succonox 18, butylated hydroxy toluene, Ionol, (2,6-di-tert butyl-p-cresol), Nonox CI (N-N di-beta napthyl-p-phenylenediamine, Santonox R (1,1' thio bis-3 methyl-6-tert-butylphenol)	solvent extraction, reaction with ethanolic $FeCl_3$ to produce Fe^{2+} which is estimated spectrophotometrically using 2-2' dipyridyl (low levels)	271
Phenolic antioxidants including Agerite Alba ((hydroquinone monobenzyl ether), Agerite Spar (styrenated phenol),	methanol or ethanol extraction coupling with diazolized p-nitroanaline-spectrophotometric finish	28
Agerite Superlite (polyalkyl polyphenol)		
Antioxidant 425, ie. (2,2' methyl-bis (6-tert butyl-4-methyl phenol))		
Antioxidant 2246, ie. (2,2'-methylene-bis (6-tert butyl-4-methyl phenol))		
Deenax, (2,6, Di-tert--butyl-p-cresol		
1-naphthol		
2-naphthol		

Table 1.21 continued

Naugawhite, (alkylated phenol)		
Nevastain A, (not disclosed)		
Nevastain B, (not disclosed		
nonyl phenol		
p-Phenyl phenol		
Polygard, (Tris (nonylated phenyl)-phosphite)		
Santovar A, (2,5-Di-tert-amyl-hydroquinone		
Santovar O, (2,3-Di-tert-butyl hydroquinone)		
Santowhite crystals, (4,4'-Thio-bis-(6-tert-butyl-m-cresol and $SC1_2$)		
Santowhite powder, (4,4'-Butylidene-bis (3-methyl-6-tert-butylphenol))		
Solux, (N-p-Hydroxyphenyl-morpholine)		
Stabilite white powder, (not disclosed)		
Styphen 1, (Styrenated phenol)		
Wingstay S, (Styrenated phenol)		
Wingstay T, (a hindered phenol)		
Amine antioxidants eg. Nonox CI (N,N' di-2-naphthyl-p-phenylene diamine)	solvent extraction spectrophotometric finish using dipyridyl	271
Succonox 18 (N-stearoyl-p-aminophenyl) Nonox CI	toluene extraction, methanol precipitation of polymer, spectrophotometric determination after reaction with hydrogen peroxide	50,51
Ultraviolet Methods		
Phenolic type Topanol OC (2,6 di-tert-butyl-4-methylphenol) Borox M (bis-(3,5-di-tert-butyl-4-hydroxy phenyl methane))	solvent extraction, thin-layer chromatography then ultra violet spectroscopy	34
Ionox 330 (1,3,5, trimethyl-2, 4, 6 tris (3,5 di-tert-butyl-4-hydroxy benzyl) benzene)	solvent extraction oxidation in $Pb0_2$ or nickel peroxide - ultra violet spectroscopy in neutral and alkaline solution	34,276
p-methoxyphenol 4,4' methylene bis-(2,6-di-tert-butyl-phenol) and Santonox R, (4,4'-thio-bis-(6-tert-butyl-m-cresol))	solvent extraction, ultraviolet spectroscopy (using bathochromic shift in sodium hydroxide solution)	38
Ionol (2,6 di-tert-butyl-p-cresol)	chloroform extraction	34,35
Santonox R, (4,4'-thio-bis-6-tert-butyl-m-cresol	ultraviolet spectroscopy	37
Irganox 1010	ultraviolet spectroscopy	41
Ionol (2,6 di-tert-butyl-p-cresol)	solvent extraction	
Santonox R, (4,4'-thio-bis (t-tert-butyl-m-cresol)	ultraviolet spectroscopy, solvent extraction	
Fluorescence and Phosphorescence Methods		
phenyl-napthylamines	solvent extractions, fluorescence method	283
Agerite D (polymeric dihydroxy)-quinone phenyl-2-napthylamine	solvent extractions, fluorescence method	45
Santonox R, (4,4'-thio-bis-(6-tert-butyl-m-cresol)) phenyl-2-naphthylamine	solvent extractions, phosphorescence method	45
Spectroscopic Techniques		
dilaurylthio-dipropionate	infrared spectroscopy of polymer film	285

Table 1.21 continued

Ionol (2,6 di-tert-butyl-p-cresol)	iso-octane extraction	41
Santonox R (4,4′-thio bis-(6'-tert-butyl-m-cresol)	infrared spectroscopy	35,287
Phenolic antioxidant	solvent extraction, various finishes including spectrophotometry, NMR analysis and mass spectrometry	36,45,291, 292,293 5
antioxidants	extraction with chloroform or hexane or diethyl ether or toluene or carbon disolphide-iso octane mixtures or cyclohexane or water followed by various instrumental finishes	6,7 299,271 29,50 9,10
Thin Layer and Paper Chromatography		
antioxidant	solvent extraction, paper chromatography using chromogenic reagents	304,305 306,307 308,309
Amine antioxidant	solvent extraction, thin-layer chromatography	310,311 312,6,299
antioxidants	solvent extraction, thin-layer chromatography	47,317
Ionol (2,6 di-t-butyl-p-cresol) Santonox R (4,4′-thio-bis-(6'-t-butyl-m-cresol)	hexane extraction, ethanol solution of residue examined by thin-layer chromatography	318 299,319
Ionol (2,6 di-t-butyl-p-cresol 2,4,6 tri-t-butyl phenol)	solvent extraction, thin-layer chromatography	27,322
Nonox range of antioxidants (amine and phenolic types) Butylated hydroxy toluene "2246" Santoflex range, (hydroquinone type) Superlite Polygard tris (2,3 trinonylphenyl) phosphite	solvent extraction, thin-layer chromatography	47
Irganox 1076 4,4′ butylidene, (2-t-butyl-5-methyl) phenol 4,4′ thio bis (6-t-butyl-m-cresol) pentaerythritol tetrabis 3,5, di-t-butyl-4-hydroxyhydrocinnamate 2,2′ methylene bis(4-methyl-6-t butyl phenol) octadecyl (3,5-di-t-butyl-4-hydroxyphenyl) acetate Ionol (2,6 di-t-butyl-p-cresol)	n-heptane extraction thin-layer chromatography	46
Column Chromatography		
antioxidants	solvent extraction, column chromatography on silica gel using a range of solvents and various detectors	29,40 49
Phenolic and amine antioxidants	solvent extraction, high performance liquid chromatography	55
dilaurylthiodipropioate	gel permeation chromatography	58
Santowhite, (4,4′ butylidene-bis-(6-t-butyl-m-cresol) butylated hydroxy toluene Santonox R, (4,4′ thio bis (6-6-butyl-m-cresol)	chloroform extraction, column chromatography chloroform (removes BHT) 1:10 aqueous methanol (removes Santowhite and Santonox R) Eluates analysed by ultraviolet spectroscopy	294
butylated hydroxytoluene Santonox R (4,4′ thio bis (6-t-butyl-m-cresol) Irganox 1076	freeze grinding, diethyl ether extraction, column chromatography with refractive index and ultraviolet detectors at column	55

Table 1.21 continued

CAO 14	outlet	
antioxidants	solvent extraction, column chromatography	65
Miscellaneous Methods		
2-t-butyl-4-methylphenol 2,6 di-t-butyl-4-methylphenol 2,2'-bis (4-methyl-(butylphenyl) methane	solvent extraction, bromometric estimation	7
Santonox R (4,4' thio bis (6-t-butyl-m-cresol)	ditto	67
dilaurylthio-dipropionate	solvent extraction, polarography	77
diorgano-sulphide Tert phosphite types distearyl thiodipropionate	oxidation with m-chlorophenoxybenzoic acid, unreacted oxidant estimated	81
dilauryl thiodipropionate 4,4' thio bis (6-tert butyl-m-cresol) 1'1' thio bis (2-naphthol) triphenyl phosphite triethylphosphite tri-p-tolyl phosphite tris (dinonyl phenyl) phosphite 2,2' thio bis (6-tert-butyl-p-cresol) triisopropylphosphite	estimated iodimetrically	3
Laser Desorption/Ionization F.T. Mass Spectrometry		
(a) Sterically hindered phenols, thioethers, phosphite, phosphonite and hindered amine types	direct analysis of polymer	398
18 antioxidants	analysis of ether extract of polymer	399
(b) Antiozonants in Polyethylene		
antiozonants	thin-layer chromatography	328,332,7, 67,77,81, 310,311, 312,98, 110,128, 47
(c) Ultraviolet Absorbers in Polyethylene		
Thin-Layer Chromatography		
benzophenone and salicylic acid types	solvent extraction, thin-layer chromatography	317
benzophenone type salicylate type benzotriazole type	solvent extraction, thin-layer chromatography for quantitative estimation	47
salicylate type substituted acryonitriles onganonickel type ultraviolet absorbers and optical brighteners	solvent extraction, thin-layer chromatography	130,142, 143,149, 163,165, 14,203,318, 246,227
optical whiteners	solvent extraction, thin-layer chromatography	237

Table 1.21 continued

Miscellaneous Methods

Tinuvin P (2'hydroxy 5'-methylphenylbenzo-triazole)) Tinuvin 326 (2-(2' hydroxy-3'-tert-butyl 5' methylethylphenyl-3 chlorobenzotriazole)) Tinuvin 327 (2-(2' hydroxy-3'-5'-di-6-butyl(phenyl)-3-chlorobenzotriazole)) Cyasorb UV531 (2-hydroxy-4-n octoxy benzophenone)	toluene extraction, glc	61
benzophenone type	solvent extraction, potentiometric titration with sodium methoxide in dimethylformamide	239
antioxidants	solvent extraction, column chromatography	65
7 UV absorbers	ether extraction-laser desorption FT ion cyclotron resonance mass spectrometry	399

(d) Antioxidants in Polypropylene

Gas Chromatography

Ionox 330 (trimethyl 2,4,6 tri(3,5 di-tert-butyl 4-hydroxybenzyl) benzene)	solvent extraction, glc	250,251
antioxidants	solvent extraction, glc	61
2,6, di-tert-butyl-p-cresol laurylthiodipropionate 2,4' methylene bis(2,6 di-tert-butylphenol) 2,2^{1} methylene bis(6-C^{1}-methylcyclphexyl)-p-cresol) 4,4^{1} thio bis(6-tert-butyl-m-cresol) 4,4^{1} butylidene bis)6-tert-butyl-m-cresol) Ionox 330 (135 trimethyl-2,4,6,tris (3,5, di-tert-butyl-4-hydroxyphenyl)phosphate octadecyl 3,5 di-tert-butyl-4-hydroxyhydrocinnamate 2-hydroxy 4-(octyloxy)benzophenone 2,hydroxy-4-methoxybenzophenone 4-dedecyloxy)-2-hydroxybenzophenone 2-(2 4 benzotriazol- 2 yl)-p-cresol 2,4 di-tert-butyl-6-(5-chloro-2 4 benzotriazol-2yl phenol	p-xylene extraction, glc	60

Thin Layer Chromatography

dilauryl thio dipropionate	hydrolysis to lauryl alcohol, chloroform-ethanol-alcohol-hexane extraction. Thin-layer chromatography then glc	252
4,4' butylidene antioxidants (2-tert-butyl-5-methyl) phenol 4,4' thio bis (6-tert-butyl-m-cresol) penterithitol tetra bis 3,5,di-t-butyl 4-hydroxyhydrocinnamate 2,2' methylene bis(4 methyl-6-tert-butyl phenol) octadecyl (3,5 di-tert-butyl-4-hydroxyphenyl)acetate 2,6 di-tert-butyl-p-cresol	n-heptane extraction thin-layer chromatography for quantitative analysis	46

Table 1.21 continued

Miscellaneous Methods

alkyl phenol and bis phenol antioxidants	volatile preconcentration, mass spectrometry	253
diorgano sulphides and tert phosphites eg. distearylthiodipropionate dilaurylthiodipropionate triphenylphosphite triethylphosphite didecylphosphite	heptane extraction. oxidation with m-chloro-peroxy-benzoic acid. unreacted oxidant estimated iodimetrically	57
phosphite antioxidants and their oxidation products	laser desorption/electron ionization FT ion cyclotron mass spectrometry	400
Hydroperoxides in Polypropylene		
hydroperoxides	iodometric method	260
(e) Antioxidants and Optical Brighteners in Polystyrene		
antioxidants phenolic acid amine type	solvent extraction, thin-layer chromatography for quantitative analysis	47
Ionol (2,6 di-tert-butyl-p-cresol)	methylene dichloride solution, glc	255
optical brightener Tinuvin P (hydroxybenzyl-benzotriazole	methylene dichloride solution, glc	255
Ionol, Tinuvin P	methylene dichloride solution, glc	255
(f) Organic Peroxide Residues in Polystyrene Gas Chromatography		
tert-butylhydroperoxide tert-pentylhydroperoxide tert-butylperoxyisobutyrate di-tert-butylperoxide di-tert-pentylhydroperoxide	solvent extraction, glc	69
di-tert-butylhydroperoxide	thermal decomposition of benzene solution, glc estimation of acetone and ethane produced	70
Thin-layer Chromatography		
dicumylperoxide	acetone extraction, thin-layer separation and iodometric estimation	68
Polarography		
benzoyl peroxide succinic acid peroxide lauroyl peroxide acetyl peroxide peroxy acetic acid methyl ethyl ketone peroxide phenyl cyclohexane peroxide di-tert-butyl perphthalate p-tert-perbenzoate	toluene extraction, polarography	264

Table 1.21 continued

Stearic acid in Polystyrene		
surface and internal stearic acid	FTIR spectroscopy	401
(g) Organotin Type Stabilizers in Poly(vinylchloride)		
Thin-layer Chromatography		
dibutyltin-dilaurate dioctyltin-dilaurate butyltintrichloride dimethyltindichloride diphenyltin dichloride hexabutylditin tributyltin laurate dibutyltin bis-(2-ethyl-hexylthioglycollate)	solvent extraction, thin-layer chromatography for identification of organotin compounds	47
dibutyltin bis(2-ethylhexyl) thioglycollate	solvent extraction, thin-layer chromatography	47
di-n-octyltin maleate	solvent extraction, thin-layer chromatography	10
stabilizers	various methods	266,269, 270,272-274
dibenzyltin bis(isooctyl-mercaptoacetate) dibenzyltin bis(butyl-mercaptoacetate) dibenzyltin bis(cyclohexylmer-captoacetate)	solvent extraction, hydrolysis to glycol, thin-layer chromatography of 3,5 dinitrobenzoates	275
Titration methods		
stabilizers	titration of carboxylate end-groups with sodium methoxide in pyridine	95,96
stabilizers	determination of sulphur by potentiometric titration with potassium periodate in glacial acetic acid	277
dialkyltin thio compounds	potentiometric titration with silver nitrate in aqueous iso-propanol. Also, identification by infrared spectroscopy	95
Miscellaneous methods		
stabilizers	heptane-acetic acid extraction various instrumental finishes	278
dialkyltin types dialkylthioglcollates dialkyltin laurylmercaptides eg. dibutyltin oxide, di-octyltin oxide, dibutyltin maleate dioctyltin laurate, dioctyltin thioglycollate	identification by infrared spectroscopy and chemical methods	95
(h) Non-tin Type Stabilizers in Poly(vinylchloride)		
Thin-layer Chromatography		
3 amino crotonic ester of 2,2' thiodiethanol	mascerate with water, acetic acid-ethanol or heptane, then thin-layer chromatography	279
phenylamine 2 phenylindal dicyandiamide	solvent extraction, thin-layer chromatography	280-282, 284

Table 1.21 continued

aminocrotonic esters		
stabilizers	solvent extraction, thin-layer chromatography	47
epoxy types	methanol extraction, thin-layer	288,289
epoxidized soyabean oil	chromatography	
epoxidized linseed oil		
isooctylepoxystearate		
2-ethylhexylepoxystallate		
butyl epoxy tallate		
butyl ester of epoxy linseed oil		
di(isoamyl)4,4,epoxytetrahydro-phthalate		
Miscellaneous methods		
stabilizers	solvent extraction, glc	246,290, 247
stabilizers	diethyl ether extraction, various instrumental finishes	12
diphenylthioureas	methanol or diethyl ether extraction,	11
2-phenylindole	spectrophotometric finish	

(I) Metal Stearate Stabilizers in Poly(vinylchloride)

metal stearates	solvent extraction, thin-layer chromatography	246
Ba	solvent extraction, column	100
Cd stearates	chromatography	
Zn		
Cd	solvent extraction, paper chromatography	101
Pb stearates		
Zn		
Cd	solvent extraction or ashing, polarography	102
Pb stearates		
Zn		
Cd	solvent extraction, infrared spectroscopy	295
Pb stearates		
Zn		
Na	solvent extraction, chemiluminescence	296
K stearates	techniques	
Ba		
Cd	ashing and elemental analysis by	99
Zn stearates and laurates	polarography	
Ba		
Na	ashing, flame photometry	297
K		
Ba stearates		
Cd		
Pb		

(j) Plasticizers in PVC

Solvent extraction methods		
eg. dioctylphthalate	diethyl ether extraction, weighing	83,298
tritolylphthalate		
plasticizers	solvent extraction, weighing	82
poly(propylene adipate)	diethyl ether then methanol extraction,	86
dioctylphthalate	weighing	

Table 1.21 continued

tritolylphthalate		
poly(propylene sebacate)	methanol extraction	292
poly(propylene adipate)		
Gas chromatography		
plasticizers	solvent extraction, glc	290,247
alkyl adipate	solvent extraction, glc	300,246,
alkyl phthalate		301,302
benzylbutylphthalate	solvent extraction, glc	303
alkyl phthalates	solvent extraction, glc	313
fatty acid esters		
dimethylphthalate	dichloromethane extraction, glc	314
dimethyl sebacate		
triacetin		
diacetin		
diethylphthalate		
adipate and phthalate type	mild pyrolysis-gas chromatography	315
plasticizers	solvent extraction, gas chromatography	316
dibutylphthalate	pyrolysis-gas chromatography	321,322
diisobutylphthalate		
bis(2-ethylhexyl)phthalate		
di-n-octyl phthalate		
alkyl phthalates	tetrahydrofuran extraction, glc (includes	89
alkyl adipates	determination of alcohol degradation	
alkyl azealates	products)	
alkyl citrates		
alkyl sebacates		
alkyl glycollates		
alkyl phosphates		
plasticizers	pyrolysis-gas chromatography of polymer	43
Thin-layer chromatography		
plasticizers	dimethoxymethane extraction, thin-layer	323
	chromatography	
phthalic acid esters	solvent extraction, thin-layer	324-340,
sebacic acid esters	chromatography for identification and	279,286,
phosphoric acid esters	determination	289
epoxy esters		
amino crotonic esters		
diphenylthio		
adipic acid type		
chlorinated paraffin waxes	solvent extraction, glc	43
tricresyl phosphate		
dibutyl succinate		
di-n-octylphthalate		
bis(2-ethylhexyl)phthalate		
tributyl phosphate		
esters and carboxylic acids and		
dicarboxylic acids		
diethyl phthalate		
dimethylsebacate		
phthalate type	solvent extraction, thin-layer	95
	chromatography and infrared identification	
Infrared and NMR spectroscopy		

Table 1.21 continued

polypropylene adipate polypropylene sebacate tritolylphosphate bisoflex 791 bisoflex 220 bisoflex 795	extraction diethyl ether then carbon tetrachloride then with methanol followed by infrared spectroscopy	82
tricresylphosphate dibutyl succinate di-n-octylphthalate bis(2 ethylhexyl)phthalate tributylphosphate esters of carboxylic acids dicarboxylic acids diethylphthalate dimethylsebacate	solvent extraction, infrared spectroscopy	43
diiso-octylphthalate	solvent extraction, NMR spectroscopy	341
Miscellaneous methods		
plasticizers	benzene extraction, measurement of saponification value	342
plasticizers	diethyl ether extraction, various instrumental finishes	12
phthalate plasticizers	extraction with boiling hydrochloric acid, polargraphy	343,344

(k) Antioxidants, Antiozonants, Plasticizers and Accelerators in Rubbers and Polyisoprenes

Spectroscopic Methods		
amine and phenolic antioxidants	ethanolic hydrochloric acid extraction, spectroscopy	13
phenyl salicylate resorcinol benzoate	diethylether extraction, spectroscopy	14,345,16
antioxidants	acetone extraction, spectroscopy	16
ketone-amine condensates 2-mercapto-benzinidazole	acetone, extraction, spectroscopy	17
antioxidants phenylnapthyl amines	solvent extraction, fluorescence spectroscopy	346,353
rubber accelerators	solvent extraction, polarography	347-353
polygard (tris-nonylated phenyl) phosphate	solvent extraction, hydrolysis to nonyl phenol, diazotrization and spectrophotometric evaluation	354
Column chromatography		
amine types of antioxidants and antiozonants	solvent extraction, gel permeation chromatography using tetrahydrofuran dilution solvent	137,138-141 329,355-358
antioxidant, plasticizers accelerators	carbon tetrachloride solution of sample put through column chromatography, ultra-violet monitoring of effluent successive elution with various solvents	31
antioxidants N-N' diethyl aniline N-ethylaniline diphenylamine N-phenyl-2-napthylamine	freeze grinding, diethyl ether extraction, column chromatography with refractive index and ultraviolet detectors	328

Table 1.21 continued

Gas chromatography		
antioxidants	solvent extraction, glc	144,145,405
amine type 1,2 dihydroxy-2,2,4 trimethyl-6-ethoxy quinoline N-isopropyl-N'-phenyl-p-phenylenediamine N,phenyl-2-napthylamine N,N'di-2-octyl-p-phenylene diamine N,N' diphenyl-p-phenylene diamine	acetone extraction, glc	83
di-2-napthyl-p-phenylene diamine poly(trimethyl-dihydroquinoline)	solvent extraction, glc	145
(l) Antioxidants, Antiozonants, Accelerators in Rubber Vulcanizates Antioxidants		
Gas chromatography		
amine and phenolic type	solvent extraction, acetylation glc	359
Column chromatography		
antioxidants and accelerators, guanidine type, thiocarbanilide mercaptobenz-thiazole-2-disulphide 2-benzthiazolyl-N' cyclohexyl sulphonamide phenyl napthylamines sym-di-beta-napthyl-p-phenylene diamine zinc dialkyldithiocarbamates formaldehyde-aniline condensation products aldol napthylamine condensation products polymerized trimethyl dihydroquinone	solvent extraction, column chromatography. column streaked with detection reagents, UV and IR examination of extracts	360-362
Thin-layer chromatography		
amine and phenolic types N-phenylnapthylamine acylated diphenylamines 4,4' dimethyoxydiphenylamine	solvent extraction, TLC for identification	311
phenolic and amine	solvent extraction, TLC	311
antioxidants	solvent extraction, TLC	363
Paper chromatography		
antioxidants phenolic and amine antioxidants	solvent extraction, paper chromatography	305 364 365
Spectroscopy		
agerite range of antioxidants stabilite alba (di-o-tolyl ethylene diamine) Santoflex 66 (N-phenyl-N'-cyclohexyl-p-phenylene diamine) Santoflex DD (6-dodecyl-1,2-dihydro-2,2,4-trimethyl-quinoline) Santoflex 13	acetonitrile extraction - then mass spectrometry, infrared spectroscopy and NMR	293

Table 1.21 continued

N-phenyl beta-napthylamine Wingstay S (styrenated phenol) 2,6, di-tertbutyl-p-cresol N,N' diphenyl-p-phenylene diamine		
(m) Antiozonants		
Thin-layer chromatography		
antiozonants	solvent extraction, TLC	311
Paper chromatography		
antiozonants	solvent extraction, paper chromatography for identification	365,366, 305,66
(n) Accelerators		
Thin-layer chromatography		
accelerators	solvent extraction, TLC	367,368
guanidines, thiuram type carbonate type eg. tetramethylthiuram monosulphide, dipentamethylene thiuram-tetrasulphide cyclic thiuram piperidinium pentamethylene-dithiocarbamate zinc (bismuth and cadmium)-dimethylthiocarbamate 2-benzothiazyl N,N' diethyl-ethiocarbonyl sulphide 2-mercaptobenzthiazole benzothiazyl disulphide	solvent extraction, TLC for identification	311
Paper chromatography		
accelerators	solvent extraction, paper chromatography	365,293, 305,364
Mass spectrometry		
2-mercapto-benzothiazole accelerator sulphenamide types eg. N-tertbutyl-2-benzothiazole sulphenamide 2-(4 morpholinothio)benzothiazole 2-(2,6-dimethyl morpholinothio)benzothiazole N,N-diisopropyl-2-sulphenamide N,N-dicyclohexyl-2-benzothiazole-sulphenamide	volatile preconcentration/mass spectrometry down to 0.2%	369
(o) Phenolic Antioxidants in Polycarbonate		
p-tert-butyl phenol	diazotization with p-nitro-aniline	370,375

Table 1.21 continued

(p) Toluene Diisocyanate in Polyurethane Foam

toluene diisocyanate	derivativization with pentafluoropropionic anhydride - gas chromatography	402

1. Crompton, T.R. and Myers, L.W., Plastics and Polymers, page 205, June (1968).
2. Jeffs, A.R., Analyst (London), **94,** 249 (1969).
3. Lorenz, O., Scheel, W. and Dummer, W., Kautschuk Gummi, 7, 273 (1954).
4. Campbell, R.H., and Wise, R.W., J. Chromatog, **12,** 178 (1963).
5. Slonaker, D.R. and Sievers, D.C., Anal. Chem., **36,** 1130 (1964).
6. Van der Heide, R.F. and Wouters, O., Lebensum Untersuch Forsch, **117,** 129 (1962).
7. Schroder, E. and Rudolph, G., Plaste Kautschuk, **10,** 22 (1963).
8. Metcalf, K. and Tomlinson, R., Plastics (London), **25,** 319 (1960).
9. Yushkevichyute, S.S. and Shlyapnikov, Yu. A., Plasticheskie Massy, No. **1,** 54 (1967).
10. Stafford, C., Plasticheskie Massy, **34,** 794 (1962).
11. Korn, O. and Woggon, H., Plaste Kautschuk, **11,** 278 (1964).
12. Zilio-Grandi, F., Libralesso, G., Sassu, G. and Sveglidao, G., Mater. Plaste. Elast., **30,** 643 (1964).
13. Brock, M.J. and Louth, G.D., Anal. Chem., **27,** 1575 (1955).
14. Wandel, M., Tengler, H., Fette, Seifen, Anstrichmittel, **66,** 815 (1964).
15. Varma, J.P., Suryanaraya, N.P. and Sircar, A.K., J. Sci. Ind. Res., India, **20,** 79 (1961).
16. Yuasa, T. and Kamiya, K., Japan Analyst, **13,** 966 (1964).
17. Varma, J.P., Suryanaraya, N.P. and Sircar, A.K., J. Sci. Ind. Res. India, **21,** 49 (1962).
18. Hilton, C.L., Anal. Chem., **32,** 1554 (1960).
19. Wims, A.M. and Swarin, S.J., J. Appl. Polym. Sci., **19,** 1243 (1975).
20. Lichtenhaler, R.G., and Ranfelt, F., J. Chromatogr., **149,** 533 (1978).
21. Liquid Chromatography Procedure for Polyolefin Additives, WAPP-100 Waters Asssoc., Milford, Mass., (1978).
22. Swaren, S.J., and Wims, A.M., J. Appl. Polym. Sci., **19,** 1243 (1975).
23. Ranfelt, F. and Lichtenhaler, R.G., J. Chromatogr., **149,** 553 (1978).
24. Schabron, J.F. and Fenska, L.E., Anal. Chem., **52,** 1411 (1980).
25. Schabron, J.F. and Bradfield, D.Z., J. Appl. Poly,. Scie., **26,** 2479 (1981).
26. Murakami, S., Tukomori, T., Tsurugi, J. and Murata, N., J. Chem. Soc. Japan, Ind. Chem. Sect., **67,** 1161 (1964).
27. Novitskaya, L. and Kararinova, N., Zhur. Analit. Khim., **28,** 1233 (1973).
28. Hilton, C.L., Anal. Chem., **32,** 383 (1960).
29. British Standard 2782, Part 4 Method 405D (1965).
30. Cornish, P.J., J. Appl Polymer Sci., **7,** 727 (1963).
31. Fiorenza, A., Bonomi, G. and Saredi, A., Mater Plast. Elast., **31,** 1045 (1965).
32. Auler, H., Asbest. Kunststoffe, **14,** 1024 (1961).
33. Kawaguchi, T., Ueda, K. and Koga, A., J. Soc. Rubber Ord. (Japan) **28,** 525 (1955).
34. Ruddle, L.H. and Wilson, J.R., Analyst (London), **94,** 105 (1969).
35. Hilton, C.L., Rubber Age, **84,** 263 (1958).
36. Burchfield, H.P., and July, J.N. Anal. Chem., **19,** 383 (1960).
37. Cieleszky, C. and Nagy, F., Z. Lebensm Unterforsch, **114,** 13 (1961).
38. Wexler, A.S., Anal. Chem., **35,** 1926 (1963).
39. Soncek, J. and Jelinkova, E., Analyst (London), **107,** 623 (1982).

40. Berger, K.G., Sylvester, N.D. and Haines, D.M., Analyst (London), **85,** 341 (1960).
41. Spell, H.L. and Eddy, R.D., Anal. Chem., **32,** 1811 (1960).
42. Miller, R.G.J. and Willis, H.A., Spectochim. Acta., **14,** 119 (1959).
43. Guichon, G. and Henniker, J., British Plastics, **37,** 74 (1964).
44. Luongo, J.P., Appl. Spectry., 4, 117 (1965).
45. Drushel, H.V. and Sommers, A.L., Anal. Chem., **36,** 836 (1964).
46. Dobies, R.S., J. Chromatography, **40,** 110 (1969).
47. Simpson, D. and Currell, B.R., Analyst (London), **96,** 515 (1971).
48. Albarino, R.V., Applied Spectroscopy, **27,** 47 (1973).
49. Van der Neut, J.H. and Maagdenberg, A.C., Plastics, **31,** 66 (1966).
50. British Standard 2782 Part 4, Method 405B (1965).
51. Crompton, T.R., Appl. Polymer Science, **6,** 538 (1962).
52. Sawada, M., Yamagi, I. and Yamashina, T., J. Soc. Rubber Ind. (Japan), **35,** 284 (1962).
53. Glavind, J., Acta. Chem. Scand., **17,** 1635 (1963).
54. Kabota, T., Kuribayashi, S. and Furuhama, T.S., Soc. Rubber Ind. (Japan), **35,** 662 (1962).
55. Majors, R.E., J. Chromatographic Science, **8,** 338 (1970).
56. Hilton, C.L., Rubber Chem. Techol., **32,** 844 (1959).
57. Kellum, G.E., Anal. Chem., **43,** 1843 (1971).
58. Coupek, J., Pokorny, S., Protivova, J., Holak, J., Korvas, H. and Poaposil, J., J. Chromatography, **65,** 279 (1972).
59. Bundergesund Heitschlatte, **16,** 155 (1973).
60. Lappin, G.R. and Zannucci, J., Anal. Chem., **41,** 2076 (1969).
61. Denning, J.A. and Marshall, J.A., Analyst (London), **97,** 710 (1972).
62. Davies, J.T. and Bishop, J.R., Analyst (London), **96,** 55 (1971).
63. Roper, J.N., Anal. Chem., **42,** 688 (1970).
64. Crompton, T.R. and Myers, L.W., British Plastics, December (1965).
65. Dengreville, M., Analusis, **5,** 195 (1977)
66. Maltese, P., Clementini, L. and Panizza, S., Mater Plaste Elastomerie, **35** 1669 (1969).
67. Morgenthalen, L.P., unpublished work.
68. Brammer, J.A., Frost, S. and Reid, W.V., Analyst (london), **92,** 91 (1967).
69. Bukata, S.W., Zabrocki, L.L. and McLaughlin, M.F., Anal. Chem., **35,** 886 . (1963).
70. Hyden, S., Anal. Chem., **35,** 133 (1963).
71. Vasil'eva, A.A., Vodzinskii Yu V., and Korshunov, I.A., Zavod Lab., **34,** 1304 (1968).
72. Barendrecht, E., Anal. Chim. Acta., **24,** 498 (1961).
73. Budke, C., Bannergee, D.K. and Miller, F.D., Anal. Chem., **36,** 523 (1964)
74. Ward, G.A., Talanta, **10,** 261 (1963).
75. Hamilton, J.W. and Tappel, A.L., J. Amer. Oil Chemists Soc., **40,** 52 (1963).
76. Elving, P.J. and Krivis, A.F., Anal. Chem., **30,** 1645 (1958).
77. Neureitter, N.P. and Bown, D.E., Ind. Eng. Chem. Prod. Res. Dev. **1,** 236 (1961)
78. Lintner, C.J., Schleif, R.H. and Higuchi, T., Anal. Chem.,**22,** 534 (1950).
79. Yamaji, O., and Yamashina, T., J. Soc. Rubber Ind. (Japan), **35,** 280 (1962).
80. Sawada, M., Yamagi, I. and Yamashina, T., J. Soc. Rubber Ind. (Japan), **35,** 284 (1962).
81. Kellum, G.E., Anal. Chem., **43,** 1843 (1971).

82. Robertson, M.M. and Rowley, R.M., British Plastics, January 26th (1960).
83. Haslam, J. and Soppet, W.W., J. Chem. Soc., **67,** 33 (1948).
84. Doehring, H., Kunststoffe, **28,** 230 (1938).
85. Thinius, K., Analytische Chemie der Plaste, Springer Verlag (1952).
86. Haslam, J., and Squirrel, D.C.M., Analyst (London), **80,** 871 (1955).
87. Wake, W.C., The Analysis of Rubber and Rubber-like Polymers, McLaren London (1958).
88. Clarke, A.D. and Bazill, G., British Plastics, **31,** 16 (1958).
89. Krisher, A., Anal. Chem., **43,** 1130 (1971).
90. Berens, A.R., Brider, L.B., Tomanek, C.M and Whitney, J.M., BP Goodrich Tyre Centre, Brecksville, Ohio. M/S circulated to members of the Vinyl Chloride Safety Association. Nov. 14th 1974 entitled "Analysis of Vinyl Chloride in PVC powder by Head-Space Chromatography".
91. Berens, A.R., Paper delivered to the 168th ACS Meeting, Atlantic City, New Jersey, Sept. 1974 entitled "The Solubility of Vinyl Chloride in Polyvinylchloride".
92. Baba, T. and Shokuhin Eiseigaku Zasshi., **18,** 6, 500 (1977); Chem. Abstr., **88,** 153342e (1978).
93. Gilbert, S.G., Giacin, J.R., Morano, J.R. and Rosen, J.D., Package Dev. Syst., **5,** 20 (1975).
94. Mansfield, P.B., Chemy, Ind., **28,** 792 (1971).
95. Udris, J., Analyst (London), **96,** 130 (1971).
96. Groagova, A. and Pribyl, M., A. Analyst Chem., **234,** 423 (1968).
97. Belpaire, F., Revie Belg. Mat. Plast., 6, 201 (1965).
98. Delves, R.B., J. Chromatography, **26,** 296 (1967).
99. Mal'kova, L.N., Kalanin, A.I. and Derepletchikova, E.M., Zhur. Analit. Khim., **27,** 1924 (1972).
100. Verfungungen and Mittalungen des Ministeriums for Gesund-heitwesen No. 2 (1965).
101. Schroeder, E. and Thinius, K., Deutsch Farben S., **14,** 146 (1960).
102. Schroeder, E. and Malz, S. Deutsch Farben Z., **5,** 417 (1958).
103. Milligen, M.B., Anal. Chem., **46,** 746 (1974).
104. Lattimer R.P., Harris, R.E., Rhoe, C.K. and Schulten, H.R., Anal. Chem., **58,** 3188 (1986).
105. Van Houwelingen, G.D.B., Analyst (London), **106,** 1057 (1981).
106. Van Houwelingen, G.D.B., Peters, M.W.M.G. and Huysmans, W.G.C., Fresenius Z. Anal. Chem., **293,** 396 (1978).
107. Tanaka, M. and Kogima, I., Anal. Chim. Acta., **41,** 75 (1968).
108. Groom, T., Babicc, J.S. and Van Leuwen B.G., J. Cell Plast., **10,** 43 (1974).
109. Horozawa, S.T. In Kolthoff, I.M. and Elving, P.J. Editors "Treatise on Analytical Chemistry" part II volume 14. John Wiley, New York, P. 23 (1971).
110. Gororyova, A., Elekt. Kabl. Tech., **23,** 207 (1970).
111. Van Lingen, R.L.M., Fresenius Z. Anal. Chem., **169,** 410 (1969).
112. Van Houwelingen, G.D.B., Albero, J.G.M. and Do Hoog, A.J., Fresenius Z Anal. Chem., **300,** 112 (1980).
113. Allen, B.J., Elsea, G.M., Keller, K.P. and Kinder, H.D., Anal. Chem., **49,** 741 (1977).
114. Hoffman, E.R., Anal. Chem., **48,** 445 (1976).
115. Squirrel, D.C.M., Analyst (London), **106,** 1042 (1981).
116. Welsch, T., Engewald, W. and Kowash, E., Plaste. u Kaut., **23,** 584 (1976).

117. Urbanski, J., Anal. Chem. (Warsaw), **22,** 749 (1977).
118. Smith, T.L. and Whelihan. B.N., Text. Chem. Color., **10,** 35 (1978).
119. Bradley, A. and Heagnay, T.R., Anal. Chem., **42,** 894 (1970).
120. Tengler, H. and Von Falkai, B., Kunststoffe, **62,** 759 (1972).
121. Cope, J.F., Anal., Chem., **45,** 562 (1973).
122. Brown, L., Analyst (London), **104,** 1165 (1979).
123. Budlyina, U.V., Marinin, V.G. and Vodzinsku, Yu V., Zavod Lab., **36,** 1051 (1970).
124. Hazzczyc, B. and Walczyk, W., Chem. Anal. (Warsaw), **20,** 885 (1975).
125. Majewski, J., Polimeri, **18,** 142 (1973).
126. Stetzler, R.S. and Smullin, C.F., Anal. Chem., **34,** 194 (1962).
127. Fritz, J.S. and Schlenk, G.H., Anal. Chem., **31,** 1808 (1959).
128. Fiorenza, M., Bonomi, G. and Saradi, A., Rubber Chem. Technol., **41,** 630 (1968).
129. Floria, J.A., Dobratz, J.W. and McClure, J.H., Anal. Chem., **36,** 111 (1964)
130. Stahl, E., Thin-layer chromatography, Academic Press, New York, pp. 349 - 352, 48 (1965).
131. Siggia, S., Hanna, J.G. and Culmo, R., Anal. Chem., **33,** 900 (1961).
132. Spirin, Yu L. and Yatsimirskaya, T.A., Vysokamol. Soedin Ser. A.15 2595 (1973).
133. Fritz, D.E., Sahil, A., Keller, H.P. and Kovat, E., Anal. Chem., **51,** 7 (1979).
134. Kaduji, I.I and Rees, J.H., Analyst (London), **99,** 435 (1974).
135. Smith, R.M. and Dauson, M., Analyst (London), **105,** 85 (1980).
136. Li Ho, F.F., Anal. Chem., **45,** 603 (1973).
137. Protivova, J. and Pospisil, S., J. Chromatography, **88,** 99 (1974)
138. Yoshikova, T. and Kimura, K., J. Appl. Polymer Sci., **15,** 2513 (1971).
139. Colburn, A.P., and Pigford, R.L., J.H. Perry (Editor) Chemical Engineers Handbook, McGraw Hill, New York, 34d Ed. p. 538 (1950).
140. Edwards, G.D., J. Polym Sci., Part. 21, 105 (1967).
141. Henrickson, J.G. and Moore, J.C., J. Polym. Sci., Part A - I, **4,** 167 (1966)
142. Randerath, K., Thin-layer chromatography, Academic Press, New York, pp. 179 - 181 (1966).
143. Robbitt, J.M., Thin-layer chromatography, Reinhold, New York, p. 180 (1963).
144. Wize, R.W., Sullivan, A.B. and Rubber Chemistry and Technology, **335,** No. 3 July - September (1962).
145. Wize, R.W., and Sullivan, A.B., Rubber Age, **91,** 773 (1962).
146. Gaeta, L.J. and Schleuter, E.W. and Altenau, A.G., Rubber Age., **47,** 101 (1969).
147. Schleuter, D.D. and Siggia, S., Anal. Chem., **49,** 2349 (1977).
148. Schleuter, P.D., Ph. D. Dissertation University of Massachusetts, Amherst Mass (1976).
149. Zijp, J.W.H., Rec. Trav. Chim., **76,** 313 (1957).
150. Frankoski, S.P. and Siggia, S., Anal. Chem., **44,** 507 (1972).
151. Kreshkov, A.P., Shvelsova, L.N. and Emelin, E.A., Sov. Plast. (Engl. Transl.) (10) **53,** (1968) Chem. Abstra., **70,** 20345q (1969).
152. Emelin, E.A., Savinov, V.M. and Sokolov, L.B., J. Anal. Chem. USSR (Engl. Transl.), **28,** 1188 (1973); Chem. Abstr., **79,** 14695u (1973).
153. Curnuck, P.A. and Jones, M.E.B., British Polym. J., **5,** 21 (1973).
154. McGowan, R.J., Anal. Chem., **41,** 2074 (1969); **42,** 942 (1970).
155. Kulikova, N.P., Shablygin, M.V. and Utervskii, LE., Khim. Volokna, **3,** 24 (1973); Chem. Abstr., **79,** 92637v (1973).

156. Teleshova, A.S., Teleshov, E.N. and Pravednikov, A.N., Vysokomol. Soedin., Ser. A., 12 2309 (1971); Chem. Abstr., **76,** 46644k (1972).
157. Gomoryova, A., Chem. Abstr., **83** 59692u (1975)
158. Stahl, E. and Dey, L.S., Kunststoffe, **64** 657 (1974), Chem. Abstr., **82** 112514v (1975)
159. Mlejnek, O. and Cyeckova, L., J. Chromatogr., **94** 135 (1974)
160. Kalinina, L.S. and Doroshina, L.I., Chem. Abstr., **79** 54100g (1973)
161. Betso, S.R. and McLean, J.D., Anal. Chem., **48** 766 (1976)
162. Skelly, N.E. and Husser, E.R., Anal. Chem., **50** 1959 (1978)
163. Husser, E.R., Stehl, R.H., Prince, D.R. and Delap, R.A., Anal. Chem., **49** 154 (1977)
164. Guthrie, J.L. and McKinney, R.W., Anal. Chem., **49** 1676 (1977)
165. Zijp, J.W.H., Red. Trav. Chim., **77** 129 (1958)
166. Parker, R.D., Dow Corning Corporation, Barry, Wales, unpublished procedure
167. Fritz, G. and Hurcht, H., Z. Anorg. Allg. Chem., **317** 35 (1962), Chem. Abstr., **57** 1443g (1962)
168. Krasikova, V.M. and Kaganova, A.N. and Lobtov, V.D., J. Anal. Chem. USSR. (Engl. Transl.), **28** 1458 (1971)
169. Voronkov, M.G. and Shemyatenkova, V.T., Bull. Acad. Sci. USSR. Div. Chem. Sci., 178 (1961) Chem. Abstr., **55** 16285b (1961)
170. Franc, J. and Placek, K., Collect. Scech. Chem. Commum., **38** 513 (1973)
171. Franc, J., Chem, Abstr., **82** 67923 q (1975)
172. Kreshkov, A.P. Shemyatenkova, V.T., Syaatsillo, S.V. and Palamarchuk, N.A., J. Anal. Chem. USSR (Engl. Transl.) **15** 727 (1960)
173. Heylmun, G.W., Bujalski, R.L. and Bradley, H.B., J. Gas Chromatogr., **2** 300 (1964)
174. Krasikova, V.M. and Kaganova, A.N., J. Anal. Chem. USSR (Engl. Transl.) **25** 1212 (1970)
175. Bissell, E.R. and Fields, D.B., J. Chromatog. Sci., **10** 164 (1972)
176. Franc, J. and Placek, K., Mikrochim. Acta. 11., **1** 31 (1975)
177. Hanson, C.l. and Smith, R.D., Anal. Chem., **44** 1571 (1972)
178. Dubiel, S.V., Griffith, G.W., Long, C.L., Baker, G.K. and Smith, R.E., Anal. Chem., **55** 1533 (1983)
179. Eckstein, Y., Dreyfuss, Anal. Chim., **52** 537 (1980)
180. Gerschchenko, Z.V., Blinov, V.F. and Zimin, Yu B., Plast, Massy, **12** 12 (1975)
181. Poharelsky, L. and Heran, J., Chem. Prumsyl., **27** 630 (1977)
182. Germanenko, E.N. and Perepletchikova, E.M., Zhur. Anal. Khim., **29** 830 (1974)
183. Mathieson, A.R., McLaren, J.V., J. Polymer Sci. A-3 2555 (1965)
184. Kathalsky, A., Shavit, N. and Eisenberg, H.J., Polymer Science, **13** 69 (1954)
185. Law, R.D., J. Polymer Science, **9** 589 (1971)
186. Connor, A.Z. and Egler, F.W., Anal. Chem., **22** 1129 (1950)
187. Francis, C.V., Anal. Chem., **25** 941 (1953)
188. Monett, W. and Horowicz, S., Anal. Chem., **52** 1529 (1980)
189. Gyenes, I., "Titration in Non-Aqueous Media", J. Iliffe Books, London, p. 261 (1967)
190. Urbanski, J., Plaste Kautsch., **15** 260 (1968)
191. Haslam, J., Hamilton, J.B. and Jeffs, A.R., Analyst (London), **83** 66 (1958)
192. Miller, P.F., Samsel, E.P. and Cobler, T.G., Anal. Chem., **33** 677 (1961)
193. Kappelmeier, C.P.A., Farbon, Zeitung, **40** 1141 (1935)
194. Kappelmeier, C.P.A., Verfkroniek, **27** 291 (1954)

195. Ferskow, F.W., Farben Zeitung, **44** 33 (1939)
196. Bandel, G., Angew Chem., **51** 570 (1938)
197. Malin, C.J., Gening, L.B. and Williams, R.F., Ind. Eng. Chim, Anal. Ed., **14** 935 (1942)
198. Malin, C.J., Tonghe, L.J., Laird, B.C. and Smith, G.D., Anal. Chem. **26** 188 (1954)
199. Viebock, F. and Brechner, C., Ber. deutsch. Chem. Ges., **63** 3207 (1930)
200. Cross, C.K. and Mackay, A.C., J. Amer. Oil Chem. Soc., **50** 249 (1973)
201. Stenmark, G.A. and Weiss, F.T., Anal. Chem., **28** 1784 (1956)
202. Motorina, M.A., Kalinina, L.S. and Metalkina, E.I., Plast, Massy., **6** 74 (1973)
203. Dobies, R.S., J. Chromatography, **35** 370 (1968)
204. Berl, E. and Hefter, K., Cellulose Chemie., **14** 67 (1933)
205. Scheddel, R.T., Anal. Chem., **30** 1303 (1958)
206. Tukuchi, T., Tonge, S. and Sigimura, Y., J. Polymer Science, A - 1, **6** 3415 (1968)
207. Cohen, M., Ind. Eng. Chem., **47** 2096 (1955)
208. Durbetaki, A.J., Anal. Chem., **28** 2000 (1956)
209. Stenmark, G.A., Anal. Chem., **29** 1367 (1957)
210. Kolthoff, J.M., Mitarb. J. Polymer Sci., **2** 199, 206, 220 (1947)
211. Boyer, R.F., J. Phys, and Colloid Chem., **51** 80 (1947)
212. Kuryatnikov, E.C., Vizyert, R.V. and Berlin, A.A., Vign L'viv Politekh. Inst. No. **57** 36 (1971)
213. McNeill, I.E., Polymer, **4** 15 (1963)
214. McGuchan, R. and McNeill, I.E., J. Polymer Science, A - 1, **4** 2051 (1966)
215. Shibata, C., Yamazaki, M. and Tabenuchi, T., Bull. Chem. Soc. (Japan), **50** 311 (1977)
216. Di Pasquale, C., Di. Iorio, G. and Capaccioli, T., J. Chromatogr., **152** 538 (1978)
217. Shanks, R.A., Pye Unicam Newsletter (1975)
218. Rohrschneider, L.Z., Analyt. Chem., **255** 345 (1971)
219. Shapras, P. and Claver, G.C., Anal. Chem., **34** 433 (1962)
220. Tweet, O., Miller W.K., Anal. Chem., **35** 852 (1963)
221. Wilkinson, L.B., Norman, C.W. and Brettner, N.P., Anal. Chem., **36** 1759 (1964)
222. Nowak, P. and Klemmett, O., Kunststoffe, **52** 604 (1962)
223. Adcock, L.H., Patra, **3** 5 (1962)
224. Ragelis, E.P. and Gajan, R.J.F., Assoc. Offic. Agric. Che., **45** 918 (1962)
225. Pfab, W. and Noffz, D., Z. Anal. Chem., **37** 195 (1963)
226. Shapras, P. and Claver, G.C., Anal. Chem., **36** 2282 (1964)
227. Van der Heide, R.F., Ernahrungforschung, **24** 239 (1966)
228. Crompton, T.R. and Myers, L.W., European Polymer Journal, **4** 355 (1968)
229. Schwoetzer, G., Z. Anal. Chem., **260** 10 (1972)
230. Podzeeva, R.M., Lukhovitskii, U.I. and Forpov, V.L., Zavod. Lab., **37** 160 (1971)
231. Crompton, T.R. and Myers, L.W., Plastics and Polymers, 205 (June 1968)
232. Novak, V., Chem. Prumsyl, **22** 298 (1972)
233. Claver, G.C. and Murphey, M.E., Anal. Chem., **31** 1682 (1959)
234. Katelin, A.I., Komleva, V.N. and Mal'kova, L.N. Metody. Anal. Kontrolya, Provoizvod. Khim. Prom - st., **11** 62 (1977)
235. Water, D.N., Proc. Conf. Raman Spectros., 5th, 500 (1976). Chem. Abstr., **87** 185223n (1977)
236. Cobler, J.G. and Samsel, E.D., SPE Transactions, April 1962

237. Jentzsch, D., Kruger, H., Lebricht, G. and Deucks, G., Gut. I.Z. Anal. Chem., **236** 96 (1968)
238. Jeffs, A.R., Analyst (London), **94** 249 (1969)
239. Romano, S.J., Renner, J.A. and Leitner, P.H., Anal. Chem., **45** 2327 (1973)
240. Armieau, V. and Costain, D., Mater. Plast. (Bucharest), **9** 606 (1972)
241. Harrington, R.C. and Keister, D.C., Paper presented at the Plastics Paper Conference of the Technical Association of the Pulp and Paper Industry in New York City, September (1966)
242. Rice, D.D. and Trowell, J.M., Anal. Chem., **39** 157 (1967)
243. Schmalz, E.O., Fraserforsch Tex. Tech., **21** 209 (1970)
244. Maltese, P., Clementini, L. and Panizza, S., Mater. Plast. Elastomeric, **35** 1669 (1969)
245. Schmalz, E.O., Anal. Abstracts, **17** 2449 (1969)
246. Schroeder, E. and Hagen, E., Plaste Kautsch, **15** 625 (1968)
247. Wandel, M., Tengler, H. and Ostromov, H., Die Analyse von Weischmachern, Springer Verlag, Berlin, Heidelberg, New York (1967)
248. Kardos, E., Kosljar, V., Hudec, J. and Ilka, P., Chemcike, Zvesti, **22** 768 (1968)
249. Buttery, R.G. and Stuckey, B.N., J. Agr. Food Chem., **9** 283 (1961)
250. Knight, H.S. and Siegel, H., Anal. Chem., **38** 1221 (1966)
251. Styskin, E.L. Gurvic, Ya.A. and Kumak, S.I., Khim Prom., **5** 359 (1973)
252. Sedlar, J., Feniokova, E. and Pac, J., Analyst, (London) **99** 50 (1974)
253. Yashikawa, T., Ushimi, K., Kimura, K. and Tamuro, N.I., J. Appl. Polym. Sci., **15** 2065 (1971)
254. Wheeler, D.A., Talenta, **15** 1315 (1968)
255. Roberts, C.B. and Swank, J.D. Anal. Chem. **36** 271 (1964)
256. Nosikov, Yu. D. and Vetchinkina, V.N., Neftckhimiya, **5** 284 (1965)
257. Helf, C. and Bockwan, D., Plate Kauteschuk, **11** 624 (1964)
258. Pocaro, P.J., Anal. Chem., **36** 1664 (1964)
259. Duvall, A.H. and Tully, W.F., J. Chromatography **11** 38 (1963)
260. Citovicky, P., Simek, I., Mikulasova, D. and Chrastova, V., Chem. Zvesti, **30** (3), 342 (1976)., Chem. Abstra., **88** 105960h (1978)
261. Long, R.E. and Guvernator, G.C., Anal. Chem., **39** 1493 (1967)
262. Seldar, J., Feniokova, E. and Pac, C., Analyst, (London) **99** 50 (1974)
263. Reed, P.R. and Warren, P.L., Tech. Paper Reg. Tech, Conf. for plast. Wnf., **11** 139 (1975)
264. Kuta, E.J. and Quackenbush, F.W., Anal. Chem., **32** 1069 (1960)
265. Mayer, H., Deut. Lebunsun Rundshau, **57** 170 (1961)
266. Thinus, K., Stabilizierung and Alterung von Plastwerkstoffen Akademie Verlaq Berlin (1969)
267. Blois, M.S., Nature (London), **181** 1199 (1958)
268. Kharasch, M.S. and Joshi, B.S., J. Org. Chem., **22** 1439 (1957)
269. Schmidt, W., Beckman Report, **4** 6 (1961), **3** 13 (1962) and **3** 13 (1962)
270. Neubert, G., Beckman Report, 15 203, 265 (1964)
271. Metcalf, K. and Tomlinson, R.F., Plastics, **25** 319 (1960)
272. Burger, K., Beckman Report, **15** 192, 280 (1963)
273. Franzen, V. and Neubert, G., Chemiber, Zeitung, Chem. Apparatur., **89** 801 (1965)
274. Berger, K.G., Rudt, U. and Mack, D., Deutsch. Lebensmittell Rdsch., **6** 180 (1967)
275. Mazur, H., Rocnz. panst. Zakl. Hig., **24** 551 (1973)

276. Haslam, J. and Willis, H.A., Identification and Analysis of Plastics, The Iliffe Group, London p. 307 (1965)
277. Stapfer, C.H. and Dvorkin, R.D., Anal. Chem., **40** 1891 (1968)
278. Schroeder, E., Pure Appl. Chem., **36** 233 (1973)
279. Mazur, H., Rocnz. panst. Zakl. Hig., **23** 263 (1972)
280. Hagen, E., Deutsch. Lebensmittel. Rdsch., **14** 158 (1967)
281. Schroder, E., Hagen, E. and Frimel, S., Deutsch Lebensmittel Rdsch., **31** 814 (1967)
282. Korn, O. and Waggon, H., Nahrung, **8** 351 (1964)
283. Parks, C.A., Anal. Chem., **37** 140 (1961)
284. Waggon, H., Kohler, U. and Korn, O., Ernahrungforschung., **11** 548 (1966)
285. Major, J. and Kocmanova, V., Chem. Prum., **17** 372 (1967)
286. Veres, L., Kunststoffe, **59** 241 (1969)
287. Szalkowski, C. and Garber, J., J. Agr. Food Chem., **10** 110 (1962)
288. Weatherhead, R.G., Analyst, (London) **91** 445 (1966)
289. Kreiner, J.G., J. Chromatography, **75** 271 (1973)
290. Kucera, J., Collect. Czechoslavakia Chem. Commun., **28** 1344 (1963)
291. Webster, P.V. and Franks, M.C., J. Inst. Petrol, London, **56** 50 (1970)
292. Wake, W.C. The Analysis of Rubber and Rubber-like Polymers, McLaren, London (1969)
293. Carlson, D.W., Hayes, M.W., Bansaw, H.G., McFadden, A.S. and Altenau, A.G., Anal. Chem., **43** 1874 (1971)
294. Campbell, R.H. and Wise, R.W., J. Chromat., **12** 178 (1963)
295. Malik, W.U., Hague, R. and PalVerma, S., Bull, Chem. Soc. Japan, **36** 746 (1963)
296. Schmidt, W., Beckman Report, **4** 6 (1961), **3** 13 (1962)
297. Schroeder, E., Hagen, E. and Zysik, M., Beckman Report, **13** 720 (1966)
298. Dohring, H., Kunststoffe, **28** 230 (1938)
299. Waggon, H., Korn, O. and Jehle, D., Nahrung, **9** 495 (1965)
300. Hasse, H., Kaut. Gummi Kunstoffe, **20** 501 (1967)
301. Zulake, J. and Guichon, G., Revista de Plastices Modernos, **13** 12 (1963)
302. Zulake, J., and Landault, C. and Guichon, G., Bull. Soc. Chim. France, 1294 (1962)
303. Cook, C.D., Elgood, E.J. and Solomon, D.H., Anal. Chem., **34** 1177 (1962)
304. Williamson, F.B. and Rubber, S., International Plastics, 148 (No. 2) 24 (1966)
305. Zijp, J.W.H., Rec. Trav. Chim., **75** 1155 (1956)
306. Zijp, J.W.H., Dissertation Technical University, Delft (1955)
307. Zijp, J.W.H., Rec. Trav. Chim., **75** 1129 (1956)
308. Auler, H., Rubber Chem. Technol., **37** 950 (1964)
309. Zijp, J.W.H. and Kautschuk, U., Gummi, **10** 14 (1957)
310. Kreiner, J.G., Rubber Chem. Technol., **44** 381 (1971)
311. Kreiner, J.G. and Warner, W.C., J. Chromatography, **44** 315 (1969)
312. Auler, J., Identification of Antioxidants, Antiozonants and Accelerators by means of Thin-layer Chromatography ETDC Aachen, Materials and Research Dept., (1967)
313. Veres, L., Kunststoffe, **59** 13 (1969)
314. Turnstall, F.I.H., Anal. Chem., **42** 542 (1970)
315. Zulaica, J. and Guichon, G., Anal. Chem., **35** 1724 (1963)
316. Yoshita, T., Toyoda Gosei Giho, **17** 64 (1975), Chem. Abastr., **88** 170952w (1978)

317. Kirchner, J.G., Thin-layer Chromatography, Interscience, New York, pp 168 - 169, 679 - 684 (1966)
318. Cumpelik, B.M., Drug Cosmet. Ind., **113** 44 (1973)
319. Sokolowska, R., Rocze panst. Zakl., Hig., **20** 661 (1969)
320. Barla, F., Muanyag Gumi, **14** 330 (1977), Chem. Abstr., **88** 1705955z (1978)
321. Takada, T., Fukui-ken Kogyo Shikenjo Nempo., **49** 66 (1974), Chem. Abstr., **87** 118636m (1977)
322. Takada, T., Fukui-ken Kogyo Shikenjo Nempo, **50** 77 (1975), Chem. Abstr., **87** 118638p (1977)
323. Stueuerle, H. and Pfab, "W., Deutsch Lebensmittel Rdsch., **65** 113 (1964)
324. Peereboom, J.W.C., J. Chromatography, **3** 323 (1960)
325. Von Nagy, F.Z., Lebensmittel Unterschung and Forschung, Part 4, 126 (1965)
326. Ligotti, I., Piazantini, R. and Bonomi, G., Hass. Chim. No. 1. **16** (1964)
327. Von Braun, D., Chimia, 19 February (1965)
328. Jaminel, F., I.C. Formaco Edizons Practica. Anno XVIII No. 12 December 1963)
329. Burns, W., J. Appl. Chem., **5** 599 (1955)
330. Schevschenko, Z.A. and Favorskaya, I.A., Vestu. Lenigr. Univ., **19** 2 (1964)
331. Ruddle, L.H., Swift, S.D., Udris, J. and Arnold, P.E. in P.W. Shallis (Editor) Proceedings of the SAC Conference, Nottingham 1955" W. Heffer and Sons td., Cambridge Press
332. Walker, W.H. and Gaenshirt, K.H., Z. Analyt. Chem., **267** 127 (1973)
333. Veres, L., Kunststoffe, **59** 12 (1969)
334. Cambell, G.E., Foxton, A.A. and Wordsall, R.L., Lab. Pract., **19** 369 (1970)
335. Stuele, W. and Pfab, W., Deutsch Lebensmittel Rundschau. Sch., **65** 113 (1969)
336. Burns, W., Polymer J. Singapore, **3** 58 (1971)
337. Von Braun, D. Kunststoffe, **52** 2 (1962)
338. Gomaryova, A., Elekrviz. Kahl. Tech., **23** 207 (1970)
339. Sokolowska, R., Rocnz. panst. Zakl. Hig. **20** 595 (1969)
340. Kula, H., Rocnz. panst. Zakl. Hig., **20** 307 (1969)
341. Mansfield, P.B., Chemy. Ind., **28** 792 (1971)
342. Hurtibese, R.J. and Latz, H.W., J. Agric. Food Chem., **18** 377 (1970)
343. Wexler, T., Jaks, I. and Jucan., Mater. Plat. (Bucharest), **9** 268 (1972)
344. Mol'kova, L.N., Kalinin, A.I. and Perepletchikova, E.M., Zh. Anal. Khim., **27** 1924 (1972)
345. Varmier, J.P., Suryanaraya, N.P., and Sircar, A.k., J. Sci. Ind. Res. India, **20** 79 1961)
346. Parker, C.A. and Barnes, W.J., Analyst (London), **82** 606 (1957)
347. Mocker, F., Kautchuk Gummi., **11** 281 (1958)
348. Mocker, F., Kautchuk Gummi., **12** 155 (1959)
349. Mocker, F., Kautchuk Gummi, **13** 92 (1960)
350. Adams, R.N., Rev. Polarog., **11** 71 (1963)
351. Zweig, A., Lancaster, E., Neglis, M.T. and Jura, W.H., J. Am. Chem. Soc., **86** 413 (1964)
352. Vodzinskii, Yu. and Semchikova, G.S., T. Po Khim I Khim Teknol., 272 (1963)
353. Vermillion, F.J. and Pearl, T.A., J. Electrochem. Soc., **111** 1392 (1964)
354. Nawakoswki, A.C., Anal. Chem., **30** 1868 (1958)
355. Coupek, J., Pekorny, S. and Pospisil, J., IUPAC Microsymposium on akromoleculres 11th Prague, September (1972)
356. Coupek, J., Pokorny, S., Jirachova, L. and Pospisil, J., J. Chromat. **75** 87 1973)

357. Coupek, J., Kalovec, M., Krivakova, M. and Pospisil, J., Angew. Makromol. hem., **15** 137 (1971)
358. Tswrugi, J., Murakaus, S. and Goda, K., Rubber Chem. Technol., **44** 857 1971)
359. LiGottil, L., Materie Plast. Elastomerie, **39** 889 (1973)
360. Bellamy, L.J., Laurie, J.H. and Pren, E.W.S., Trans. Inst. Rubber Industry, 15 1947)
361. Mann, J., Trans. Inst. Rubber Industry, **27** 232 (1951)
362. Parker, C.A. and Berriman, J.M., Trans. Inst. Rubber Industry, **28** 279 (1952)
363. Kueda, I. and Fernandez, G., Revta. Plast., **24** 82 (1973)
364. Burger, V.L. Rubber Chem. Technol., **32** 1452 (1959)
365. Miksch, R. and Prolsi, O., Bummi. Asbest., **13** 250 (1960)
366. Gaczynski, R. and Stephen, M., Przemylek Chem., **38** 9 (1959)
367. Freiner, J.G., Rubber Chem. Technol., **44** 381 (1971)
368. Yuasa, T. and Kamiya, K., Bunseki Kagsku, **13** 966 (1964)
369. Hilton, A.S. and Altenau, A.G., Rubber Chemistry and Technology, **46** 1035 (1973)
370. Kalinina, L.S. and Doroshina, L.I., Metody Ispyt. Kontr. Issled. Mashinostroit, Mater., **3** 5 (1973), Chem. Abstr., **80** 54100 (1974)
371. Baumgartner, C.E., Analytical Chemistry, **59** 2716 (1987)
372. Kullik, E., Kalyurand, M., Laumberg, M., Laboratory Practice, P73 Jan - Feb (1987)
373. Deans, J.R., Journal of Chromatography, **289** 43 (1984)
374. Heinson, E., Analytical Applied Pyrolysis, **2** 131 (1980)
375. Alexander, R., The Analytical Report, Peskin Elmer Ltd., Beaconsfield, Bucks., U.K., No. 7, February (1985)
376. Rudewics, P., Munsen, B., Analytical Chemistry, **58** 358 (1986)
377. Vargo, J.D., Olsen, K.L., Analytical Chemistry, **57** 672 (1985)
378. Peltonen, K., Analyst (London), **111 819** (1986)
379. Schutjes, C.P.M., Journal of Chromatography, **279** 269 (1983)
380. Risby, T.H., Yergey, I.A., Socca, I.I., Analytical Chemistry, **54** 2228 (1982)
381. Meinzelaar, H.L.C., Havenkamp, I., Hileman, F.D., in "Pyrolysis Mass pectrometry of Recent and Fossil Biomaterials", Elsevier Amsterdam (1982)
382. Hmelnitskii, R.H., Mikashenko, I.M., Brodskii, E.S. in "Pyrolysis Mass spectrometry of Polymers", Khimija Moscow (1980)
383. Sammerrai, D., Al Najjar, H., Salim, W., Analyst (London) **110** 1267 (1985)
384. Mourey, T.H., Smith, G.A., Analytical Chemistry, **56** 1773 (1984)
385. Balke, S.T., Patel, R.D. Advance Chemistry Ser. No. 203 281 (1983)
386. Lattimer, R.P., Harman, D.J., Analytical Chemistry, **52** 1808 (1980)
387. Snyder, L.R., Glajch, J.K., Kirkland, J.J., Journal of Chromatography, **3** 157 1983)
388. Simak, P., Makromol Chemie Makromal Symposium, **5** 61 (1986)
389. Dickie, R.A., Hammond, J.S., de Vries, J.E., Holubka, J.W., Analytical Chemistry, **54** 2045 (1982)
390. Kim C. Y., Dodge, A.I., Lai, S., Kawasaki, A., Analytical Chemistry, **54** 232 1982)
391. Ernes, D.A., Hashumaker, O., Analytical Chemistry, **55** 408 (1983)
392. Joseph, K.T., Browner, R.F., Analytical Chemistry, **52** 1083 (1980)
393. Ludwig, F.J., Analytical Chemistry, **56** 2081 (1984)
394. Utterbach, D.J., Millington, D.S., Gold, A., Analytical Chemistry, **56** 470 (1984)
395. Gauer, R.K., Sharma, P. Gupta, K.C., Analyst (London), **114** 1147 (1989)

396. Hagman, A., Jacobsson, S., Analytical Chemistry, **61** 1202 (1989)
397. Di Noto, V., Saccon, M., Bresadola, S., Zanetti, R., Analyst (London), **115** 041 (1990)
398. Johlman, C.L., Wilkins, C.L., Hogan, J.D., Donovan, T.L., Laude, D.A., Youssefi, M.J., Analytical Chemistry, **62** 1167 (1990)
399. Asamato, B., Young, J.R., Citerin, R.J., Analytical Chemistry, **62** 61 (1990)
400. Xiang, X., Dahlgren, J., Enlow, W.P., Marshall, A.G., Analytical Chemistry, **64** 2862 (1992)
401. Kumar, T., Analyst (London) **115** 1319 (1990)
402. Jedrzejczat, K., Gaind, V.S., Analyst (London), **118** 149 (1993)
403. Hogan, J.D., Laude, D.A., Analytical Chemistry, **64** 763 (1992)
404. Liang, Z., Marshall, A.G., Westmoreland, D.G., Analytical Chemistry, **63** 815 (1991)
405. Kirkland, J.J., Dilks, C.H., Analytical Chemistry, **64** 2836 (1992)
406. Bletsos, I.V., Hercules, P.M., Vanleyen, D., Benninghoven, A., Karakatsanis, C.G., Riech, J.N. Analytical Chemistry, **61** 2142 (1989)
407. Poli, J.B., Schure, M.R., Analytical Chemistry, **64** 896 (1992)
408. Via, J.C., Brauer, C.L., Taylor, L.R., Analytical Chemistry, **66** 603 (1994)
409. Imasaki, T., Hozumi, M., Ishibashi, N., Analytical Chemistry, **64** 2206 (1992)
410. Jones, G.G., Pauls, R.E., Willoughby, R.C., Analytical Chemistry, **63** 460 (1991)
411. Kirkland, J.J., Reuenter, S.W., Analytical Chemistry, **64** 904 (1992)
412. Vincenti, M., Pelizzetti, E., Guarini, A., Costanzi, S., Analytical Chemistry, **64** 1879 (1992)
413. Northrop, D.M., Martire, D.E., Scott, R.P.W., Analytical Chemistry, **64** 16 (1992)
414. Hancock, D.O., Synovec, R.E., Analytical Chemistry, **60** 2812 (1988)
415. Bowden, M., Donaldson, P., Gardiner, D.J., Birnie, J., Gerrard, D.L., Analytical Chemistry, **63** 2915 (1991)
416. Lockmiller, C.H., Breiner, S.J., Kock, M.N., Elomaa, M.A., Analytical Chemistry, **63** 1685 (1991)
417. Tindall, G.W., Perry, R.L., Little, J.L. Spaugh, A.J., Analytical Chemistry, **63** 1251 (1991)
418. Bartle, K.D., Boddington, T., Clifford, A.A., Cotton, N.J., Analytical Chemistry, **63** 2371 (1991)
419. Bletsos, I.V., Hercules, D.M. Van Leyen, D., Hagenhoff, B., Niehuis, E., Benninghoven, A., Analytical Chemistry, **63** 1953 (1991)
420. Kirkland, J.J., Dilks, C.H., Rementer, S.W., Analytical Chemistry, **64** 1296 (1992)
421. Bletsos, I.V., Hercules, D.M., Van Leyen, D., Hagenhoft, B., Niehuis, E., Benninghoven, A., Analytical Chemistry, **63** 1953 (1991)
422. Okada, T., Analytical Chemistry, **62** 734 (1990)
423. Okada, T., Analytical Chemistry, **62** 327 (1990)
424. Hagenhoff, D., Benninghoven, A., Barthel, H., Zoller, W., Analytical Chemistry, **63** 2466 (1991)
425. Okada, T., Analytical Chemistry, **63** 1043 (1991)
426. Kirkland, J.J., Dilks, C.H., Rementer, S.W., Analytical Chemistry, **64** 1295 (1992)
427. Benincasa, M.A., Giddings, J.C., Analytical Chemistry, **64** 790 (1992)
428. Fowler, D.E., Johnson, R.D., Van Leyn, D., Benninghoven, A., Analytical Chemistry, **62** 2088 (1990)

429. Van Breeman, R.B., Huang, C.H., Bumgardner, C.L., Analytical Chemistry, **63** 2577 (1991)
430. Guarini, A., Guglielmetti, G., Vencenti, M., Guarda, P., Analytical Chemistry, **64** 790 (1992)
431. Petro, M., Svec, F., Gitsov, I., Frechet, J.M.J., Analytical Chemistry, **68** 315 (1996)
432. Schriemer, D.C., Li, L., Analytical Chemistry, **68** 2721 (1996)
433. Jackson, C., Larson, B., McEwen, C., Analytical Chemistry, **68** 1303 (1996)
434. Dey, M., Castero, J.A., Wilkins, C.A., Analytical Chemistry, **67** 1575 (1995)
435. Thielking, H., Kuliche, W.M., Analytical Chemistry, **68** 1169 (1996)
436. Tuminello, W.H., Bletsos, I.V., Davidson, F., Weigert, F.J., Simpson, N., Slinn, D., Analytical Chemistry, **67** 1955 (1995)

CHAPTER 2

COPOLYMERS AND CONDENSATION PRODUCTS

2.1 STYRENE-BUTADIENE

2.1.1 Functional Groups

Most of the published work on the determination of functional groups, not unexpectedly, has been carried out on copolymers. This is because the determination of a functional group that is specific to one of the copolymer constituents is the key to the determination of the monomer ratios in the copolymer.

2.1.1.1 Unsaturation

2.1.1.1a Iodine Monochloride Procedure. Crompton and Reid[1] have described procedures for the separation of high impact polystyrene, ie. styrene-butadiene copolymer, into a solvent soluble function, ie. free rubber plus rubber grafted polystyrene plus copolymerized rubber and a gel fraction and for estimating total unsaturation in the two separated fractions. To separate a sample into gel and soluble fractions it was first dissolved in toluene. Only gel remains undissolved. Methanol is then added, which precipitates the polystyrene-rubber graft, ungrafted rubber and polystyrene. Any styrene monomer, soap or lubricant remain in the liquid phase, which is separated from the solids and rejected. The addition of toluene to the solids dissolve all the polymeric material with the exception of the gel. The toluene solubles are separated from the solid gel by centrifuging and made up to a standard volume with toluene. The gel is then dried in vacuo and weighed.

Both the gel and toluene soluble fractions are reserved for determination of unsaturation by the iodine monochloride method. To determine unsaturation in styrene-butadiene rubbers with good accuracy using the iodine monochloride procedure it was found necessary to contact the sample with chloroform for 15 hours before reaction with iodine monochloride. Even with a 30 hour reaction period, a constant iodine value (ca. 20) is obtained only when the sample size is 0.05g or less, ie. a five-fold excess of iodine monochloride reagent. The solid gel, separated from a high impact polystyrene by the solvent extraction procedure, is completely insoluble in chloroform and in the iodine monochloride reagent solution. A contact time with chloroform of 90 hours with a 75 hour reaction period with reagent is required.

Crompton and Reid[1] used these procedure to study the distribution of rubber added in several laboratory preparations of high impact polystyrene containing 6% of a styrene-butadiene rubber and 94% styrene, ie. theoretical 4.1% butadiene. The results in Table 2.1 show the way in which the added unsaturation of 4.1% butadiene distributes between the gel and soluble fractions: the butadiene content of the separated gel remains fairly constant, in the 20 - 25% region, regardless of the quantity of gel present in the sample. As the gel content increases, therefore, so more of the rubber becomes incorporated into the gel and less remains as free rubber or soluble graft. The recovered

Table 2.1 - Distribution of butadiene between soluble and gel fractions obtained from polystyrenes containing different amounts of gel

Gel content of sample (wt%)	Butadiene content isolated gel (wt%)	Soluble graft butadiene content A (calculated on original sample) (wt%)	Gel butadiene content B (calculated on original sample) (wt%)	Total butadiene content (A + B) (calculated on original sample) (wt%)	Amount of original rubber un-saturation in the sample $C=\frac{(A+B) \times 100}{4.1}$ %
	-	3.5	-	3.5	85
4.7	19.5	2.8	0.0	3.7	90
5.6	16.2	2.9	0.9	3.8	93
8.9	23.3	1.5	2.1	3.6	88
11.8	20.0	1.5	2.4	3.9	95

From Crompton and Reid[1] John Wiley, New York

unsaturation lies mainly in the 90 - 95% region, indicating that loss of unsaturation due to grafting or cross-linking reactions occurs only to the extent of some 5 - 10%.

The iodine monochloride method has been used for a variety of polymers. These polymers include those which are highly unsaturated, such as polybutadiene and polyisoprene[3-6] and polymers having low unsaturation such as butyl rubber[7] and ethylene propylene diene terpolymer. Considerable work has been done investigating the side reactions of iodine monochloride with different polymers[7]. These side reactions are substitution and splitting out rather than the desired addition reaction.

2.1.1.1b Infrared Spectroscopy. Albert[2] has compared determinations of butadiene in high impact polystyrene by an infrared method and by the iodine monochloride method described by Crompton.[1] The infrared method is based on a characteristic absorbance in the infrared spectrum associated with the trans configuration in polybutadiene. Since different grades of high impact polystyrene may contain elastomers with different trans contents, calibration curves based on the standard rubber are not always suitable for analyzing these products. The results obtained by the two methods for several high impact grades are compared in Table 2.2. The rubber content of high impact polystyrene (sample 1) determined by titration is lower than the value obtained by the infrared method. This is expected in inter-polymerized polymers because of cross-linking which reduces the unsaturation of the rubber. The other polymers (except sample 3) appear to contain Diene 55 type rubber since reasonable agreement was obtained between the iodine monochloride and infrared methods. High impact polystyrene 3, however, must contain a polybutadiene of high cis content to explain the low (1.2% w) amount of rubber found by the infrared method compared to the 9.0% found by the titration method.

It is seen that the infrared and iodometric methods are complimentary to each other.

Near infrared spectroscopy has been used to determine the microstructure and composition (cis 1:4, trans 1.4 and 1.2 butadiene contents) of polybutadiene and styrene butadiene copolymers.[127]

Table 2.2 - Rubber content of high-impact polystyrenes (based on PBD)

Sample	Polybutadiene %w (iodine monochloride method)	Polybutadiene %w (IR method)
Standard: 6.0%w diene 55	6.2	-
Standard: 12%w diene 55	12.2	-
Standard: 15%w diene 55	14.8	-
High-impact polystyrene 1	8.6	9.7
High-impact polystyrene 2	5.6	5.8
High-impact polystyrene 3	9.0	1.2
High-impact polystyrene 4	11.2	11.4
High-impact polystyrene 5	5.8	5.9

2.1.2 Monomers and Other Volatiles

Gas chromatography is used extensively for the determination of residual monomers and other volatiles in copolymers. In this technique the polymer is dissolved in a suitable, non-interfering solvent, and the solution to which a suitable internal standard is added, injected directly into the gas chromatograph. Head space analysis techniques, discussed later in section 2.3.2 (method 57)[25] have also been applied to the determination of monomers in copolymers.

In a variant of the direct gas chromatographic technique, the polymer containing an internal standard is dissolved in a suitable solvent, the polymer is precipitated by the addition of water and the aqueous solvent phase is injected into the gas chromatograph. However, in general, direct injection of a solvent solution of the polymer is to be preferred. This technique has the disadvantage of leading to deposition of polymeric material in the injection port of the gas chromatograph which would lead eventually to blockage and necessitate frequent cleaning. To overcome this the injection port can be fitted with a replaceable glass liner loosely packed with glass fibre as illustrated in Fig. 2.1. When a solution of polystyrene is injected into the liner, polymer is retained by the glass fibre and volatile components are swept on to the chromatographic column by the carrier gas.

Using this technique, Crompton and Myers[8] have described a technique (Method 53) for determining styrene monomer and a wide range of other volatiles in conventional polystyrene and styrene-butadiene copolymers.[15]

Shanks[9] has determined residual butadiene and styrene in polymers with an analytical sensitivity of 0.05 to 5 ppm by analysis of the equilibrated headspace over polymer solutions and determined acrylonitrile, alpha-methyl styrene and styrene monomers by headspace analysis over heated solid polymer samples.

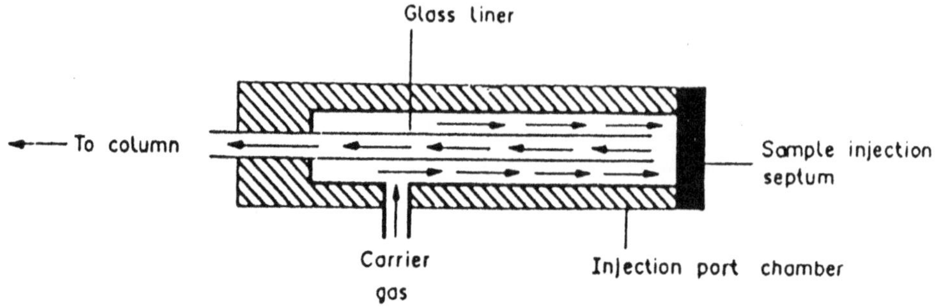

Figure 2.1. Injection port glass liner fitted to F. & M. Model 1609 gas chromatograph. The glass liner measures 60 mm x 40 mm o.d x 2 mm and is very loosely packed with glass fibre.

2.1.3. Additives

Many of the procedures described in Chapter 2 for the determination of additives in polymers can be applied without modification to the analysis of copolymers (Methods 31 - 33).

2.1.3.1 Stearic Acid and Sodium Stearate.

A method is described in Method 54 for the determination of stearic acid and sodium stearate in styrene butadiene.

2.1.3.2 Tertiary Phosphite Antioxidants.

Nawakowski[10] has described a colorimetric method for determining Polygard (trisnonylphenylphosphite) based on hydrolysis to nonyl phenol, followed by coupling with p-nitro benzene-diazonium fluoroborate and colorimetric estimation at 550 nm.

2.1

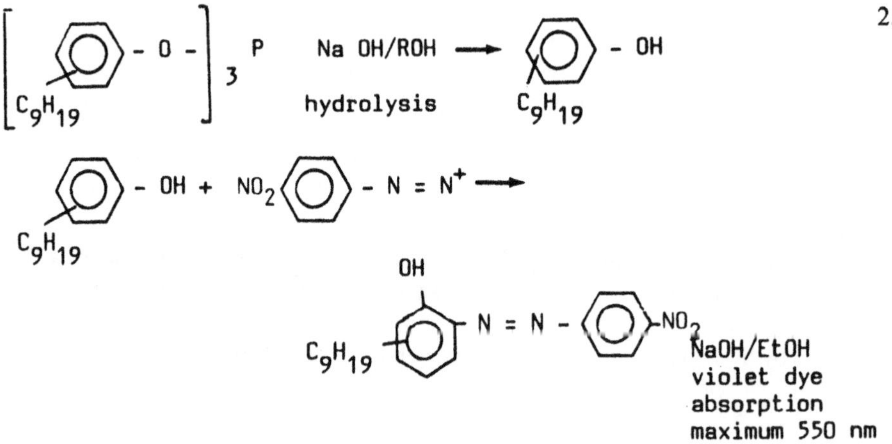

Various other phenolic antioxidants produced dyes under these conditions, viz. Wingstay S, Agerite Superlite and Nevastain A.

Brandt[11] has described an alternate method for the determination of Polygard in styrene butadiene latices which utilizes the bathochromic shift in the spectrum of phenols resulting from the formation of phenolate ions in alkaline solution. The latex is flocculated by the addition of acid and the antioxidant extracted with iso-octane. Polygard in iso-octane has an ultraviolet spectrum with a peak at 273 nm in neutral solution. By adding a strong base (tetrabutylammonium hydroxide) the Polygard is hydrolysed and the peak is shifted to 296 nm. The difference in absorbance at 299 nm between the neutral and alkaline solutions is directly proportional to the amount of Polygard present. By use of this bathochromic shift, interference of nonphenolic impurities is eliminated and a background correction factor is not required.

The bathochromic shift method has the advantage of being more specific than the diazotization method. Results obtained by this procedure agree well with these based on direct determination of elemental phosphorus (Table 2.3).

2.1.3.3 Antioxidants and Stabilizers.

Hilton and Altenau[12] and Hayes and Altenau[13] used mass spectrometry to qualitatively identify volatile antioxidants in sheet samples of synthetic styrene-butadiene rubbers and rubber type vulcanizates. They extracted the polymer with acetone in a Sohxlet apparatus, removed excess solvent and dissolved the residue in benzene. Substances identified and determined by this procedure include N-phenyl-ß-napthylamine, 6-dodecyl-2,2,4-trimethyl 1,2-dihydroquinolines, trisnonyl-phenyl-phosphite, isobutylene - bisphenol A reaction product, 2-mercaptobenzothiazole sulphen-amide (accelerator) N-cyclohexyl-2-benzothiazole sulphenamide, N-tert-butyl-2-benzo-thiazole sulphenamide, 2-(4-morpolinothio) benzothiazole, 2-(2,6-dimethyl-morphal-inothio) benzothiazole, N,N'-diisopropyl 2-benzothiazoles, 2-mercaptobenzo-thiazole and N,N'-dicyclohexyl-2-benzothiazole sulphamide.

2.1.4 Fractionation

Styrene-butadiene copolymers have been fractionated by high performance liquid chromatography.[54,72]

2.2 STYRENE-ACRYLONITRILE

2.2.1 Characterization

Cortes et al[131] have described an on-line coupled microcolumn size exclusion chromatographic - pyrolysis - gas chromatographic system for the characterization of styrene-acrylonitrile copolymers.

2.2.2 Monomers

Residual amounts of styrene and acrylonitrile monomers usually remain in manufactured batches of styrene-acrylonitrile copolymers. As these copolymers have a potential use in the food packaging field, it is necessary to ensure that the content of both of these monomers in the finished copolymer is below a stipulated level.

In a polarographic procedure[14] for determining acrylonitrile (down to 2 ppm) and styrene (down to 20 ppm) monomers in styrene-acrylonitrile copolymer, the sample is dissolved in 0.2 M tetramethyl ammonium iodide in dimethyl formamide base electrolyte

Table 2.3 - Determination of Polygard: comparison of ultraviolet with perchloric acid methods

SBR latex type 6101	Percentage Perchloric acid (phosphorus) method		Ultraviolet method	
Sample 1	1.31	1.30	1.18	1.30
Sample 2	1.26	1.28	1.24	1.36
Sample 3	1.23	1.24	1.36	1.32
Sheet rubber				
Type 1019	1.25	1.28	1.04	1.06
Type 1503	1.62	1.63	1.53	1.60
Type 1018	1.59	1.56	1.57	1.58
Type 1022	1.27	1.31	1.16	1.14

From Brandt[11] American Chemical Society

and polarographed at start potentials of -1.7 v and -2.0 v respectively for the two monomers (Method 55).

Titrimetric methods[16] (Method 56) have been described for the determination of styrene and acrylonitrile monomers in styrene-acrylonitrile. These methods unfortunately have the disadvantage of being more subject to interference than the polorographic method described above.

Methods have been described for determining styrene and acrylonitrile (dimethyl formamide extraction-polarography)[17] and styrene, acrylonitrile and butadiene monomers (solution in dimethyl formamide-gas chromatography).[18] Acrylonitrile has been determined by polarography[19] and by gas chromatography[20] of a dimethyl formamide solution. The polarographic method[19] also determines methacrylonitrile. Styrene monomer has been determined by a polarographic method.[17]

2.2.3 Fractionation

Styrene acrylonitrile copolymers have been fractionated by high performance liquid chromatography using the precipitation technique.[76]

2.3 VINYLCHLORIDE BUTADIENE ACRYLONITRILE STYRENE 2-ETHYLHEXYL ACRYLATE COPOLYMERS

2.3.1 Characterization

^{13}CNMR spectroscopy has been used to elucidate structural detail of vinyl copolymers.[89]

2.3.2 Monomers

In head space procedures a solution of the polymer in a suitable solvent is placed in a closed container and allowed to equilibriate at a controlled temperature so that volatile monomers or other impurities dissolved in the polymer solution partition between the solution and the gas phase. Subsequent analysis of the gas phase enables the concentration of monomers to be calculated. In a variant of this method, the solid polymer is allowed to equilibriate with the head space gas.

Although the solid head space method discussed later provides about 10-fold more sensitivity than the solution head space method (assuming a 10% sample solution) the solid method may be applied only to sample systems where equilibration with the head space is rapid and complete. For example, residual styrene monomer in polystyrene does not reach equilibrium with the head space after 20 hours[21,22] and thus may not be determined by the solid head space method. Furthermore, even if equilibrium between the solid and head space is obtained, the partition coefficient must also be determined for the component of interest in each type of sample matrix. The solution head space approach is applicable to a much wider range of samples than the solid approach. When working with sample solutions, head space equilibrium is more readily attained and the calibration procedure is simplified.

Greater sensitivites and shorter analysis times were obtained using the head space analysis methods than were possible by the direct injection of polymer solutions into a gas chromatograph (Table 2.4).

Various methods have been described for the determination of styrene monomer in polystyrene by solution head space analysis.[23,24]

Steichen[25] has discussed a modified solution approach for the gas chromatographic determination of residual vinyl chloride, butadiene, acrylonitrile, styrene and 2-ethylhexyl acrylate monomers in their associated polymers by head space analysis (Method 57).

2.4 ETHYLENE OXIDE-PROPYLENE OXIDE-GLYCEROL CONDENSATES

2.4.1 Functional Groups

2.4.1.1 Hydroxy Groups by Direct Injection Enthalpimetry.

This technique involves the reaction of a small portion of sample with a large excess of acetic anhydride under conditions where reaction is rapid (less than 1 s), the change in temperature associated with the reaction, dT, is recorded using a thermistor bridge. Under conditions of constant heat capacity, dT should be directly proportional to the number of reactive groups per unit mass of the sample. The main advantages of the technique are, firstly, that it is relatively simple to operate and secondly, that only a few minutes are required in order to perform an analysis.

Kaduji and Rees[26] and others[27] employed direct injection enthalpimetry to determine the hydroxy value of glycerol - alkylene oxide polyethers and butane-1, 4-diol-adipic acid polyesters (Method 58).

Direct injection enthalpimetry has great potential as a method for determining the hydroxyl values of polyethers and polyesters. The method is rapid, the temperature rises for two samples and a standard being recorded in duplicate in about 10 minutes.

The methods described above determine total hydroxy groups and are applicable to many types of hydroxy containing polymers including polyethylene glycol, polypropylene glycol and glycol/alkylene oxide condensates.

In many pratical situations, it is necessary to be able to distinguish between primary and secondary hydroxyl groups in polymers. Thus, the reaction product of a glycerol

Table 2.4 - Comparison of Quantitation Limited* for Residual Monomers Using Conventional and Head-Space GC Methods

Monomer	Boiling point	Direct solution injection[a]	Solution head-space	Modified solution head-space
Vinylchloride	-13°C	1-2 ppm	0.05 ppm	-[b]
Butadiene	-4°C	5 ppm	0.05 ppm	-[b]
Acrylontirile	76°C	10 ppm	0.50 ppm	-[b]
Styrene	145°C	10 ppm	20.00 ppm	1 ppm
2-ethylhexyl acrylate	214°C	200 ppm	1000.00 ppm	5 ppm

*The quantitation limit is defined as the monomer concentration necessary to produce a peak at least three times the baseline noise or 3% of full scale.

[a] Injection of a 10% polymer solution into a gas chromatograph.

[b] A 2- to 3-fold increase in monomer peak height resulted from the injection of water into the polymer solution. A baseline disturbance due to elution of water negated any real improvement in detection limit for these monomers. From Pfab and Noffz[22] American Chemical Society

-ethylene oxide condensate with propylene oxide would contain both types of hydroxy group.

2.2

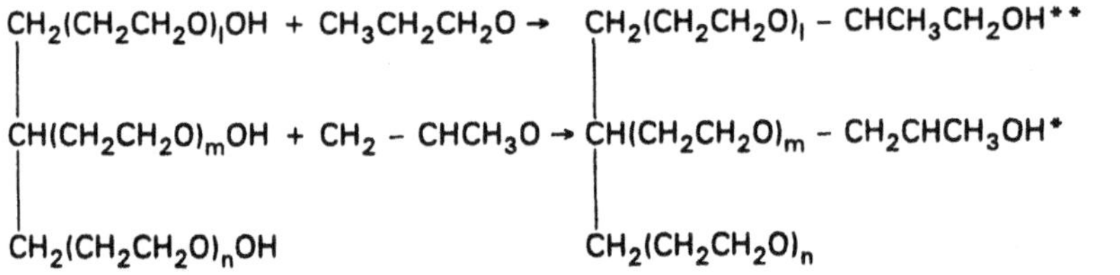

*secondary hydroxy, **primary hydroxy

Reaction rate differences of primary and secondary hydroxy groups with phenyl isocyanate are the basis of a kinetic method for carrying out this determination[28] (Method 59).

2.4.1.2 Alkoxy Groups.

A method for determining alkoxy groups is based on thermal degradation of alkoxy groups to the corresponding olefin which is determined by gas chromatography. This latter method is typified by a procedure[29,30,31] for the determination of ethoxy groups in ethylene oxide-propylene oxide condensates in which the sample is pyrolysed in an evacuated vial at 360 - 410°C and the resulting olefin mixture analysed by gas chromatography.

$CH_2CH_2OCH\ CH_3\text{-}CH_2O{=}CH_2{=}CH_2{+}CH_3CH{=}CH_2$ 2.3

A curve relating per cent ethylene as a function of the ethylene oxide content of ethylene oxide-propylene oxide condensate is linear up to about 50% ethylene oxide, then turns sharply upward to an ethylene content of 38.6% for pure polyethylene glycol.
The relative contents of ethylene oxide and propylene oxide in polyethylene - polypropylene glycols has been determined using combined pyrolysis - gas chromatography calibrated with polyethylene glycol and polypropylene glycol standards.[32]

2.5 ETHYLENE OXIDE - PROPYLENE OXIDE - POLYHYDRIC ALCOHOL AND ETHYLENE OXIDE - PROPYLENE OXIDE AMINE CONDENSATES

2.5.1 Functional Groups

2.5.1.1 Oxyethylene and Oxypropylene

Ethylene oxide and propylene oxide adducts of polyhydric alcohols and amines are widely used as polyethers in the production of polyurethane foams by reaction with diisocyanate. The physical properties of the foams depend to a certain extent on the chemical structure of these polyethers, so it is very important to establish a method for the identification of the base compounds and for the determination of the proportions of their oxyethylene and oxypropylene groups.
Mathias and Mellor[32] split the polyethers with hydrobromic acid - acetic acid to give bromo compounds, which were analysed by gas chromatography.

$$-\,CH_2\text{-}CH_2O + 2HBr \quad = \quad Br\,CH_2CH_2Br + H_2O$$

$$-\,CH\,CH_3\text{-}\,CH_2O + 2HBr \quad = \quad BrCH(CH_3)\text{-}CH_2Br + H_2O \qquad 2.4$$

In this way, the content of oxyethylene groups, and therefore, the original polyhydric alcohols, can be determined. Stead and Hindley[33] modified this method and obtained good results for the determination of the oxyethylene group contents of ethylene oxide-propylene oxide copolymers. The determination of the content of oxyethylene groups in these copolymers can also easily be carried out by nuclear magnetic resonance spectrometry[32,34] without chemical splitting of the ether linkage. However, it is difficult to identify the base compounds by this method. A number of methods for the cleavage of ethers have been studied but few were applied to the identification of the base compounds of the polyurethane polyethers.

Dowle and Campbell[44] pryolysed alcohol-polyethylene oxide condensates in the presence of hydrogen bromide and examined the products by gas chromatography. Several mixed anhydrides of carboxylic and sulphonic acids, as proposed by Karger and Mazur[35] act as reagents for the cleavage of ether linkages. This is particularly true of acetic anhydride toluene-p-sulphonic acids, which is not only a powerful reagent for the cleavage of ether linkages but is also an active acetylating agent. For example, when the propylene oxide adduct of glycerol is treated with this reagent, the polyether is split, thus giving glycerol triacetate and propylene glycol diacetates, which are easily identified by gas chromatography.

$$
\begin{array}{l}
CH_2-(CH_3CHCH_2O)_nOH \\
| \\
CH-(CH_3CHCH_2O)_nOH \quad + \quad 6(CH_3CO)_2O \quad = \\
| \\
CH_2-(CH_3CHCH_2O)_nOH
\end{array}
\qquad (2.5)
$$

$$
\begin{array}{l}
CH_2OOCCH_3 \\
| \\
CHOOCCH_3 \quad + \quad 6CH_3COOCHCH_3-CH_2OOCCH_3 \quad + \quad 3CH_3COOH \\
| \\
CH_2OOCCH_3
\end{array}
$$

Cervenka and Merrall[37] have investigated the application of acidic dehydration of ethylene oxide-propylene oxide condensates in bromonapthalene in the presence of p-toluene sulphonic acid to the elucidation of the molecular structure and monomer sequence of these polymers. Gas chromatography was used to determine dehydration products.

Studies on poly(ethylene glycol) and poly(propylene glycol) homopolymers showed that dioxane and its derivatives (eg. methyl dioxone 1.4) are not the only reaction products. Dehydration of poly(ethylene glycol) gave three products; that of poly(propylene glycol) six products. The majority of them were identified.

Cervenka and Merrall[37] conclude that results on homopolymers, their blends and model copolymers of different chain architectures demonstrate that acidic dehydration is capable of distinguishing ethylene oxide-propylene oxide copolymers of different structures, giving correct absolute values of overall monomer contents and also ranking polyols according to their degrees of randomness.

Tsuji and Kounishi[36] extended this method to the identification of base compounds and the determination of their oxyethylene and oxypropylene group contents (Method 60).

2.6 ETHYLENE PROPYLENE-DIENE TERPOLYMERS

2.6.1 Functional Groups

2.6.1.1 Trace Unsaturation.

Polymerization of ethylene and propylene results in a saturated copolymer. In order to vulcanize this rubber, some unsaturation has to be introduced. This is commonly done by adding a few percent of non-conjugated diene (termonomer) such as dicyclopentadiene, 1,4-hexadiene, or ethylidene norborene during the polymerization. Since only one of the double bonds of the diene reacts during polymerization, the other is free for vulcanization. The amount of unsaturation left in the ethylene propylene diene terpolymer is of great interest because the vulcanization properties will be affected.

Sewell and Skidmore[38] used time averaged NMR spectroscopy at 60 MC/sec to identify low concentrations of nonconjugated dienes introduced into ethylene-propylene copolymers to permit vulcanization. Although infrared spectroscopy[39] and iodine monochloride unsaturation methods[40] have been used to determine or detect such dienes, these two methods can present difficulties. Identification of the incorporated third monomer is not always practical in infrared spectroscopy at the low concentrations involved and in the high-resolution NMR spectra of these terpolymers the presence of unsaturation is not usually detected, since the signals from the olefinic protons are of such low intensity that they become lost in the background noise. The spectra obtained by time averaged NMR are usually sufficiently characteristic to allow identification of the particular third monomer incorporated in the terpolymer. Moreover, as the third monomer initially contains two double bonds, differing in structure and reactivity, the one used up in copolymerisation may be distinguished from the one remaining for subsequent use in vulcanization. Therefore, information concerning the structure of the remaining unsaturated entity may be obtained. Table 2.5 shows the chemical shifts of olefinic protons of a number of different third monomers in the copolymers.

The cyclooctadiene and dicyclopentadiene terpolymers have olofinic protons with the same chemical shift, 4.55 ppm and so these cannot be differentiated by this technique but may be distinguished by the use of iodine monochloride. The hexadiene type of terpolymer may be identified by its olefinic resonance at 4.7 ppm. These three monomers have what appears as a single olefinic resonance in the terpolymer. On the other hand, the two norbornadiene types of monomer each show two characteristic resonances. In the methylene norbornene terpolymer the olefinic resonances arise from two protons, each giving a separate signal, whereas in the ethylidene norbonene terpolymer there is only one proton, the signal of which appears as a doublet. In view of these considerations it is more difficult to detect the olefinic resonance in the latter instance.

Altenau et al[41] applied time averaging NMR to the determination of low percentages of termonomers such as 1:4 hexadiene, dicyclopentadiene and ethylidene norbornene in ethylene-propylene termonomers. They compared results obtained by NMR and the iodine monochloride procedure of Lee et al.[42] The chemical shifts and splitting pattern of the olefinic response were used to identify the termonomer.

Table 2.6 compares the amount of termonomer found by the NMR method of Altenau et al[41] and by iodine monochloride procedures.[42,45] The termonomers were identified by NMR and infrared.

Table 2.5 - N.M.R. determination of non-confugated olienes in ethylene propylene copolymers

Third Monomer	Chemical Shift ppm
Cyclooctadiene 1,5	4.55
Dicyclopentadiene	4.55
1:4 hexadiene	4.7
Methylene norbornene	5.25 and 5.5
Ethylidene norbornene	4.8 and 4.9

From Sewell and Skidmore[38]. John Wiley, New York

Table 2.6 - Determination of termonomer in ethylene-propylene diene terpolymers - comparison of methods (data shown as weight % terpolymer)

NMR	Lee[42] Koltoff and Johnson	Kemp and Peters[45]	Termonomer
7.3	3.0		dicyclopentadiene
1.1	1.6		1,4-hexadiene
5.7	5.9	9.0	ethylidene norbornene
2.8	3.6	4.5	ethylidene norbornene
1.7	2.3	4.8	ethylidene norbornene
4.6	5.4	6.0	ethylidene norbornene

From Altenau[41] American Chemical Society

Table 2.6 shows that the data obtained by the NMR method agrees more closely with the Lee et al iodine monochloride method[42] than with the iodine monochloride method of Kemp and Peters.[45] The difference between the latter two methods is best explained on the basis of side reactions occurring between the iodine monchloride and polymer because of branching near the double bond.[42] The reason for the difference between the NMR and the Lee et al[42] methods is not clear. The reproducibility of the NMR method was 10 $\pm$ to 15%.

Infrared spectroscopy has been used for the determination of trace unsaturation in ethylene-propylene-diene terpolymers.[46] Determination of extinction coefficients for the various terpolymers is required if quantitative work is to be done.

2.7 ACRYLATE COPOLYMERS

2.7.1 Functional Groups

2.7.1.1 Carboxyl Groups.

Most methods for the determination of carboxyl groups in polymers are based on titration techniques. The following copolymers, acrylic acid-itaconic acid,[47] acrylic acid-ethyl acrylate[48] and maleic acid-styrene[49] have been studied. High frequency titration has been applied[50] to the analysis of itaconic acid-styrene copolymers. High frequency titration gives a precise location of the inflection points related to the polymer carboxyl groups and is a sensitive method for the determination of the freedom of the copolymer samples from monobasic acid impurities (comonomer acids), since mixtures of copolymer acids with monobasic and dibasic acids show definite inflection points that can be related to the individual carboxylate species present.

Johnson et al[51] found that in titrating copolymers of methylacrylate and methacrylic acid with standard base to determine carboxyl groups, a number of deficiences were encountered (eg. the presence of up to 5% water and unreacted monomers both lead to underestimations of the acid content, also titration was not applicable to polymers of molecular weight over one million or high acid content because of a tendency to reprecipitate during the titration). For this reason they investigated the application of proton NMR by using the integral of the ester methoxy protons and combining this result with the total integral for CH_2 and CH_3 protons (the overlap between CH_2 and CH_3 resonances was enough at 100 MHz to prevent separate determination of these integrals), the copolymer composition could be ascertained. A problem was that the reaction solvents (toluene and hexane) and comonomers had resonances that overlapped those of the CH_2 and CH_3 protons of the copolymers introducing considerable inaccuracy in the total CH_2 CH_3 integral. For this reason they investigated the applicability of ^{13}C NMR. Because of the greater spectral dispersion and narrower resonance lines obtained with ^{13}C NMR relative to proton NMR problems associated with resonance overlap can be resolved. Excellent agreement was obtained between carboxyl values obtained by this procedure and conventional titration in 1:1 ethanol:water with standard potassium hydroxide to the phenol phthalein end-point over in the acid content range 13% to 100%.

Sharp and Paterson[52] have described a pyrolysis - gas chromatographic - mass spectrometric procedure (Method 61) for the determination of 1 - 10% of copolymerized acrylic acid and methacrylic acid in acrylic polymers.

Nissen et al[53] have described a method for carboxyl groups in poly(ethylene terephthalate). Hydrazinolysis led to formation of terephthalomonohydrazide from carboxylated terephthalyl residues to provide a selective analysis for carboxyl group viz ultraviolet absorbance at 240 nm.

Titrimetry, NMR spectroscopy and pyrolysis gas chromatography each give different information in the analysis of carboxyl groups. Frequently more than one technique has to be applied to obtain the required information.

2.7.1.2 Ester Groups.

Most methods for the determination of ester groups in polymers are based on the following procedures:

(a) Saponification
(b) Zeisel procedures, based on hydriodic acid
(c) Pyrolysis - gas chromatography
(d) Physical methods, eg. infrared spectroscopy and nuclear magnetic resonance spectroscopy.

2.7.1.2a Ester Groups Saponification Methods. Ester groups occur in a wide range of polymers, eg. poly(ethylene terephthalate) and in copolymers such as, for example, ethylene-vinyl acetate, acrylic acid-vinyl ester, methyl acrylate vinyl ester and poly(methyl acrylate). The classical chemical method for the determination of ester groups, namely, saponification, can be applied to some types of polymers. For example, copolymers of vinyl esters and esters of acrylic acid can be saponified in a sealed tube with 2 M sodium hydroxide. The free acids from the vinyl esters were determined by potentiometric titration or gas chromatography. The alcohols formed by the hydrolysis of the acrylate esters can be determined by gas chromatography. Vinyl acetate-ethylene copolymers can be determined by saponification with 1 M ethanolic potassium hydroxide at 80°C for 3 hours and back titration with standard acid[55,70,71] or by saponification with p-toluene sulphonic acid and back titration with standard acetic acid.[58,59]

Poly(methyl acrylate) can be hydrolyzed rapidly and completely under alkaline conditions, on the other hand the monomer units in poly(methyl methacrylate) are resistant to hydrolysis[56] although benzoate end-groups react readily.[57] Only about 9% of the ester groups in poly(methyl methacrylate) react even during prolonged hydrolysis; hydrolysis of polymethyl acrylate was complete in 30 minutes. Although only about 9% of the ester groups in polymethyl methacrylate homopolymers are hydrolyzed by alcoholic sodium hydroxide, this proportion is increased by the introduction of comonomer units into the polymer chain. Thus, saponification techniques should be applied with caution to polymeric materials.

2.7.1.2b Ester Groups Zeisel Procedures. Hydrolysis using hydriodic acid has been used for the determination of the methyl, ethyl, propyl and butyl esters of acrylates, methacrylates or maleates[60] and the determination of polyethyl esters in methyl methacrylate copolymers.[62,63] First the total alcohol content is determined using a modified Zeisel hydriodic acid hydrolysis.[61] Secondly, the various alcohols after being converted to the corresponding alkyl iodides are collected in a cold trap and then separated and determined by gas chromatography. Owing to the low volatility of the higher alkyl iodides the hydriodic acid hydrolysis technique is not suitable for the determination of alcohol groups higher than butyl alcohol. This technique has also been applied to the determination of alkoxy groups in acrylate esters.[60]

Anderson et al[64] have used combined Zeisel reaction - gas chromatography to analyse acrylic copolymers. Acrylic esters were cleaved with hydriodic acid and gas chromatography was used for analysing the alkyl iodides so formed (Method 62).

2.7.1.2c Ester Groups Pyrolysis - Gas Chromatography. Barrall et al[65] have described a pyrolysis-gas chromatographic procedure for the analysis of polyethylene-ethyl acrylate and polyethylene-vinyl acetate copolymers and physical mixtures thereof. They used a specially constructed pyrolysis chanber as described by Porter et al.[66] The pyrolysis chromatogram of poly(ethylene-vinyl acetate) contains two principal peaks. The first is methane and the second acetic acid. The pyrolysis chromatogram of poly(ethylene-ethyl acrylate) at 475°C shows one principal peak due to ethanol. Table 2.7 shows the analysis of 0.05 gram samples of poly(ethylene-ethyl acrylate) and poly(ethylene-vinyl acetate) obtained at a pyrolysis temperature of 475°C.

Pyrolysis - gas chromatography - mass spectrometry has been used to identify ester groups in acrylic copolymers.[68]

Ziesel reduction followed by gas chromatography or pyrolysis followed by gas chromatography have the advantage of being more specific for the determination of ester groups than standard saponification procedures.

2.7.1.2d Ester Groups Physical Methods. Infrared spectroscopy has been applied to the determination of free and combined vinyl acetate in vinyl chloride-vinyl acetate copolymers.[67] This method is based upon the quantitative measurement of the intensity

Table 2.7 - Pyrolysis results on physical mixtures of poly (ethylene-ethyl acrylate) and poly (ethylene-vinyl acetate)

Mixture	Acetic acid found	wt% calcd	Ethylene found	wt% calcd[a]	Oxygen found	wt% calcd
50% PEEA-1 and 50% PEVA-2	9.10	9.05	2.65	2.62	7.88	8.25
33.3% PEEA-2 and 66.6% PEEVA-3	12.15	12.33	0.75	0.70	7.33	7.49

[a]Calculated from results for acetic acid and ethylene content for individual samples on weight per cent basis. From Blackwell[65], American Chemical Society

of absorption bands in the near infrared spectral region arising from vinyl acetate. A band at 1,613 cm^{-1} (6.20 u) vinyl group enables the free vinyl acetate content of the sample to be determined. A band at 4651 cm^{-1} (2.15 u) is characteristic for the acetate group and arises from both free and combined vinyl acetate. Thus, the free vinyl acetate content may be determined by difference at 2.15 u. Polymerized vinyl chloride does not influence either measurement.

The vinyl acetate content of films of ethylene-vinyl acetate copolymers can be determined by methods based on the measurement of absorbances at 1639 and 1389 cm^{-1} (6.10 and 7.20 u)[70] and at 1245 cm^{-1} (8.03 u) and 1743 cm^{-1} (5.73 u).[71,74] The acrylate salt in acrylate salt-ethylene ionomers has been determined from the ratio of absorbances at 1560 cm^{-1} (6.41 u) (asymmetric vibration of the carboxylate ion) and 1380 cm^{-1} (7.25 u).[69]

Anderson et al[64] have described an infrared procedure for distinguishing between copolymerised acrylic and methacrylic acids in acrylic polymers containing more than 10% of the acid. This method is based on precise measurement of the wavelength of the carboxylic acid absorption maximum at about 1700 cm^{-1} (5.88 u). The identification of the acid in compositions containing less than 10% of acid was hitherto not possible, except when unpolymerised acid residues can be separated from the polymer.[73]

Haslam[74] has discussed infrared methods for the determination of ester groups in acrylic copolymers. NMR spectroscopy has been used for the determination of isophthalate in poly(ethylene terephthalate isophthalate) dissolved in 5% trichloroacetic acid. The NMR spectra of these polymers were measured on a high resolution NMR spectrometer at 80°C. A singlet at 7.74 ppm is due to the four equivalent protons attached to the nucleus of the terephthalate unit. The complicated signals which appear to 8.21, 7.90, 7.80, 7.35, 7.22 and 7.10 ppm are due to the four protons attached to the nucleus of the isophthalate unit. The content of the isophthalate unit can be calculated from the integrated intensities of these peaks.

NMR has also been used to determine ethyl acrylate in ethyl acrylate-ethylene and vinyl acetate-ethylene copolymers.[75] Measurements were made on 10% solutions in diphenyl ether at an elevated temperature. Resolution improved with increasing temperature and lower polymer concentration in the solvent. This technique has also been used to identify ester groups in acrylic copolymers[74] and copolymers.

Campbell[77] has described an isotope dilution derivative method for the determination of vinyl acetate in vinyl acetate - vinyl chloride copolymers.

2.7.1.3 Epoxy Groups.

Swaraj and Ranby[79] used infrared spectroscopy to determine epoxy groups in methylmethacrylate - glycidyl methacrylate copolymers. The analysis was performed on dried potassium bromide pellets containing 0.5 mg sample in 200 mg potassium bromide. The peaks at the wavenumbers 907 (11.02 u) and 1717 cm^{-1} (5.82 u) are the most suitable ones for analysis of epoxy and carbonyl groups respectively. Using the "base line density" method the values of the absorbances at 907 and 1717 cm^{-1} wavenumbers were determined. The average values of the absorbances, their ratio and the glycidylmethylacrylate mole fraction determined chemically are presented in Table 2.8.

The absorbance ratio at 907 cm^{-1} vs 1717 cm^{-1} is linearly related to the glycidyl methylacrylate content in the copolymer and can be expressed by the following equation:

$$R = 0.250\ X_G + 0.033$$

where R is the absorbance ratio at 907 and 1717 cm^{-1} and X_G is the mole fraction of glycidylmethacrylate containing the epoxy group in the copolymer.

2.7.2 Fractionation and Molecular Weight

High performance liquid chromatography has been applied to the fractionation of styrene-methyl methacrylate[78,95,107,109,113] and styrene n-butyl methacrylate[116] copolymers. Separations were achieved on silica gel using 1,2 dichloroethane-chloroform as eluting solvents and an ultraviolet detector.[129]

The chemical composition and molecular weight distribution of high conversion polystyrene - methyl methacrylate (54 - 85% methyl methacrylate) has been determined by liquid adsorption and size exclusion chromatography.[130]

2.8 ACRYLAMIDE COPOLYMERS

2.8.1 Functional Groups

2.8.1.1 Alkoxy Groups.

The Ziesel reaction has been extensively studied and has been used successfully for the determination of alkoxy groups in cellulosic materials.[80-82] Quantitative cleavage of the alkoxy groups in polymers is routinely obtained. However, as discussed in section 2.7.1.2 on the determination of ester groups, hydriodic acid also cleaves any ester linkages on the polymer backbone, giving a positive interference.

Anderson et al[83] identified and determined the etherifying alcohols present in thermosetting acrylamide interpolymers of the type shown below via alcohol exchange when interfering ester linkages are present in the polymer backbone (Method 64). No interference is encountered from alkyl esters present in the sample.

$-(CH_2\text{-}CHCONHR'')x - (CH_2\text{-}CRCOOR^1)y\text{-}(CH_2\text{-}CHPh)z$ 2.6

where R′ = H,CH_3,C_2H_5,C4H9 or C_8H_{17} and R″ = H, $CH_2OC_4H_9$ or CH_2OH

Table 2.8 - Analytically determined mole fraction of GMA in the MMA-GMA copolymers and the infrared absorbanced at 907 (11.02 micron) and 1717 cm^{-1} (5.82 micron)

Expt. no.	Mole fraction of GMA in the co-polymer deter-mined chemically	A, 5.82 micron	A, 11.02 micron	Absorbance ratio A, 11.02 micron / A, 5.82 micron
R50	0.218	1.366	0.128	0.093
R51	0.394	0.911	0.120	0.131
R52	0.584	0.629	0.115	0.182
R61	0.623	0.921	0.177	0.192
R53	0.706	0.955	0.204	0.213

From Swarej and Ranby[79]. American Chemical Society

2.9 OCTADECANE-MALEIC ANHYDRIDE COPOLYMERS

2.9.1 Functional Groups

2.9.1.1 Anhydride Groups.

Van Houwelingen[84] discussed the determination of anhydride groups in a resin derived from octadecene-1 and maleic anhydride.

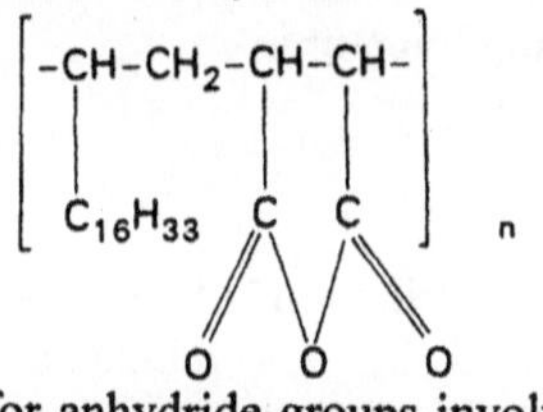

2.7

The common method for anhydride groups involving reaction with an excess of aniline and subsequent backtitration of the excess[85] is unsuitable, as the reactivity of the anhydride group is low. Even after hydrolysis with aqueous pyridine (containing 40% v/v of water) in a Parr bomb at 150°C for 4 hours, anhydride groups are still seen in the infrared spectra.

A suitable method for determining the anhydride group is titration with aqueous potassium hydroxide in pyridine after previous esterification of the carboxyl group with diazomethane. This esterification is carried out in diethyl ether methanol (9 + 1). After methylation, which takes about 10 m for 0.5g of sample, the solvents are removed by evaporation and a portion of the derivatised polymer is dissolved in pyridine and titrated. In the infrared spectra of the resin before and after methylation it can be seen that the absorption band of the acid group at 1710 cm^{-1} (5.84u) disappears and a carbonyl band of the ester at 1740 cm^{-1} (5.74u) is formed. The acid content of the sample is found from the difference in titres of an unmethylated and a methylated product.

2.10 EPOXY RESINS

2.10.1 Volatiles

Thermogravimetric analysis coupled with gas chromatography and mass spectrometry has been used to study the nature of the volatiles produced during the curing of epoxy resins.[117]

2.10.2 Functional Groups

2.10.2.1 Epoxy Groups.

Peltonen et al[120] used infrared spectroscopy to determine epoxy resins and their degradation products.

Further methods for the determination of functional groups, volatiles and additives in copolymers are reviewed in Tables 2.9 - 2.11.

2.11 TERTOCTYLPHENOL - ETHYLENE OXIDE CONDENSATES

2.11.1 Oligomers

Wallingford[132] carried out an oligomeric separation of ionic and non-ionic ethoxylated polymers by capillery gel electrophoresis. The oligomeric distribution of several sulphated and phosphated alkyl phenol ethoxylate surfactants was determined on commercial grade cross linked polyacrylamide gel columns. Non-ionic detergents and polyethylene glycol oligomers reacted with phthalic anhydride to provide a charge to the molecule thereby facilitating the detection of polyethylene glycol oligomers from ethylene glycol to species with 120 monomer units. A linear relationship was found between molecular weight and migration time.

2.11.2 Molecular Weight

The molecular weight distribution of tert octyl phenol - ethylene oxide condensates has been determined by a combination of high performance liquid chromatography and laser desorption Fourier Transform ion cyclotron resonance mass spectrometry.[128]

Table 2.9 - Functional Groups in Copolymers

			Ref
Nitrile	Polystyrene	Dye partition method	86
Unsaturation	styrene-butadiene methylmethacrylate terpolymer	Raman spectroscopy	87
Unsaturation	acrylonitrile-butadiene styrene terpolymers	NMR	88
Unsaturation	ethylene-propylenediene terpolymer	NMR	38
Vinyl	vinylchloride-vinylidene chloride	NMR	90,91
Vinyl	styrene-divinyl benzene	Pyrolysis-mass spectrometry	92

Table 2.10 - Monomers and Volatiles in Copolymers

Monomer or volatile	Copolymer	Method	Ref
Divinylbenzene	Styrene-divinylbenzene	Pyrolysis - mass spectrometry	93
Styrene	Polyester laminates Styrene butadiene acrylics	Polarographic	94
Styrene and butadiene	Styrene-butadiene	Head space analysis and gas chromatography of DMF solution	18, 96
Benzene and toluene	Rubber adhesives	Gas chromatography	97
Styrene	Latex	Gas chromatography	98, 99
Styrene, dialkyl phthalate,	Polyester resin mouldings mouldings	Gas chromatography	100, 101
d-limonene benzaldehyde methyl acetate, toluene	Plasticized vinyl chloride copolymers	Permeability of polymer films to organic vapours mass spectrometry	126
Styrene, ethyl acrylate	Styrene ethyl acrylate	Radioactive tracer methods	102, 103
Styrene, methacrylic acid	Styrene methacrylic acid	Mercurimetric	104, 105
Ethylene glycol, dimethyl tereph-thalate, dimethyl isophthalate, dimethyl adipate	Isophthalic acid-terephthalic acid-ethylene glycol	Gas chromatography	106
Ethyl acrylate, styrene, vinyl acetate	Styrene - ethyl acrylate	Distill in presence of toluene, glc	103
Methyl acrylate, ethyl acrylate, 2-ethylhexyl acrylate, butyl	Mixed polyacrylates solution, glc		108

Table 2.10 continued

acrylate, ethyl acrylate, vinyl propionate			
2-ethylhexyl acrylate, vinyl acetate, butyl acrylate, methyl acrylate, methacrylic acid	Mixed polyacrylates	Solution in propyl acetate-cyclohexanol, glc. Solution in isopropanol-glc	110, 111
Methyl methacrylate, ethyl acrylate, styrene	Styrene - acrylate and styrene methacrilate	Solution, glc	112
Vinyl acetate, 2-ethyl hexyl acrylate	Vinylacetate - 2-ethylhexyl-acrylate	Solution in propyl acetate, glc	108

Table 2.11 - Determination of Additives in Copolymers

Additive	Copolymer	Method	Ref
Dilauryl thiodipropionate	Ethylene - vinyl acetate and acrylonitrile - butadiene-styrene	Hydrolysis to lauryl alcohol hydrolysis with chloroform-ethanol-hexane then TLC or GLC	121
Santonox	Tetramethylene glycol-dimethyl acrylate	High performance liquid chromatography	114
Polygard (tris-nonylated phenyl) phosphite	Styrene butadiene	Methanol or ethanol extraction, coupling with diazotized-p-nitro-aniline, spectrophotometric finish	115
Polygard (tris-nonylated phenyl) phosphite, Agerite Supealite NevastainA	Styrene-butadiene	Solvent extraction, hydrolysis to nonyl phenol, diazotization and spectro-photometric evaluation	10
Polygard (tris-nonylated phenyl) phosphite	Styrene-butadiene	Solvent extraction - addition of sodium hydroxide and measurement of batho chromic shift spectrometry	11
Various types Phenyl ß napthylamine WingstayS (styrenated phenols), Polygard	Styrene-butadiene	Solvent extraction Colormetric tests	118
Amine acid phenolic types		Solvent extraction, thin layer chromatography for quantitative analysis	118, 119
N-phenyl-ß napthylamine		Volatile preconcentration mass spectrometry	17
6 dodecyl-2,2,4-tri methyl 1,2 dihydro-quinaline, trisnonyl phenyl phosphate, dilaurylthio dipropionate		Hydrolysis to lauryl alcohol, chloroform-ethanol-hexane extraction and thin-layer chromatography then glc	121
Alkylated cresols (2,6-di-t-butyl-p-cresol), amine type (N-phenyl-2		Ethanol extraction, glc	122

Table 2.11 continued

napthylamene), Subst'd p-phenylene diamine type, Santoflex 13 sec heptyl phenyl-p-phenylene diamine			
Additives and low molecular weight compounds		Solvent extraction - column chromatography	123
Accelerators and antioxidants	Rubbers	Infrared spectroscopy	124
Amine antide-gradients	Rubbers	Gas chromatography	1
Accelerators and antioxidants	Rubbers	Column chromatography	43
Antioxidants phosphite type, (bis (2,4 di-tert butyl phenol), pentaery thritol diphosphite) and its diphosphate oxidation product also distearylpent-aerythritol diphosphite	Acrylonitrile-butadiene-styrene terpolymer and polyethylene terephthalate	Laser desorption electron impact FT ioncyclotron mass spectrometry	125

REFERENCES - CHAPTER 2

1. Crompton, T.R. and Reid, V.W., J. of Polymer Science, Part A, **1** 347 (1963)
2. Albert, N.K., Woodbury Research Laboratory, Shell Chemical Co., Woodbury, U.S.A. Private Communication, October (1966)
3. Basch, A., and Lewin, M., J. Poly., Sci. Polym., **11** 1707 (1973)
4. Dollimore, D. and Holt, B., J. Polym. Sci. Polym. Phys. Ed., **11** 1703 (1973)
5. Varma, S.D. and Narasimhan, V., J. Appl. Polym. Sci., **16** 3325 (1972)
6. Funt, J.M. and Magill, J.H., J. Poly. Sci. Polym. Phys. Ed., **12** 217 (1974)
7. Kokta, B.V. Valade, J.L. and Martin, W.N., J. Appl. Polym. Sci., **17** 1 (1973)
8. Crompton, T.R. and Myers, L.W., European Polymer Journal, **4** 355 (1968)
9. Shanks, R.A., Scan., **6** 20 (1975)
10. Nawakowski, A.C., Anal. Chem., **30** 1868 (1958)
11. Brandt, J.J., Anal. Chem., **33** 1390 (1961)
12. Hilton, A.S. and Altenau, A.G., Dublin Chemistry and Technology, **46** 1035 (1973)
13. Hayes, M.W. and Altenau, A.G., Rubber Age, **102** 59 (1970)
14. Crompton, T.R. and Buckley, D., Analyst (London), **90** 76 (1965)
15. Mino, G., J. Polymer Science, **22** 369 (1956)
16. Roy, S.S., Analyst (London), **102** 302 (1977)
17. Claver, G.C. and Murphey, M.E., Anal. Chem., **31** 1682 (1959)
18. Shapras, P. and Claver, G.C., Anal. Chem., **36** 2282 (1964)
19. Mayo, F.R., Lewis, F.M. and Walling, C., J. Amer. Chem. Soc., **70** 1529 (1948)
20. Doak, K.W., J. Amer. Chem. Soc., **70** 1525 (1948)
21. Wise, R.W. and Sullivan, A.B., Rubber Chemistry and Technology, **35** 684 July/Sept (1962)
22. Pfab, W. and Noffz, D., Anal. Chem., **37** 195 (1963)
23. Rohrschneider, L., Z. Anal. Chem., **255** 345 (1971)
24. Shanks, R.A., Pye Unicam Newsletter (1975)
25. Steichen, R.J., Anal. Chem., **48** 1398 (1976)
26. Kaduji, I.I. and Rees, J.H., Analyst (London), **99** 435 (1974)
27. Bark, L.S. and Grime, J.K., Analyst (London), **97** 911 (1972)
28. Crompton, T.R. unpublished work
29. Newmann, E.W. and Nadeau, H.G., Anal. Chem., **10** 1454 (1963)
30. Swan, W.B. and Dux, J.P., Anal. Chem., **33** 654 (1961)
31. Zeman, T., Novak, L., Mitter, L., Stekla, J. and Holendova, O., J. Chromatography, **119** 581 (1976)
32. Mathias, A. and Mellor, N., Anal. Chem., **38** 472 (1966)
33. Stead, J.B. and Hindley, A.H., J. Chromat., **42** 470 (1969)
34. Konishi, K. and Kanoh, Y., Japan Analyst., **15** 1110 (1966)
35. Karger, M.H. and Mazur, Y., J. Amer. Chem. Soc., **90** 3878 (1968)
36. Tsuji, K. and Kouishi, K., Analyst (London), **99** 54 (1974)
37. Cervenka, A. and Merrall, C.H., J. Chem. Phys., **62** 120 (1975)
38. Sewell, R.P. and Skidmore, D.W., J. Polymer Science, A - I, **6** 2425 (1968)
39. Cooper, W., Eaves, D.E., Tunnicliffe, M.E. and Vaughan, G., European Polymer J., **1** 121 (1965)
40. Tunnicliffe, M.E., Mackillop, D.A. and Hank, R., European Polymer J., **1** 259 (1965)
41. Altenau, H.G., Headley, L.M., Jones, S.O. and Rensaw, H.C., Anal. Chem., **42** 1280 (1970)

42. Lee, T.C., Kolthoff, I.M. and Johnson, E., Anal. Chem., **22** 995 (1950)
43. Darker, C.A. and Berriman, J.M., Transactions of the Institution of the Rubber Industry, **28** 279 December (1952)
44. Dowle, C.J., Campbell, W.C., Analyst (London), **113** 1157 (1988)
45. Kemp, A.R. and Peters, H., Ind. Eng. Chem. Anal. Ed., **15** 52 (1943)
46. Hank, R., Rubber Chem. Technol., **40** 936 (1967)
47. Crisp, S., Lewis, B.G., Wilson, A.D.J., Dental. Res., **54** (6) 1238 (1975)
48. Tan, J.S. and Gasper, S.P., Macromolecules, **6** (5) 741 (1973)
49. Garrett, S.R. and Guile, R.L., J. Am. Chem. Soc., **74** 4533 (1951)
50. Douglas, J., Timnick, A. and Guile, R.L., J. Polymer Sci., A - I 1609 (1963)
51. Johnson, D.E., Lyeria, J.R., Horikawa, T.T. and Pederson, L.A., Anal. Chem., **49** 77 (1977)
52. Sharp, J.L. and Paterson, G., Analyst (London), **105** 517 (1980)
53. Nissen, D., Rossbach, V. and Zahn, H., J. Appl. Polymer Sci., **18** 1953 (1974)
54. Sato, H., Takeuchi, H., Suzuki, S., Tanaka, Y., Makromol Chemie Rapid Commun. **5** 719 (1984)
55. Munteanu, D. and Savu, N., Rev. Chim. (Bucharest), **27** (10) 902 (1976) Chem. Abstr. **86** 121973 d (1977)
56. Bevington, J.C., Eaves, D.E. and Vale, R.L., J. Polymer Sci., **32** 317 (1958)
57. Bevington, J.C., Trans. Faraday Soc., **56** 1762 (1960)
58. Aydin, V., Kaczmar, B.U. and Schulz, R.D., Angew Makromol. Chem., **24** 171 (1972)
59. Lenkrath, G., Gummi Asbest. Kunststoffe, **29** 585 (1976)
60. Miller, D.L., Samsel, E.P. and Cobler, J.G., Anal. Chem., **33** 677 (1961)
61. Samsel, E.P. and McHard, J.A., Ind. Eng. Chem. Anal. Ed., **14** 750 (1942)
62. Haslam, J., Hamilton, J.B. and Jeffs, A.R., Analyst (London), **83** 66 (1958)
63. Haslam, J. and Jeffs, A.R., J. Anal. Chem., **7** 24 (1957)
64. Anderson, D.G., Isakson, K.E., Snow, D.L., Tessari, D.J. and Vandeberg, J.T., Anal. Chem., **43** 894 (1971)
65. Barrall, E.M., Porter, R.S. and Johnson, J.E., Anal. Chem., **35** 73 (1963)
66. Porter, R.S., Hoffman, A.S. and Johnson, J.F., Anal. Chem., **34** 1179 (1962)
67. Helmuth, J., Poly. Vehromarium Plast., **3** 7 (1973)
68. Squirrel, D.C.M., "Automatic methods in Volumetric Analysis" Hilger and Watts, London, p. 94 - 99 (1964)
69. Munteanu, D. and Toader, M., Mater. Plasst. (Bucharest), **13** 97 (1976)
70. Major, J. and Sodenka, J., Chem. Prumsyl., **25** 601 (1975)
71. Sirynk, A.G. and Bulgakova, R.A., Vysokomol Soedin, Ser. B., **19** 152 (1977)
72. Glockner, M., Van der Berg, J.H.M, Mellerink, N.L.T., Scholte, T.G., Konigsvelt, R. Journal of Chromatography, **317** 615 (1984)
73. Puchalsky, C.B., Anal. Chem., **51** 1343 (1979)
74. Haslam, J., Willis, H.A. and Squirrel, D.C.N., Identification and Analysis of Plastics, 2nd edition, Iliffe, London, pp 57 - 63 (1972)
75. Porter, R.S., Nicksic, S.W. and Johnson, J.F., Anal. Chem., **35** 1948 (1963)
76. Glockner, G., Van der Berg, J.H.M., Meijerink, N.L.J., Scholte, T.G., Konigsvelt, R, Makromolecules, **17** 962 (1984)
77. Campbell, D.R., Anal. Chem., **47** 1477 (1975)
78. Phillips, M.A., J. Chem. Soc. (London), 2393 (1928)
79. Swaraj, S. and Ranby, B., Anal. Chem., **47** 1428 (1975)
80. Kretz, R., Fresenius, Z. Anal. Chem., **176** 421 (1960)
81. Steyermark, A., J. Assoc. Off. Angric. Chem., **38** 367 (1955)

82. Ehrlich, Rogozinsky, S. and Patchornick, A., Anal. Chem., **36** 840 (1964)
83. Anderson, D.G., Isakson, K.E., Vandeberg, J.T., Jao, M.Y.T., Tessari, D.J. and Afremow, L.C., Anal. Chem., **47** 1008 (1975)
84. Van Houwelingen, G.D.B., Analyst (London), **106** 1057 (1981)
85. Gyenes, I., "Titration in Non-aqueous Media" Iliffe Books, London p 261 (1967)
86. Kanjilal, C., Mitra, B. and Palit, S.R. Makromol. Chem., **178** 1707 (1977)
87. Sloane, H.J. and Branstone-Cooke, R., Appl. Spectroscopy, **27** 217 (1973)
88. Turner, R.R. and Carlson, D.W., Altenau, U., unpublished work
89. Cheng, H.N., Bennett, M.A., Analytical Chemistry, **56** 2320 (1984)
90. Chujo, S., Satoh, S., Ozeka, T. and Nagai, E., J. Polymer Sci., **61** 512 (1962)
91. Chujo, S., Satoh, S. and Nagai, E., J. Polymer Sci., A - 2 895 (1964)
92. Svob, N. and Flajsman, F., Croat. Chem. Acta., **42** 417 (1970)
93. Ymasa, T. and Kamiya, K., Japan Analyst., **13** 966 (1964)
94. Novak, V. and Siedl, J., J. Chem. Prumsyl., **28** 186 (1978)
95. Tanaka, T. Omoto, M. Donkai, M. Inagaki, H.I. Makromolec Science, B17 211 (1980)
96. Shanks, R.A., Scan. **6** 20 (1975)
97. Takashima, S. and Okada, F., Himeji, Kogyo Daigaku Kenkyu Hokoku, **12** 34 (1960)
98. Brodsky, J., Kunststoffe, **51** 20 (1961)
99. Nelson, F.M., Eggertsen, F.T. and Holst, J.J., Anal. Chem., **33** 1150 (1961)
100. Holtmann, R. and Souren, J.R. Kunststoffe, **67** 776 (1977)
101. Yamaoka, A. and Matsui, T., Himeji Kogyo Daigaku Kenkyu Hokoku, **30A** 101 (1977); Chem. Abstr., **89** 7000w (1978)
102. Acosta, J.L. and Sastre, R., Rev., **29** (224) 212 (1975)
103. Tweet, O. and Miller, W.K., Anal. Chem., **35** 852 (1963)
104. Kreshkov, A.P., Balyatinskaya, L.N. and Chesnokova, S.M., Tr. Mosk. Khim-Tekhnol. Inst., **70** 146 (1972); Chem. Abstr., **80** 19144 (1974)
105. Kreshkov, A.P., Balyatinskaya, L.N. and Chesnokova, S.M., Zh. Anal. Khim., **28** (**8**), 1571 (1973)
106. Oprea, N. and Pogorevic, A., Rev. Chim. Bucharest, **25** 244 (1974)
107. Mori S.. Uno, Y. Analytical Chemistry, **59** 90 (1987)
108. Schwoetzer, G., Z. Anal. Chem., **10** 260 (1972)
109. Mori, S., Uno, Y., Analytical Chemistry, **58** 303 (1986)
110. Shiryaev, B.V. and Kozmenkhova, E.B., Zavod. Lab., **38** 1303 (1972)
111. Mel'nikova, S.L., Tishehenko, U.T. and Sazonenko, U.V., Lakokras Mater, Ikh. Primen., **4** 56 (1977)
112. Rosenthal, R.W., Schwartzman, L.H., Greco, N.P. and Proper, P.J., J. Org. Chem., **28** 2835 (1963)
113. Danielwicz, M., Kubin, M., J. Applied Polymer Science, **26** 951 (1981)
114. Pasteur, G.A., Anal. Chem., **49** 363 (1977)
115. Hilton, C.L., Anal. Chem., **32** 383 (1960)
116. Balke, S.T., Separation Purification Methods, **11** 1 (1982)
117. Peltonen, K., Analyst (London), **111** 819 (1986)
118. Parks, C.A., Anal. Chem., **33** 140 (1961)
119. Simpson, D. and Currell, B.R., Analyst, (London), **96** 515 (1971)
120. Peltonen, K., Pfaffi, P., Itkonen, A., Analyst (London), **110** 1173 (1985)
121. Sedlar, J., Feniokova, E. and Pac, J., Analyst, **99** 50 (1974)
122. Gaeta, L.J., Schleuter, E.W. and Altenau, A.G., Rubber Age, **101** 47 (1969)

123. Fallick, G.J., Talarico, P.C. and McGough, R.R., Soc. Plast. Eng., Tech. Pap., **22** 574 (1976)
124. Mann, J., Transactions of the Institution of the Rubber Industry, **27** 249, October (1951)
125. Xiang, X, Dahlgren, J. Enlow, W.P., Marshall, A.G., Analytical Chemistry, **64** 2862 (1992)
126. Ton, J.C., Rulf, D.C., Analytical Chemistry, **62** 592 (1990)
127. Miller, C.E., Eichinger, B.E., Gurley, T.W., Hermiller, J.G., Analytical Chemistry, **62** 1778 (1990)
128. Liang, Z., Marshall, A.G., Westmoreland, D.G., Analytical Chemistry, **63** 815 (1991)
129. Mori, S., Analytical Chemistry, **62** 1902 (1990)
130. Mori, S., Analytical Chemistry, **60** 1125 (1988)
131. Cortes, H.J., Jewett, G.L. Pfeiffer, C.D., Martin, S., Smith, C., Analytical Chemistry, **61** 961 (1989)
132. Wallingford, R.A. Analytical Chemistry, **68** 2541 (1996)

CHAPTER 3

DETERMINATION OF ELEMENTS

As many of the methods for determining elements are applicable to various types of polymers and copolymers, they are grouped in this chapter and not listed separately under the polymer type listed in chapters 1 and 2.

In addition to carbon and hydrogen and possibly oxygen, many polymers contain other metallic and non-metallic elements in concentrations ranging from less than one part per million up to several tens of percentage.

3.1 MINOR METALLIC CONSTITUENTS

3.1.1 Catalyst Residues

These are usually either residues left in the polymer after manufacture, such as aluminium, titanium or vanadium in low pressure polyolefins, or are adventicious metallic impurities left in the polymer after manufacture. As in many cases, trace metals can adversly effect the chemical properties such as oxidative stability of the polymer, it is important to control the level of such materials left in the final polymer. Methods are reviewed below.

3.1.1.1 Iron, Aluminium and Titanium. see Method 65

3.1.1.2 Zinc.

The zinc content of chlorinated polymers has been determined by fusing the sample with sodium peroxide followed by a standard complexometric titration of the resulting solution.[1]

3.1.1.3 Aluminium.

Trace amounts of aluminium in polyethylene have been determined by gas chromatography of volatile aluminium complexes with trifluoroacetone or pivaloyltrifluoroacetone.[2]

3.1.1.4 Aluminium and Vanadium.

In Methods 66 and 67 are described procedures for ashing polyolefin samples and ethylene propylene rubbers[3] and the determination of aluminium and vanadium.

3.1.1.5 Chromium.

Traces of chromium in polyolefins originating from Phillips type polymerization catalysts have been determined by spectrophotometric, atomic absorption spectrometric and neutron activation analysis procedures.[35]

3.1.1.6 Silica.

Commonly, nowadays, the active catalyst (based on chromium, titanium or vanadium) used in high density polyethylene manufacture is adsorbed on to a highly porous silica support. Determination of the silica catalyst support content of the final polymer gives an assessment of the economic productivity of the reactor, ie. its output of polyethylene per g of catalyst and also enables the very low concentration of active catalyst metal in the polymer to be calculated

Battiste et al[4] have described three methods for the determination of silica catalyst supports in polyethylene, (Method 68).

3.1.1.7 Sodium and Lithium.

Another source of traces of metals in polymers are neutralizing chemicals added to the final stages of manufacture to eliminate the effects of acidic catalyst remnants on polymer processing properties (eg. mold corrosion) and on polymer physical properties, (eg. hygroscopicity due to residual chloride ion).
Procedures for the determination of sodium and lithium in polyolefins are described in Methods 69 and 70.

3.1.2 Adventicious Metallic Impurities

Atomic absorption spectrometry is a useful technique for the determination of traces of heavy metals in polymers. Generally, the polymer is ashed at a maximum temperature of 450°C to avoid losses of elements by volatilization, then the ash is digested with warm nitric acid prior to spectrometric analysis, (Method 71).

Deacon et al[36] have described a high performance liquid chromatographic method for determining down to 30 ppm of eight divalent metals in polymeric sealants.

Henn[5] has reported on a flameless atomic absorption technique with solid sampling for determining trace amounts of iron, copper and chromium in polymers such as polyacrylamide with a detection limit of approximately 0.01 part per million.

The advantages of atomic absorption techniques as opposed to spectrophotometric analysis for the determination of metals are that the former are amenable to multielement analysis and can be automated. There is some evidence[6,7] that ashing polymers in silica crucibles rather than platinum can lead to up to 10% losses of elements such as copper by adsorption within the silica matrix to produce a compound that is not extractible by subsequent acid leaching. This does not occur when ashing is carried out in platinum. If silica crucibles are used then a magnesium oxide ashing aid should be employed as is demonstrated in the method for determining down to 0.1 ppm of copper in polyolefins, (Method 72).

Another method for avoiding losses of metals during ashing is the low temperature controlled decomposition technique using active oxygen. This method has been studied in connection with the determination of trace metals in PVC, polypropylene and polyethylene terephthalate.[8]

X-ray fluorescence spectrometry has several advantages over other methods. The analysis is non-destructive, specimen preparation is simple, measurement time is usually less than for other methods and x-rays interact with elements as such, ie. the intensity measurement of a constituent element is independent of its state of chemical combination. However, the technique does have some drawbacks, eg. absorption effects of other elements present, for instance, the carbon and hydrogen of the polyethylene matrix and excitation of one element by x-rays from another, eg. cadmium and selenium mutually affect one another.

Cook et al[9] studied the determination of chromium, manganese, iron, cobalt, nickel, copper and zinc in polybutadiene, polyisoprene and polyester resins. The samples were ashed and the ash dissolved in nitric acid prior to x-ray analysis. No separation schemes are necessary and concentrations as low as 10 ppm can be determined without inter-element interference. Many investigators have obtained much higher recoveries using various ashing aids such as sulphuric acid,[7] elemental sulphur[10,11] magnesium nitrate[7,14] and benzene and xylene sulphonic acids[12] than by dry ashing.

Leyden[15] used x-ray fluorescence spectrometry to determine metals in acid digests of polymers. The aqueous solutions were applied to filter paper discs. He found that recoveries of metals by the x-ray technique was 101 - 110% compared to 89 - 94% by chemical methods of analysis.

3.2 METAL CONTAINING ADDITIVES

In addition to traces of metals due to catalyst remnants, metal containing processing chemicals and adventicious impurities, polymers might also contain higher concentrations of metals. Examples of this include metallic stearate heat stabilizers in PVC and fire retardant additives such as antimony trioxide in acrylics and polyolefins. Such elements can be determined by atomic absorption procedures such as that described in Method 71.

3.2.1 Antimony

Ogure[16] has described a method for the determination of antimony trioxide in polypropylene powder by extraction with aqueous hydrochloric acid followed by atomic absorption spectrometry.

3.2.2 Arsenic.

Korenaga[17] has described an atomic absorption method for the determination of trace amounts of arsenic in acrylic fibres containing antimony oxide fire retarding agent, (Method 73).

3.2.3 Tin

Udris[18] and Belpaire[19] described various procedures for the quantitative analysis of organotin stabilizers.

Tin can be also determined in PVC by polarography and atomic absorption spectrometry[30] and in PVC foils by x-ray fluorescence analysis.[31]

3.2.4 Zinc.

Radovici et al[20] reported on a polarographic method for the micro-determination of zinc in polyesters.

3.3 ORGANIC AND INORGANIC PIGMENTS

Organic and inorganic pigments are used for colouration of polymers, polymer films and polymer coatings. Vapour phase ultraviolet absorption spectrometry at 200 nm has been used[21] to identify such pigments. In this method powdered samples are directly vaporized in the heated graphite atomizer. Thermal ultraviolet profiles of organic pigments show absorption bands between 300 and 900°C, while profiles of inorganic pigments are characterized by absorption bands at temperatures above 900°C. Temperature, relative intensity and width of the bands allow the identification of the pigments. The technique shows fast acquisition of thermal ultraviolet profiles (2 - 3 min for each run) good repeatability and wide thermal range (from 150 to 2300°C). The method has been applied to a variety of polymers, (Method 74).
Hsu and Marshall[32] identified dyes in solid poly(methylmethylacrylate) by laser desorption Fourier Transform Ion Cyclotron Resonance Mass Spectrometry. Accurate mass measurements identified several red and orange dyes in untreated poly(methmethacrylate) at an order of magnitude lower (0.1%) than obtained by infrared attenuated total reflectance spectroscopy (1 - 2%).

3.4 NON-METALLIC ELEMENTS

As in the case of metals, non-metallic constituents such as halogens, nitrogen and sulphur can occur in polymers in amounts ranging from a few parts per million (eg. traces of chlorine catalyst remnants in polyolefins) up to tens of percent (eg. chlorine in PVC, nitrogen in polyamides).

A variety of techniques have been used for the determination of non-metallic elements, ranging from modern non-destructive techniques such as x-ray fluorescence spectroscopy to more classical techniques based on digestion of the sample followed by an appropriate analytical finish. Generally, the modern techniques have the advantages of speed of analysis, specificity and sensitivity but are costly to purchase. For many applications the classical techniques will suffice. A particularly interesting and rapid technique is that based on oxygen flask combustion. In this, a small weighed portion of polymer wrapped in a paper container is supported in a platinum holder and burnt in a sealed flask in pure oxygen. An absorbing solution in the flask collects the combustion products which are then subsequently analysed by an appropriate spectrophotometric or titrimetric technique.

3.4.1 Halogens

Three basic types of methods exist for the determination of chlorine and other halogens in polymers. These are based (i) on fusion of the polymer in a ground form with sodium carbonate, followed by subsequent determination of the sodium halide (Method 75), (ii) combustion of the polymer in an oxygen flask containing an absorbing solution followed by determination of halide ion (Methods 76 - 78)[2] and (iii) non-destructive methods based on x-ray fluorescence spectrometry of the polymer.

Both silver nitrate (Methods 75 and 77) and mercuric nitrate (Method 78) have been used to titrate halides following sample digestion.

Mittenberger and Gross[24] determined chlorine in PVC by fusion with sodium peroxide followed by silver nitrate titration of the chloride produced. Tanaka and Morikawa[25] described a semimicro technique for the determination of total chlorine in PVC using a semimicro method based on the Schoniger oxygen flask and Fajans method.

Morett[33] has described a 13C NMR method for the determination of chlorine in polystyrene. Qi and Pickup[37] used x-ray emission analysis to determine the ratio of chlorine to sulphur present in copolymers based on poly (3-methyl thiophene).
Johnson and Leonard[26] have described a method (Method 79) for the determination of fluorine based on decomposition of the sample by oxygen flask combustion followed by spectrophotometric determination of the fluoride produced by a spectrophotometric procedure involving the reaction with the cerium III complex of alizarin complexan (1,2-dihydroxy-anthraquinone 3-ylmethylamine N,N-diacetic acid).

3.4.2 Sulphur

An oxygen flask combustion procedure is described in Method 80 for the determination of sulphur in amounts from 50 ppm upwards in polyolefins and other polymers.

Sodium peroxide fusion in a metal bomb has also been used[27] to decompose polymers prior to the determination of down to 50 ppm sulphur. In the procedure described in Method 81 sodium is removed from the fusion product by means of a cation exchange resin and the sulphate is determined by titration with standardised 0.005M barium perchlorate. Chlorine, fluorine and nitrogen in amounts up to 2 mg. in the sample are without serious effect on the determination of sulphur. The effect of larger amounts of fluorine can be suppressed by the addition of boric acid.

The method described in Method 82 for the determination of macro amounts of sulphur in polymers is based on combustion of the sample in an oxygen filled flask over a solution of hydrogen peroxide. The amount of sulphuric acid produced is determined by titration with 0.005 M or 0.0005 M barium perchlorate the equivalence point being obtained photoelectrically. The use of a photoelectric method of end-point detection overcomes the difficulties associated with visual end-point detection described in Methods 80 and 81 as it makes the assessment of end-point independent of invididual operators observation of colours.

Inductively coupled plasma atomic emission spectrometry[13,34] has been used to determine sulphur in polymers.

3.4.3 Nitrogen

Various methods are available for the determination of nitrogen in polymers. The first method (Method 83) based on kjeldahl digestion - boric acid titration, is designed to determine very low concentration (50 - 100 ppm) of nitrogen in polymers and as such is suitable for the determination of low concentrations of nitrogen containing additives in polymers (eg. 0.25% Nonox CI antioxidant in polyolefins). Method 84 also based on kjeldahl digestion - boric acid titration, is suitable for the determination of much higher concentrations of nitrogen (1 - 90%) in nitrogen containing copolymers. Method 85 has a wide range of application (0.002 - 75% nitrogen) and is based on kjeldahl digestion with a spectrophotometric indophenol blue finish for the determination of the ammonia produced on digestion of the polymer. The final method (Method 86) is based on micro dumas combustion and is capable of determining nitrogen in the range 1 - 50% in copolymers. This method is available in a commercially produced automated version (eg. the Perkin Elmer CHN Analyser).

Hernadez[29] has described an alternative procedure based on pyro-chemiluminescence which he applied to the determination of 250 to 1500 ppm nitrogen in polyethylene. In this technique the nitrogen is the sample is subjected to oxidative pyrolysis to produce nitric oxide. This when contacted with ozone produces a metastable nitrogen dioxide molecule which, as it relaxes to a stable state, emits a photon of light. This emission is measured quantitatively at 700 - 900 nm.

3.4.4 Phosphorus

Two oxygen flask methods are described in Method 87 and 88 for the determination of phosphorus in polymers. The first is applicable in the range 0.01 to 2% phosphorus and the second in the range 2 to 13% phosphorus.

A further method for the determination of organic phosphorus is based on digestion of the sample with concentrated sulphuric - perchloric acids, followed by a spectrophotometric finish (Method 89).

Phosphorus has been determined[28] in thermally stable polymers by mineralizing with a nitric-perchloric acid mixture and subsequent titration with lanthanum nitrate or by photometric determination of the phosphomolybdenum blue complex.

REFERENCES - CHAPTER 3

1. Bahrani, M.L., Chakravarty, N.K. and Chopra, S.C., Indian J. Technol., **13** 576 (1975)
2. Sokolov, D.N., Nesterenko, G.N. and Golubeva, L.K., Zabod. Lab., **39** 939 (1973)
3. Smith, A.J., Anal. Chem., **36** 944 (1964)
4. Battiste, D.R., Butler, J.P., Cross, J.B. and McDaniel, M.P., Anal. Chem., **53** 2232 (1981)
5. Henn, L., Anal. Chim. Acta., **73** 273 (1974)
6. Gorsuch, T., Analyst, (London), **87** 112 (1962)
7. Gorsuch, T., Analyst, (London), **84** 135 (1959)
8. Narasaki, H. and Umezawa, IK., Kobunshi Kagaku, **29** 438 (1972)
9. Cook, W.S., Jones, D.O. and Altenau, A.G., Canadian Spectroscopy, **13** 64 (1968)
10. Bergmann, J.S., Ehrhart, C.H., Grantelli, L. and Janik, L.J., 153rd National ACS meeting, Miami Beach, Florida (April 1967)
11. Rowe, W.A. and Yates, K.P., Anal. Chem., **35** 368 (1963)
12. Shott, J.E., Jr., Garland, T.J. and Clark, R.O., Anal. Chem., **33** 506 (1961)
13. Caselta, B., Di Pasquale, G., Sottientini, A., Atomic Spectroscopy, **6** 62 (1985)
14. Gamble, L.W. and Jones, W.H., Anal. Chem., **27** 1456 (1955)
15. Leyden, D.E., Lennox, J.C. and Pittman, C.U., Anal. Chim. Acta., **64** 143 (1973)
16. Ogure, H., Bunseki Kagaku, **24** 197 (1975)
17. Korenaga, T., Analyst (London), **106** 40 (1981)
18. Udris, J., Analyst (London), **96** 130 (1971)
19. Belpaire, F., Revue. Belg. Mat. Plast., **6** 201 (1965)
20. Radovici, A., Radovici, R. and Ciornei, T., Mater. Plast. (Bucharest), **11** (7) 360 (1974); Chem. Abstr., **83** 11093x (1974)
21. Tittareli, P., Zerlia, T., Colli, A. and Ferrari, G., Anal. Chem., **55** 220 (1983)
22. Falcon, J.Z., Lowe, J.L., Gacta, L.J. and Altenau, A.G., Anal. Chem., **47** 171 (1975)
23. Colson, A.F., Analyst (London), **35** January (1969)
24. Mitterberger, W.D. and Gross, H., Kunststofftecknick, **12** (7), 176; (8) 219; (9) 281; (10) 277; (11) 303; (1973)
25. Tanaka, Y. and Morikawa, T., Kagaku to Kogyo (Osaka), **48** (10) 387 (1974); Chem. Abstr., **82** 98702w (1974)
26. Johnson, C.A. and Leonard, M.A., Analyst (London), **111** 101 (1961)
27. Colson, A.S., Analyst (London), **113** 791 (1963)
28. Kalinina, L.S., Nikitina, N.I., Motorina, M.A. and Sedova, I.V., Plast. Massy., **5** 66 (1976)
29. Hernandez, H.A., International Laboratory 84 September (1981)
30. Fassy, H., Lalet, P., Chem. Analyst, **52** 1281 (1970)
31. Havranek, E., Bumbalova, A., Kapisinka, V., Chemicky Prum., **20** 536 (1970)
32. Hsu, A.T., Marshall, A.G., Analytical Chemistry, **60** 817 (1988)
33. Morrett, S., Horowicz, Analytical Chemistry, **52** 1529 (1980)

34. Di Pasquale, G., Caselta, B., Atomic Spectroscopy, **5** 209 (1984)
35. Haukka, S., Analyst (London), **116** 1055 (1991)
36. Deacon, M, Smyth, M.R., Leonard, R.G., Analyst (London), **116** 897 (1991)
37. Qi, Z., Pickup, P.G., Analytical Chemistry, **65** 696 (1993)

CHAPTER 4

COMPOSITIONAL ANALYSIS

Functional group analysis and the determination of monomer unit concentrations can be used to determine the monomer composition and hence the empirical formula of a copolymer or terpolymer. Thus, if a styrene-butadiene copolymer contains 6% (6/56 moles) bound butadiene and 94% (94/104 moles) bound styrene, then the empirical formula is:

$$Sty \frac{56}{6} Bu \frac{94}{104} \quad = \quad Sty_{8.44} Bu$$

ie. there are 8.44 moles of styrene per mole of butadiene in the copolymer.

4.1 STYRENE BUTADIENE COPOLYMERS

The compositional analysis of styrene butadiene copolymers is discussed in Section 2.1.1.

Some other examples are discussed below of the application of these principles to copolymer and terpolymer analysis.

4.2 ACRYLONITRILE-BUTADIENE - STYRENE TERPOLYMERS

Acrylonitrile can be determined via a determination of organic nitrogen by the Kjeldahl procedure (Section 3.4.3). Styrene units can be determined by infrared spectroscopy. Butadiene units can be determined by the iodine monochloride procedure (Section 2.1.1). Some typical results obtained on an ABS terpolymer are presented below:

	Analysis of original polymer	Analysis of chlorobenzene soluble fraction (56%) of original polymer	Analysis of isolated-chlorobenzene insoluble gel (44%) content of original polymer
Butadiene (B)	20.5	30.8	10.3
Acrylonitrile (A)	23.8	22.1	27.6
Styrene (S)	54.0	48.0	62.0
	____	____	____

Total	98.3	100.9	99.9
Empirical formula	$AB_{0.813}$ $S_{1.155}$	$AB_{1.320}$ $S_{1.108}$	$AB_{0.353}$ $S_{1.146}$

Oxidation with osmium tetroxide has been used as the basis for a method of determining bound butadiene in acrylonitrile - butadiene - styrene terpolymers.[1]

4.3 NATURAL RUBBER, STYRENE-BUTADIENE RUBBER AND ETHYLENE-PROPYLENE-DIENE TERPOLYMER IN CURED STOCKS

Krishen[2] has described a procedure for the determination of these monomer units (Method 90). He quantitatively analysed the gaseous pyrolysis products from natural rubber, styrene-butadiene rubber and ethylene-propylene diene terpolymer rubber by gas chromatography. He showed that the 2-methyl-2-butene peak was linear with the natural rubber content of the sample. Styrene-butadiene rubber was determined from the peak area of the 1,3-butadiene peak. The ethylene-propylene-terpolymer content was deducted from the 1-pentane peak area of the pyrolysis products.

Miller et al[61] used transmission spectroscopy in the near infrared region (1100 - 2500 nm) to determine cis-1,4 butadiene, trans-1.4 butadiene and 1.2 butadiene units in butadiene rubber and styrene-butadiene rubber in bulk and in carbon tetrachloride solution.

4.4 STYRENE ISOPRENE COPOLYMERS

Prime[62] coupled a thermogravimetic analyser and an atmospheric pressure chemical ionization tandem triple quadruple mass spectrometer system to carry out evolved gas analysis on isoprene-styrene copolymers thereby obtaining information on comonomer ratios.

Thermogravimetric analysis has also been coupled[63] with gas chromatography /mass spectrometry and with gas chromatography to obtain information on styrene-isoprene block copolymers.

4.5 ALKENE COPOLYMERS

4.5.1 Ethylene-Propylene Copolymers

An infrared spectroscopic method for the determination of bound ethylene in ethylene propylene copolymers is described in Method 91. This method utilizes the absorbance of the 732 cm^{-1} infrared absorption band attributed to (gamma)$_r$ $(CH_2)_3$ groups characteristic of an ethylene unit isolated between two head to tail propylene units;

$$-CHCH_3 - \underline{\overline{CH_2-CH_2-CH_2}}-CHCH_3-CH_2- \quad 4.1$$

Dekmesian and Morioka[64] applied Fourier Transform infrared spectroscopy to the fractions obtained by gel permeation chromatography of ethylene-propylene rubbers.

Various other workers have studied the application of infrared spectroscopy to the determination of propylene groups in ethylene-propylene copolymers. This work is summarized in Table 4.1.

^{13}C NMR spectroscopy has also been applied to the determination of the ethylene propylene ratio in these copolymers.[50]

Carbon 13 NMR has proved to be an excellent technique for analysis of sequence distributions and comonomer contents in ethylene-propylene copolymer.[18 - 26] These analyses are particularly straightforward if one of the monomer units is present at a level of 94% or greater because the other monomer will then occur primarily as an isolated unit.

Paxton and Randall[27] used carbon-13 NMR (Method 92) and infrared spectroscopy (see Method 91) to measure the concentration of ethylene is ethylene-propylene copolymers. These polymers contained greater than 95% propylene, with the ethylene units present as isolated entities between two head-to-tail propylene units. These workers point out that most infrared bands used for determining copolymer compositions are sensitive to sequences of both monomers. This infrared method for compositional analysis can be calibrated if: (a) known standards of similar consitiution to the copolymers being analysed are available and (b) assignments and behaviour of the calibration bands are well established; preferably the absorptivities of these bands should be relatively independent of the position of monomer units in the chain. Thus, quantitative infrared analysis of copolymers depends primarily on the standards employed whose composition can be determined directly and reliably. Paxton and Randall[27] used 13 C-NMR (Method 92) to provide such reference standards for the less time-consuming infrared measurements by Method 91. Because it is reletively inexpensive and easy to operate for copolymer analysis they showed that an excellent correlation is obtained between 13 C-NMR and infrared results on a series of ethylene-propylene copolymers containing greater than 95% wt% propylene.[28]

4.5.2 Ethylene-Butene Copolymers

A method[29] is discussed below for the determination of the composition of an ethylene-butene-1 copolymer containing up to about 10% butene. This technique has been applied to the pyrolysis gas chromatography of ethylene-butene copolymers. Pyrolysis were carried out at 410°C in an evacuated gas vial and the products swept into the gas chromatograph. Under these pyrolysis conditions, it is possible to analyse the pyrolysis gas components and obtain data within a range of about 10% relative. The peaks observed on the chromatogram were methane, ethylene, ethane, combined propylene and propane, isobutane, 1-butene, trans-2-butene, cis-2-butene, 2-methyl-butane and n-pentane.

A straight line relationship exists between the amount of ethylene produced on pyrolysis and the amount of i-butene in the ethylene - butene copolymer. The y intercept of 16.3% ethylene should represent that amount of ethane which would result from a purely linear polyethylene. An essentially unbranched Phillips-type polyethylene polymer yielded 14.5% ethylene, which is fairly close to the predicted 16.3%.

Table 4.1 - Application of Infrared Spectroscopy to the Determination of Propylene in Ethylene-Propylene Copolymers

Wavelength	Related to	Comments	Ref
7.25 (CCl_4 solution)	7.25 methyl groups	High scatter in results	3-5
Ratio 13.95/8.7 (solid film)	Propylene content in range 30 - 50% propylene	Results adversly affected by polymer crystallinity	6-8
Ratio 7.25 band (solid diecast film) and product of absorbance by the halfwidth of the 6.85 hand	Propylene content	Calibrated versus C14 labelled ethylene or propylene copolymers by plotting ratio 7.25/6.55 versus % propylene weight fraction	9
7.25 and ratio 8.6/2.32 (CCl_4 solution)	7.25, methyl groups	Copolymers containing 30% propylene	10
Ratio 1.692/1.764 (CCl_4 solution)	1.692 CH3 groups 1.764 CH2 groups	Applicable to copolymers containing 15 - 52% propylene	11
13.7 and solid film scanned at 180°C	13.70-13.89 crystalline phase 13.89 amorphous phase	Calibrated versus known composition copolymers prepared with C14 labelled ethylene	40 15-17
Ratio 11.00/11.25 pyrolysis of film at 450°C	11.00 vinyl groups 11.25 vinylidine groups	Ratio 11.00/11.25 varies with propylene content of copolymer copolymers ctg >8% propylene	14 12,13

In this method, accurately prepared standard copolymers are required for calibration purposes. Physical blends of the two homopolymers, polyethylene and polybutene-1 will not suffice as these behave differently in the methods to the true copolymers. An excellent method for preparing such standards is to copolymerize blends of ethylene and 14C labelled butene -1 of known activity. From the activity of the copolymer determined by scintillation counting its butene -1 content can be calculated.

Numerous methods have been applied to the analysis of 1-butene-olefin copolymers, including pyrolysis gas chromatography,[29,30,32] differential thermal analysis[32] x-ray crystallography[34] and melting-point fractionation.[35] The most widely adopted technique is infrared spectrometry[34 - 38] either using a heated cell to eliminate the effects of crystallinity, or more simply by scanning a film of known thickness and comparing the absorbance to those of standards. Infrared is the easiest method to run and the least demanding in equipment. It requires, however, a set of standards for calibration of the instrument used. These standards have in the past been frequently obtained by copolymerization of 14-C tagged monomer and radioassay.[14,39] This technique introduces additional manipulations and the possibility of isotope effects and is time-consuming.

Fisch and Dannenberg[41] have reported that for the analysis of propylene butene-1 copolymers containing up to 11% bound propylene, satisfactory calibration standards can be obtained using 13-C NMR analysis.

Other NMR methods for the determination of the composition of olefins copolymers are tabulated in Table 4.2.

4.5.3 Ethylene -Alpha Olefin Copolymers

Hatfield et al[69] applied melt state MAS ^{13}C NMR to the determination of comonomer types and composition in ethylene-alpha olefin copolymers. Melt state ^{13}C NMR with magic angle spinning and dipolar coupling was used for this analysis and gave the advantages of reduced analysis time and the ability to analyse samples which were not amenable to solution state NMR. Hatfield et al[69] examined five different ethylene -alpha olefin copolymers; one cross linked ie. insoluble. Good agreement was obtained between melt and solution NMR methods.

4.6 OTHER ETHYLENE COPOLYMERS

4.6.1 Ethylene-Vinyl Acetate Copolymers

Jones and McClelland[65] have demonstrated the application of transient infrared emission spectroscopy (TIRES) to quantitative compositional analysis of ethylene-vinyl acetate copolymers. Standard errors are less than 1%.

4.6.2 Ethylene - Tetrafluoroethylene Copolymers

Morelli et al[66] used evolved gas mass spectrometry to carry out compositional analysis of ethylene-tetrafluoroethylene copolymers. The technique responded to the HF^+ signal observed in the evolved gas analysis mass spectrum.

4.7 ACRYLIC COPOLYMERS

4.7.1 Styrene-Acrylate Copolymers

Anderson et al[51] determined acrylate units and styrene units in copolymers of acrylic acid or methacrylic acid with styrene, vinyl chloride and acrylamide by a combination of techniques. Acrylate groups were determined by carrying out a Ziesel reaction with hydrogen iodide to convert acrylate groups to alkyl iodides which were determined by gas chromatography. This method is discussed in further detail in the earlier section on the determination of acrylate groups (Section 2.7.1., Method 62).

Styrene units were determined,[51] as described in Method 93, by infrared spectroscopy at 700 cm^{-1} (14.28u) which is the phenyl ring out of plane mode. This frequency provides specificity, freedom from interferences and an absorption that is directly proportional to the styrene content.

Cheng-Yu Wang et al[70] applied pyrolysis gas chromatography to the determination of the structure of styrene-n-butyl acrylate copolymers. The number average sequence length which reflects monomer arrangement was calculated using formulae that incorporate pure trimer peak intensities and hybrid trimer peak intensities.

The degree of structure ie. number average sequence lengths also composition were determined for styrene-n-butyl acrylate copolymers and compared to those obtained for homogenous ie. non-structured (random) copolymers.

Table 4.2 - NMR Methods for Determining Copolymer Composition

Polymer	Determined	Reference
Ethylene-propylene	Propylene	42
Ethylene-propylene	alpha- olefins	43
4-methyl-isopentene-1-pentene	4-methyl-1-pentene	44
Butene-propylene	Butene	45
4-methylpentene-1-pentene	4-methyl pentene	46
Propylene-butene	Propylene	45, 47, 48
Propylene-vinyl chloride	Propylene	49

4.7.2 Styrene-Methacrylate Copolymers

Evans et al[52] have described techniques employing pyrolysis - gas chromatography, (Method 95) proton NMR (Method 94) and carbon analysis for the determination of styrene and methacrylate units in styrene - methylmethacrylate and styrene-*n*-butyl methacrylate copolymers. Agreement between the three independent methods is excellent. The comparison stresses the complementary nature of all three methods. The pyrolysis gas chromatography method possesses advantages such as simplicity and rapidity.

Various other workers[53-58] have used NMR to determine methyl-methyl-methacrylate in styrene - methmethacrylate copolymers.

Cheng Yu Wang and Smith[71] used pyrolysis gas chromatography to carry out quantitative analysis and microstructure determination of styrene-methyl methacrylate copolymer. Compositional analysis was achieved by measurement of monomer peak intensities. Due to the poor stability of methyl methacrylate oligomers, neither methyl methacrylate dimers or trimers were detected under normal pyrolysis conditions. The number average sequence length for styrene units was determined from pure and hybrid trimer peak intensities. The number average sequence length for methyl methacrylate was determined using formulae that incorporate composition and the number average sequence length of styrene units.

4.7.3 Methylmethyacrylate - Methacrylic Acid Copolymers

In titrating copolymers of methyl methacrylate and methacrylic acid with standard base to determine composition, a number of deficiences have been encountered. For example, there were two common sources of "contamination" that gave rise to underdermination of the acid content: 1), the copolymers tended to be hygroscopic and hence, could on occasion contain up to 5% absorbed moisture and 2), they could, on occasion, retain solvents and/or monomers. Additionally, the method has been found to be inapplicable to copolymers of high molecular weight ($MW > 10^6$) and/or high acid content (> 60% acid) because of the tendency of such systems to reprecipitate during the titration procedure. Because of these limitations, Johnson et al[59] examined the applicability of Fourier Transform ^{13}C NMR to the compositional analysis of methylmethacrylate acid copolymers, (Method 96). In pyridine solutions of these copolymers, the resonances arising from acid carboxyl and ester carbonyl carbons are sufficiently resolved to allow the determination of relative integrals.

4.7.4 Butyl Methacrylate-Methyl Methacrylate Copolymers

Transient infrared emission spectroscopy (TIRES)[65] and transient infrared transmission spectroscopy (TIRTS)[67] have both been applied to quantitative compositional analysis of butyl methacrylate methyl-methacrylate copolymers. Standard errors are less than 1%.

4.7.5 Acrylic Acid-Methacrylic Acid Copolymers

Sharp and Patterson[68] have described a pyrolysis - gas chromatographic procedure (Method 61) for the determination of up to 10% of copolymerized acrylic acid and methacrylic acid in acrylic copolymers.

4.7.6 Functional Group Analysis of Acrylic Copolymers

The determination of carboxyl, ester and epoxy groups in various acrylic copolymers is discussed in Section 2.7.1. In many cases such analyses enable copolymer analysis to be carried out.

4.7.7 Vinyl Chloride-Vinylidene Chloride Copolymers

Chang Yu Wang and Smith[72] used pyrolysis - gas chromatography to elucidate the composition of and carry out structural studies on vinyl chloride-vinylidene chloride copolymers. The number average sequence length, which reflects monomer arrangement in the copolymer, were calculated using formulae that incorporate pure trimer and hybrid trimer peak intensities.

Due to the difference in reactivity between vinyl chloride and vinylidene chloride monomers the structure of the polymer was further investigated on the basis on the percentage of grouped monomers (ie. the number average sequence length for vinyl chloride and vinylidene chloride repeat units). The results obtained for compositional analysis achieved by this method and by H NMR were in excellent agreement. In the method 2.5 mg of sample was pyrolysed in a quartz tube, equilibriated for 5 minutes at 180°C, then pyrolysed at 700°C for 20 seconds using a pyroprobe CD5190 with platinum coil. Gas chromatography was carried out using a flame ionization or mass spectrometric detector.

4.8 HEXAFLUOROPROPYLENE - VINYLIDENE FLUORIDE COPOLYMERS

Two methods have been described for determining the compositional analysis of these copolymers, one based on high resolution continuous and Fourier Transform 19 F NMR and the other on pyrolysis - gas chromatography,[33] (Method 97).

4.8.1 19 F NMR

When it is desired to measure the composition of a single component in a mixture, it is necessary to relate the component resonance to that of another compound, an internal standard, which is of known chemical composition and has been added in known weight to a known weight of unknown. Brame and Yeager[31] used dichlorobenzotrifluoride as an internal standard in the continuous wave method for determining the compositional analysis of both repeat units in hexafluoro-propylene-vinylidene fluoride copolymers. This work demonstrated the utility of the Fourier transform NMR method in quantitative analysis of the copolymer in relation to results obtained by continuous wave 19 F NMR and proton NMR.

The lines observed are attributed to the following:

CF_3 group (delta = -70 to -75), CF_2 groups (delta = -90 to -120) and CF group (delta = -180 to -185).

The results of the determination on a copolymer sample are given in Table 4.3. The value obtained is in excellent agreement with those obtained by mass balance.

4.8.2 Pyrolysis - Gas Chromatography.

Blackwell[33] used a Curie point pyrolyser to carry out quantitative analysis of monomer units in polyhexafluoropropylene - vinylidene fluoride (Method 97). The polymer composition is calculated from the relative amounts of monomer regenerated and the trifluoromethane (CHF_3) produced during pyrolysis. The exact mechanism by which trifluoromethane is produced during pyrolysis is not known but it is presumed that the free trifluoromethyl group is cleaved from the polymer backbone. The trifluoromethyl group then extracts a proton from the polymer chain to form trifluoromethane.

4.2

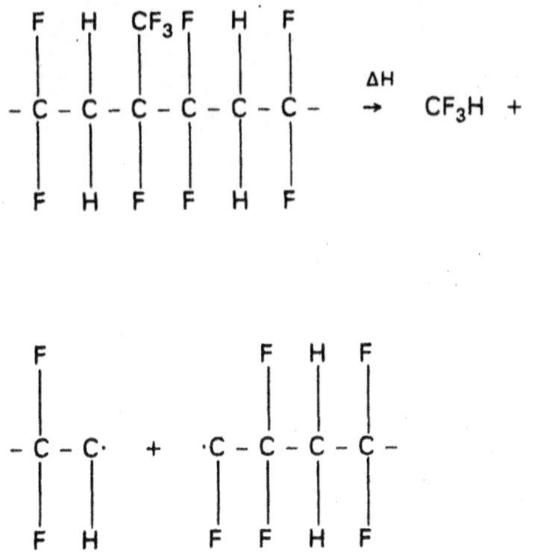

Table 4.3 - HFP/VF_2 Copolymer Compositions

Sample	Mass balance, wt % VF_2	FT-NMR	
		wt% VF_2	wt% HFP
A	48 ± 2	46.6 ± 1.1[a]	53.4 ± 1.2[a]

[a] Precision at 95% confidence level. From Brame and Jaeger[31]. American Chemical Society

REFERENCES - CHAPTER 4

1. Kranz, D., Dingles, K. and Wentling, P., Angew Makromol. Chem., **51** 25 (1976)
2. Krishen, A., Anal. Chem., **44** 494 (1972)
3. Natta, G., Mazzanti, G., Valvassori, A., and Pajaro, A., Chim. Ind. (Milan), **29** 733 (1957)
4. Drushell, H.V., and Iddings, F.A., Anal. Chem., **35** 28 (1963)
5. Drushell, H.V. and Iddings, F.A., Division of Polymer Chemistry 142nd meeting American Chemical Society Atlantic City, September (1962).
6. Wei, P.E., Anal. Chem., **33** 215 (1961)
7. Gossl, T., Makromol Chem., **42** 1 (1961)
8. Corish, P.J., Anal. Chem., **33** 1798 (1961)
9. Tosi, G., and Simonazzi, T., Die Angewandte Makromolecular Chemie., **32** 153 (1973)
10. Ciampelli, G., Bucci, G., Simonazzi, A. and Santambrogio, A., La Chimica L'Industria, **44** 489 (1962)
11. Bucci, G. and Simonazzi, T., La Chemica et L'Industria, **44** 262 (1962)
12. Drushel, H.V., Crit. Rev. Anal. Chem., **1** 161 (1970)
13. Tosi, C. and Ciampelli, G., Adv. Polym. Sci., **12** **88** (1973)
14. Brown, J.E., Tyron, M. and Mandel, J., Anal. Chem., **35** 2173 (1963)
15. Fraser, G.V., Hendra, P.J., Walker, J.H., Cudby, M.E.A. and Willis, H.A., Makromol. Chem., **173** 205 (1973)
16. Tosi, C., Makromol. Chem., **170** 231 (1973)
17. Popov, U.P. and Duvonov, A.P., Zhur. Prikl. Spektrosk., **18** 1077 (1973)
18. Porter, N., Nicksie, O. and Johnson, P., Anal. Chem., **35** 1948 (1963)
19. Wilkes, C.E., Carmen, C.J. and Harrington, R.A., J. Polymer Sci., **43** 237 (1973)
20. Carmen, C.J., Harrington, R.A. and Wilkes, C.E., Macromolecules, **10** 536 (1977)
21. Ray, G.J., Johnson, P.E. and Knox, J.R., Macromolecules, **10** 773 (1977)
22. Randall, J.C., Macromolecules, **11** 33 (1978)
23. Crain, Jnr., W.O., Zambelli, A. and Roberts, J.D., Macromolecules, **4** 330 (1971)
24. Carman, C.J. and Wilkes, C.E., Rubber Chem. Technol., **44** 781 (1971)
25. Schaefer, J., Makromolecules, **4** 107 (1971)
26. Grant, D.M. and Paul, E.G., J. Amer. Chem. Soc., **86** 2984 (1964)
27. Paxton, J.R. and Randall, J.C., Anal. Chem., **50** 1777 (1978)
28. Saunders, J.M. and Komoroski, K., Makromolecules, **10** 1214 (1977)
29. Newmann, E.W. and Nadeau, H.G., Anal. Chem., **10** 1454 (1963)
30. Verdier, J.C. and Guyot, A., Macromol. Chem., **175** 1543 (1974)
31. Brame, E.G. and Jeager, F.W., Anal. Chem., **48** 709 (1976)
32. Barrall, E.M., Porter, R.S. and Johnson, J.F., J. Appl. Polym. Sci., **9** 3061 (1965)
33. Blackwell, J.T., Anal. Chem., **48** 1883 (1976)
34. Turner Jons, A., J. Polym. Sci. Polym. Lett. Ed., **3** 591 (1965)
35. Slonaker, D.F., Combs, R.L. and Coover, Jr., H.W., J. Macromol. Sci., Chem., **1** 539 (1967)
36. Huff, T., Bushman, C.J. and Cavender, J.W., J. Appl. Polym. Sci., **8** 825 (1964)
37. Tosi, C., Lachi, M.P. and Pinto, A., Macromol. Chem., **120** 225 (1968)
38. Wegemer, N.J., J. Appl. Polym. Sci., **14** 573 (1970)
39. Lomonte, J., J. Polymer Sci. Polym. Lett. Ed., **1** 645 (1963)

40. Lamonte, J.N., Tirpak, G.A., J. Polym. Science **A2** 705 (1964)
41. Fisch, M.H. and Dannenberg, J.J., Anal. Chem., **49** 1408 (1977)
42. Yur'eva, F.A., Gusanova, F.O., Pant-yanskii, A.E., Seidov, N.M. Mamedova, V.M., Abasov, A.T. and Malova, H.G., Yysakomol Soedin Sea A., **19** 2401 (1977)
43. Popov, V.P. and Duvanova, A.P., Zh. Prikl. Spektrosk, **22** 1115 (1975)
44. Manett, S. and Horowicz, S., Anal. Chem., **52** 1529 (1980)
45. Barrall, E.M., Porter, R.J. and Johnson, J.F., J. Chromatogr., **11** 177 (1963)
46. Manlus, G.G., J. Polymer Sci., **8** 137 (1965)
47. Barrell, E.M., Porter, R.S. and Johnson, J.F., Anal. Chem., **35** 73 (1963)
48. Porter, R.S., Hoffman, A.S. and Johnson, J.F., Anal. Chem., **34** 1179 (1962)
49. Kubic, I., Singbar, M. and Navratil, Y., Petrochemia., **17** 10 (1977)
50. Cheng, H.M., Anal. Chem. **54** 1828 (1982)
51. Anderson, D.G., Isakson, K.E., Snow, D.L., Tessari, D.J., and Candeberg, J.T., Anal. Chem., **43** 894 (1971)
52. Evans, D.L., Weaver, J.L. Mukherji, A.K. and Beatty, C.L., Anal. Chem., **50** 857 (1978)
53. Nishoika, A., Katoy Ashikari, N., H. Polymer Sci., **62** 510 (1962)
54. Nishoiks, A., Katoy Mitsuoka, H., J. Polymer Sci., **62** 59 (1962)
55. Bovey, F.A., J. Polymer Sci., **62** 197 (1962)
56. Ito, K., and Yashimata, Y., J. Polymer Sci., **83** 625 (1965)
57. Harwood, H.J. and Ritchey, W.M., J. Polymer Sci., **83** 419 (1965)
58. Kranz, D., Dinges, K. and Wendling, P., Anal. Makromol. Chem., **51** (1) 25 (1976)
59. Johnson, D.E., Lyeria, J.R., Horikawa, T.T. and Pederson, L.A., Anal. Chem., **49** 77 (1977)
60. Cortes, H.J., Hewett, G.L., Pfeiffer, C.D., Martin, S., Smith, C. Analytical Chemistry, **61** 961 (1989)
61. Miller, C.E., Eichinger, B.E., Gurley, T.W., Hermiller, J.G. Analytical Chemistry **62** 1778 (1990)
62. Prime, R.B., Shushan, B. Analytical Chemistry, **61** 1195 (1989)
63. McClennan, W.H., Buchanan, R.M., Arnold, N.S., Divorzanski, J.P., Menzelaar, H.L.C. Analytical Chemistry, **65** 2819 (1993)
64. Dekmezian, A.H., Marioka, T. Analytical Chemistry, **61** 458 (1989)
65. Jones, R.W., McClelland, J.F. Analytical Chemistry, **62** 2074 (1990)
66. Morelli, J.J., Grayson, M.A., Wolf, C.J. Analytical Chemistry, **61** 802 (1989)
67. Jones, R.W., McClelland, J.F. Analytical Chemistry, **62** 2247 (1990)
68. Sharp, J.L., Patterson, G. Analyst (London) **105** 517 (1980)
69. Hatfield, G.R., Killinger, W.E., Zeigler, R.C. Analytical Chemistry, **67** 3082 (1995)
70. Cheng Yu Wang, F., Gerhart, B.B., Smith, P.B. Analytical Chemistry, **67** 3536 (1995)
71. Cheng Yu Wang, F., Smith, P.B. Analytical Chemistry **68** 3033 (1996)
72. Cheng Yu Wang, F., Smith, P.B. Analytical Chemistry, **68** 425 (1996)

CHAPTER 5

ADDITIVE MIXTURES

Generally speaking, the methods described in Chapters 1 and 2 for the determination of additives in polymers refer to particular known additives or, at least, to a particular class of additives such as phenolic antioxidants. Frequently, however, the analyst is faced with the problem of dealing with unknown additives in polymers, either single additives or, more commonly, mixtures of additives. He has first to identify the additive then determine it, usually in a mixture with other additives, which may interfere in the determination of the additive in which he is interested.

A variety of procedures have been evolved for handling this situation. Naturally, since mixtures may be involved, chromatographic techniques feature strongly in the repertoire of analytical methodology available. These include gas chromatography, high performance liquid chromatography, liquid column chromatography, supercritical fluid chromatography and thin layer chromatography. Any of these separation techniques might be coupled with an identification technique such as mass spectrometry and infrared spectroscopy or, indeed, in some instances, these latter techniques may be applicable without a prior chromatographic separation. Generally, however, chromatography will be required.

A further technique is based on pyrolysis or photolysis of the polymer additives followed by gas chromatography and/or mass spectrometry of the products produced in order to facilitate their identification and, consequently, the identity of the original additive. First, however, and certainly in the case of chromatographic techniques, it is necessary to quantitatively separate the additives from the polymer and to determine the elements present in the polymer and in extracts. These aspects are discussed below.

It is advisable when commencing the analysis of a polymer for unknown additives to determine first its content of various non-metallic and metallic elements as discussed in Chapter 3. Any element found to be present must be accounted for in the subsequent examination for, and identification of, additives. Hence, elemental analysis reduces the possibility of overlooking any additive which contains elements other than carbon, hydrogen and oxygen. The analytical methods used to determine elements should be sufficiently sensitive to determine about 10 ppm of an element in the polymer, ie. should be able to detect in a polymer a substance present at 0.01% and containing down to 10% of the element in question.

This requirement is met for almost all the important elements by use of optical emission spectroscopy and x-ray fluorescence spectrometry and other classical methods discussed in Chapter 3. X-ray fluorescence spectrometry is applicable to all elements with an atomic number greater than 12. Using these two techniques, all metals and non-metals down to an atomic number of 15 (phosphorus) can be determined in polymers at the required concentration (Cook et al,[1] Hous and Silverman,[2] Mitchell and O'Hear[3] and Bergmann et al[4]). Nitrogen is determinable at these levels by micro Kjeldahl digestion techniques.

5.1 SEPARATION METHOD FOR ADDITIVE MIXTURES

Squirrel[5] has described general analysis schemes for the examination for the presence of additives and process residues in PVC, polyolefins and acrylics. Foreknowledge of the types of additives present is not required in these schemes, which are therefore very useful when examining polymers of unknown composition. In the schematics shown in Figures 5.1 to 5.3, the major points to note are as follows. Where identification, particularly of minor organic components is required, then some separation from the plastic compound is often necessary. Special care and specialised techniques are required when dealing with laminates and surface-coated films. For major components the separation is made quantitatively and the analysis is completed by various techniques. For volatile components, separation, identification and quantification can often be carried out in one analytical process.

5.1.1 Solvent Extraction Methods

The main extraction procedures used are summarised in Table 5.1 and although most of them can be carried out on material cut to a particle size of less than 2 mm diameter, it is often advantageous to produce material of a smaller size and with a larger surface area to mass ratio. This is conveniently done by grinding at the temperature of liquid nitrogen using an efficient and easily cleaned cutter mill.

The extraction of additives strongly adsorbed or chemisorbed on the polymer filler matrix must be carefully watched by the analyst, as a change of the method of manufacture of, for example, the filler or in the method of compounding the plastic formulation, can also markedly alter the degree of adsorption being produced and hence invalidate an established quantitative extraction procedure. The use of "stronger" extraction reagents can cause complications at the measurement stage and hence each system must be carefully screened and frequently checked. Polymer extraction procedures using organic solvents do not extract all types of organic additives from polymers, also many inorganic compounds and metal inorganic compounds (eg. calcium stearate) are insoluble. The presence of metals will have been indicated in the preliminary examination of the polymer. Most types of organic polymer additives, however, can be readily extracted from polymers with organic solvents of various types. The first stage is to solvent-extract the total additives from the polymer in high yield and with minimum contamination by low molecular polymer. Extracts should be used for analysis without delay as they may contain light or oxygen sensitive compounds. When delay is unavoidable, storage in actinic glassware under nitrogen in a refrigerator minimises the risk of decomposition.

Total internal plus external additives can be extracted from low and high density polyethylene and polystyrene by procedures involving solution or dispersion of the polymer powder or granules (3 g) in cold redistilled sulphur-free toluene (50 - 100 ml), followed in the case of polyethylene by refluxing for several hours. Rubber-modified polystyrene does not completely dissolve in toluene if it contains gel. Methyl ethyl ketone or propylene oxide are alternative suitable solvents for polystyrene. Dissolved polymer is then reprecipitated by the addition of methyl alcohol or absolute ethanol (up to 300 ml) and polymer removed by filtration or centrifuging. The additive-containing extract can then be gently concentrated to dryness as described previously. Alternative procedures for the extraction of polyethylene and polypropylene involve refluxing with chloroform for 6 hours or contacting with cold diethyl ether for 24 hours or soxhlet extraction with diethyl ether, methylene dichloride, chloroform or carbon tetrachloride for

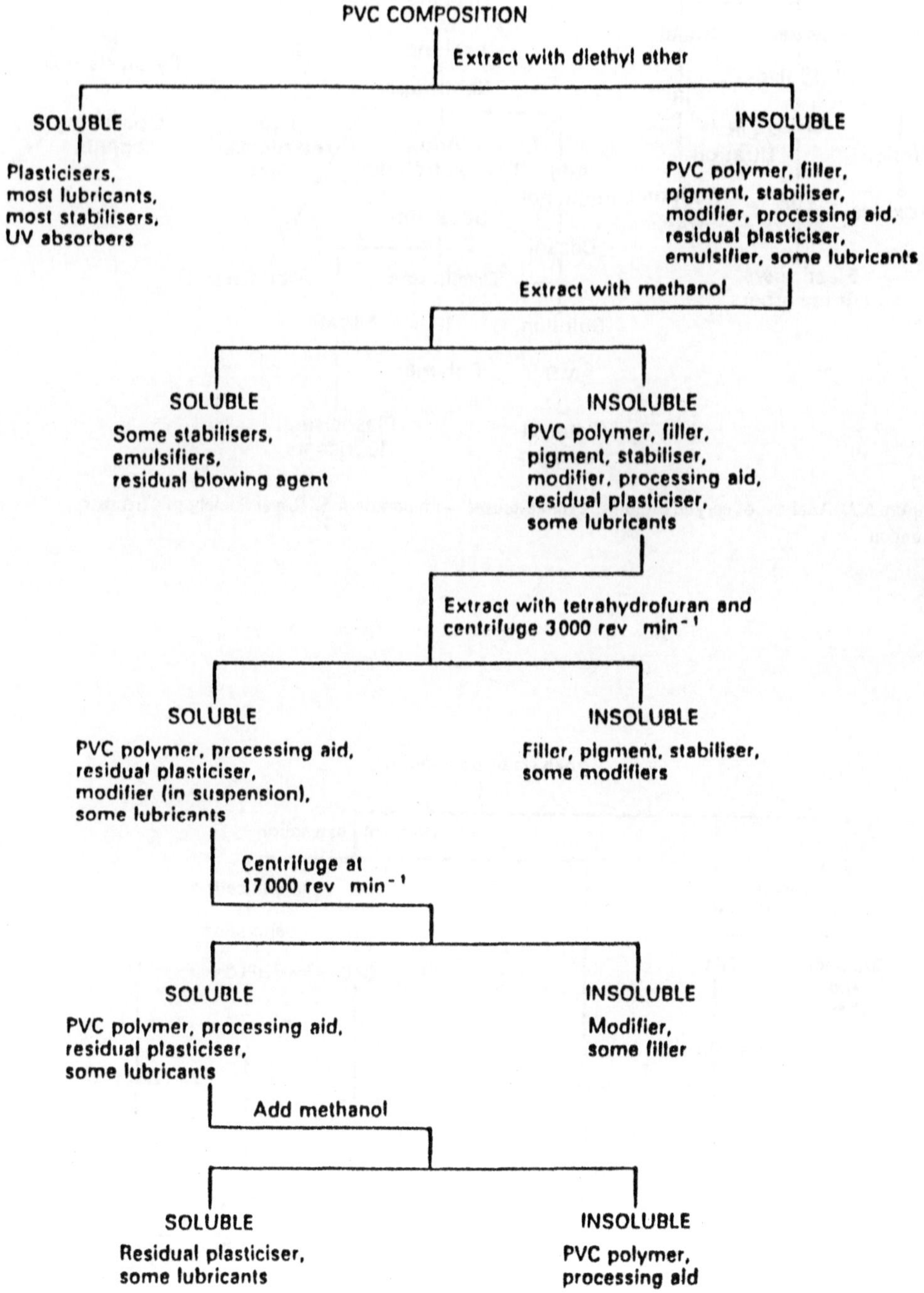

Figure 5.1. Analysis of PVC compositions. From Squirrel[5] with permission. Royal Society of Chemistry, London

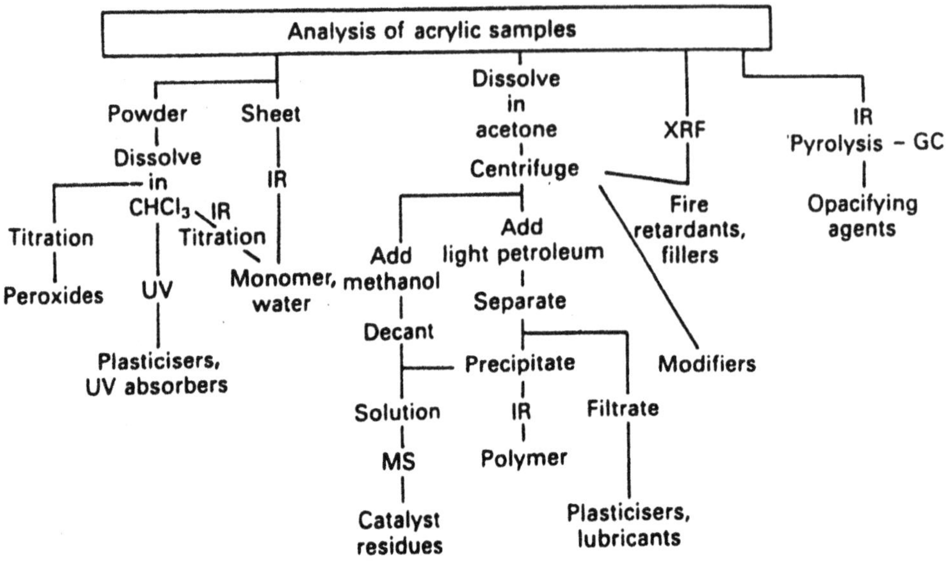

Figure 5.2. Analysis of acrylic samples. From Squirrel[5] with permission. Royal Society of Chemistry, London

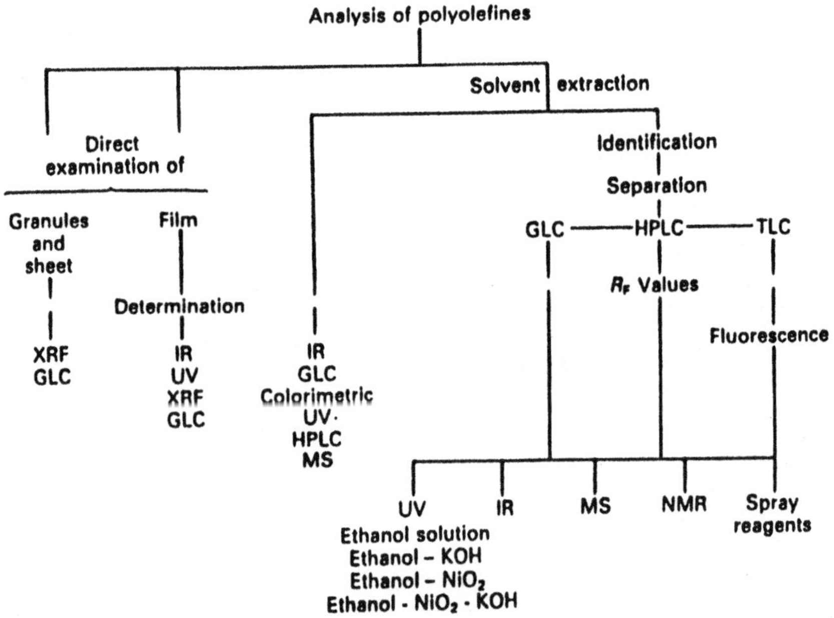

Figure 5.3. Analysis of polyolefines. From Squirrel[5] with permission. Royal Society of Chemistry, London

6 - 24 hours followed by concentration of the extract. Methylene dischloride is a particularly good solvent for polypropylene extraction because of its high volatility. In addition to additives, most solvents also extract some low molecular weight polymer with subsequent contamination of the extract. To overcome this, Slonaker and Sievers[16] have described a procedure for obtaining polymer-free additive extracts from polyethylene based on low temperature extraction with n-hexane at 0°C. This procedure is also applicable to polypropylene and polystyrene.

5.1.2 Solution Precipitation Methods

This method of extraction involves dissolution of the organic phase of the plastic composition in a suitable (sometimes hot) solvent, followed by precipitation of the polymeric constituents, often in a finely divided form in suspension with the inorganic fillers, by cooling or by means of another solvent in which the analyte compound is also soluble. This method is labour intensive but very effective and if it does not completely release any chemisorbed constituents from the polymer, for example filler matrix, it will often leave them in a form very vulnerable to attack by the analytical reagent(s) finally used in the determination.

5.1.3 Vacuum Thermal Displacement Extraction Method

These procedures are used extensively for the direct isolation or release of volatile components from a polymeric matrix and may involve the combined use of vacuum and heat, as for example in the mass spectrometer direct insertion probe or during dry vacuum distillation. Alternatively, the volatiles may be swept from the heated sample by a flow of inert gas for concentration by freeze trapping and/or collection on to a solid adsorbent prior to thermal or solvent desorption for gas-chromatographic or mass spectrometric examination

5.2 CHROMATOGRAPHIC METHODS OF SEPARATION

When the above procedures for preliminary isolation of the analyte materials from the polymer matrix are completed, further separation is often required for identification and determination.. Four forms of chromatography are generally used; gas chromatography (section 5.2.1), thin layer chromatography (section 5.2.2), high performance liquid chromatography (section 5.2.3) and more recently supercritical fluid chromatography (section 5.2.4).

5.2.1 Gas Chromatography

Gas chromatography in all its forms with appropriate detectors and, when necessary, temperature programming, heart cutting and back-flushing techniques, is used extensively for volatile components.

Headspace methods are used extensively for the determination of residual monomers and other residues in polymer compositions after dissolution or dispersion in a suitable solvent and equilibration in a sealed vial at constant temperature prior to chromatography of the headspace gas. For samples in the form of fine powders or thin films, the technique can be applied directly to the solid and liquid samples. (Tables 5.2 and 5.3).

Table 5.1 - Examples of Extraction Methods

Extraction Method	Polymer or Compound	Solvent(s)	Additives or Contaminant Extracted
Single solvent (Soxhlet)	PVC	Diethyl ether	Plasticisers
Single solvent (reflux)	Polyethylene	Chloroform-1,1,1-trichloroethane	Antioxidants
Solvent + reagent (reflux)	Polypropylene	Ethylene dichloride-trichloroacetic acid	Chemisorbed amides/amines
	Polyester	Methanol-Karl Fischer reagent	Water
Solution/precipitation	Polyolefines	Toluene-methanol, xylene-methanol	UV absorbers, antioxidants, slip agents
	Acrylics	Acetone-light petroleum	Plasticiser, lubricants
Steam/solvent distillation	Packaging films	Water - diethyl ether	Odour and taint-forming additives
Vacuum/thermal extraction	Nylon		Water
	Fluorocarbon polymers		Process fumes

The attraction of gas chromatography lies in its ability to simultaneously separate, identify and estimate sub milligram quantities of complex mixtures. There are certain drawbacks to the technique. Retention times are no more specific for gas chromatography than are R_f values for thin-layer and paper chromatography. Day-to-day reproducibility of retention times is not good for some types of instruments, especially when operating at high column temperatures. Since many polymer additives are not very volatile, low stationary-phase loadings have to be used in order to reduce retention times to a sensible value, which means that large areas of uncoated solid support appear which in turn lead to bonding with compounds such as phenolic and amine antioxidants. This results in distortion of peaks and in lengthening of retention times. There are measures which can be taken to meet these difficulties. Thus, relative retention times are often more reproducible than unadjusted retention times. Non-volatile sample components can be converted into more volatile derivatives such as tri-methyl-silyl ethers. The solid support can also be treated with hexamethyldisilazane, for example, to reduce the number of active sites on the column material available for bonding. The formation of derivatives before chromatography does, however, in some applications lead to sample loss and can often lead to the appearance of spurious peaks on the chromatogram. Nevertheless, despite these limitations gas chromatography has found numerous applications in the identification, characterization and estimation of additives in polymers.

Gas chromatography is particularly useful for the characterisation of more volatile substances such as mixtures of fatty acids or of alcohols, esters or hydrocarbons which are used in polymer formulations. It has the advantage that it provides more information regarding the carbon number distribution of each of these types of compounds than can be obtained by infrared spectroscopy alone. Similarly, if the polymer is known or suspected to contain dicarboxylic acids, polyol esters, polyhydric alcohols or alcohol esters of phthalic, sebacic, adipic or azelaic acids, then frequently, information on the type of compound present can be obtained by comparing under the same conditions the retention time of the unknown with those of known authentic substances.

Table 5.2 - Some Examples of Gas Chromatographic Procedures Applied to Liquids

Solids (direct examination)	Packaging films	and printing solvent residues, residual monomers	
"		Polyethylene	Catalyst carrier,
	Granular polymers	Polypropylene	process residues,
		Ethylene-vinyl acetate co-polymers	ethylene, vinyl acetate, acetic acid
"		PVC and co-polymers	Monomers, moisture
	Polymer powders	Poly(ethylene terephthalate	acetaldehyde
"	Foodstuffs	Biscuits, cakes and crisps	Migratory trace monomers
Solids (indirect examination)	Polymer solutions (normally 10%)	PVC compositions, ABS	Residual monomers

Table 5.3 - Some Examples of Gas Chromatographic Procedures applied to Liquids

Liquids Foodstuffs	Cooking oils, fruit squashes, water, wine, spirits	Migatory trace monomers
Liquids	Solution extracts of absorption tubes (atmospheric monitoring)	Vinyl chloride in dichloromethane, acrylonitrile in benzyl alcohol, benzene in ethylene carbonate
Liquids	Latices	Butadiene, styrene, acrylate monomers, methacrylate monomers dimers and co-dimers
Liquids Biological samples	Whole blood	Vinyl chloride from PVC transfusion sachets, trichloroethylene, trichloroethanol
	Urine	Trichloroethanol, trichloroacetic acid

Some typical applications of gas chromatography to the determination of antioxidants in polymers are tabulated in Table 5.4

5.2.2 Thin Layer Chromatography

For additives extracted from polyolefins, usually with diethyl ether, the extract is refluxed with ethanol and the solution is decanted from the insoluble residual polymer. On cooling, additives such as dilauryl and distearyl thiodipropionate separate out and are identified by infrared examination. A 30 ul volume of the ethanolic solution is then spotted on to a thin Kieselgel 60 TLC plate and eluted with a suitable solvent, usually 98.5 + 1.5 toluene-ethyl acetate. The eluted plate is dried and sprayed with colour-developing reagents and the spots are examined. If the spots are to be submitted to mass spectrometric examination, methanolic iodine is used as the colour-developing reagent as this does not over-complicate the mass spectrometry.

The normal spray procedure uses a 0.5% solution of 2,6-dibromo-p-benzoquinone-4-chlorimine in ethanol followed, after drying, by a 0.5% borax solution. After this spray the plate is dried at 120°C for 5 minutes to develop the colours.

Table 5.4 - Separation of Phenolic Antioxidants - Gas Liquid Chromatographic Techniques

Substances separated	Stationary phase	Column ^{0}C	Other details	Ref
2,5-Di-t-butyl-p-cresol, (2-hydroxy-5-methyl-phenyl) benzotriazole	25% LAC-2R/446 (adipate ester) + 2% H_3PO_4 on Chromosorb	135	H_2 carrier gas, F.I.D. error $\pm$ 1%	6
2,6-Di-t-butyl-p-cresol, 2, 6-di-t-butyl-phenyl	10% Apiezon N on Celite 545	164	H_2 carrier, F.I.D. 10-3 M in presence of others can be detected	7
2,4,6-Tri-t-butyl phenol, Diphenylamine 2,6-Di-t-butyl-p-cresol	Apiezon		F.I.D.	8
Halogenated bis-phenols	10% DC-710 Silicone oil on Chromoport 80-100 mesh	225-250	12 in glass column ¼ in o.d. Carrier: 130 ml He/min	9
Low b.p. phenols	Capillery column coated with 10% xylenol phosphate	125	F.I.D.	10
Phenols and 5-t-butyl derivatives	Silicone oil 550-Carbo-wax 400 (3:2)	200	Mean deviation 0.4%	11
Phenols and cresols	5% W/W of various phosphate esters of phenols	110	120 cm x 4.5 mm column Pye-Argon Chromatography	12
Ionox 330	a) 20% DC-710 Silicone oil on Chromosorb b) 2% SE30 Silicone gum on Chromosorb mesh	200-300 in. 10 min	a) 12 x 3/16in column b) 12 x 1/16in stainless steel column	13
Low molecular weight phenol	Silicone-coated capillary column		Converted to trime-thyl silyl esters before chromatography	14
2,6-Di-4-methylphenol	20% SE30 on HMDS-treated 60 mesh Chromosorb W	200	E.C. detector	15

Some typical applications of thin-layer chromatography to the determination of one type of additive, namely antioxidants, in polymers are tabulated in Table 5.5 (Rf values) and Table 5.6.

5.2.2.1 Thin-layer Chromatography followed by Infrared, Ultraviolet and Mass Spectrometry.

This technique, described in Method 98 is applied when it is required to identify mixtures of additives in polymers. Firstly, it is necessary to separate total additives from the polymer by a suitable solvent extraction procedure as discussed in Section 5.1. The total extract is the subject to thin-layer chromatography to separate it into individual additives which are then separately removed from the plate by scraping off bands of the plate coating. The individual portions of plate coating are then each separately extracted with a suitable low boiling solvent prior to examination by one or more of the afore-mentioned spectroscopic techniques.[69]

Table 5.5 - Thin-layer Chromatography of Polymer Additives

Additive	R_f value* Solvent 1	Solvent 2	Solvent 3	Solvent 4	Colour of spot
Topanol OC	1.00	1.00	1.00	1.00	Pale yellow centre with pink outer
DLTDP	0.40	1.00	0.05	0.30	Yellow-brown
DSTDP	0.45	1.00	0.10	0.35	Yellow-brown
Distearyl disulphide	1.10	1.05	1.20	1.20	Bright yellow
DTB glycol ester	0.00	0.75	0.00	0.00	Brown centre & mauve outer
Ionox 330	1.05	1.05	0.80	0.80	Pinkish brown
Irganox 259	0.30	1.05	0.00	0.65	Brown
Irganox 288	0.25	1.05	0.00	0.60	Brown
Irganox 1010	0.30	1.10	0.00	0.60	Brown
Irganox 1076	0.85	1.05	0.60	0.95	Brown
Nonox WSP	0.85	0.95	0.45	0.35	Bright yellow centre with pink outer
Polygard	0.30, 0.15	0.80, 0.70	0.10	0.05	Blue and red
Santonox R	0.35	0.80	0.10	0.00	Purple
Tinuvin 326	1.05	1.05	0.95	1.05	Yellow
Tinuvin 327	1.05	1.05	1.05	1.10	Yellow
Tinuvin 328	1.00	1.00	1.00	1.05	Yellow
Topanol CA	0.10	0.60	0.00	0.00	Brown
UV 531	0.75	1.00	0.35	0.80	Blue
Boechst D55	0.85, 0.75 0.55, 0.35 0.30, 0.20	0.90, 0.85	0.55, 0.45 0.25, 0.15 0.10, 0.00	1.10, 0.45 0.25, 0.15	Blue and red
Oleamide	0.00	0.00	0.00	0.00	These additives give only a very faint brown coloration
Erucamide	0.00	0.00	0.00	0.00	
Ethomeen T12	0.00	0.00	0.00	0.00	
Stearic acid	0.00	0.00	0.00	0.00	

* R_f values quoted to the nearest 0.05. Solvent 1 = toluene - ethyl acetate (98.5 + 1.5); solvent 2 = toluene - isopropanol (88 + 12); solvent 3 = toluene - light petroleum (b.p. 60-80 °C) (1 + 1); and solvent 4 = cyclohexane - toluene - methanol (88 + 10 + 2).

Dohrman [68] discusses preparative-layer chromatography which, can separate considerably larger quantities of compounds on plate layers up to 2 mm thick and 100 cm x 20 cm in area.[29-42]

5.2.3 High Performance Liquid Chromatography

This has now become a very valuable tool for plastics analysis,[73-77] particularly in the additive field. This technique can be used with both reverse-phase and adsorption columns and isocratic and gradient elution.

One of the difficulties of column chromatography is the problem of identifying the fractions in which the separated compounds are concentrated. This can be achieved by the laborious process of examining all the fractions, for example, by infrared or ultraviolet spectroscopy or by evaporating to dryness and weighing the residues; or by the less laborious process of monitoring the effluent as it leaves the chromatographic column so that solute-containing fractions from the fraction collector can be picked out from the fractions which do not contain any substances. Several types of effluent monitors are available, based on the measurement of the ultraviolet absorption, conductivity, etc.

Table 5.6 - Separation of Antioxidants - Thin-layer Chromatographic Methods

Substances separated	Stationary phase	Mobile phase	Detection	Ref
Phenolic antioxidants	Silica Gel G	Methanol-cyclohexane (1:24)	30% Molybdoph-osphoric acid + amonia vapour	16
Organo-tin stabilizers	Not stated	Acetic acid-isopropyl ether (1.5:98.5)	20% Molybdoph-osphoric acid + amonia vapour	17
Antioxidants	Not stated	Light petroleum-ethyl acetate (9:1)	a) Ethanolic 2,6-dichloro-p-benzo quinone-4 chlor-imine + 2% aq. $Na_2B_40_7$ b) Diatzotized p-nitroaniline	18
Organic stabilizers	Kieselgel G	Ethanol-free chloroform		19
Phenolic antioxidants	Polyamide powder	Methanol-water (3:2) or methanol-carbon tetrachloride (1:9)	Diazotized sulph-anillic acid	20
Phenyl salicylate, Resorcinol benzoate	Kieselgel G	Dichloromethane or isoproyl ether-light petroleum (40-60%) (7:3)	Ultraviolet	21
BHA,2,6-di-t-butyl-p-cresol	Silica gel	Chloroform	20% Molybdophos -phoric acid + amonia vapour	22
Antioxidants	Polyamide powder	Methanol-acetone -water (6:1:3)	Diazotized sulpan-ilic acid or moly-bdophosphoric acid	23
Antioxidants	Kieselgel G		alpha, alpha'-Diphenyl-beta-picryl hydrazyl (free radical)	24
Antioxidants	Alumina + 5% plaster of paris on microscope slide	Petrol-dioxan (10:1)	5% Ethanol, phos-phomolybdic acid	25
Antioxidants	Silica gel	Acetone, chloroform, benzene, carbon tetra-chloride or binary mixtures		26
Antioxidants	a) 10% starch in polyamide powder b) 10% PVC in polyamide	Methanol-acetone-water (3:1:1) light petro-leum-benzene-acetic acid DMF (40:40:20:1)		27
Antioxidants	Silica Gel G	Benzene	0.5% $Fe_2(SO_4)_3$ in sulphuric acid + 0.2% $K_4Fe(CN)_6$ 1:1	28

These have the disadvantage of being too specific for dealing with mixtures of compounds of unknown type. For example, compounds which do not either absorb in the ultraviolet or ionize would be missed using these detectors. The most useful general purpose monitors are those based on the measurement of refractive index and on thermal effects. The latter operates on the principle that as each separated compound moves down the column it is accompanied by heat of absorption and desorption due to interaction between solute molecules and the stationary phases. These heat pockets (ie. separated compounds) are detected by a thermistor at the column outlet and recorded on a strip chart which can be operated in conjunction with a fraction collector. Thus, separated fractions can be readily located and bulked if necessary for further examination. The most recent development in liquid chromatography, namely high pressure liquid chromatography, combines the advantages of built-in detectors with improved resolution in the separation of mixtures due to improved column packings and operation at an elevated pressure. A number of high-efficiency liquid chromatographic supports are now available. These include Zipax (DuPont's CSP support)[46,48,49] Corasil I and II[47-49] and Durapak (Waters Associates).[49] With the exception of Durapak, these materials in the micron particle range, consist of particles with a solid core and thin porous coating. This unique combination gives very high coefficients of mass transfer. Durapaks consist of conventional liquid phases, such as ß,ß'oxy-dipropionitrile, chemically bonded to a rigid porous bead. Textured glass beads for liquid-liquid chromatography similar to those reported for gas chromatography[44] have been developed by Corning.[45]

Fig. 5.4 shows a chromatogram of a mixture of antioxidants and ultra-violet absorbers. A reversed-phase system is again preferred because it can be washed clean with methanol but this time an isocratic solvent system is used. With the variable wavelength detector, the optimum wavelength can be set for the type of compounds being examined.

Majors[51] has described a high performance liquid chromatographic system for the determination of antioxidants in polyethylene and plasticizers in PVC, (Method 99).[30-41,52]

5.2.4 Supercritical Flow Chromatography[78-79]

Raynor et al[71] separated additives in complex mixtures by capillery supercritical flow chromatography and deposited the separated products in potassium bromide discs prior to measurement of their infrared spectra by FTIR spectroscopy. Supercritical flow chromatography coupled with F.T. infrared spectroscopy (SFC-FTIR) has been used to provide quantitative information on and characterization of a range of polymeric surfactants.[80] Bartle et al[50] and Chester[70] also investigated this technique.

5.3 MASS SPECTROMETRY

Like other forms of molecular spectroscopy, mass spectrometry[84,85] may be used as a "fingerprint" technique to identify the components of additive systems extracted from polymers. The strengths of mass spectrometry are high sensitivity and the ability to distinguish between closely related compounds of differing relative molecular mass, eg. the various alkyl thiodipropionates used as synergistic stabilisers in polyolefins and the ultraviolet absorbing benzotriazole derivatives. Often it is not necessary to separate the components before examination as some separation may be achieved by careful variation of the sample probe temperature to produce in effect, a fractional distillation of the components. The presence, however, of large amounts of low relative molecular weight

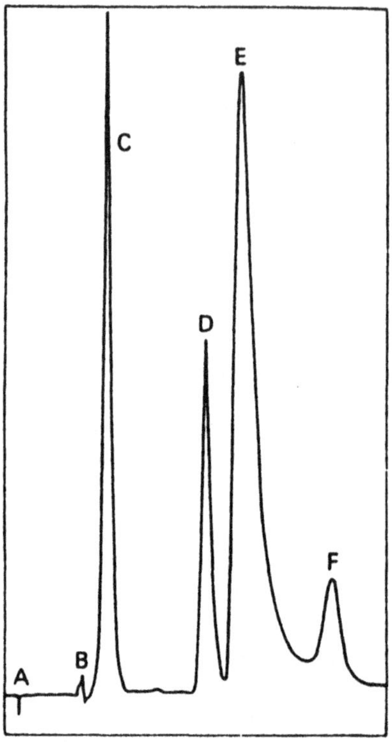

Figure 5.4. Determination of additives from polypropylene by HPLC. Chromatographic conditions: column Spherisorb ODS (5pm), 100 x 5 mm; solvent, 90% V/V methanol: flow-rate 40 cm^3h^{-1}; detection, absorbance at 280 mm and 0.2 a.u.f.s. A, injection (10pl); B, solvent front; C, Topanol CA, 10.12 mg per 50 cm^3; D, Tinuvin 326, 4.96 mg per 50 cm^3; and F, Irganox 1010, 10.10 mg per 50 cm^3.

polymers such as polyethylene and polypropylene can cause interference by producing a high hydrocarbon background extending to several hundred relative atomic mass units. In such instances thin-layer chromatographic separation can be used as a clean-up procedure.

Blestos[72] produced time of flight secondary ion mass spectra of thick films of poly(dimethylsiloxane) and fluorocarbon polymers containing additives in the range m/z up to 4500. The spectra obtained were characteristic of the polymer and its additives or any surface contamination.

5.3.1 Static Secondary Ion Mass Spectrometry (SSIMS)[86]

Rudewicz and Manson[53] have reported a method for the determination of additives in polypropylene without prior separation by solvent extraction or precipitation techniques, (Method 100)[54]. The additives are vaporized from polypropylene samples in a heatable glass probe under chemical ionization conditions using a 1.1% ammonia in methane reagent gas mixture. The dominant ion in this mixture, NH_4 + is a low energy reagent ion that reacts with the additives to give very simple spectra of $(M + H)^+$ or (M +

$NH_4)^+$ ions and little fragmentation. The detection of additives in a hydrocarbon matrix is very selective.

5.3.2 Liquid Chromatography - Mass Spectrometry

The information obtained by absorbance detection when coupled with liquid chromatography is usually not specific enough to allow the qualitative identification of compounds present in complex polymer samples with any degree of certainty. With the use of mass spectrometric detection Vargo and Olson[55] showed that characteristic mass spectral data can be obtained which greatly aid in elucidating the identity of unknown compounds. They used mass spectrometric detection in series with absorbance detection to identify or characterise antioxidants and ultra-violet light stabilizing additives in plastic materials, which were separated by liquid chromatography, (Method 101).

5.3.3 Gas Chromatography - Mass Spectrometry

Polymers contain trace amounts of residues of the organic catalyst used in their preparation and the identification of these is often necessary. The use of gas chromatography is conjunction with mass spectrometry is required in order to separate the complex mixture of components that are extracted. For example, tetra-methylsuccinodinitrile has been detected in extracts of polymers prepared using azobisiso-butyronitrile catalyst. Substantial losses of tetramethylsuccinodinitrile occur in the evaporation of methanolic solutions which explained earlier difficulties in detecting residues of this catalyst. Even without concentration of the polymer extract it was possible to achieve a lower limit of detection of 20 ppm in the polymer.

A further family of catalysts often used are peroxides (eg. benzoyl or lauroyl peroxide), these produce acids as residues which may be detected by mass spectrometry or by methylation of the evaporated extract prior to gas chromatography - mass spectrometry examination.

The analyst in the plastics industry may be required to trace the cause of odour and taint produced in foodstuffs packaged in plastic materials. This provides good examples of the use of high sensitivity gas chromatography - mass spectrometry in the identification of such compounds. Two methods have proved useful for the concentration of these components.

(A) Where the sample is a sealed plastic bag the headspace gas from the bag (or many bags) is withdrawn through an adsorption tube normally packed with Porapak Q. The trapped organic species are then thermally desorbed for gas chromatography - mass spectrometry examination. This has enabled residual printing and coating solvents in the bag headspace to be identified as causes of odour and taint.

(B) For containers and foods, the Likens and Nickerson combined solvent extraction - steam distillation procedure[56] has proved useful. The advantage of this form of extraction is that large amounts of sample may be extracted with a small volume of organic solvent prior to further concentration by evaporation. In this procedure the evaporation of the organic solvent can concentrate trace amounts of solvent impurities to a significant proportion of the final extract and redistilled solvent. Therefore, distilled rather than deionised water is used as deionized water can contain trace organics from the resin bed. A blank extraction is always carried out. Extracts can be examined by gas chromatography - mass spectrometry.

5.4 ADDITIVE DEGRADATION TECHNIQUES

5.4.1 Photolytic Degradation - Gas Chromatography

Juvet et al[57] showed that characteristic reaction products occur during photolytic degradation and polymer additives may on this basis be identified without separation from the polymer samples. The excellent reproducibility reported for photolytic degradation of liquid samples indicated the possibility of quantitative determination of polymer additives (Method 102). They developed equations to predict the shape of calibration curves expected for both trace level additives and those present at higher concentration. Photolytic degradation has been shown to yield simple and more reproducible decomposition patterns due to greater control of input energy and the more predictable manner in which compounds decompose photolytically.

Since photolytic analysis does not require separative steps and can be performed with samples weighing less than a milligram, it provides a rapid, practical approach for the determination of additives in polymeric materials.

5.4.2 Pyrolysis - Gas Chromatography - Mass Spectrometry

Pyrolysis - gas chromatography has been used for many years for the characterisation of plastics materials, particularly when they are of an intractable nature owing to cross-linking or are very heavily filled. The technique has now been extended to include mass spectrometry and is of particular value when minor components need to be identified. An example has been described by Sharp and Paterson[58] for the identification of small amounts (1- 10%) of copolymerised unsaturated acids in acrylic polymers. The method can be summarised as follows:

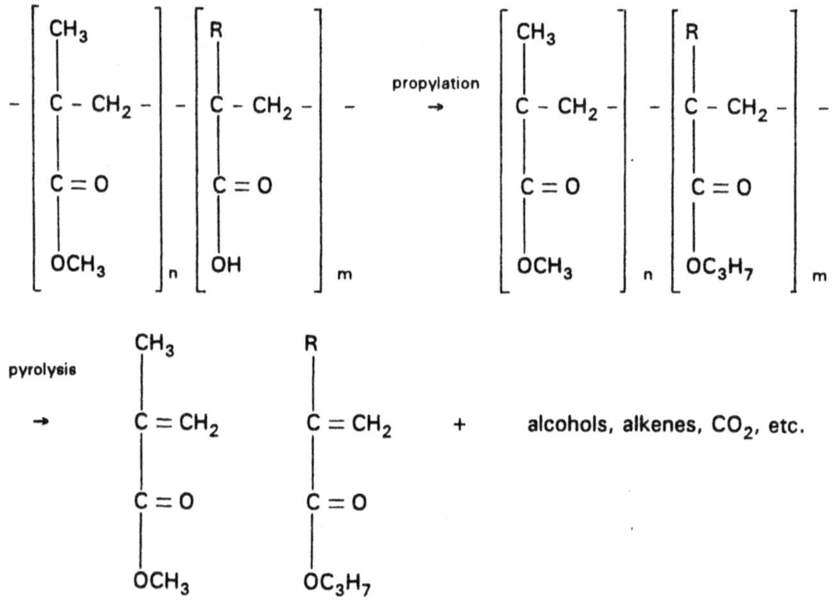

5.1

The copolymerised acid is propylated by treatment of the sample with Propyl 8 reagent, the resultant polymer is pyrolysed and the propyl ester of the acid, if present, is identified by gas chromatography - mass spectrometry. By this procedure copolymerised acrylic or methacrylic acid has been identified in terpolymers with (a) butyl acrylate and styrene, (b) methyl methacrylate and ethyl acrylate and (c) ethylene and propylene. A methyl methacrylate - alpha-methylstryene - maleic acid terpolymer, when examined by this proylation - pyrolysis procedure, yielded dipropyl fumarate and a smaller amount of dipropylmaleate.

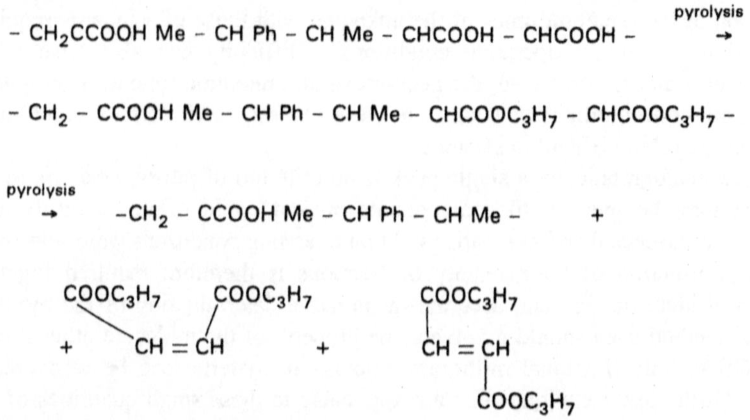

5.2

5.5 INFRARED SPECTROSCOPY

5.5.1 Infrared Laser Raman Spectroscopy[81-83]

Although both of these spectroscopic methods have a wide use in their own right, the example given below demonstrated well the complementary value of the two methods, taking advantage of the fact that elements of high atomic number, eg. antimony and bromine, have relatively more intense Raman spectra but the lighter elements show up clearly in the infrared spectra.

An example of the application of these techniques is the identification of additives in PVC. When a sample of PVC was examined by infrared spectroscopy the strongest bands (9.8 and 14.9 um) were due to a talc-type material and bands of medium intensity were assigned to polypropylene and possibly antimony trioxide (13.4 um). Additional weak bands in the 7.3 - 7.7 um region were possibly due to decabromodiphenyl ether. In the Raman spectrum, however, the strongest bands (250 and 185 cm^{-1} shift) confirmed the presence of antimony trioxide and some bands of medium intensity confirmed the presence of decabromodiphenyl ether (doublet at 140, triplet at 220 cm^{-1}-1 shift) and polypropylene (800, 835, 1150, 1325, 1450 and 2900 cm^{-1}-1 shift). The silicate bands that obscured the regions of the infrared spectrum were not observed in the Raman spectrum.

Examples of other spectroscopic techniques that have been applied to the identification of additives in polymers are given in Table 5.7.

5.5.2 Laser Desorption - Fourier Transform Techniques

These relatively new techniques viz laser desorption/ionization Fourier transform mass spectrometry and fast atom bombardment[87] and laser desorption Fourier transform ion cyclotron resonance mass spectrometry[88] have been applied to the determination of non volatile polymer additives (thioester, phosphite, phosphonate and hindered amine antioxidant types) and antioxidants, ultraviolet absorbers and amide waxes.

5.5.3 Gas Chromatography - Infrared Spectroscopy

The direct application of gas chromatography to the identification and determination of unknown additives in polymers has its limitations. Amongst these are the fact that retention times are purely relative and not always exactly reproducible; even when the chemical class of the sample constituents is known, identification by comparison of the retention times of the unknown with those of standards requires exact reproduction of column operating conditions. Difficulty can also arise when non-symmetrical peaks are produced; the peak-maximum retention time of a component then depends on concentrations. Peaks can also overlap and certain combinations of substances are often difficult to separate.

Chromatographically a single peak is no criterion of purity, since more than one substance may be present, the components present could often be resolved if their presence was suspected and alternative column operating conditions were selected.

Confirmation of homogeneity of fractions is therefore required together with unequivocal identification and accurate quantitative determination of the product. The analytical method used should involve some property of the molecule other than boiling point. Often, only fractional milligram amounts of material can be recovered from a column. Of the few techniques that are applicable to these small quantities of material, mass and infrared spectroscopy are particularly suited.

The use of fraction collecting techniques in conjunction with gas chromatography is now well established[59-65] and is an attractive proposition in additive identification problems, despite the fact that little published work has yet appeared on the application of this technique to polymer additives. In this technique the separated compound as it emerges from the gas chromatographic column is swept by the carrier gas through a cold trap where it condenses. The material in the trap is then either transferred to an infrared gas cell for examination in the vapour phase, or is transferred as a liquid to a suitable micro cell or may be condensed on a cold surface as a solid for examination by conventional spectroscopic techniques.

Anderson[66] has described in detail a technique for collecting individual products separated from mixtures by gas chromatography followed by their identification by vapour phase infrared spectroscopy using a double beam spectrometer. He claimed that the technique would identify and determine to within $\pm$ 2% amounts as low a 5 umole of all substances having a boiling point up to about 175°C. Heated gas cells enable liquids of higher boiling point to be examined. Naturally, as the infrared spectrum of a compound in the liquid and vapour form are different, it is necessary to compare spectra obtained by the above technique with spectra of standard vapours. Generally speaking, the additive constituents of polymers have boiling points which are appreciably higher than 200°C and hence cannot be handled in infrared gas cells. For such substances, infrared examination as a liquid or a solid, as is discussed in Method 103 is more relevant.

Volatile constituents are, however, sometimes encountered, viz expanding agents, plasticizers, lubricants, adhesives, solvents, monomers and degradation products of additives or of the polymer itself and infrared gas cell techniques can be of value in the examination of gas chromatographic fractions containing these types of substances.

Haslam et al[67] have studied in detail the collection of volatile and liquid fractions emerging from a gas chromatographic column and their subsequent identification by infrared spectroscopy. They applied these techniques to various polymeric materials encountered in the plastics industry (Method 103).

Table 5.7 - Other Techniques Applied in Plastics Analysis

Technique	Principal use
X-ray diffraction	Identification of crystalline additives in situ (eg hydrates), Identification of polymer phases
NMR	Identification of additives after extraction. Calibration of infrared methods (structure). Identification of "bound" additives Wide-line NMR for liquid - solid phase blends
ESCA	Surface composition studies, eg presence of slip additive and antistatic agents at polymer surfaces
MOLE	Particulate contamination at or near film surfaces
SIMS	Ultra-surface analysis: catalyst studies
Spectrofluorescence	Degradation studies: surface oxidation; study of scintillation additives

REFERENCES - CHAPTER 5

1. Cook, W.S., Jones, D.O. and Altenau, M.G., Canadian Spectroscopy, **13** 64 71 (1968)
2. Houk, W.W. and Silverman, L., Anal. Chem., **31** 1069 (1959)
3. Mitchell, B.J. and O'Hear, H.G., Anal. Chem., **34** 1621 (1962)
4. Bergmann, J.S., Ehrhart, C.H., Granatelli, L. and Janck, J.L., 153rd National American Chemical Society Mtg., Miami Beach, Florida, April (1967)
5. Squirrel, D.C.M., Analyst (London), **106** 1042 (1981)
6. Roberts, C.B. and Swank, J.D., Anal. Chem., **36** 271 (1964)
7. Harland, W.G., Khadr, M.M. and Peters, R.H., Polymer, **15** 81 (1974)
8. Mehta, A. and Wunderlich, B. Makromol. Chem., **175** 977 (1974)
9. Pocoro, P.J., Anal. Chem., **36** 1664 (1964)
10. Freedman, R.W., and Charlier, G.O., Anal. Chem., **36** 1880 (1964)
11. Duvall, A.T. and Tully, W.F., J. Chromatogr., **11** 38 (1963)
12. Brooks, V.T., Chemy. Ind. 1317 (1959)
13. Knight, H.S. and Siegel, H., Anal. Chem., **38** 1221 (1966)
14. Grant, D.W. and Vaughan, G.H., Gas Chromatography p. 305, Butterworths, London (1962)
15. Bosch, K., Beitr. Gerichtl. Med., **33** 280 (1975)
16. Slonaker, D.F. and Sievers, D.C., Anal. Chem., **36** 1130 (1964)
17. Neubert, G., Z. Anal. Chem., **203** 265 (1964)
18. Van der Heide, R.F. and Wouters, O., Lebensun Untersuch. Forsch., **117** 129 (1962)
19. Van der Heide, R.F., Z. Lebensum. Untersuch. Forsch., **124** 198 (1964)
20. Davidek, J. and Pokorney, J., Rev. Univ. Ind. Santander, **4** 11 (1962)
21. Wandel, M. and Tengler, H., Flete. Seifen. Anstrichmittel, **66** 815 (1964)
22. Ishikaway, S. and Katsun, G., Bitamin, **30** 203 (1964)
23. Davidek, J. and Pakorny, J., Z. Lebensum Untersuch. Forsch, **115** 113 (1961)
24. Waggon, H., Korn, O. and Jehle, D., Nahrung, **9** 495 (1965)
25. Cook, D.C., Nash, N.G. and Flanagan, H.R., J. Amer. Chem. Soci., **75** 6242 (1953)
26. Yuasa, T. and Kamiya, K., Japan Analyst, **13** 966 (1964)
27. Van der Heide, R.F., Maagdenberg, A.C. and Van der Neut, J.H., Chem. Weekblad, **61** 440 (1965)
28. Copius-Peereboom, J.W., Nature, (London), **204** 748 (1964)
29. Raynor, M.W., Davies, I.L., Bartle, K.D., Williams, A., Chalmers, J.D., Cook, B.W., European Chromatography News, **1** 18 (1987)
30. Van der Neut, J.H. and Maagdenberg, A.C., Plastics, **31** 66 Jan (1966)
31. Frew, N.M., Johnson, C.G., Bromund, R.M. (Chappenter B.A. and Sevenants, M.R. Editors). In "Supercritical Fluid Extraction and Chromatography" ACS Symposium Series No. 366 American Chemical Society pp 208 - 228 (1988)
32. Waggon, H. and Jehle, D., Die Nahrung, **4** 495 (1965)
33. Braun, D., Chimica Ind., **19** 77 (1965)
34. Wandel, M. and Tengler, H., Kunststoffe, **55** 11 (1965)
35. Haalpaap, H., Chem. Org. Tech., **35** 488 (1963)
36. Braun, D. and Geenen, I.T., J. Chromatography, **7** 56 (1962)
37. Knappe, E. and Peteri, D., Z. Analyst Chem., **188** 184 (1961)

38. Lynes, A.J., Chromatography, **15** 108 (1964)
39. Lane, E.S., J. Chromatography, **18** 426 (1965)
40. Crump, G.B., Anal. Chem., **362** 2447 (1964)
41. Dallas, H.S.J., J. Chromatography, **17** 267 (1965)
42. Ritter, F.J. and Meyer, G.M. Nature (London), **193** 941 (1962)
43. Shahl, E., Laboratory Practice, June (1964)
44. McDonnell, H.L., Anal. Chem., **40** 221 (1968)
45. Haulein, A., Corning Glass Works, Personal Communication (1970)
46. Whipple, E.B. and Green, P.J., Macromoleccules, **6** 38 (1973)
47. Waters Associates "New Developments in Chromatography" No. 1 (1970)
48. Kirkland, J.J., Anal. Chem., **41** 2118 (1969)
49. Kirkland, J.J., J. Chromatogr. Sci., **7** 7 (1969)
50. Bartle, K.D., Davies, I.L., Raynor, M.W. (Hunt, B., Holden's Editors) In "Size Exclusion Chromatography", Blackie, Glasgow (1988)
51. Majors, R.E., J. Chromat. Sci., **8** 338 (1970)
52. Scott, C.D., Johnson, W.T. and Walker, V.E., Anal. Biochem., **32** 182 (1969)
53. Rudewicz, P. and Munson, B., Anal. Chem., **58** 358 (1986)
54. Spreen, R.C., University of Delaware, Newark, D.E. (1983)
55. Vargo, J.D. and Olsen, K.L., Anal. Chem., **57** 672 (1985)
56. Likens, S.T. and Nickerson, G.B., Proc. Amer. Soc. Brow-Chem., 5 (1964)
57. Juvet, R.S., Smith, L.S. and Li, K.P., Anal. Chem., **44** 49 (1972)
58. Sharp, J.L. and Paterson, G., Analyst (London), **105** 517 (1980)
59. Haslam, J., Soppet, W. and Willis, H.A., Appl. Chem., **1** 112 (1951)
60. Liberti, A., Costa, G. and Pauluzzi, E., Chim. and Ind., **38** 674 (1956)
61. Kendall, D.N., Appl. Spectroscopy, **7** 179 (1953)
62. Cachia, M., Southwart, D.W. and Davison, W.H.T., J. Appl. Chem., **8** 291 (1958)
63. Bellamy, L.J. and Williams, R.L., Paper presented at International Symposium on Microchemistry, Birmingham, August (1958)
64. Martin, A.J.P., 1st International Gas Chromatographic Symposium, Instrument Society of American, Michigan State University (1957)
65. Kuelemans, A.I.M. and Kwantes, A., 4th Worl Petroleum Congress, Rome (1955)
66. Anderson, D.M.W., Analyst, (London), **84** 50 (1959)
67. Haslam, J., Jeffs, A.R. and Willis, H.A., Analyst (London), **86** 44 (1961)
68. Dohmann, K., Laboratory Practice July (1965)
69. McCoy, R.N., Fiebig, E.C. Anal. Chem., **37** 593 (1965)
70. Charter, T.L., Burkes, L.J., Delaney, T.E., Innis, D.P., Owens, G.D., Dinkstoa, J.D., (Charpentier B.A., Sevenants, M.R. Editors) In Supercritical Extraction and Chromatography ACS Symposium Series No. 366, American Chemical Society, pp 144 - 160 (1988)
71. Raynor, M.W., Bartle, K.D., Davies, I.L., Williams, A., Clifford, A.A., Analytical Chemistry, **60** 427 (1988)
72. Bletsos, C.V., Hercules, D.M., Magill, J.H., Analytical Chemistry, **60** 824 (1988)
73. Kirkland, J.J., Analytical Chemistry, **64** 1239 (1992)
74. Kirkland, J.J., Dilks, C.H., Rementer, S.W., Analytical Chemistry, **64** 1295 (1992)
75. Kirkland, J.J., Rementer, S.W., Analytical Chemistry, **64** 904 (1992)
76. Barth, H.G., Boyes, B.E., Analytical Chemistry, **62** 381R (1990)
77. Boehm, R.E., Martire, D.E., Analytical Chemistry, **61** 471 (1989)
78. Wallace, J.C., Krieger, M.S., Hites, R.A., Analytical Chemistry, **64** 2655 (1992)
79. Moulder, R., Bartle, K.D., Clifford, A.A., Analyst (London), **116** 1293 (1991)

80. Jenkins, J.L., Kaplan, M., Simmonds, M.R., Davidson, G., Healy, M.A., Poliakoff, M., Analyst (London) **116** 1305 (1991)
81. Bowden, M., Donaldson, P., Gardiner, D.J. Birnie, J., Gerrard, D.L., Analytical Chemistry, **63** 2915 (1991)
82. Rapp, T.L., Kavalchyk, W.K., Davis, K.L., Todd, E.A., Liu, K.L., Morris, M.D., Analytical Chemistry, **64** 2434 (1992)
83. Windig, W. Guilment, J. Analytical Chemistry, **63** 1425 (1991)
84. Harrington, P de B, Street, T.E., Vourkees, K.J., Radicatidi Brozolp F., Odom, R.W., Analytical Chemistry, **61** 715 (1989)
85. Fowler, D.E., Johnson, R.D., Van Leyen, D., Benninghoven, A., Analytical Chemistry, **62** 2088 (1990)
86. Feld, H., Leute, A., Zurmuehlen, R., Benninghoven, A., Analytical Chemistry, **63** 903 (1991)
87. Johlman, C.L., Wilkins, C.L., Hogan, J.D., Donovan, T.L., Lande, D.A., Youseffi, M.J., Analytical Chemistry, **62** 1167 (1990)
88. Asamoto, B., Young, J.R., Citerin, R.J., Analytical Chemistry, **62** 61 (1990)

CHAPTER 6

PRELIMINARY QUALITATIVE IDENTIFICATION OF POLYMERS

The polymer analyst working in industry is frequently asked to identify a polymer or at least to give an opinion as to the type of a polymer. This might arise as a result of an interest in the materials being used by competitors or manufacturers of components or packaging materials. Obviously, if a detailed examination of a polymer or copolymer and its additive system is required then other chapters of this book should be referred to. However, if a quick opinion is all that is required then some simple physical and chemical tests and the preliminary qualitative identification or fingerprinting approach discussed in this chapter might suffice.

Simple Physical and Chemical Tests

Physical and chemical tests considered here are elemental analysis, saponification number, solvent solubility, ignition tests, colour reactions and burning/heating tests.

6.1 PRELIMINARY REMOVAL OF ADDITIVES FROM POLYMERS

Chemically pure high polymers are rarely used in practive. They are usually mixed with additives. As these additives influence the reactions of the plastic it may be necessary to remove them by one of the methods listed below before attempting to identify the polymer by the methods listed above under Physical and Chemical Tests.

A. Fractional precipitation (Table 6.1.1)

(a) The polymer is first completely diluted in a good solvent. The resulting plastic solution is stirred into at least 10 times its amount of a solvent which acts as a precipitant for the dissolved plastic but which combines completely with the first solvent. The precipitated plastic is separated from the solvent mixture by centrifuging or filtration and is pulverzed and repeatedly extracted with the precipitant/solvent or redissolved in a good solvent and precipitated. This treatment provides additive-free polymer. (See Table 6.1.1a)

(b) The plastic is dissolved in a solvent which dissolves the plastic at higher temperatures; the plastic separates out on cooling while the additives remain in solution. (See Table 6.1.1b)

(c) Plastic dispersions can often be separated by nullifying the action of the emulsifier and/or dispersing agent. The following methods can be used according to the type of emulsifier. (See Table 6.1.2)

1. Alteration of the solvent phase, ie. addition of a further solvent to coagulate the polymer. (Table 6.1.2a)

2. Precipitation of the emulsifier by the addition of acids or a counter interfacially active ion. (Table 6.1.2b)
3. Freezing the dispersion out with dry ice. (Table 6.1.2c)
4. Subjection of the dispersion to dialysis, whereby the water-soluble inorganic salts and relatively low-molecular emulsifiers are diffused through cellophane membranes so that the protective colloid and the polymerizate can now be easily separated quantitatively by centrifuging. (Table 6.1.2d)

B. Separation by Fractional Extraction. (Table 6.1.3)

Fractional extraction can be used in three ways, depending on the substance under examination:

(a) Finely shredded solid polymers are extracted in a Soxhlet apparatus with a set of solvents of increasing solvent power. Some of the extracts will contain polymer and some will contain the additives. (See Table 6.1.3a)

(b) To ensure that the polymer particles are fine enough to allow the components to be extracted to diffuse sufficiently quickly, a solution or dispersion of the polymer is poured over an inert carrier, such as silica gel or sterchamol, the solvent is allowed to evaporate from the solid phase and the carrier is then extracted in a heatable column with solvents of increasing solvent power to isolate a solution of the polymer. (Table 6.1.3b)

(c) The components of a polymer are distributed between two non-miscible liquids in separating funnels or distribution apparatus with the correct selection of solvents, one phase will contain the additives and the other will contain the polymer. (Table 6.1.3c)

C. Separation by Diffusion. (Table 6.1.4)

The polymer is precipitated from a solution or dispersion in a glass vessel as a thin film on the inner wall. Then a solvent which is a non-solvent towards the polymer is added in several stages. This solvent will dissolve certain additives, eg. plasticizers leaving an additive free film of polymer on the inner wall of the vessel. (Table 6.1.4)

D. Separation by Dialysis or Electro-Dialysis. (Table 6.1.5)

This method has only been used for separating low-molecular salts and interfacialy active substances from protective colloids and high polymer polymers and for fractionating pure polymers according to their molecular weight.

Some examples of these procedures are given in Table 6.1.

6.2 CLASSIFICATION OF POLYMERS

6.2.1 Classification By Elements

An undoubtedly incomplete classification of polymers based on elemental classification and saponification number is provided in Table 6.2. Note that copolymers are not included in this list.

A cautionary note is that, in addition to the polymer itself, the polymer additive system may contain elements other than carbon, hydrogen and oxygen. The detection of an element such as nitrogen, sulphur, halogens, phosphorus, silicon or boron in a polymer

Table 6.1 - Examples of Separating Procedures

Plastic	Operation	Aim or separating procedure
6.1.1 Fractional precipitation procedures (precipitation by addition of non-solvent for polymer to solvent solution of polymer)		
Polyvinyl chloride	Stirring of a concentrated solution of tetrahydrofuran into methanol	Plasticizers, emulsifiers remain in solution, PVC and its co-polymers are precipitated
Polyacrylate	Stirring the solution in acetone into 20 times the amount of water	Emulsifiers, watersoluble resins and salts remain in solution, polymerizate is precipitated
Polyacrylate	Water is added to the solution in acetone until faint turbidity occurs, acetone then distilled off under vacuum	Emulsifiers, watersoluble resins and salts remain in solution, polymerizate is precipitated
6.1.1a Fractional precipitation (thermal precipitation of polymer from solution at low temperature)		
Polyethylene	Dissolve in enough benzene to make a clear solution appear when warmed then cool	Polymerizate is precipitated additives soluble in cold benzene, paraffin, waxes, resins remain in solution
6.1.2 Emulsion breaking methods		
6.1.2a Addition of solvent		
Polymethacrylate-dispersions	Dispersion is stirred into 20 times its amount of methanol or isopropanol	Polymerizate is precipitated emulsifier and any protective colloid remain in solution
6.1.2b Emulsifier precipitation by acid addition		
Rubber or butadiene	Dispersion is broken by acidifying as long as soap is used as the emulsifier	After acidifying, fatty or resin acid is extractable with ether; latex is precipitated
6.1.2c Freezing out dispersion by cardice addition		
Polyvinyl acetate or silicone oil dispersions	Action of emulsifier destroyed by cold	Polymerizate is precipitated, can now be filtered off or centrifuged off
6.1.2d Application of dialysis		
Polyvinyl acetate ester or polyacrylate dispersions	Salts and low molecular emulsifiers are dialyzed	In the dialyzate, salts and emulsifiers, in the dialyzate residue, protective colloid polymerizate
6.1.3 Fractional extraction procedures		

Table 6.1 continued

6.1.3a Solvent extraction of shredded plastic		
Polyvinyl chloride	Consecutive extractions with CCl_4, ether, benzene, CH_2 CCl_2, tetrahydrofuran	Isolation of stabilizers and plasticizers, recognition of copolymers or poly blends
6.1.3b Solution or dispersion of plastic passed over inert carrier on which plastic absorbs, solvent evaporation from carrier, then desorbtion from carrier with strong solvents		
Polyethylene	Consecutive extraction of silica with gasoline, methanol, $CHCl_3$, water, methanol or acetone, then benzene or toluene (perhaps warm)	In the gasoline: waxes; in CH_3OH: emulsifiers; in $CHCl_3$; montan waxes in water: cellulose derivatives and polyvinyl alcohol; in benzene/toluene: polyethylene
6.1.3c Plastics components distributed between two non-miscible liquids		
Polyvinyl acetate	Extracted from aqueous methanol solutions or dispersions with pentane/ ether mixtures	Isolation of plasticizers
6.1.4 Separation by diffusion		
Polyvinyl acetate	Polymer precipitated with ligroin or ligroin/ether mixtures	Isolation of the plasticizers, can also be used for traces
6.1.5 Separation by dialysis or electrodialysis		
Polyacrylate	see 6.1.2d	

is indicative that the elements originate in the polymer and not the additive system if the element is present at relatively high concentrations such as several per cent. This is highlighted by the example of a high density polyethylene which might contain 0.2 - 1% chlorine originating from polymerization residues and PVC homopolymer which contains more than 50% chlorine.

Useful rapid tests for elements which can be used qualitatively or quantitatively are tabulated in Table 6.3.

Haslam et al[7] have adopted the oxygen flask combustion technique to the qualitative detection in polymers of nitrogen, chlorine, bromine, iodine, fluorine, phosphorus and sulphur in amounts down to 0.25%, (see Method 104).

6.2.2 Classification By Solubility

This technique is of limited value, especially when applied to copolymers. Solubility does not only depend on the type of the monomer base products but also to a large extent on the degree of polymerization (ie. on the molecular weight) on the branching and cross-linking of the molecules and on the steric configuration of the polymerizate. Very efficient solvents can however be distinguished from the less efficient ones for different plastics.

Table 6.2 - Elemental Analysis and Saponification Number

I Principal elements C,H

Aliphatic	Aromatic
Polyethylene	Polystyrene
Polypropylene	Polyindene
Polyisobutylene	Polyxyleneyls
Polybutadiene	Polymer petroleum fractions
Polyisoprene	
Natural rubber	
Butyl rubber	

II Principal elements C,H,O

Saponification number = O	Saponification number = < 200	Saponification number = > 200
Regenerated cellulose	Natural resin	Cellulose acetate
Polyvinyl alcohol	Modified phenoplasts	Cellulose butyrate
Phenoplasts		Cellulose acetate butyrate
Phenol-furfural resins		Polyvinyl acetate and its co-polymers
Xylene-formaldehyde resins		Polyvinyl propionate
Cellulose ether		Polyacrylate
		Polymethacrylate
Polyvinyl ether		Alkyd resins
Coumarone resins		Polyester
Polyglycols		Polycarboxylic acid anhydride
Polyvinyl acetals		Polycarbonic ester
Polymer aldehydes		
Polyketones		
Epoxy resins		

III Principal element, halogen

Polymer halogen olefins	Rubber derivatives	Others
Polyvinyl chloride (PVC)	Chlorinated rubber	Colophen resins
PVC co-polymers	Rubber hydrochloride	Chloronaphthalenes
Polyvinylidene chloride	Chlorinated synthetic rubber	Chlorinated paraffins
Poly-2-chlorobutadiene		
Polychlorostyrene		
Polytetrafluoroethylene		
Polytrifluorochloroethylene		
Polyvinyl fluoride		

IV Principal elements N or N and O

Polyacrylic and polyvinyl compounds	Basic components of formaldehyde aminoplasts	Others
Polyacrylonitrile	Urea	Nitrocellulose
Polyacrylamide	Ethylene urea	Polyamides
Polymethacrylamide	Propylene urea	Polyurethanes

Table 6.2 continued

Polyvinylidene cyanide	Dicyandiamide	Polyureas
Polyvinyl pyridine	Melamine	Phenoplasts and epoxy resins cured with amines
Polyvinyl pyrrolidone	Acetylene diurea	
	Glyoxylureides	
	Anilines	

V Principal elements, S in addition to O

Products of synthesis	Modified natural products
Polyethylene polysulphide	Vulcanized rubber
Polydiethylether-polysulphides	Sulphurized stand oils
Polythioether	

VI Principal element, silicon

Silicon oils and rubbers
Silica acid esters

VII Principal elements, nitrogen and sulphur

Thiourea formaldehyde resins
Sulphonamide resins
Polyacrylamides, polyacrylimides

VIII Principal elements, halogen and sulphur

Sulphochlorinated polyethylene and its vulcanizates
Polychlorobutadiene, vulcanized with compounds containing sulphur

IX Principal elements, nitrogen, sulphur and phosphorus

Casein condensates

X Principal elements, phosphorus, nitrogen and halogen

Polyphosphoric nitrile chloride

Co-polymers usually dissolve in a greater number of solvents than homopolymers, eg. poly(vinyl chloride) is only very slightly soluble in acetone or methylene chloride, in which co-polymers with vinyl acetate or acrylates easily dissolve. Attempts have failed to separate plastics according to their solubility in the same way in which cations are separated in inorganic qualitative analysis. Table 6.4 summarizes the solubility of plastics.

6.2.3 Classification By Colour Reactions

Colour reactions for particular homopolymers are listed in Table 6.5. These tests are rather laborious and are perhaps best regarded as confirmatory tests for polymer identifications achieved by other means.

Table 6.3 - Methods of Determination of Traces of Various Elements in Polymers

Element	Procedure	Ref	Interferences
Sulphur up to 30 mg	Combusted in oxygen-filled flask over dilute hydrogen peroxide solution Potentiometric titration of sulphuric acid with N/100 sodium hydroxide or photometric titration of sulphate with N/100 barium perchlorate	1,2	Chloride, fluoride, phosphate, nitrogen, boron and metals all interfere Up to 2 mg chlorine, fluorine, nitrogen, boron and metals do not interfere in the determination of 1 mg sulphur. Phosphorus (up to 2 mg) interferes in the determination of sulphur (1 mg) but this interference can be overcome using the procedure of Colson (1,2)
Chlorine or bromine	Combustion as above Chloride titrated potentiometrically with N/100 silver nitrate in presence of nitric acid in acetone		
Chlorine or bromine or iodine	Combustion as above Halide titrated with mercuric nitrate	3	Up to 8 mg phosphorus, fluorine, sulphur, do not interfere in determination of 2 mg chlorine
Phosphorus, 0.1 g polymer	Digested with sulphuric acid/perchloric acid Digest diluted with ammonium vanadate/molybdate added. Yellow phosphovandmolybdate complex evaluated at 430 nm	4	No interference by sulphur, chlorine, fluorine, nitrogen
Nitrogen	Digestion carried out on 0.5-1.0g polymer a) Kjeldahl digest made alkaline and distilled into 4% boric acid. Ammonia estimated by acid titration b) Spectrophotometric estimation at 630 nm of phenol indophenol derivative	6	No interference by chlorine and nitrogen
Fluorine	30 mg polymer combusted in an oxygen filled flask over distilled water Reacted with buffered alizarin complex/cerous nitrate, blue colour produced evaluated at 610 nm	5	No interference by large excesses of sulphur, chlorine, phosphorus and nitrogen
Silicon	30 mg polymer in gelatin capsule and combusted with sodium peroxide, sucrose and benzoic acid in 22 ml capacity Parr Bomb		No interference by sulphur, halogens, phosphorus, nitrogen and boron
Boron	0.1g polymer digested with concentrated nitric acid in a sealed ampoule to convert organoboron compounds to boric acid Digest dissolved in methyl alcohol and boron estimated flame-photometrically at 5.9.5nm	6	No interference by chlorine and nitrogen

Table 6.4 - Summary of the Solubility of Plastics

Plastics	Typical solvents	Typical non-solvents
Polyethylene	Benzene, xylene tetrahydronaphthalene	Gasoline, alcohols, esters hydrocarbon halides
Polypropylene	Xylene, trichloroethylene chloroform	Alcohols, gasoline, esters
Polyisobutylene	Gasoline, ether	Alcohols, esters
Polybutadiene	Benzene	Gasoline, alcohols, esters ketones
Polyisoprene	Benzene	Gasoline, alcohols, esters ketones
Natural rubber	Hydrocarbons and hydrocarbon halides	Gasoline, alcohols, esters ketones
Polystyrene	Dioxane, benzene, chlorohydrocarbons	Gasoline, alcohols
Regenerated cellulose	No solvent	
Polyvinyl alcohol	Water, formamide alcohols	Gasoline, aromatic hydrocarbons
Phenolic plastics uncured	Alcohol, ketones halides	Gasoline, hydrocarbon
Phenolic plastics cured	Insoluble and often infusible	
Cellulose methyl ether	Ethylene chlorohydrin, water dilute solution of caustic soda	Aliphatic and aromatic hydrocarbons
Cellulose ethyl ether	Methylene chloride/methanol	Water, aliphatic and aromatic hydrocarbons
Cellulose benzyl ether	Acetone, butanol	Water, aliphatic and aromatic hydrocarbons
Polyvinyl methyl ether	Water, methanol	Gasoline
Polyvinyl ethyl ether	Gasoline, aromatic hydrocarbons, hydrocarbon halides alcohols, esters, ketones	Water
Polyvinyl butyl ether	Aromatic and hydrocarbon halides, ketones	Methanol, ethanol
Polyglycols	Water, hydrocarbon halides, alcohols	Gasoline
Polyvinyl formal	Tetrahydrofuran, ketones esters	Methanol, aliphatic hydrocarbons
Polyvinyl acetal	Tetrahydrofuran, ketones esters	Methanol, aliphatic hydrocarbons
Polyvinyl butyral	Tetrahydrofuran, ketones, esters, chloroform-methanol (9+1)	Methanol, aliphatic
Polymer aldehydes	Ketones, esters	Gasoline, alcohols
Polymer ketones	Ketones, esters	Gasoline, alcohols
Epoxy resins, uncured	Alcohol, esters, ketones, dioxane	Water, hydrocarbons
Natural resins	Alcohol, benzene, hydrocarbon halides, esters, ether	Gasoline
Modified phenoplasts, accord. to cure	Ethanol	Aliphatic and hydrocarbon halides
Polymethacrylates	Esters, aromatic and chlorinated hydrocarbons, ketones, dioxane	Aliphatic hydrocrbons, alcohols, ether
Polyacrylates	Aromatic hydrocarbons, hydrocarbon halides, esters ketones, tetrahydrofuran	Gasoline

Table 6.4 continued

Polyvinyl acetate	Aromatic and chlorinated hydrocarbons, methanol, ketones	Gasoline
Polyesters	Phenols, halogen phenols, nitrohydrocarbons, benzyl alcohol	Hydrocarbons, alcohols
Alkyd resins	Aromatic hydrocarbons, hydrocarbon halides, hydrocarbons, lower alcohols, ketones, esters	Hydrocarbons
Cellulose esters	Ketones, esters	Aliphatic hydrocarbons
Polyvinyl chloride	Tetrahydrofuran	Hydrocarbons, butyl acetate alcohols
Polyvinylidene chloride	Tetrahydrofuran, dioxane ketones, butyl acetate	Alcohols, hydrocarbons
Polyvinyl chloride copolymer with vinyl acetate	Tetrahydrofuran, methylene chloride	Hydrocarbons, alcohols
Acrylic esters	Tetrahydrofuran, methylene chloride	Hydrocarbons, alcohols
Maleic esters	Tetrahydrofuran, methylene chloride	Hydrocarbons, alcohols
Chlorinated rubber	Carbon tetrachloride, ketones esters, tetrahydrofuran	Aliphatic hydrocarbons
Rubber hydrochloride	Ketones	Aliphatic hydrocarbons, carbon tetrachloride
Neoprene	Toluene, hydrocarbon halides	Alcohols, esters
Trifluorochloroethylene	none	
Tetrafluoroethylene	none	
Polyamides	Formic acid, phenols trifluoroethanol	Alcohols, esters, aliphatic and aromatic hydrocarbons
Cellulose nitrate	Lower alcohols, acetic esters, ketones, ether/ alcohol (3:1)	Ether, benzene, hydrocarbon halides
Polyacrylonitrile	Dimethyl formamide, nitrophenol, ethylene carbonate	Alcohols, esters, ketones hydrocarbons
Polyacrylamide	Water	Alcohols, esters, hydrocarbons

6.2.4 Classification By Burning/Heating Tests

These tests are of limited value although Bird[33] has described a simple thermal test for the identification of plastic films, (Method 105). The above tests are of limited value nowadays becaused of the wide range of polymers now being manufactured and certainly are probably of no value in the case of copolymers. Conventional low impact polystyrene is soluble in hot toluene, whereas high density polyethylene or propylene have little or no solubility in this solvent. However, if the polystyrene contains some copolymerized butadiene, as occurs in the case of high impact polystyrenes, then due to the presence of crosslinked gel, the polymer would not completely dissolve in hot toluene. So even in the case of simple polymers solubility tests are of limited value and for them to provide any useful information required detailed knowledge. Polystyrene on the other hand, unlike the polyolefins when it is held in a flame, due to its aromatic nature

Table 6.5 - Colour Reactions to Show Structures and Functional Groups

Plastics	Reagent	Reaction	Functional group Responsible for Colour Reaction	Ref
Polyvinyl alcohol	Boric acid/iodine	Greenish blue	Hydroxy	8,9
Cellulose derivatives	Alpha-naphthol/H_2SO_4	Reddish violet	Hydroxy	10
	Aniline acetate	Turns filter paper red	Acetyl	11
	Anthrone	Violet	Acetyl	12
Polyvinyl ether	Solution in acetic anhydride + conc. H_2SO_4	Blue-violet	Ether	13-15
	Dichloroacetic acid	Blue	Ether	9
Coumarone resins	Bromo-glacial acetic acid with solution of $CHCL_3$	Turns perm-anently red	Phenolic	16-17
Phenolic resins	Diazotized 2-nitro-4-chloroaniline	Red	Phenolic	18
	2,6 dibromoquinone	Blue	Phenolic	19
	$FeCI_3$	Green	Phenolic	20
	Millon's reagent	Violet	Aldehyde	11
Furfural resins	Aniline acetate	Turns filter paper red		21,22
Epoxy resins	Nitration reactions		Epoxy	21,22
	Mercuric sulphate	Precipitate	Epoxy	23
	Paraformaldehyde/ conc. H_2SO_4	Bright blue	Epoxy	24,25
	Reaction with pyridine or quinoline bases	Blue		13,14
Natural resins	Acetic anhydride/H_2SO_4	Reddish violet		5
Polymethacry-lates	Reaction of depolymeriz-ates with conc. HNO_3 Zn power	Blue colouring which can be removed with $CHCI_3$		29
HCHO amino plastics	Carbazole/H_2SO_4	Dark blue		
Hydrocarbon halides	Solution in pyridine + methanolic KOH	Red-brown	Halogen	11,30,32

will burn with a smoky flame. However, if it is a non-flame grade it will not burn. PVC will, when burnt, produce an acrid odour of hydrogen chloride, so will poly(vinylidene chloride) and many copolymers containing vinyl chloride or vinylidene chloride. If the polymer contained fluorine, smelling the odours produced on combustion might be a hazardous operation as it would be in the case of acrylonitrile and polyurethane polymers which, in these circumstances, are likely to produce hydrogen cyanide. Density measurement, ie. whether the polymer sinks or floats in water is another simple parameter that can be observed. It will distinguish polyethylene with a density of less than one from a highly chlorinated polymer with a density of greater than one. However, if the polyethylene contains lead stearate or iron oxide filler, its density may well exceed unity.

Enough has been said to indicate that with the exception of elemental analysis, in many instances simple physical tests just do not provide definitive information.

Qualitative or quantitative examination for the presence of elements and the pyrolysis gas chromatograph or infrared fingerprinting approach, as discussed below, do however provide more useful information and indeed in many cases will lead to a successful resolution of the problem.

6.3 QUALITATIVE IDENTIFICATION BY PYROLYSIS-GAS CHROMATOGRAPHY

Pyrolysis is an analytical technique whereby complex involatile materials are broken down into smaller volatile constituent molecules by the use of very high temperatures. Polymeric materials lend themselves very readily to analysis by this technique. Providing that the pyrolysis conditions are kept constant, a sample should always degrade into the same constituent molecules. Therefore, if the degradation products are introduced into a gas chromatograph, the resulting chromatogram should always be the same and a chromatogram uniquely characteristic of the original sample should be obtained.

In one technique a pyrolyser probe is inserted into a purpose designed adaptor which is installed in the injector unit of the gas chromatograph. Different types of adaptor are available for packed or capillary column work. Correct selection of the appropriate adaptor is a pre-requisite for optimum system performance. The choice of packed or capillary pyrolysis gas chromatography is generally a matter of personal requirements. The type and range of samples to be analysed, the complexity of pyrogram required and the length of the analysis - all will play a significant role in decision making.

For the general examination of a wide range of rubbers or plastics, packed column pyrolysis is more than adequate. However, for comparison of one batch of rubber with another batch of the same rubber, the detail obtained from capillery pyrolysis gas chromatography will probably give the best results because minor details can be compared. Applications for pyrolysis are vast and include all types of synthetic polymers, rubbers and plastics, as well as latexes, paints and varnishes; in fact, almost any sample that contains involatile organic material that can be contained in a tube or coated onto a platinum ribbon so that it may be pyrolysed.

Pyrolysis-gas chromatographic procedures have generally followed one of two basic patterns. In the flash technique, a sample is deposited on an electrically heatable filament, or placed in a boat in a furnace, and the gas chromatographic carrier stream used to transport the pyrolyzate directly to the column. The alternative procedure involves heating of the polymer in a separate enclosure, trapping the off-gases and admitting the pyrolyzate to the chromatograph after a given collection interval. Each approach has certain advantages and the type of information desired from pyrolysis - gas chromatograms should therefore, dictate the sampling method to be employed.

6.3.1 Filament Pyrolysis

For qualitative identification a thin film of sample is usually coated onto Nichrome or platinum spiral or is placed in a small boat so that the weight of the residue remaining after heating can be determined. Although the filament may catalyze the degradation, Jones and Moyles[34] showed that with 30 - 30 ug samples the pyrograms obtained with Nichrome, platinum or gold-plated platinum filaments are identical. The power supply to the filament is generally controlled by a variable transformer and timed by a stopwatch, but exact measurement of the filament temperature is difficult. More elaborate automatic time and voltage controls have been suggested.[35,36] If desired, the pyrolyzer temperature can also be manually programmed to obtain more rapid equilibrium and the remove the pyrolysis products from the heated zone immediately after they are formed.[37] For quantitative studies of the mechanism and the kinetics of polymer degradation where accurate analysis of the volatile and non-volatile reaction products obtained at a certain temperature and under closely controlled conditions is required, it is preferable to employ preheated tube furnaces than to refine the design of the filament-type pyrolyser. Despite the fact that the filament-type pyrolyzer does not allow optinum control of degradation conditions the pyrograms are entirely satisfactory for identification purposes when polymers of known composition are available for comparison.

Giacobbo and Simon[38] have described a very useful pyrolysis unit for microgram samples. The material coated on a small ferromagnetic wire is pushed by means of a magnet into the pyrolysis capillary which is surrounded by an induction coil. Using a frequency of 450 Khz the high frequency induction oven will heat an iron wire of 0.6 mm diameter to the Curie temperature in 2 x 10^{-3}s. During the heating time, which can be controlled from 0.06 s to several seconds, the temperature of the wire remains at an approximate constant maximum (of a fixed frequency). The pyrolysis temperature can be varied by choosing a ferromagnetic conductor with a suitable Curie point temperature. With a reactor capillary of 0.6 mm and a wire of 0.5 mm diameter and 1 cm length the carrier gas will pass through the reactor in 5 x 10^{-3}s, assuming a flow rate of 10 cm^3/min.

May et al[39] have used this Curie point filament pyrolyser to produce pyrolysis - gas chromatograms for various polymers (Method 106).

Voigt[40-42] employed the platinum filament pyrolyser. This unit is attached directly to the gas inlet of the gas chromatograph for the examination of ethylene - propylene copolymers.

Pyrograms will differentiate between random copolymers and block polymers or polymer mixtures.[35,40,45,46] Presence of foreign monomer may interrupt chain transfer processes involved in the degradation. Similar products result but their quantities as determined from peak heights are different.

Systematic identification of plastics was attempted by Nelson et al[47] by pyrolyzing 0.2 - 0.5 mg samples for 10 s at 650 - 750^{0}C in an argon gas chromatograph using a 4 ft column of 5% silicone oil on Chromosorb W. Similar plastics were distinguished from pyrograms obtained at two different temperatures.

Groten[48] used a platinum filament type pyrolyser and six-way gas sampling valve pyrolyser in conjunction with an isothermal gas chromatograph by heating of a platinum coil at 950° C for 26 s. For the 150 different polymers investigated, the individual members of a group generally gave pyrograms that allowed unambiguous identification of the original material. The attainment of mazimum temperature is quite rapid under all conditions and is fairly insensitive for variations in carrier gas flow rate.

6.3.2 Combustion Furnace Pyrolysers

Earlier pyrolysis equipment tended to be somewhat complex in design and operation. Thus, Cox and Ellis[49]described a micro-reactor pyrolyser which they applied to large numbers of polymers. Temperatures increases of 700 - 1000°C were used in order to completely pyrolyse 0.1g samples of the polymers. The pyrolysis products were collected for 15 min then swept onto a gas chromatographic column equipped with a flame ionization detector. This type of equipment has now been displaced by the more recently described equipment as discussed later.

Kolb and Kaiser[50] pyrolyzed 4 mg of polyethylene in an oven furnace at 1000°C for 10 s. The first peak of the repeated groups of triplet peaks of the pyrogram was identified as the normal paraffin.

Cieplinski et al[51] have investigated extensively the thermal degradation products of polyolefins in an oven furnace using a chromatograph equipped with a differential flame ionization detector. Samples (50 - 70 mg) were pyrolyzed in a reaction chamber as described by Ettre and Veradi[52] and the volatile reaction products were passed through a linearly programmed temperature, open tubular (Golay) column. On pyrolysis at 690°C a large number of triplet peaks up to C22 were obtained. The last peak in each triplet group was identified as the normal paraffin.

Comparison of pyrograms of high and low density polyethylenes indicates significant differences for peaks above C21, with the relative amounts of the two major peaks being reversed. Thus, for the low density polyethylene the last peak corresponding to the _n_-paraffin is always larger, whereas for high density type polyethylene the first peak predominates. The pyrograms of polymethylene obtained by Ciepielsky et al[51] are greatly dependent on the pyrolysis temperature.

Two procedures have been described, (I) flash filament pyrolysis (Section 6.3.1) and (ii) heating the polymer in a micro-reactor and collecting the total volatiles produced, (Section 6.3.2). In yet another more recent approach, the pyrolyser probe method discussed below, the polymer is placed in a quartz tube which is then inserted into a platinum coil heater element. The coil heater is switched on to release the pyrolysis products which are then directly swept into the gas chromatograph.

Chemical Data Systems UK supply a range of pyrolyser probes which are discussed in Method 107. These probes can be used in conjunction with any suitable gas chromatograph.

A large number of materials can be analysed by packed column pyrolysis gas chromatography. The chromatograms obtained are relatively simple and the analysis time comparatively short. These chromatograms are therefore especially suitable for qualitative polymer identifications. Relatively large amounts of sample can be introduced onto the column and therefore minor differences between batches of similar sample may be observed. Figs 6.1(a) and (b) are examples of two types of acrylic material, both analyzed using the same conditions. Note the similarity of the peak pattern produced from the identical thermal degradation of the two samples, but also note the relative difference in the size of the peak at 12.8 min and also the presence of the peak at 5.04 min in the modacrylic chromatogram. The results obtained from this type of analysis are qualitative. Quantitative analysis is difficult as the identities of the separated degradation products are usually unknown. Packed column analysis of pyrolysis products generally allows observation of the more volatile constituents produced by the thermal degradation.

The chromatograms obtained from using capillary columns to separate the pyrolysis products are generally complex. Separation of all the more volatile fragments is not possible without cryogenic cooling of the oven. Under standard chromatographic conditions it is the high boiling degradation products that are readily separated as can be seen from fig. 6.2(a) and (b). Therefore, it is the latter part of the chromatogram that is used for identification. Because it is the higher boiling degradation products that are being observed, a number of polymers give very similar pyrograms. Thus the pyrograms obtained for polyethyl acrylate and polyvinylacetate are almost identical. The only observable differences are the presence of the two additional peaks in the polyvinyl-acetate pyrogram at 8.74 minutes and 12.79 minutes. Therefore, capillary column pyrolysis gas chromatography is slightly less useful than packed column pyrolysis as a

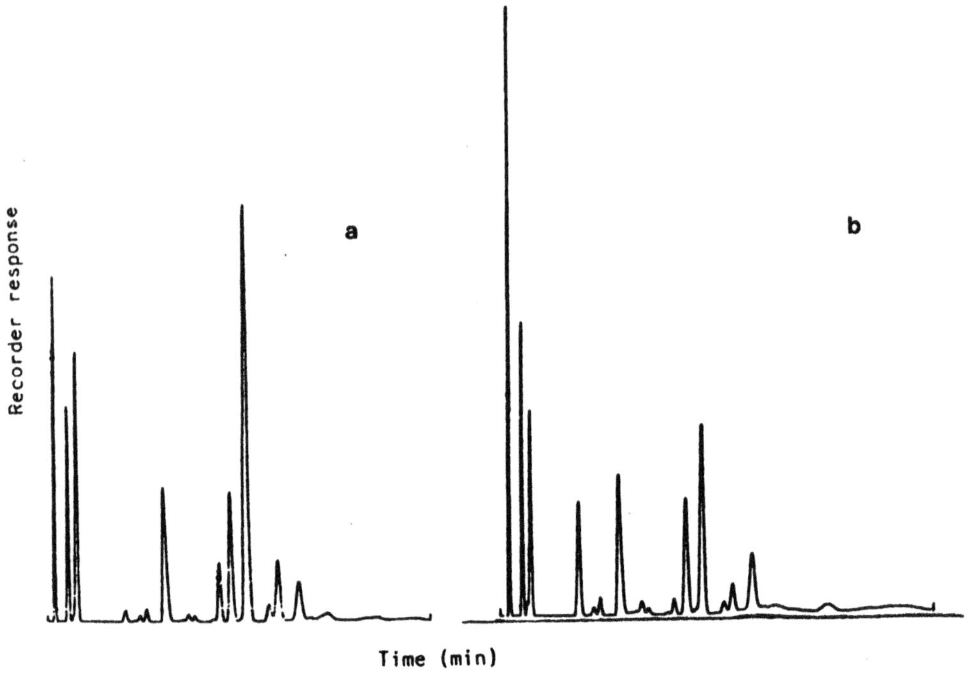

Figure 6.1. Pyrolysis - gas chromatography of (a) modacrylic resin; (b) acrylic resin.
Instrument: P-E8310 series
Column: 1m Porapak QS
Conditions: 50 C (2.0 mins) 15 deg C min 1 230 C (15 mins)
Pyrolysis: 950 C for 10 sec

strict identification technique. However, the complexity of the pyrograms produced allows minute details to be observed. Analysis times used in capillary column pyrolysis are two to three times longer than those for packed column pyrolysis gas chromatography; runs are generally in the order of 60 minutes in length. The high split ratio and very small amounts of sample required to obtain good results with capillary column pyrolysis gas chromatography mean that there is an obvious advantage in using this technique when only very minute amounts of sample are available for analysis.

In the chemical data systems - filament coil type pyrolyser, Pyro-probe 122.the polymer samples are held in a quartz tube that is inserted into the platinum coil. If pyrolyses are to be conducted at slow rates the pyrolyser is interfaced to a sample concentration (Model 320, Chemical Data Systems) which collected the pyrolysates on a Tenax filled trap. In this way polymer samples can be processed for minutes or hours at slow heating rates and the pyrolysates collected for a single gas chromatographic analysis. When the pyrolysis is complete, the trap is pulse heated and backflushed with the carrier gas. The desorbed pyrolysates are transferred to a gas chromatograph (Model 3700 Varian or equivalent) equipped with a 50 m x 0.25 mm i.d. SE-54 capillary column (Quadrex).

A 6-:1 split was established at the injection port of the gas chromatograph and the column was programmed from 50 to 280° at a rate of 40°C min^{-1}. A flame ionization detector is used and the area data obtained from a recording integrator.

Wampler and Levy[53] used this apparatus to study the effects of slow heating rates on the products formed during the pyrolysis of polyethylene. They showed that both the pyrolysis temperature and the rate at which that temperature is achieved have significant effects on the formation of pyrolysates from a solid polymer.

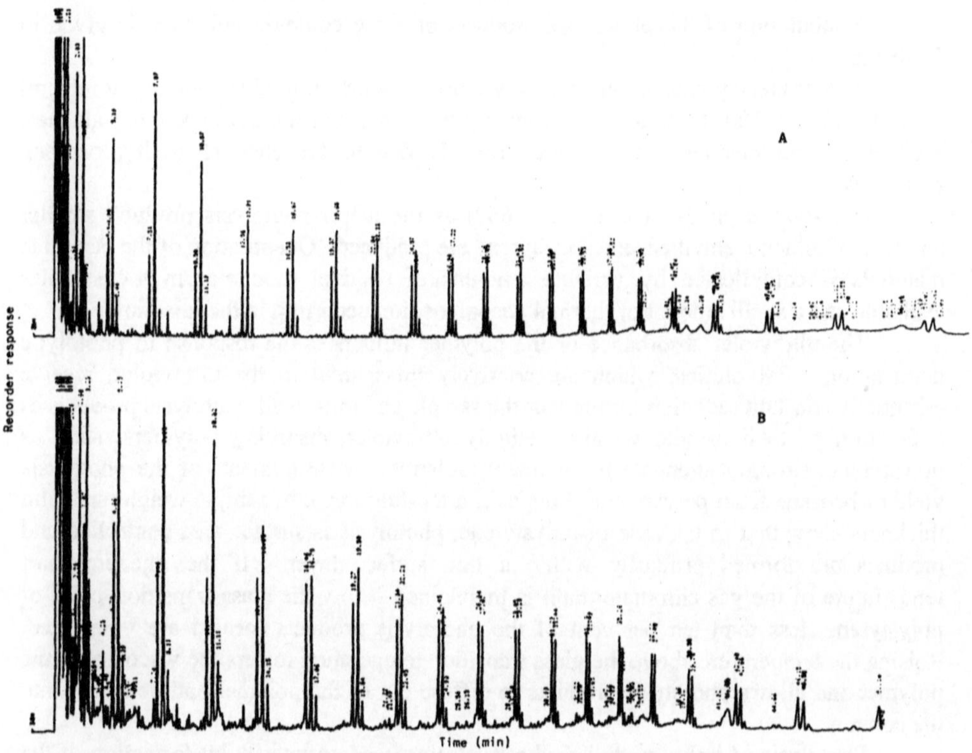

Figure 6.2. Pyrolysis - gas chromatography of (a) polyethylene; (b) polypropylene - polyethylene copolymer.
Instrument: P-E 8310 Series with split/split less injector.
Column 25, SE30 Capillary (0.25 mm i.d.)
Conditions: 50 C (2.0 mins) 5 deg C min 1 230 C (after last peak)
Pyrolysis: 850 C for 10 sec.

Pyrolysis coupled with simultaneous multidetector gas chromatography has been applied to studies of the structure of end-groups in polymethylmethacrylate.[86]

6.4 LASER PYROLYSIS - GAS CHROMATOGRAPHY

The advantages claimed for this technique[54,55] include rapid heating and cooling of the sample and relatively simple fragmentation patterns, (Method 108).

6.5 PHOTOLYSIS - GAS CHROMATOGRAPHY

Juvet et al[56] have used this technique for the preparation of pyrograms of polymers. They claim that this technique yields considerably more simple and reproducible decomposition patterns than filament, furnace and Curie point pyrolysers and claim that this is due to greater control of energy input and a more predictable manner in which the polymers decompose photolytically. Photolysis is carried out on a pure thin film of the polymer by irradiation using a medium pressure mercury source. The photolysis products are swept onto a gas chromatogram to produce a pattern characteristic of the polymer.

A tabulation of the photolysis products of some common polymers is given in Table 6.6.

Polyethylene yields on photolysis primarily a series of products eluting at integral multiples of 1 = 100; these are undoubetedly n-alkanes and the corresponding alkenes. Products at intermediate I values are probably due to branched chain hydrocarbon fragments.

Polystyrene yields benzene (I = 660) as the major photolysis product, smaller amounts of toluene, ethylbenzene and styrene are produced. Quantitation of the last three materials is complicated by variable amounts of residual monomer in the samples available and a small amount of thermal decomposition occurring in the injection port.

The ultraviolet absorbance of the polymer influences the response to photolytic degradation. Polyolefins, which are relatively transparent in the ultraviolet, receive essentially constant radiation throughout the sample and thus yield photolysis products as a function of total sample weight. Highly ultraviolet absorbing polymers, such as polystyrene, strongly attenuate the incident radiation. Measurements of the photolysis yield of benzene from polystyrene films over a ten-fold range of sample weight and film thickness show that in the case of polystyrene, photolysis is surface area controlled and products are formed primarily within a thin surface layer. If the injection port temperature of the gas chromatograph is maintained below the glass transition point of polystyrene, less than ten per cent of the photolysis products formed are volatilized. Raising the temperature above the glass transition temperature lowers the viscosity of the polymer and allows photolysis products to diffuse out of the polymer and be swept into the column.

Photolysis of poly (methyl methacrylate) proceeds primarily by formation of the monomer, corresponding to the product observed at I = 720. As in the case of polystyrene, small amounts of residual monomer and monomer formed from thermal decomposition were observed in non-irradiated sampled. Methanol and methyl formate, which have also been reported as photolysis products are not separated on SE-30 but analysis on Carbowax 20 M give two peaks corresponding to these substances.

6.6 PYROLYSIS - MASS SPECTROMETRY

Hughes et al[57] and Meuzelaar[58] have shown that pyrolysis - mass spectrometry has considerable potential for the characterisation and discrimination of natural and synthetic polymers. A Curie-point pyrolyser, combined with a dual flame detector gas chromatograph and V.G. Micromass 12 F mass spectrometer were used by these workers. The pyrolyser was mounted on the front of the gas chromatograph and was purged with helium at a flow rate of 5 cms min^{-1}. Samples were pyrolysed at 610°C (heating time 15 s) and the pyrolysate was swept into an empty 45 cm x 6.35 mm o.d. x 2 mm i.d. glass column maintained at 200°C in the gas chromatograph. A make-up flow of helium (10 cm^3 min^{-1}) was introduced at the end of the column by using a Swagelok stainless-steel 1/4 - 1/16 in reducing union with a length of 1/16 in o.d. stainless steel tubing brazed into it. The mixed outlet flow was passed through a length of glass-lined stainless-steel microbore tubing to the jet separator of the mass spectrometer. The mass spectrometer was operated under standard electron impact conditions; electron energy 70 eV, emission current 100 uA, accelerating voltage 4 kV and a source temperature of 240°C. The mass range (25 - 200) was scanned at 3 s per decade with a magnet re-set time of 3 s. About 40 - 50 scans of the pyrolysate were made with acquision, storage and processing of spectra carried out with a V.G. 2040 data system, supplemented with an X - Y plotter.

Table 6.6 - Photolysis Products of Common Polymers

Polymer	Retention Index I on SE-30	Rel. peak area[a]
	< 500	1.0
	604 ± 12	0.1
	680 ± 8	0.02
	702 ± 6	0.1
	765 ± 5	0.02
	800 ± 5	0.15
	876 ± 5	0.01
	900 ± 5	0.07
Polystyrene	< 500	0.02
	660 ± 10	1.0
	784 ± 8)	
	830 ± 8)	ca. 0.5[b]
	886 ± 6)	
Poly(methylmethacrylate)	< 500	1.0
	720 ± 8	0.5
	1525 ± 5	0.2
	1550 ± 5	0.2
Poly(tetrafluoroethylene)	< 500	...

[a] Irradiation time, 30 minutes. [b] Total of three incompletely resolved peaks. From Juvet[56] American Chemical Society

The system was used to study a wide range of polymeric materials. The sensitivity is sufficiently high to allow samples of 5 ug or less to give adequate electron-impact spectra, but in the chemical ionisation mode larger samples are necessary. Mass pyrograms are usually characteristic of the sample type and frequently allow discrimination between samples of similar composition.

The advantages of pyrolysis - mass spectrometry over pyrolysis - gas chromatography for generating information about polymeric materials are its speed, sensitivity, ease of producing data that can be computer processed and the elimination of the variables associated with gas chromatography. A major disadvantage of pyrolysis - mass spectrometry is that a complex mixture is produced by a combination of pyrolysis and electron - impact fragmentation, which makes a mass pyrogram more difficult to interpret than the chromatograms produced in pyrolysis - gas chromatography, in which only a pyrolytic breakdown stage is involved.

Jackson and Walker[59] studied the applicability of pyrolysis combined with capillary column gas chromatography mass spectrometry to the examination of phenyl polymers (eg. styrene-isoprene copolymer) and polymer like phenyl ethers (eg. bis(m-(m-phenoxy phenoxy)phenyl)ether). They examined the effect of varying parameters affecting the nature of products formed and relative product distribution in routine pyrolysis. These parameters include the effects of pyrolysis temperature rise times, pyrolysis temperatures up to 985°C and pyrolysis duration. Temperature rise time (0.1 to 1.5 s) is not a critical factor in the Curie point pyrolysis of a styrene-isoprene copolymer, either with regard to the nature of the products formed or their relative distributions. Additionally, the variation of pyrolysis duration or hold time (2.0 to 12.5 s) at a fixed Curie temperature reflected no change in the nature of components formed; however changes in product distributions were observed. Variations in Curie temperature at a fixed pyrolysis duration produced drastic changes in product distributions such as a three-fold change in isoprene dimer formation in styrene-isoprene copolymers; however, temperature variance did not change the nature of the products formed. Bis(m-(m-phenoxy Phenoxy)phenyl)ether) produced two primary pyrolysis products, diphenylether and dibenzofuran. In the procedure a 10% w/v solution of the polymer in a volatile solvent (benzene) is prepared. The pyrolysis Curie point temperature wire is dipped 0.25 in into the polymer solution and the wire placed in a vacuum oven at 75 to 80°C for 30 min to remove the solvent.

A pyrogram of the copolymer (isoprene-styrene) resulting from a 10 s pyrolysis at 601°C yields product distributions similar to the sum of the two constituent product distributions. For example, when the polymer polyisoprene is pyrolyzed, C_2, C_3, C_4, isoprene and $C_{10}H_{16}$ dimers are produced. When polystyrene is pyrolyzed, styrene and aromatic hydrocarbons are the products. The copolymer product distribution and relative area basis resemble the two individual polymer product distributions.

Mattern et al[44] carried out laser mass spectrometry on polytetrafluoroethylenes. They found a fragmentation mechanism common to each fluoropolymer yields structurally relevant ions indicative of the orientation of monomer units within the polymer chain. A unique set of structural fragments distinguished the positive ion spectra of each homopolymer, allowing identification.

Chin-An-Hu[63] has described a pyrolysis mass spectrometric method for polymer characterisation.

Qian et al[91] carried out a rapid identification of polymers using a technique based on ion source direct pyrolysis mass spectrometry and library searching. Polymers were pyrolysed using a coiled filament designed for desorption chemical ionization/desorption electron ionization applications. Pyrolysis products were ionized at 70 ev electron impact. This yielded highly reproducible spectra characteristic of the polymer. Using these techniques and library searching a comprehensive library of 150 polymers was developed.

6.7 TYPICAL POLYMER PYROGRAMS

6.7.1 Filament and Furnace Pyrolysis

Samples can be pyrolyzed at a series of temperatures between 400 and 1000°C and the change in degradation behaviour as evidenced by the (1) "appearance temperatures" of various peaks and (2) relative abundance of products as a function of temperature noted.[60,61] Changes of the most characteristic breakdown products versus temperature can also be plotted. For homopolymers the temperatures at which degradation products are first obtained may be more distinctive than the retention times of the products. The utility of the pyrolysis technique is that the chemist can assemble a collection of pyrograms of polymers which are of particular interest and use them to identify unknown polymers. For example, Figure 6.3 shows pyrograms obtained by furnace and filament methods of pyrolysis via a range of synthetic elasomers. An example of a sophisticated recent application of pyrolysis gas chromatography is the work of Denig[62] Figure 6.4(a) and (b) who compares the pyrograms of PVC and poly(vinylidene chloride). The high quality of these pyrograms is self-evident and makes it possible to distinguish between the two polymers. The technique also distinguished between a physical mixture of poly(ethylene acrylate) and poly(isobutyl acrylate) and a poly(ethylacrylate)-poly(isobutyl acrylate) copolymer.

6.7.2 Curie Point Pyrolysis

In Figure 6.5 are shown some pyrograms obtained using a Curie Point pyrolyser in conjunction with a Pye 104 Model 64 gas chromatograph.[39] The polymers were not pre-treated in any way and they were pyrolysed in the solid phase as received. The pyrograms are those for a linseed pentaerythritol-o-phthalate alkyd (Figure 6.5(b)), an acrylic emulsion (Figure 6.5 (c)) and a vinylchloride (95 per cent)-vinyl acetate (5 per cent copolymer (Figure 6.5(a)). A solid phase Poropak Q column was used.

May et al[64] have published a full reference collection of Curie point pyrograms.

6.7.3 Laser Pyrolysis

Some typical laser pyrograms are reproduced in Figure 6.6.[54,55]

6.8 POLYMER IDENTIFICATION BY INFRARED SPECTROSCOPY

Infrared spectra of thin films of polymer in the region up to 4000 cm^{-1} are characteristic of the polymer. The spectra can be run either as potassium bromide discs (1.5 - 2 ug polymer per 400 mg potassium bromide) or as polymer film of varying thickness up to 12 micron using a sodium chloride prism under the following conditions:[67]

Adjustment 0 to 100 cent T at 2.00 u
Gain: 3.9 - 4.0 (on the brink of aperiodicity)
Time constant of the amplifier: 2
Speed of the optical attenuator: 1100
Resolution (split programme): 927
Scanning speed: 1 x 10 + scan program (about 11 min from 1.25 to 15.25 u)
Speed supression: 8
Nernst current: 0.30 to 0.35 A
Wavelength region scanned: 1.25 to 15.25 u

Table 6.7 tabulates data on the infrared absorptions occuring in polymers. It should be emphasised that any conclusions reached following examination of infrared spectra should be confirmed by other techniques such as elemental analysis and pyrolysis - gas chromatography.

Denchant[68] has discussed in detail the infrared spectra of 35 polymers and Hippe and Kerste[69] developed an algorithm for the infrared identification of vinyl polymers. Pyrolysis followed by infrared spectroscopy is a particularly useful technique for application to polymers which are rendered opaque or completely non-transparent by the presence of pigments or fillers.

Leukoth[70] used this technique to determine the chemical composition of polymers and rubbers containing high proportions of pigments or fillers.

Alexander[71] has described a method for obtaining spectra of thin films of polymer which are free of interference fringes. The method is based on measuring a transmission/reflection spectrum of the polymer using a specular reflectance accessory on the infrared instrument. The film is placed on a reflectance accessory with a mirror above it. The radiation transmitted through the film is returned by the mirror and so is measured together with the radiation reflected by the film.

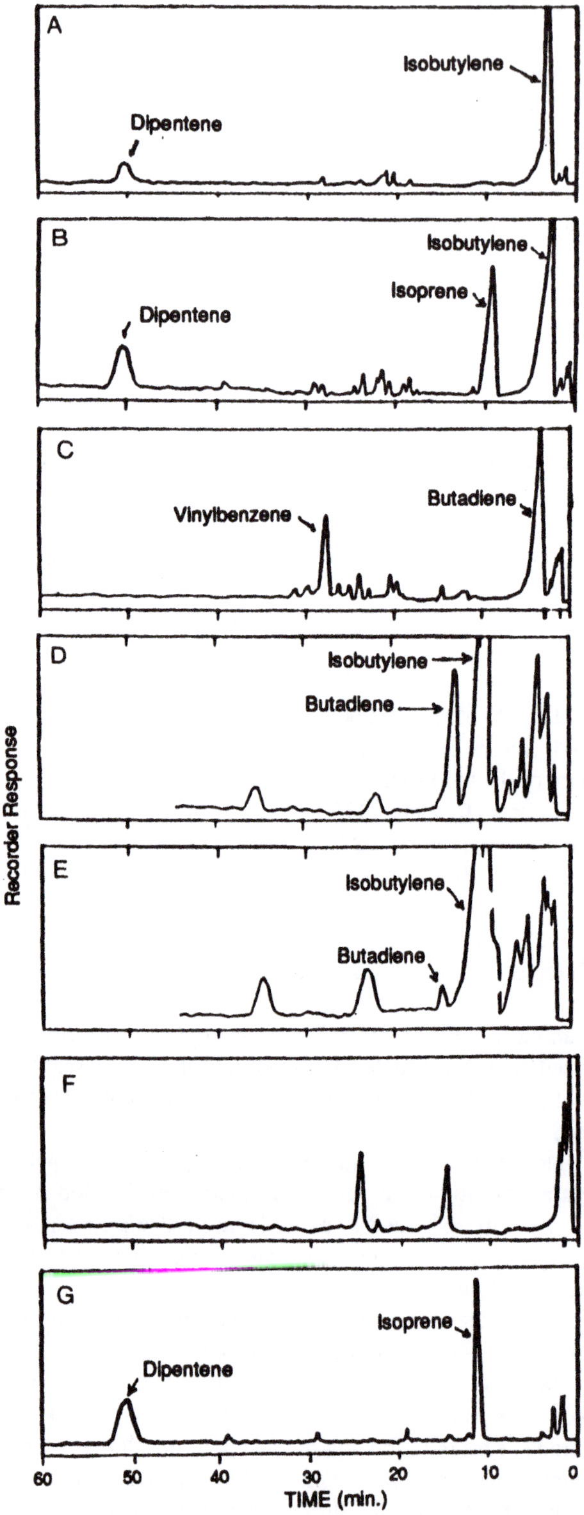
A
Isobutylene
Dipentene
B
Isobutylene
Isoprene
Dipentene
C
Vinylbenzene
Butadiene
D
Isobutylene
Butadiene
E
Isobutylene
Butadiene
F
G
Isoprene
Dipentene
Recorder Response
60
50
40
30
20
10
0
TIME (min.)

Figure 6.3 Pyrolysis - gas chromatograms for

(a) natural rubber
(b) isobutylene - isoprene elastomers
(c) mixture of natural rubber and isobutylene - isoprene elastomer
(d) cis-butadiene elastomer
(e) 1:1 mixture of cis-butadiene and isobutylene - isoprene elastomer
(f) 1:9 mixture of cis-butadiene and isobutylene - isoprene elastomer
(g) Chloroprene elastomer

Osland[26] has described a heated press for the preparation of plastic films for analysis by infrared spectroscopy. The press can produce films as thick as 500 um and of reproducible thickness. Earlier workers have used the hot press film method, where the sample is heated until molten, pressed to a thin film and allowed to solidify. Sampling handling accessories are now available with which films of constant thickness can be prepared. Alternatively, the sample material can be dissolved in a suitable organic solvent and a film cast onto glass or a cell window.

These techniques are adequate for identifying the bulk polymer but polymer blends can contain many other components which must be identified or quantified and for this purpose they are generally unsuitable, the reason being that they cannot produce thick films or reproducible film thicknesses.

Other approaches to obtaining the infrared spectrum of polymers are surface reflectance technique[27,28] and infrared microscopy.[66]

Gardalla and Grobe[28] compared attenuated total reflectance and photoacoustic sampling for surface analysis of polymer mixtures by Fourier transform infrared spectroscopy. They show that analysis by attenuated total reflectance is more suitable for smooth surfaces and is faster. Photoacoustic methods have shallower sampling depths than attenuated total reflectance but the latter technique is applicable over a range that is more controllable.

Fox[65] employed Fourier Transform infrared spectroscopy for the identification of traces of poly(dimethylsiloxane).

Brako and Wexler[31] have described a useful technique for differentiating the presence or absence of functional groups such as hydroxyl, carboxylic acid or ester in polymers containing small percentage components of such groups. Films of latices or polymers are subjected to chemical treatment which results in marked changes in the infrared spectrum which can be associated with the disappearance of a functional group or its replacement by another functional group. Infrared data may be readily interpreted negatively so that one may definitely preclude the presence of hydroxyl, carbonyl, amine, amide, nitrile, ester, carboxylic, aromatic, methylene, tertiary butyl and terminal vinyl groups, if the corresponding group vibrations are absent in the infrared spectrogram. More difficult is the assignment of functional groups where multiple or several alternate possibilities exist as in the mixture of a carboxylic and keto group or in the assignment of a band to an olefinic group.

When applying infrared spectroscopy to the preliminary qualitative identification of polymers it is desirable to assemble a library of standard spectra. Razinger et al[43] have described a minicomputer procedure for digitization, maintenance and retrieval of an infared spectral collection.

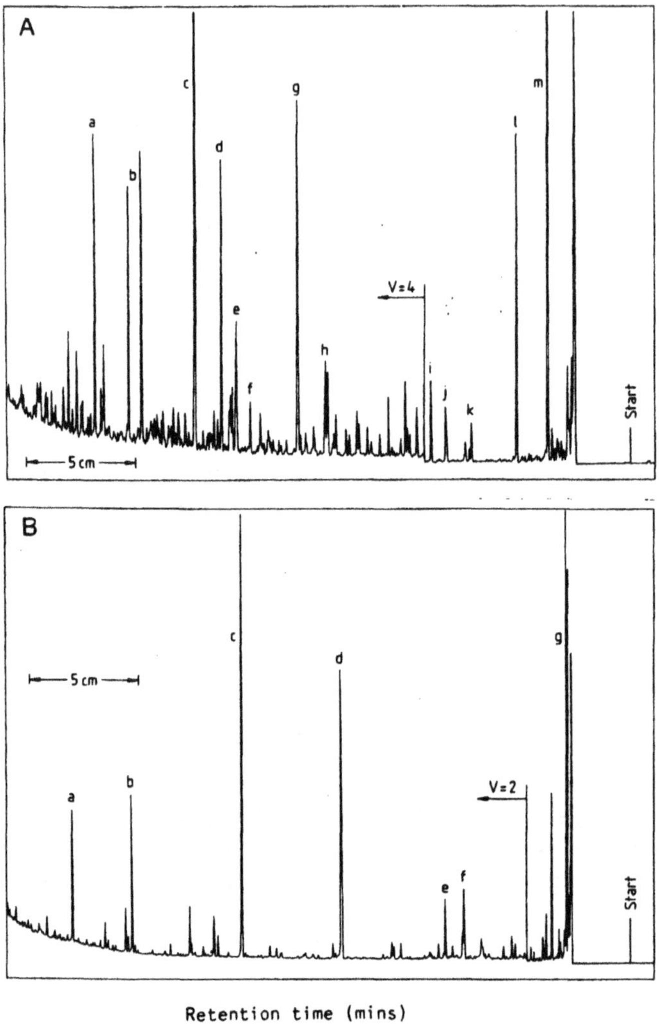

Figure 6.4

(A) Pyrolysis - gas chromatogram of PVC (a) biphenyl, (b) methyl naphthalene, (c) naphthalene, (d) methylindene, (e) tetralin, (f) methyl indene, (g) indene, (h) indane, (I) styrene, (j) o-xylene, (k) ethylbenzene, (l) toluene, (m) benzene.

(B) Pyrolysis - gas chromatography of polyvinylidene chloride, (a) tetra-chlorostyrene, (b) trichlorostyrene, (c) 1,3,5 trichlorobenzene, (d) m-dichlorobenzene, (e) trichlorobutadiene, (f) chlorobenzene, (g) vinyldene-chloride. From Denig with permission.[62] Carol Hauser Verlag Munich

6.9 POLYMER IDENTIFICATION BY LASER DESORPTION - ION MOBILITY SPECTROMETRY

In this technique the pulse from a Nd/YAG laser ablates the surface of the polymer generating molecular fragments in the vapour phase that are representative of the target material. These vapour phase components are drawn into the ion mobility spectrometer source where they are ionized. Both positive and negative ions are formed

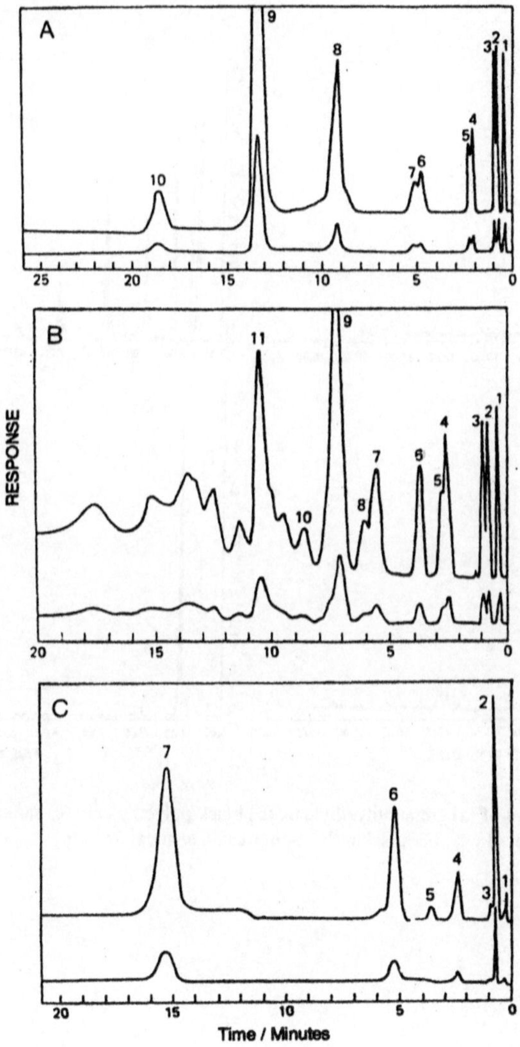

Figure 6.5

(A) Pyrogram for vinyl chloride (95 per cent) - vinyl acetate (5 per cent) copolymer. Peak 1, methane; 2, ethylene; 3, ethane; 4, propylene; 5, propane; 6, n-butene; 7, n-butane; 8, acetic acid; 9, benzene and 10, toluene. Amount pyrolysed 41 ug.

(B) Pyrogram for a linseed pentaerythritol o-phthalate alkyd. Peak 1, methane; 2, ethylene; 3, ethane; 4, propylene; 5, propane; 6, acetaldehyde; 7, n-butene; 8, n-butane; 9, acrolein; 10, allyl alcohol and 11, methacrolein. Amount pigmented paint pyrolysed 14 ug.

(C) Pyrogram for an acrylic emulsion. Peak 1, methane; 2, ethylene; 3, ethane; 4, methanol; 5, acetaldehyde; 6, ethanol and 7, methyl acrylate. Amount pyrolysed 7 ug.

From May et al with permission.[39] Royal Society of Chemistry, London

and the spectrometer can operate in either mode. The ions are formed and the spectrometer can operate in either mode. The ions are resolved into a pattern or signature characteristic of the polymer. Simpson et al[72] examined PVC, Nylon 66, acrylonitrile-butadiene-styrene terpolymer, polyethylene terephthalate and polyethylenes by this tech-

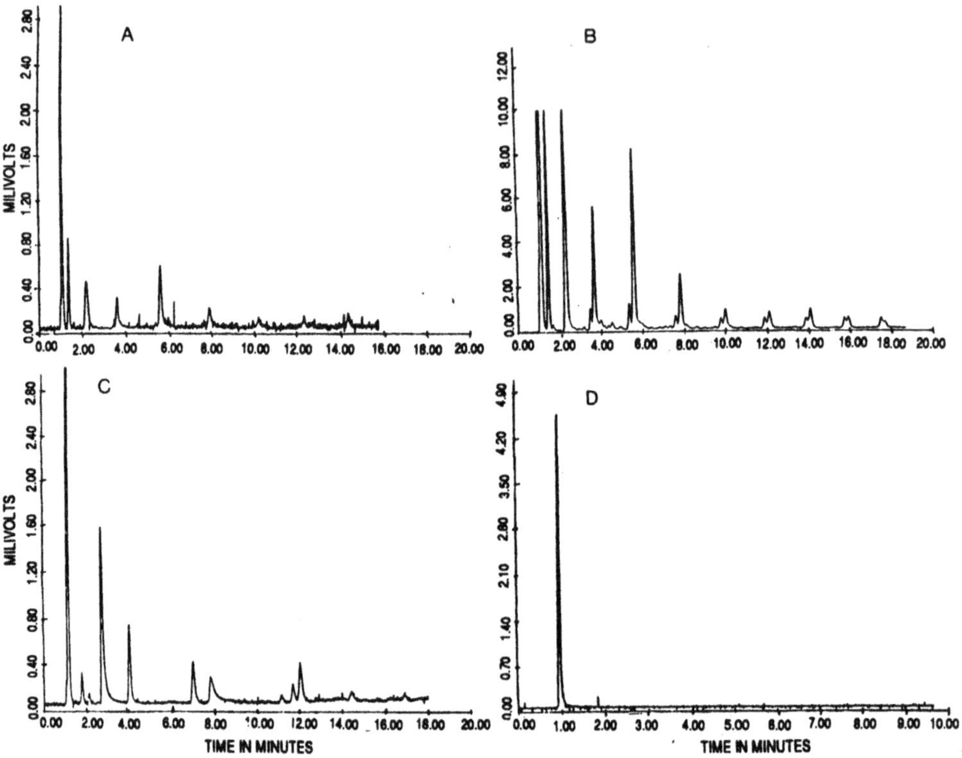

Figure 6.6
Lazer pyrogram of (a) green polyethylene, (b) black polyethylene, (c) saran, (d) teflon. From Folmer et al with permission.[54,55] American Chemical Society

nique. Similar materials such as high and low density polyethylenes could be differentiated.

6.10 OTHER MASS SPECTROMETRIC TECHNIQUES

Perfluoroethylenes have been characterized by desorption chemical ionization and tandem mass spectrometry[83] Fourier transform ion cyclotron resonance mass spectroscopy has also been applied to the identification of polymers, eg. polyethylene glycols.[84] Comparative complimentary plasma desorption mass spectrometry/secondary ion mass spectrometry has been applied to the identification of oligomers of various polymers including polyethylene glycol, polytetrafluoroethylene, polycarbonate, polyacrylates, polyethylene terephthalate and siloxanes.[85]

Shick et al[92]used radio frequency powered glow discharge atomization/ionization mass spectrometry (Rf GDMS) for the characterisation of bulk polymers. Fingerprint mass spectra were obtained for a series of polytetrafluoroethylene based polymers by looking at the differing base peaks and relative peak intensities. The technique exhibited excellent discharge stabilization and internal stability characteristics with a relative

Table 6.7 - Some Infrared Spectral Lines of Polymers

1. Polyacrylonitrile	C=N, 4.47 Aliphatic C - H, 3.41, 3.49 C = O stretching, 5.78 Others 6.01, 6.90, 3.34, 3.41
2. Polyformaldehyde	C - H stretching, 4.90, 6.80, 7.0, 7.23, 8.08, 9.15 Others, 3.34, 3.41, 11.15 (strong)
3. Polyisobutylene	Aliphatic CH) 3.35 - 3.50) 6.70 - 6.90 Others, 7.20, 7.32, 8.15 Doublet at 10.56, 10.87
4. Polyester resin	Aliphatic CH, 3.30, 3.40 C = O stretching, 5.80 (strong) C = O stretching ester, 7.65 - 8.03 Others, 6.09, 6.15, 6.50, 6.61, 6.70, 7.65 and 12.90 and 14.35 (characteristics)
5. Buna Rubber	Aliphatic CH, 3.40 Double bond, 5.90 Others, 6.10, 6.25, 6.69, 6.90 - 6.96 Strong, 10.36, 11.0, 13.22, 14.33
6. Vinylite polymers (Union Carbide) copolymerized vinyl chloride - vinyl acetate	Vinylite VMCH aliphatic CH, 3.40 C = O stretching, 3.65, 5.76 H O H binding absorbed water, 6.16 C - O stretching, 7.00, 7.32, 7.56, 8.1 Others, 9.17, 9.28
7. Vinylite VAGH	Aliphatic CH, 3.45, 3.56 C = O stretching, 5.79 Adsorbed water, 6.16 C - O stretching, 7.02, 7.55, 8.05 Others, 9.16, 10.42 On comparing with above spectra for Vinylite VMCH it is seen that there are sufficient differences to make positive identification possible in the region 7.0-7.5 and 9.5-10
8. Vinylite XYHL	The difference between this and the spectra of Vinylite VMCH and Vinylite VAGH is the strong absorption at 8.73
9. Silicone oil	Methyl in Si-CH3, 7.95, 9-10, 12.6 Si-O-Si, 12. 6 Aliphatic CH, 3.36, 3.43 - vibrations
10. Methyl cellulose	OH stretching, 2.90 Aliphatic CH stretching, 3.42, 3.51 Adsorbed water, 6.15 (similar spectrum to cellulose)
11. Ethyl cellulose	Splitting of aliphatic CH stretching vibration at 3.37-3/50 Others, 6.94, 7.27 (strong), 10.90, 11.37 (weak)
12. Carboxyl methylcellulose sodium salt	Resembles spectrum of cellulose
13. Carboxy methyl cellulose (free)	C = O stretching, 5.57 (strong) Carboxy, 3-4
14. Nitrocullulose	R-O - NO2, 6.05, 7.80 (characteristic)
15. Phenol formaldehyde (asbestos filled)	C-H groups, 3.3 (weak) Double bond plus C-H absorption, 6.25 double peak -Si O (from asbestos), 9.5-10.0
16. Melamine formaldehyde (asbestos filled)	N-H groups 2.75 Si O (from asbestos), 9.5-10.0
17. Polycarbonate	OH, 2.90, 6.12 (adsorbed water) CH adjacent to double bond, 3.28 (aromatic CH) Aliphatic C-H, 3.37, 3.48 C = O stretching, 5.88 (strong) CH, 6.9, 7.10-7.37

Table 6.7 continued

	Carbonyl ester, 8.0
	1.4 disubstitution, 12.1, 13.8
18. Polyethylene glycol	OH, 2.90 weak
	Aliphatic C - H, 3.45 (strong), 6.75-7.50
	- CH2 - O-CH2 band, 9
19. Polypropylene glycol	CH adsorption, 2.90
	Triple C-H, 3.35 - 3.50
	CH, 6.9
	CH_3, 7.3 (strong)
	C-O-C, 9 (strong)
	Others, 10.8, 11.6, 12.0
20. Polyvinylidene chloride	Aliphatic C-H, 3.4-3.5
	C = O absorption, 5.75 (impurity)
	CH, 7.10, 7.31
	C-CI, 13.35 (usually 13.3-14.3)
21. Chlorinated polypropylene	Absorbed water, 2.93, 6.16
	Aliphatic C-H, 3.40, 6.96, 7.25
	Methyl C-CI 13.15, 13.65, 12.75 (probably)
22. Polyvinylacetal	OH, 2.91
	Aliphatic CH, 3.4
	C = O ester, 5.76
	Methyl groups, 7.28
	- CH_2 - O - CR, 8.8
23. Polyacrylamide	C = O, 6.00
	C = N (probably), 4.5
	OH (moisture), 2.95
24. Cellulose propionate	Moisture, 2.9, 6.13
	Triple C - H adsorption, 3.4
	Ester carbonyl, 5.7, 8.6
	Others (characteristic), 7.25-8.5, 11.4, 12.45, 13.45
25. Cellulose acetate butyrate	Characteristic absorptions, 10.5, 11.1, 12.0, 12.6, 13.4, 14.5
26. Butyl rubber	Aliphatic Ch, 3.4 (strong), 6.8
	Two methyls attached to carbon, 7.19, 7.31
	CH adjacent to aliphatic double band, 10.55, 10.85
27. Nitrile rubber	CH adjacent to double band, 10.3 (strong), 3.25, 6.00, 6.10, 6.25 (strong)
	Aliphatic C - H, 3.40, 3.49, 6.9
	C = N, 4.46
28. Styrene-butadiene rubber	Polystyrene bands, 3.25, 3.30, 3.35
	(C - H adjacent to double band)
	Monosubstituted benzene ring, 5.6, 13.2, 14.33
	C - H adsorption, 10.35
29. Styrene-butadiene acrylonitrile	Mositure, 2.92, 6.12
	Aromatic C - H, 3.28, 3.31, 6.28 and C - H adjacent to double bond
	Aliphatic C - H, 3.41, 3.50, 6.88, C = N, 4.47
	Monosustituted benzene ring, weak 13.2, 14.35
30. Ethylene-propylene copolymer	Closely resembles polyethylene spectrum plus double bonds, 10.3
	Aliphatic C - H, 3.4
	CH adjacent to double bond, 3.3
	Methyl groups, 7.25
31. Ethylene-propylene-diene terpolymer	Closely resembles spectrum of ethylene propylene copolymer.
	Addition spectral lines
	Aliphatic double bands, 3.3 (weak) 10.65
32. Epoxy resin (araldite)	Hydroxy, 2.9, 9.04
	C - H, 3.27, 3.34, 6.20, 6.30
	Split methyl group, 7.20, 7.32
	1,4-di-substituted benzene ring, 5.6 (weak) 12.3
	Aliphatic C - H, 3.39, 3.43, 3.45,6.84

Table 6.7 continued

33. Coumarone-indene resin	Adsorbed water, 2.92 Aromatic C - H, 3.31, 6.49 Substituted benzene ring, 13.4, 14.3 Aliphatic C - H, 3.42, 3.50, 6.90
34. Polyvinylacetate	C = O, 5.75 Ester, 8.0, 9.8
35. Polyvinyl butyral	Characteristic absorptions, 7.25, 8.05, 8.9, 10.0
36. Polytetrafluorethylene	CF2, 8.4 Others, 8.4, 4-5
37. Polyethylene terephthate (Terylene)	OH stretching, 2.82 C = O, 2.90, 5.80 H stretching, 3.25 CH2 stretching, 3.35, 3.43 CH2 deformation, 7.08, 7.20 CH deformation, 7.90 C-O-C stretching, 8.51, 9.59 C-C, 6.80 Others, 5.80, 6.19, 11.45, 13.77, 6.65, 7.30, 7.45, 8.86, 9.10, 9.80, 10.3, 11.15, 11.45, 11.80, 12.62, 13.75
38. Polymethlmethacrylate	Ester carbonyl, 5.75 C-O-C stretching, 6.70-6.95, 8.30-8.80 Other, 10.00, 10.40
39. Cellulose (cellophane)	Water Absorption, 2.8 - 3,4 OH stretching, 3.4 CH stretching, 3.4 C = O stretching, 5.8 COO - ion stretching, 6.28 CH2 deformation, 7.0 CH deformation, 7.3, 7.6, 11.1 OH deformation, 7.5, 8.9, 15.4 C - O stretching, 8.6 OH bending, 9.4 C - OH stretching, 9.8 C - OH, 8.9 C - O, C - C stretching, 10.0 Other absorption, 9.0, 6.1
40. Cellulose acetate	Adsorbed water, 2.9 Ester carbonyl, 5.75 Other absorption, 3.5-5.35, 8.0, 7.5-10.0
41. Melamine	Characteristic absorptions, 3.0, 6.0-6.5, 9.7, 12.3
42. Urea formaldehyde	Characteristic absorptions, 6.0-6.2, 6.9, 8.0, 10.0
43. Natural rubber	C = C, 6.08 CH2 and CH3 deformation, 7.0, 7.25, 12.00 Other absorption, 7.5 - 11

standard deviation of less than 5%. In-depth profiles of polymers were obtained for copper layers deposited on a PTFE substrate. The following polymers were examined;

PTFE
PTFE - perfluoromethylvinyl ether copolymer
PTFE - hexafluoropropylene - polyvinylidene fluoride terpolymer.

It was observed that side chain structure had a dramatic effect in the fingerprint mass spectra.

6.11 EXAMINATION OF POLYMER SURFACES

Polymer films for certain specific applications such as food wrapping film consist of laminates of very thin layers of two or more different polymers. Various techniques have been adopted for the identification of the different layers in such laminates by examining each side of the film separately.

Techniques that have been applied include time of flight static secondary ion mass spectrometry (SSIMS),[73-76] low energy ion scattering spectrometry, x-ray photoelectron microscopy[77] scanning electrochemical microscopy (SECM),[78] laser ionization mass spectrometry,[79] surface enhanced infrared reflection spectroscopy,[80] Fourier Transform infrared spectroscopy[81] and nuclear magnetic resonance spectroscopy.[82]

Time of flight static secondary ion mass spectroscopy (SSIMS) has been applied to perfluorinated polymers,[88] polystyrene,[89] polyacylacrylates (including poly cyclohexylmethacrylate, polybenzyl methacrylate, polyphenyl methacrylate, poly n-hexyl methacrylate, poly n-butyl methacrylate, polymethylmethacrylate, poly n-propyl methacrylate, polyisopropyl methacrylate and poly secbutyl methacrylate). Blends of polystyrene and polyvinyl chloride,[87] bisphenol A and polystyrene,[73] polycarbonate and polystyrene[73] and tetramethyl bisphenol A and polycarbonate[73] have also been studied by this technique.

Scanning electrochemical microscopy (SECM) has been applied to polymethylmethacrylate, polystyrene and polyethylene glycol.[78] Surface enhanced infrared reflection microscopy was applied to polyacrylonitrile, polybutadiene and styrene resins[80] whilst Fourier transform infrared spectroscopy was applied to polyimides.[81] Finally, nuclear magnetic resonance spectroscopy has been applied to the examination of the surfaces of films of polyethylene, Suryln and ethylene-vinyl acetate copolymer.[82]

Schreimer and Li[93] carried out a surface analysis of bulk polymers using laser induced photoelectron ionization with laser desorption in a time of flight mass spectrometer. In this technique the carbon dioxide laser beam is focussed on the polymer surface to induce fast thermal dissociation and/or photo dissociation of the polymeric materials. The resulting neutral species are then ionized using photoelectron ionization (PEI). The PEI process involves a laser metal interaction in which a low power pulsed ultraviolet laser beam is directed to an appropriate metal surface exterior to the source region of a Wiley-McLaren time of flight mass spectrometer. The photoelectrons generated are accelerated as a narrowly distributed beam by the fringing fields of the source region. The electron beam, travelling in a direction almost parallel to the extraction grid is used to ionize the desorbed neutral molecules that are entrained into the ionization region. The technique was applied to the following polymers;

polystyrene
poly (2-vinyl pyridine)
polystyrene-2-vinyl pyridine
polethylene glycol
poly (methylmethacrylate)

Plots of signal intensity versus m/z all gave readily interpretable mass spectra suitable for structural analysis and chemical identification. The technique was also used for end-group analysis. Results compared well with those obtained by other mass spectrometric methods. The detection sensitivity was high, eg. a limit of detection of less than 100 a mole per laser shot for polyethylene glycol.

Lin et al[94] carried out an analysis of photoablation products resulting from polymer materials following supersonic beam/multi-photon ionization/time of flight mass spectrometry. They examined the photo-ablation products obtained in this technique and compared them with those obtained by thermal decomposition.

The high sensitivity provided by supersonic beam spectrometry permitted the detection of minor species, thus styrene monomer is produced from poly alpha methyl-styrene by cleavage of a methyl group and by proton rearrangement. Because the ablation is carried out at high temperatures it is possible to obtain results on thermally stable polymers such as poly-p-methyl styrene. The technique was applied to polystyrene foam, styrene-butadiene copolymers and acrylonitrile-butadiene-styrene terpolymers.

REFERENCES - CHAPTER 6

1. Colson, A.F., Analyst (London), **88** 26 (1963)
2. Colson, A.F., Analyst (London), **88** 791 (1963)
3. Colson, A.F., Analyst (London), **90** 35 (1965)
4. Salvage, T. and Dixon, J.F., Analyst (London), **90** 24 (1965)
5. Johnson, C.A. and Leonard, H.A., Analyst (London), **86** 101 (1961)
6. Yoshizaki, T., Anal. Chem., **35** 2177 (1963)
7. Haslam, J., Squirrell, D.C.M., Analyst (London), **86** 239 (1961)
8. Mano, E.B., Anal. Chem., **32** 291 (1960)
9. Winterscheidt, H. and Seifen, O., Fettes Wachse, **80** 239 71 (1954); **81** 16 (1965)
10. Hinteraider, A., Fette U Seifen, **51** 367 (1944)
11. Hummel, D., Kunststoff, Rundschau, **5** 85 (1958)
12. Samsel, E.P., Anal. Chem., **29** 574 (1957)
13. Liebermann, C., Chem. Berichte, **17** 1884 (1984)
14. Morawski, T., Chemiker Zeitung, **12** 1321 (1888)
15. Storch, L., Ber Osterruch Ges. 3 Forderung d. chem. Ind., **9** 93 (1887)
16. Kappelmeier, C.P.A., Farber. Zeitung., **40** 1141 (1935), **41** 161 (1936); **42** 561 (1937)
17. Moir, S., In Matiello, J.J. Protect. and Decorat. Coatings Bd. Vol 105, J. Wiley & Sons, New York (1956)
18. Gibbs, H.D. and Matiello J.J., Protect and Decorat. Coatings, Bd. Vol. 105, J. Wiley & Sons, New York, (1956)
19. Swann, M.H., and Weil, D.C., Anal. Chem., **28** 1463 (1956)
20. Berl. E.U.K., Hefter. Cellulosechemie., **14** 67 (1933)
21. Foucry, M.J., Peintures, pigments, Verniz., **30** 925 (1954)
22. Rudd, H.W. and Zonveld, J.J. J Oil Chem. Assoc. **39** 314 (1956)
23. Swann, M.H. and Esposito, G.G., Anal. Chem., **28** 1006 (1956)
24. Gunter, F.A. u Mitarb. Anal. Chem., **23** 1835 (1951)
25. Lohmann, H., J. Prakt. Chem., **135** 57 (1939)
26. Osland, R., Laboratory Practice, **37** 73 (1973)
27. Siewierski, S. and Koziowski, W., Polimery, **18** 221 (1973)
28. Gardella, J.A. and Grobe, G.L., Anal. Chem., **56** 1169 (1984)
29. Heermann, P. in Agster, A. Farberei u textil-chem., Untersuchungen Berlin - Gottinger-Heidelberg, Springer, S-139 (1956)
30. Wessler, H.J. Polymer Science, **11** 235 (1953)
31. Brako, F.D. and Wexler, A.S., Anal. Chem., **35** 1944 (1963)
32. Wechsler, H., J. Polymer Sci., **11** 233 (1953)
33. Bird, V., Plastics Technology, **40** September (1963)
34. Jones, C.E.R. and Moyles, A.F., Nature (London), **191** 663 (1961)
35. Barlow, A.R., Lehrle, S. and Robb, J.C., Polymer, **2** 27 (1961)
36. Cogliane, J.A., Rev. Sci. Instr., **34** 439 (1963)
37. Dalnogare, S., Paper presented at 147th meeting American Chemical Society, Philadelphia. Abstract of papers p 148 (1964)
38. Giacobbo, H. and Simon, W., Pharm. Acta. Helv., **39** 162 (1964)
39. May, R.W., Pearson, E.F., Porter, J. and Scothern, M.D., Analyst (London), **98** 364 (1973)
40. Voigt, J., Kunststoffe, **54** 2 (1964)
41. Voigt, J., Kunststoffe, **51** 18 (1961)
42. Voigt, J., Kunststoffe, **51** 314 (1961)

43. Razinger, M., Penca, M. and Zupan, S., Anal. Chem., **53** 1107 (1981)
44. Mattern, D.E., Tyan Lin, F., Hercules, D.M., Anal. Chim., **56** 2762 (1984)
45. Strassburger, J., Brauer, G.M., Tryon, M. and Farziati, A.F., Anal. Chem., **32** 454 (1960)
46. Bombaugh, K.H., Cook, C.E. and Clampitt, B.H., Anal. Chem., **35** 1834 (1963)
47. Nelson, D.F., Yee, J.L. and Kirk, P.L., Microchem. J., **6** 225 (1962)
48. Groten, B., Anal. Chem., **36** 1206 (1964)
49. Cox, B.C. and Ellis, B., Anal. Chem., **36** 90 (1964)
50. Kolb, A., Kaiser, K.H., J. Gas Chromatography, **2** 233 (1964)
51. Cieplinski, E.W., Ettre, L.S., Kolb, B. and Keinmer, G., Z. Anal. Chem., **705** 357 (1964)
52. Ettre, K. and Varadi, P., Anal. Chem., **35** 69 (1963)
53. Wampler, T.P. and Levy, E.J., Analyst (London), **111** 1065 (1986)
54. Folmer, O.F., Anal. Chem., **43** 1057 (1971)
55. Folmer, O.F., and Azarraga, L.V., J. Chromatogr. Sci., **7** 665 (1969)
56. Juvet, R.S., Smith, J.L.S. and Pang Li, K., Anal. Chem., **44** 49 (1972)
57. Hughes, J.C., Wheals, B.B. and Whitehouse, M.H., Analyst (London), **102** 143 (1977)
58. Meuzelaar, H.L.C., Posthumus, M.A., Kistemaker, P.G. and Kiskinaker, J., Anal. Chem., **45** 1546 (1973)
59. Jackson, M.T. and Walker, J.Q., Anal. Chem., **43** 74 (1971)
60. Lehman, F.F. and Branch, G.M., Anal. Chem., **33** 673 (1961)
61. Fiorenza, A. and Bonomi, G., Rassanga. Chimica., **5** 197 (1963)
62. Denig, R., Kunststoffe, **75** 8 (1985)
63. Chin-An Hu, J., Anal. Chem., **53** 942 (1981)
64. May, R.W., Pearson, E.F., Scothern, M.D., Pyrolysis-Gas Chromatography, Analytical Sciences Monograph, SAC, London (1973)
65. Fox, P., Analyst (London), **114** 445 (1989)
66. Young, P.H., Spectroscopy International, **1** 50 (1989)
67. Visapaa, A., Tekn. Kemian Aikakauslechti, **19** (1962) 121 - 137; The State Institute for Technical Research, Report Series IV - Chemistry, **42** (1962); Perkin-Elmer Application & News Bulletin 1963 no. 1, 1 - 8
68. Dechant, J., "Ultraospektroskopische Untermethungen on Polymerern", Akadamie Verlag, Berlin, 516 pp (1972)
69. Hippe, Z. and Kerste, A., Bull. Akad. Pol. Sci. Ser. Sci. Chim., **21** 395 (1973)
70. Leukoth, G., Gummi. Asbest. Kumst., **27** 794 (1974)
71. Alexander, R., The Analytical Report, Perkin-Elmer Ltd., Beaconsfield, Bucks, U.K. No. 7 February (1985)
72. Simpson, M., Anderson, D.R., McLeod, C.W., Coake, M. Analyst (London), **118** 449 (1993)
73. Thompson, P.M., Analytical Chemistry, **63** 2447 (1991)
74. Gardella, J.A., Pireaux, J.J., Analytical Chemistry, **62** 645A (1990)
75. Chilkoti, A., Ratner, B.D., Briggs, D., Analytical Chemistry, **63** 1612 (1991)
76. Chilkot, A., Ratner, B.D., Briggs, D., Analytical Chemistry, **65** 1736 (1993)
77. Gardella, J.A., Analytical Chemistry, **61** 589A (1959)
78. Lee, C., Bard, A.J., Analytical Chemistry, **62** 1906 (1990)
79. Pallix, J.B., Schule, U., Becker, C.H., Huestis, D.L., Analytical Chemistry, **61** 805 (1989)
80. Nishikawa, Y., Fujiwara, K., Ataka, K., Osawa, M., Analytical Chemistry, **65** 556 (1993)

81. Barner, B.J., Green, M.J., Saez, E.T., Corn, R.M., Analytical Chemistry, **63** 55 (1991)
82. Hatfield, G.R., Analytical Chemistry, **61** 1870 (1989)
83. Guarini, A., Guglielmetti, G., Vincenti, M., Guardia, P., Marchionni, G., Analytical Chemistry, **65** 970 (1993)
84. Hogan, J.D., Laude, D.A., Analytical Chemistry, **64** 763 (1992)
85. Feld, H., Leute, A., Zurmuhlen, R., Benninghoven, A., Analytical Chemistry, **63** 903 (1991)
86. Ohtani, H.O., Luo, Y.F., Nakashima, Y., Tsukahara, Y., Tsuge, S., Analytical Chemistry, **66** 1438 (1994)
87. Cox, B.D., Park, M.A., Kaercherr, G., Schweikert, E.A., Analytical Chemistry, **64** 843 (1992)
88. Bletsos, I.V., Hercules, J.M., Fowler, D., Van Leyen, D., Benninghaven, A., Analytical Chemistry, **62** 1275 (1990)
89. Hittle, L.R., Proctor, R.A., Hercules, D.M., Analytical Chemistry, **66** 108 (1994)
90. Zimmerman, P.H., Hercules, D.M. Benninghaven, A., Analytical Chemistry, **65** 983 (1993)
91. Quan, K., Killinger, W.E., Casey, M., Nicol, G.R., Analytical Chemistry, **68** 1019 (1996)
92. Shick, C.R., De Palma, P.A., Marcus, R.K., Analytical Chemistry, **68** 2113 (1996)
93. Schriemer, D.C., Li, L., Analytical Chemistry, **68** 250 (1996)
94. Lin, C.H., Murata, Y Imasaki, T., Analytical Chemistry, **68** 1153 (1996)

PART 2

ANALYTICAL METHODS

Various detailed methods of analysis, many previously unpublished, are described in Chapter 7

CHAPTER 7.1

POLYMER ANALYSIS

Methods 1 - 52

METHOD 1 - DETERMINATION OF WATER AND VOLATILES IN POLYALKENES. GAS CHROMATOGRAPHIC METHOD.

1.1 SUMMARY

A method for the determination of down to 0.01% of water and organic volatiles in polyethylene and polypropylene.

1.2 APPARATUS

The instrument is shown diagrammatically in Figure 7.1 and consists essentially of a sample tube, forming an external loop connected to a gas chromatograph.

1.3 METHOD

The solid polymer is placed in a gas chromatograph sample loop which is then heated under controlled conditions and the volatiles swept by helium carrier gas on to a gas chromatographic column prior to determination of water and volatiles.

1.4 EXPERIMENTAL PROCEDURE

The sample loop (Figure 7.1) can be isolated and the sample heated to the required temperature. After an initial heating period the volatile constituents liberated from the sample are "flushed" on to the chromatographic column by a flow of carrier gas through, or over the sample and the required components separated and determined quantitatively. A pneumatic switch valve located in the chromatograph oven to prevent the condensation of volatile constituents within the valve and a split heater mounted on a horizontal travel in a plane at right angles to the sample tube, are essential parts of the apparatus. The instrument is semi-automatic.

The carrier gas flows through the copper sulphate in tube D, which imparts a constant amount of water (about 3 p.p.m. w/v) to the helium. The "wet" carrier gas pre-

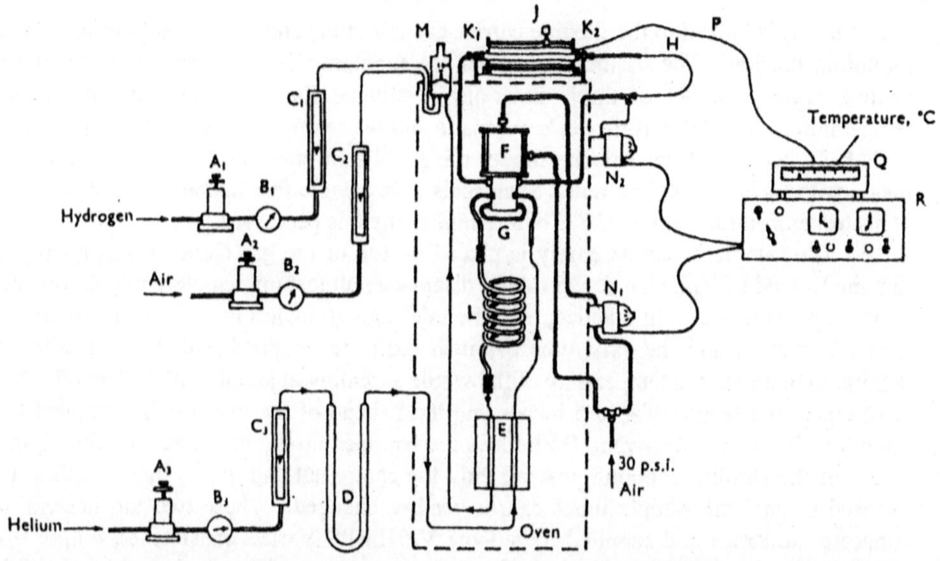

A_1, A_2 and A_3 = Edwards VPC I pressure controller

B_1, B_2 and B_3 = Pressure gauges 0 to 30 p.s.i.

C_1, C_2 and C_3 = Rotameter-type flow gauges

D = Clear plasticised PVC tubing (5 foot long × $\frac{1}{4}$ inch bore), packed with copper sulphate crystals, $CuSO_4.5H_2O$, >44 mesh

E = Katharometer

F = Pneumatic sample valve (Pye Cat. No. 12900); fitted with P 9904 change-over block

G = Internal loop

H = External loop made in part of 18-gauge stainless-steel capillary tubing

J = Sample split heater

K_1 and K_2 = Straight reducing couplings, captive seal type, for $\frac{3}{8}$ to $\frac{1}{8}$ inch o.d. tubing (Drallim, Cat. No. L/50/D/B)

L = Chromatographic column

M = Flame-ionisation detector

N_1 and N_2 = Electrically actuated 3-port pilot valves (Martinair, Type 557C/1Z)

P = Nickel - chromium/nickel - aluminium thermocouple embedded in the split heater

Q = Electronic temperature controller, proportional type

R = Sample-valve time-delay unit, containing three synchronous timers and a semi-conductor, proportional energy controller for the sample heater

Figure 7.1. Details of general purpose instrument for gas chromatographic determination of water and volatiles in polymers.

vents the gas-flow lines from "drying out". Dry pipework tends to adsorb moisture, which can then be desorbed, thus leading to spurious results. The determination of water in a sample in unaffected as the "wet" carrier gas flows continuously through both the reference and analysis cells of the katharometer. The copper sulphate crystals are sieved and any fine powder rejected. The copper sulphate tube is changed when two thirds of its length has visibly changed colour.

The pneumatic sample value F, operates merely as a switch valve, directing the carrier gas flow either around the internal loop G, or the external loop H. The pilot valve, N_1 operates samples valve F. The sample split heater J shown in more detail in Figure 7.1 consists of a cylindrical aluminium block, 6 inches long with a 2 mm hole through the centre. The block is split axially and the two halves hinged. Each half of the block contains two cartridge - heater elements, each 5.5 inches long, $^3/_8$ inch o.d. and one half contains a thermocouple pocket to accommodate the thermocouple P. The cartridge heaters are supplied by a semi-conductor energy controller contained in R, which is controlled by a galvanometer, two-position, temperature controller Q. The heater is mounted on a horizontal travel in a plane at right angles to the sample tube. A jig for mounting the sample tubes is also part of the heater assembly. The $^3/_8$ inch coupling, K_1 is brazed to its mounting bracket, which is rigidly attached to the heater base. This coupling accurately centres with the central hole through the aluminium block. The coupling, K_2 is brazed to the flexible carrier gas inlet tube and rests loosely in the second mounting bracket. The $^3/_8$ inch couplings are supplied with neoprene, or butyl rubber captive seals. Neither of these materials is suitable for continuous use at elevated temperatures, especially as the seals are made and broken many times a day. Alternative seals have been cut from silicone rubber tubing, 9 mm bore and 2 mm wall and fitted inside with $^3/_8$ inch coupling nuts. Such seals give trouble free operation for 2 or 3 days at heater temperatures up to 350°C before replacement is necessary.

The sample heater assembly is placed on top of the gas chromatograph oven so that the L-bend of $^1/_8$ inch o.d. S/S tubing disappears almost immediately into an opening on the top of the oven. In practice, both the inlet tube (6 inches long, $^1/_4$ inch o.d. and $^1/_8$ inch i.d. copper) and the exist tube ($^1/_8$ inch o.d.) are wrapped with heating tape and lagging to maintain the temperature of the whole assembly at about 100°C. The inlet tube is wrapped to a length of 4 to 5 inches and the L-bend of the exit tube is wrapped to a point 3 inches inside the oven. Both tubes are wrapped up to, and including, the $^1/_4$ inch thread of the Drallim coupling, leaving only the centre nut and the $^3/_8$ inch coupling nut exposed so that the sample tubes can be readily changed. These two tap heaters are connected in series and supplied by a Type V-5HMPS Variac transformer, output 0 to 270 V (2A). Without these heaters, volatile constituents condense at the cold ends of the sample tube and are swept away comparatively slowly on switching the carrier gas through the sample. This leads to diffuse and unsymmetrical peaks in the chromatogram.

The apparatus shown in Figure 7.1 is fitted with katharometer and flame ionisation detectors.

Disposable sample tubes are made by cutting 9 mm o.d. heat-resistant glass tubing into 6 $^7/_8$ inch ($\pm$ 1/16) lengths. The cut ends of the tubing are fire-polished in a flame. Such tubes will contain from 1.5 to 5 g of sample, depending on the polymer and its physical form.

N_2 acts as a pressure release valve to the external loop H. The sample valve time-delay unit R, contains three synchronous timers: (1) a variable 3 to 60 synchronous process timer (Chronoset Type CF.D. Robinson & Co.): (ii) a miniature, fixed cycle synchronous timer and (iii) a variable 30 s to 12 min synchronous process timer. Timers (i) and (ii) are linked, and are initiated by a push button. Timer (iii) operated independently and again is started by a second push button. The unit also incorporates a re-set button that re-sets timers (i) and (iii).

The oven contains the column, the katharometer and the pneumatic switch valve so that no condensation can take place in the switch valve. Other items of the gas chromatographic equipment not shown in Figure 7.1 include the katharometer bridge unit, the flame ionisation amplifier and the respective potentiometric recorders for the two detectors (1 mV, f.s.d.)

The chromatographic columns for the determination of water in different polymers consist of 10 per cent w/w stationary phase on polytetrafluoroethylene powder (I.C.I. "Fluon" CD4).

To examine a polymer for water and organic volatiles, the small glass sample tube is fitted with a small quartz-wool plug at about 1 inch from one end, and these are stored in an oven until just before use. Transfer several tubes from the oven to a desiccator containing phosphorus pentoxide and allow to cool. (The desiccant should be examined at weekly intervals and renewed when necessary). Use the tubes from the desiccator as required. Weigh a sample tube and pour the sample (powder or granule) into the tube. Re-weigh and insert a second quartz-wool plug with tweezers. These plugs are pre-prepared, oven-dried and stored in the desiccator. Place the sample tube in the heater jig between the couplings and tighten the couplings.

Set the heater to the appropriate temperature for the sample under examination. Operate the first push button on unit R (Figure 7.1). This starts timer (i) set for 15 s and at the same time activates pilot valve N_1, which operates the pneumatic switch valve, allowing helium carrier gas to flow around loop H, through the sample tube (still at room temperature) for 15 s and sweep the sample free from air. At the end of this 15 s purge period the pneumatic switch valve closes, thus isolating the sample. After a further 2 s delay (fixed timer (ii)) the second pilot valve, N_2, is activated for 2 s allowing the excess of helium pressure in the closed sample loop to escape to atmosphere via the tee-piece and capillary tubing attached to N_2. This valve sequence follows automatically on operating the first push button, and its completion is denoted by an indicator light. Open the split heater, move it over the sample and close. Operate the second push button for timer (iii). The sequence described below then follows.

Set timer (iii) (Figure 7.1) for 5 min. Open the split heater, move it forward on its travel and close it over the sample tube. Operate the push button for timer (iii). After the set time of 5 min the pilot valve, M_1, is triggered which operated the pneumatic switch valve, F, thus allowing helium carrier gas to flow around the external loop, H, through the sample tube and flushing the liberated volatile constituents on to the gas chromatographic column. The carrier gas remains routed around the external loop, H, until the re-set button is pressed. It is then diverted through the internal loop, G.

Two chromatograms are obtained. The first, resulting from the 15 s purge of helium through the cold tube, shows a peak for air followed by a small peak for water. This is developed while the sample is being heated for 5 min.

The second chromatogram, after the heating period, shows the liberated water followed by a peak for any organic volatiles present. Carry out a blank determination in the same manner on an "empty" tube containing two quartz-wool plugs only. Add the values for the two water peaks resulting from each determination and correct the sample figure for the blank value.

To calibrate for polyalkenes the polymer is heated above its "melting point" to obtain the total moisture; a horizontal sample tube is, therefore, used. The powder, provided it is not packed too tightly, contracts on melting and sinks to the bottom of the tube, thus allowing a free passage for the carrier gas above the sample. In this case calibration is best carried out with barium chloride dihydrate crystals, weighed into empty sample tubes. Barium chloride loses its water of crystallisation, 15.75 per cent w/w at 115°C. Chromatograms obtained in this manner are similar to those obtained on samples, ie. a small water peak after the heating period when the water is released from the crystals.

If calibration is carried out by direct injection of water on to a quartz-wool plug in an "empty" tube with a Hamilton syringe, 75 to 90 per cent of this water is purged on to the column during the 15 s period. Chromatograms are, therefore, obtained that are quite unlike those obtained for a sample, although the total amount of water should be unchanged.

Weigh about 12 mg of barium chloride crystals into an empty sample tube and obtain chromatograms as indicated under Procedure. Repeat with 9. 6. 3 and 1 mg amounts of barium chloride crystals. Add the values for the two peaks obtained for water in each determination and correct for the blank. Construct a calibration graph relating milligrams of water to the corrected total peak height (or peak area) measured from the recorder chart.

The peak on the plot, obtained by heating polypropylene powders in the presence of air, corresponds to the visible melting point of the polymer, the temperature at which the polymer begins to flow. Above this temperature the amount of water falls off before rising again at still higher temperatures. This fall off is presumably due to the sudden contraction of the surface area exposed to the oxidising atmosphere as the polymer flows. No corresponding peak is obtained with granular samples (powder that has been compounded, extruded and cut into small pellets). The continuous-line plots obtained by the second method indicate that all the moisture is not released from the polymer until the "melting point" is approached in the case of powder samples, and is exceeded in the case of granular samples. The heater temperature used for the determination of moisture in polypropylene samples was, therefore, chosen as 240°C as this temperature is applicable to both powder and granular samples.

The gas chromatographic conditions recommended for the determination of moisture in polypropylene are: column, 5 foot ($^1/_4$ inch o.d.), stainless-steel tubing, packed with 10 per cent w/w "Ucon" 50 HB 2000 fluid on "Flucon" CD4 powder; oven temperature, 100°C; and helium inlet pressure, 6 p.s.i.

1.5 DISCUSSION OF RESULTS

Low density polyethylene contains only very low concentrations of water (0.01 - 0.05% w/w) whilst polypropylene contains 0.02 to 0.5% water w/w.

At this high sensitivity and with lower amounts of water, blank determinations can be somewhat erratic with a reproducibility of 150% unless great care is taken with the storage of the dried tubes. For example, a tube that has been stored overnight in a desiccator over phosphorus pentoxide will give a much lower blank than a tube taken from the desiccator when the desiccator has been opened and closed several times. Plots relating the temperature of the sample to the amount of moisture liberated per gram of sample shows that 150°C is the minimum temperature that should be used and that, in the presence of air, oxidation, giving rapid formation of water, starts at 170°C. A heater temperature of 200°C was used for polyethylene.

Within the above concentration ranges, the reproducibility given by the gas chromatographic method is 15 $\pm$ 5 per cent relative.

METHOD 2 - DETERMINATION OF WATER IN POLYALKENES. SEMI-AUTOMATED KARL FISCHER TITRATION METHOD.

2.1 SUMMARY

This method is specific for water in polyalkenes. Other volatile constituents in the polymer do not interfere.

2.2 APPARATUS

Apparatus is shown in Figures 7.2 and 7.3 and component parts are listed in Table 7.1.

The instrument operates by means of a standard Karl Fischer dead-stop indicator circuit, with a moving coil relay replacing the microammeter. The moving coil relay operates as a switch, in conjunction with a second relay, to deactivate (at the end-point) and to activate (when water is present) an automatic titrant dispenser, which dispenses a given volume of titrant into the titrand each time it is activated.

METHOD 3 - DETERMINATION OF SANTONOX R PHENOLIC ANTIOXIDANT IN POLYETHYLENE. 2,2′-DIPYRIDYL SPECTROPHOTOMETRIC METHOD.[2]

3.1 SUMMARY

A method for the determination of various types of antioxidants in polyethylene as exemplified by the determination of Santonox R phenolic antioxidant, (Note 1).

3.2 APPARATUS

Miscellaneous glassware. Round-bottomed flasks 100 cm^3, vertical condensers, volumetric flasks 100 cm^3 and 10 cm^3 (black painted), pipettes 25 cm^3 and 1 cm^3, filter funnel 3 in. diameter.

Spectrophotometer. Unicam SP 600 or equivalent instrument. Water bath thermostatted at 25 $\pm$ 0.5°C.

3.3 REAGENTS

Toluene, "Analar" redistilled

Ethanol, Absolute,

2,2′-dipyridyl, 0.5%; dissolve 0.5g 2,2′dipyridyl in 100 ml absolute alcohol.

Ferric chloride 0.2%: dissolve 0.2g ferric chloride hexahydrate "Analar" in 100 ml absolute ethanol, (Note 3)..

To avoid photochemical reduction of this reagent by sunlight, store in an amber glass bottle which is covered with black paper. Renew the reagent daily.

3.4 METHOD

The method involves oxidation of the antioxidant under controlled conditions with an absolute ethanol solution of ferric chliride:

$$\text{A reduced} + Fe^{3+} = \text{A oxidized} + Fe^{2+}$$

This reaction is followed by reaction of ferrous iron produced with 2,2′-dipyridyl to form a coloured complex the intensity of which is proportional to the concentration of antioxidant present in the polymer.

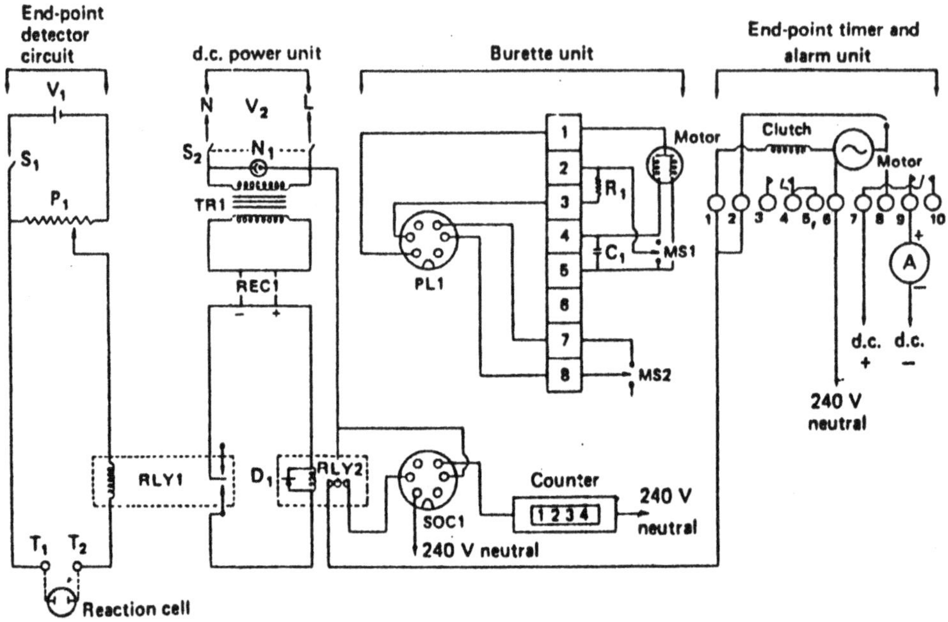

Figure 7.2 Semi automatic Karl Fischer water apparatus, Burette valve construction.

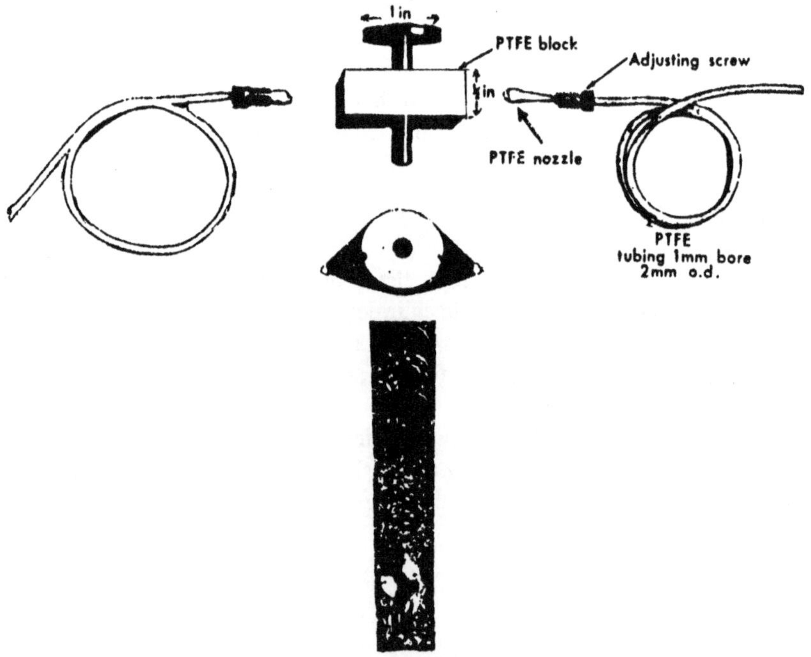

Figure 7.3 Semi automatic Karl Fischer water apparatus, Burette valve construction.

Table 7.1 - Components List (Instrument assembly by ICAM Ltd. Northop, Mold, Flintshire)

Item	Description		Manufacturer
S1	Switch SPDT	)	
S2	Switch SPDT	)	
T1 and T2	Insulated terminals	)	Radiospares Limited
N1	Panel neon clear 240V	)	P.O. Box 2BH
TR1	Transformer Hygrade 240V	)	4 - 8 Maple Street
	50 cycles 2 x 6.3V	)	London, W.1.
REC1	Rectifier Rec 20	)	
RLY2	Relay Type 1. 12V d.c. 120 ohms	)	
D1	Diode 10 DE Type REC50A	)	
P1	Potentiometer Model A 10 turn 500 ohms and Duo-Dial Model RB		Beckman, Glenrothes Scotland
RLY1	S170 d.c. relay make at 90 micro-amps resistance 3300 ohms Spec S170/1/457		Sangamo Weston Enfield, Middlesex
PL1 and SOC1	6-pin plug and socket Part No. P194		A.F. Bulgin, Barking Essex
MS1 and MS2	Microswitch Type HA1		Crouzet Ltd, Brentford, Middlesex
V1	1.5V d.c. battery		
V2	240V a.c. supply		
A	Audible alarm Bleeptone 12V d.c.		A.P. Beeson Ltd, Hove, Sussex
Burette Unit	Fisons automatic dispenser includes R1 4700 ohm resistance and C1 0.4 uF condenser		Fisons Ltd Loughborough
PTFE valve block			ICAM Ltd, Northop, Mold, Flintshire
Counter	Reset vending counter Part No. KK1441		Veeder-Root, Croyden
Timer	Chronoset CF 0-36 minutes direct clutch model		Technical Representations Ltd, Stockport, Cheshire
Case	Type DA 40168		Bedco Ltd, Harpenden, Herts.

3.5 EXPERIMENTAL PROCEDURE

Accurately weigh 1 g of sample into a 100 ml round-bottomed flask. Into the flask pipette 25 ml toluene (Note 2) and connect the flask to a vertical condenser. Heat for 90 min on a boiling water bath with occasional swirling. Low density (ie. high pressure) polyethylenes usually completely dissolve during the reflux. High density polyethylenes (ie. Zeigler low pressure) either completely dissolve or disperse sufficiently to allow full extraction of Santonox into the solvent phase.

Whilst the toluene solution is still hot pour 25 - 30 ml absolute ethanol down the condenser to precipitate dissolved polyethylene. Leave the flask to cool, stopper and shake well (refilter if the solution in the volumetric flask is cloudy). Wash the filter paper and polymer with absolute ethanol into the 100 ml volumetric flask until the liquid level reaches the 100 ml mark. Shake the flask contents well.

In a thermostatted ($25 \pm 0.5^{\circ}C$) water bath clamp two dry 10 ml volumetric flasks painted with several layers of black paint (Note 3). Ensure that the flasks are almost completely immersed in water. In one flask pipette 10 cm^3 of the polyethylene extraction solution (sample flask). Into the other flask pipette 10 ml 25:75 (v/v) toluene: absolute ethanol.

Into each flask pipette 0.5 cm^3 of 2,2′-dipyridyl reagent and then 1.0 cm^3 of iron (III) chloride reagent. Start a stopwatch, stopper both flasks and mix well. After 57 min remove both flasks from the water bath and pour into two 1 cm glass spectrophotometer cells (avoiding exposure to direct light as much as possible) and transfer the two cells to the spectrophotometer. Measure the optical density of the sample solution against the blank solution (in the comparison cell) at 60 ± 1 min after starting the stopwatch. Evaluate the solutions under the following spectrophotometric conditions.

Instrument: visible spectrophotometer.
Cells: 1 cm glass.
Blank solution: fill comparison cell with reagent blank as referred to in text.
Wavelength: 520 nm.
Temperature: 25 ± 0.5°C.

3.5.1 Calibration of Method.

Weigh out accurately 0.03g Santonox R into a 100 cm^3 volumetric flask. Make up to the 100 cm^3 mark with 25:75 v/v toluene: absolute ethanol and shake thoroughly to completely dissolve the solid (using a warm water bath to assist solution if necessary). Pipette 5 cm^3 of this solution into a further 100 cm^3 volumetric flask and make up to 100 cm^3 with 25:75 v/v toluene: ethanol. Prepare the dilutions indicated below and proceed with the colour development and measurement of optical density as described above.

To prepare a calibration graph plot optical density against the corresponding weights of Santonox R present in the 10 cm^3 volumetric flask.

	Volume of 0.015 mg/cm^3 Santonox solution/cm^3	Volume of 25:75 v/v toluene:ethanol/cm^3	Mass of Santonox mg
1.	0 (blank)	10	0.00
2.	2	8	0.03
3.	4	6	0.06
4.	6	4	0.09
5.	8	2	0.12
6.	10	0	0.15

3.6 TYPICAL RESULTS

Convert the optical density obtained from the polyethylene extract to milligram Santonox R by means of the calibration graph. Calculate the Santonox R content of the polyethylene as follows:

$$\text{Santonox R (\% w/w)} = \frac{1000}{W}$$

N = Weight (g) of Santonox R in 10 cm^3 polyethylene extract obtained by referring determined optical density to the calibration graph.

W = Weight (g) of polyethylene sample taken for analysis.

A typical application of the procedure is given below, viz. the determination of down to 0.01% of Santonox R (4,4′-thiobis-3-methyl-6-tert-butyl phenol) in polyethylene. As this procedure determines Santonox R only in its reduced form it does not include any Santonox R which may be present in the oxidized form in the original polymer, for example produced by atmospheric oxidation of the additive during polymer processing at elevated temperatures. Total reduced plus oxidized Santonox R can be determined by ultraviolet spectroscopic procedures, for example, the differential procedure described later (Method 6) and oxidized Santonox can then be obtained by difference from the two methods. Alternatively, total unoxidized plus oxidized Santonox R can be determined by the direct ultraviolet spectroscopy.

3.7 DISCUSSION OF RESULTS

Down to 0.01% Santonox in polymer can be determined by this procedure with an accuracy of ± 5%.

3.7.1 Note 1. Compounds which Interfere in the 2,2′Dipyridyl/Ferric Chloride Method.

The following compounds are known to interfere in the procedure for determining Santonox R. The described method may indeed, be used to determine these substances in polyethylene.

Succonox 18	N-stearoyl-p-amino phenol
B.H.T.	Butylated hydroxy toluene
D.C.P.	Di-cresylol propane
Ionol	2,6-di-tert-butyl-p-cresol
Nonox CI	N,N′-di-8-naphthyl-p-phenylene diamine

Compounds used as light stabilizers in polyalkenes do not usually interfere in the described procedure as they are not capable of reducing iron to the ferrous state.

3.7.2 Note 2. Composition of Solvent in which Colour Development is Carried Out.

In the procedure recommended by Metcalf and Tomlinson[2] the polyethylene sample is heated with 10 cm^3 toluene and then made up to 100 cm^3 with ethanol (ie. final solution contains 10% ethanol). This procedure was applied to some polyethylene samples and gave lower Santonox R contents than expected.

The Metcalf procedure was modified by refluxing the polyethylene with 25 cm^3 toluene and making up to 100 cm^3 with ethanol (ie. 25% toluene in final extract instead of 10% toluene as in the Metcalf method).

3.7.3 Note 3. Photochemical Reduction of Iron (III) Chloride Reagent.

Alcoholic solutions of iron (III) chloride are reduced photochemically at an appreciable rate upon exposure to daylight. Reaction of the polyethylene extract with ferric chloride and 2,2′-dipyridyl reagents is carried out, therefore, in black painted volumetric flasks. The iron (III) chloride reagent bottle must also be protected from daylight.

METHOD 4 - DETERMINATION OF PHENOLIC ANTIOXIDANTS IN POLYETHYLENE. P-NITROANILINE COUPLING - SPECTRO-PHOTOMETRIC PROCEDURE.

4.1 SUMMARY

A method for the determination of phenolic antioxidants in amounts down to 0.01% in polyethylene.

4.2 APPARATUS

Underwriters extraction cup
Volumetric flasks 100 ml
Visible spectrometer
Pipettes, 10 ml

4.3 REAGENTS

Ethyl alcohol, 95% sodium hydroxide, reagent grade, 4M; 160 g of sodium hydroxide per litre (distilled water).

o-Nitroaniline, melting point 146 - 147°C.

Sodium nitrite, analytical Reagent grade.

Coupling agent, 2,800 g of p-nitroaniline dissolved in 10 cm^3 of hot concentrated hydrochloric acid and diluted with water to 250 cm^3. After cooling to room temperature, the volume of liquid is adjusted to exactly 250 cm^3. A second solution is made containing 1.44 g sodium nitrite in exactly 250 cm^3 of distilled water. Both of the above solutions are reputed to be stable indefinitely.

25 cm^3 of each of these solutions are pipetted into separate 100 cm^3 beakers and are chilled in ice to below 10°C. The contents are mixed by combining the solutions and pouring them back and forth from one beaker to the other. Pure nitrogen is bubbled through the mixture and it is allowed to warm to room temperature. Finally, add 10 mg urea to destroy excess nitrous acid.

4.4 METHOD

The method is applicable to the determination of a wide range of antioxidants, see Table 7.2.

4.5 EXPERIMENTAL PROCEDURE

2.000 $\pm$ 0.020 g of film or powdered sample is accurately weighed and wrapped with extraction cloth which has been previously extracted to remove sizing, etc. The sample is placed in an Underwriter's extraction cup and extracted for 16 hr with 95%

Table 7.2 - Composition and Absorptivity Date for Phenolic Antioxidants

Antioxidant	Composition	Absorptivity A max-A 700	Wavelength max. mu
AgeRite Alba	Hydroquinone monobenzyl ether	31.48	565
AgeRite Spar	Styrenated phenol	44.06	548
AgeRite Superlite	A polyalkyl polyphenol	23.40	560
Antioxidant 5	Not disclosed	18.81	585
Antioxidant 425	2,2′-Methylene-bis (6-tert-butyl 4-methylphenol)	22.30	585
Antioxidant 2246	2,2′-Methylene-bis (6-tert-butyl 4-methylphenol)	20.60	578
Deenax	2,6,Di-tert-butyl-p-cresol	Does not couple	
Ionol	2,6,Di-tert-butyl-p-cresol	Does not couple	
1-Naphthol	1-Naphthol	120.1	598
2-Naphthol	2-Naphthol	115.1	540
Naugawhite	Alkylated phenol	8.20	580
Nevastain A	Not disclosed	12.44	550
Nevastain B	Not disclosed	6.62	550
Nonyl phenol	Nonyl phenol	36.25	538
p-Phenyl phenol	*p*-Phenyl phenol	80.80	548
Polygard	Tris (nonylated phenyl) phosphite	Must be hydrolysed before it will couple	
Santovar A	2,5-Di-tert-amyl-hydroquinone	Colour too weak	
Santovar O	2,5-Di-tert-amyl-hydroquinone	Colour too weak	
Santowhite crystals	4,4′-Thio-bis (6-tert-butyl-2-methyl phenol)	78.84	565
Santowhite MK	Reaction product of 6-tert-butyl m-cresol and SCl_2	66.94	560
Santowhite powder	4,4′-Butylidene-bis (3-methyl-6-tert-butyl phenol)	Colour too weak	
Solux	N-*p*-Hydroxyphenyl-morpholine	Colour too weak	
Stabilite white powder	Not disclosed	Colour too weak	
Styphen 1	Styrenated phenol	22.61	558
Wingstay S	Styrenated phenol	50.82	545
Wingstay T	A hindered phenol	10.27	590

From Hilton[2]. American Chemical Society

ethanol or methanol. The alcohol extract is transferred to a 100 cm^3 volumetric flask, cooled to room temperature and brought to the mark with the extraction solvent. A 10 cm^3 aliquot is transferred to a 100 ml volumetric flask. Two millilitres of coupling reagent are added. The solution is thoroughly mixed and 3 cm^3 of 4M sodium hydroxide solution are added. The solution is then brought to the mark with 95% ethanol or methanol.

The absorption spectrum is determined from 400 - 700 nm using a suitable spectrophotometer with quartz cells. The colour formation is complete by the time the solution is brought to the mark and is stable for at least 2 h. Ethyl alcohol is used in the reference cell unless the alcohol extract is strongly coloured. In this case, the reference solvent is taken to be a 10 cm^3 aliquot of the ethyl alcohol extract diluted to 100 cm^3 with ethyl alcohol. The absorbance readings are plotted on semilogarithmic graph paper. The per cent antioxidant is calculated using the equation developed for the antioxidant concerned.

4.6 DISCUSSION OF RESULTS

The method is capable of determining down to 0.01% antioxidant in polyethylene with an accuracy of $\pm$ 5%.

METHOD 5 - DETERMINATION OF IONOL PHENOLIC ANTIOXIDANT IN POLYETHYLENE. ULTRAVIOLET SPECTROSCOPY.

5.1 SUMMARY

A method for the determination of Ionol antioxidant in amounts down to 0.01% in polyethylene.

5.2 APPARATUS

Ultraviolet spectrophotometer, recording type.
Sample cells, quartz - 1 cm path length.
Miscellaneous glassware.
Heating mantle, electrical, 250 cm^3 capacity, with variable contol.

5.3 REAGENTS

Chloroform - AR
Ionol CP - dried by vacuum desiccation for 4 h before use.

5.4 METHOD

The absorption maximum occuring at 278 nm of a chloroform extract of the polyethylene is measured and the Ionol content of the sample then deduced from an appropriate calibration curve prepared with pure Ionol.

5.5 EXPERIMENTAL PROCEDURE

5.5.1 Extraction.

Weigh 20.0g of polymer sample into a 250 cm^3 round-bottomed flask and add 50 cm^3 of chloroform. Connect the water-cooled condenser, place on the electric heating mantle and bring gently to boiling point. Reflux for 30 min at a moderate rate. Allow to cool and then carefully decant the chloroform solution into a 100 cm^3 volumetric flask (filter if necessary) and stopper, add a further 40 cm^3 of chloroform to the round-bottomed flask and carry out a second extraction of the polymer. When cool, filter the contents into the volumetric flask containing the first extract. Wash the residue with sufficient chloroform to make up to the mark. Shake the flask to ensure homogeneity.

Record the ultraviolet spectrum of the chloroform extract against a chloroform blank from 250 to 310 nm using 1 cm cells.

Measure the absorbance of the Ionol absorption peak at 278 nm. Determine the concentration of Ionol present in the extract by reference to the prepared calibration curve.

5.5.2 Calibration.

Make up accurately a 0.05% w/v solution of Ionol CP in 250 cm^3 chloroform. Dilute this stock solution to obtain Ionol CP solutions of 0.02, 0.01, 0.025, 0.0003% w/v.

Record the ultraviolet spectrum of each of the above solution against a chloroform blank from 250 to 310 nm using 1 cm cells. Measure the absorbance of the absorption peak at 278 nm by drawing a base-line from approximately 260 to 300 nm.

Plot a calibration curve of absorbance against concentration of Ionol CP in chloroform.

5.5.3 Calculation.

Calculate the concentration of Ionol present in the polymer by means of the following expressions:

% w/w Ionol CP in polymer = 5 x (% w.v Ionol in chloroform extract)

5.6 DISCUSSION OF RESULTS

The method is capable of determining down to 0.01% Ionol in polyethylene with an accuracy of ± 5%. Any other substance in the polymer which absorbs between 250 and 310 nm will interfere.

METHOD 6 - DETERMINATION OF SANTONOX R PHENOLIC ANTIOXIDANT IN POLYETHYLENE. ULTRAVIOLET SPECTROSCOPY.

6.1 SUMMARY

A method for the determination of Santonox R phenolic antioxidant in amounts down to 0.01% in polyethylene.

6.2 APPARATUS

Apex Mill
Spectrophotometer
Water bath
1 cm silica cells
Volumetric flasks, 50, 100, 1000 ml
Condensers with ground glass neck
Filter funnels

Pipettes, 5, 10, 20 ml
Graduated cylinders, 25 ml
Filter paper, Green's No. 802 or similar

6.3 REAGENTS

Santonox R, 4,4'-thio-bis-(6-tert-butyl-3-methylphenol). Standard Santonox R solution: dissolve 0.1000 g Santonox R in cyclohexane in a 1000 cm^3 standard flask and dilute to the graduation mark.

Cyclohexane - spectroscopic grade.

Sodium hydroxide pellets - Analar, 42 gl^{-1}

6.4 METHOD

Santonox R is extracted from a finely ground sample with boiling cyclohexane. The extract is then shaken with aqueous sodium hydroxide. Santonox R forms a sodium salt in the alkali layer. This layer is then examined on an ultraviolet spectrophotometer. The base line absorbance is calculated and referred to a previously prepared calibrated graph relating weight of Santonox R to base line absorbance. Ionol does not interfere in the determination

6.5 EXPERIMENTAL PROCEDURE

6.5.1 Analysis of sample.

Grind a representative sample in the Apex Mill. Weigh about 1 g of the sample into a 100 cm^3 round-bottomed flask and add 20 cm^3 cyclohexane. Fit a condenser to the flask and allow the cyclohexane to reflux on a water bath for 1 h. Wash down the condenser with 20 cm^3 cyclohexane, remove the flask from the bath, cool to room temperature and shake well. Filter through a No. 802 filter paper into a 100 cm^3 separating funnel. Wash the filter paper with a further 10 cm^3 cyclohexane. Add 25 cm^3 freshly prepared sodium hydroxide solution, shake for 3 min then allow to settle. Run off the aqueous layer into a 100 cm^3 standard flask. Repeat the extraction with another 2 x 25 cm^3 portions of alkali, make the extract up to 100 ml with alkali and mix well. Read the optical density in 1 cm silica cells at the following wavelengths, 335, 266 and 236 nm and use sodium hydroxide solution as a blank. Calculate the baseline absorbance from the following equation.

$$A_{BL} = A_{266} - A_{335} - 0.684\,(A_{236} - A_{335})$$

Where A_{BL} = base line absorbance.

A_{335}, A_{266}, A_{236} = absorbance at 335, 266 and 236 nm respectively.

Read the concentration of Santonox from the calibration graph.

6.5.2 Calibration.

Put accurately measured portions of 5, 10, 15 and 20 cm^3 standard Santonox R solution into 50 cm^3 standard flasks and dilute to the graduation mark. Transfer each into a 100 cm^3 separating funnel and dilute to graduation mark. Transfer each into a 100 cm^3 separating funnel and add 25 cm^3 freshly prepared sodium hydroxide solution. Shake for 3 min, allow to settle, then run the lower layer into a 100 cm^3 standard flask. Repeat the extraction with another 2 x 25 cm^3 portions of alkali. Make up the volume to 100 cm^3 and mix thoroughly. Read against sodium hydroxide in 1 cm silica cells at the following wavelengths - 335, 266, 236 nm. Calculate the base line absorbance as above. Plot the base line absorbance against the concentration of Santonox R in mg/100 cm^3.

6.5.3 Calculation of Results.

Calculate the Santonox R content as follows:

$$\text{Weight \% Santonox R} = \frac{B}{W}$$

where B = Santonox R content of extract in mg/100 ml
W = weight of sample (g)

Note 1. The optical density of the cyclohexane should not exceed 0.15 at 247 nm.

Note 2. The sodium hydroxide must be Analar - and carbonate free. Carbonate adhering to the pellets may be removed by washing the pellets with distilled water.

6.6 DISCUSSION OF RESULTS

The method is capable of determining down to 0.01% Santonox R in polyethylene with an accuracy of ± 5%. Any other substance in the polymer which absorbs between 236 and 335 nm in sodium hydroxide extract will interfere. Ionol does not interfere.

METHOD 7 - DETERMINATION OF 2,6-DITERT-BUTYL-4-METHYLPHENOL AND 4 SUBSTITUTED 2,6-XYLENOL PHENOLIC ANTIOXIDANTS IN POLYPROPYLENE. DERIVATIVE ULTRAVIOLET SPECTROSCOPY.[4]

7.1 SUMMARY

Derivative ultraviolet spectroscopy is used to determine down to 0.01% 2,6 ditert-butyl-4-methylphenol and 4 substituted 2,6 xylenol phenolic antioxidants in poly-propylene.

7.2 APPARATUS

A spectrophotometer with a microcomputer differentiation module and a digital output or equivalent can be used for measurements of UV absorption spectra. The second derivative - difference spectra were measured under the following conditions: response, 2 s; slit width, 2 nm; and scan speed, 10 nn $min.^{-1}$ The samples were placed in quartz cells of 10 mm path length.

7.3 REAGENTS

Heptane - UV grade, free from aromatics, purified by treatment with concentrated sulphuric acid and distilled. Check purity by UV spectroscopy.

Propan-2-ol. Analytical reagent grade, distilled and absence of interference checked by UV spectroscopy.

Solution of potassium hydroxide in propan-2-ol. This was prepared by saturating propan-2-ol with potassium hydroxide at elevated temperature. After cooling, the solution was filtered and standardised by titration with 0.1 M hydrochloric acid using phenolphthalein as indicator. The solution was stored in the cold and dark. The solution was prepared freshly every month.

2,6-xylenol (substituted in the 4 position by a ca. C_{45} hydrocarbon chain) and 2,6-Di-*tert*-butyl-4-methylphenol. Heptane solutions of both these phenolic antioxidants were stable for 2 months under laboratory conditions.

7.4 METHOD

In non-alkaline medium the two antioxidants have virtually identical ultraviolet spectra. In alkaline medium the spectra are sufficiently different to permit the two substances to be determined without mutual interence.

The principle of the 4-substituted 2,6-xylenol method is as follows. First, the phenolic component without steric hindrance is determined by means of second derivative - difference spectroscopy of the bathochromically shifted, basified extract using the non-basified extract as a reference. The second derivative spectral amplitudes of a non-basified extract are additive for both antioxidants. The contribution of 4-substituted 2,6-xylenol to the total second derivative amplitude is then subtracted and the residual corrected amplitude corresponds to the 2,6-di *tert*-butyl-4-methyl phenol content.

7.5 EXPERIMENTAL PROCEDURE

7.5.1 Extraction of Phenols from Polypropylene Sheet.

Samples of 0.5 mm thick 50 x 50 mm polypropylene sheets were weighed and refluxed for 3 h in 50 ml of boiling heptane.

7.5.2 Preparation of Solutions for Spectroscopic Measurements.

The heptane extract was transferred quantitatively into a 50 cm^3 calibrated flask and made up to the mark with heptane (solution A). The following solutions were then prepared.

Solution of phenolics (solution B). Two millilitres of propan-2-ol plus 8 cm^3 of solution A.

Solution of phenolates (solution C). Two millilitres of a solution of potassium hydroxide in propan-2-ol (0.005 mol cm^3) plus 8 cm^3 of solution A.

Blank (solution D). Two millilitres of propan-2-ol plus 8 cm^3 of heptane.

7.5.3 Calibrations and Calculations.

Solutions containing 2,6-*tert*-butyl-4-methyl phenol in the concentration range c_1 = 30 - 150 mg dm^3 in a heptane-propan-2-ol mixture (4 + 1 v/v) are measured using the second derivative of the absorption spectra. The peak amplitude, D_1, at wavelengths of 285 and 292 nm is read off (Figure 7.4) as a function of phenol concentration. The calibration constant, $K_1 = D_1/C_1$ was found to be constant (10 x 10^{-4} mg^{-1}) (standard deviation (SD) = 0.13 x 10^{-4} mg^{-1}).

The maxima of the second derivatives of the absorption spectra of the two components do not exhibit identical positions. Therefore, calibration was carried out using the method of standard additions in order to eliminate the absorbance due to the 2,6 xylenol component. A mixture of both components with concentrations C_{-1} (2,di-*tert*-butyl-4-methylphenolic) and C_2 (2,6-xylenol) in the range 30 - 120 mg dm^3 was prepared. The peak amplitude, D_t, at wavelengths of 285 - 292 nm was measured in the form of the second derivative of the absorption spectrum of the mixture (Figure 7.4). The calibration constant, $K_3 = (Dt - C_1K_1)/C_2$, was found to be 5.2 x 10^{-4} mg^{-1} dm^3 (SD = 0.17 x 10^{-4} mg dm^{-3} from nine measurements at two constant levels of 2,6-di-*tert*-butyl-4-methylphenol concentrations (32 and 48 mg dm^{-1}), using known additions of 2,6 ditertbutyl-4-methylphenol. Table 7.3 shows the results for C_1 = 32 mg dm^{-3}. The derivative amplitude, D_3 for 2,6 xylenol is a linear function of its concentration and passes through the origin.

The 2,6-xylenol content is determined using the second derivative - difference absorption spectrum of its phenolate as the peak amplitude, D_2, at wavelengths of 302.0 - 312.5 nm. The calibration constant, $K_{-2} = D_2/C_2$, was found to be 5.71 x 10^{-4} mg^{-1} dm^3 (SD = 0.07 x 10^{-4} mg^{-1} dm^3) calculated from six measurements). The positions of the maxima on the second derivative curves are influenced by the mode of derivation (electronic, microcomputer) and by the setting of the spectrophotometer controls.

7.5.4 Calculation of the Antioxidant Concentrations in Extract.

The 4-substituted 2,6-xylenol concentration is calculated from the phenolate spectrum of the extract:

$$C_2 = D_2/K_2$$

The 2,6-di-tert-butyl-4-methylphenol concentration is calculated from the phenol spectrum of the extract:

$$C_1 = (D_t - C_2K_3)/K_2$$

To calculate the concentrations of the antioxidants in the polymer, the mass of the polypropylene sample, the volume of the extract and its dilution must be known.

7.6 DISCUSSION OF RESULTS

The relative standard deviation of this procedure at the 0.1 - 0.2% w/w additive in polymer level is between 7 and 8%.

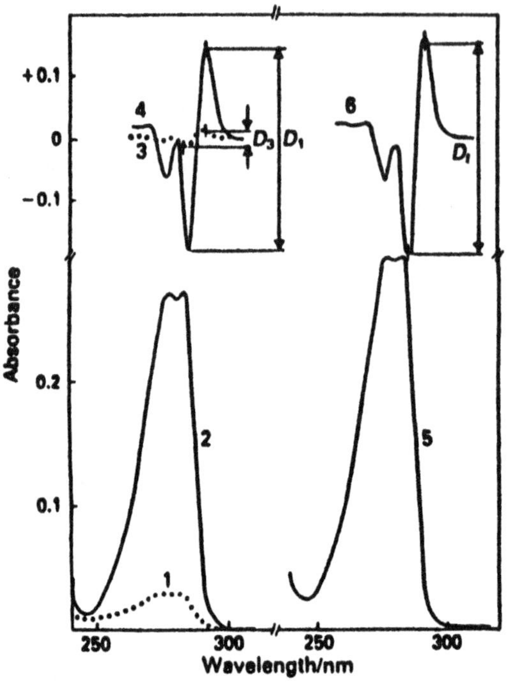

Figure 7.4 - Normal and derivative spectra of antioxidants. Concentration of each oxidant, 30 mg l^{-1}. 1. Normal spectrum of Chemantox AO-49 (4 substituted 2.6 xylenol). 2. Normal spectrum AO-4K 2.6 ditert butyl-4-methyl phenol. 3. Second-derivative spectrum of Chemantox AO-49. 4. Second-derivative spectrum of a mixture of Chemantox AO-49. 5. Normal spectrum of a mixture of Chemantox AO-49 and AO-4K. 6. Second derivative spectrum of a mixture of Chemantox AO-49 and AO-4K. Conditions for measurement of normal spectra, slit width 2 nm, scan speed, 40 nm min^{-1}. Conditions for measurement of second-derivative spectra: d (wavelength), 4 nm, slit width 2 nm, scan speed 10 nm min^{-1}. Concentrations correspond to extraction of antioxidants from 1 g of a stabilised sheet in 50 ml of heptane. From Soncek and Jelinkova with permission.[4] Royal Society of Chemistry, London.

Table 7.3 - Data for calculation of K_3 and for checking the linearity of the product K_3C_2

AO-4K: c_1/mg l^{-1}	AO-49: c_2/mg l^{-1}	Dt absorbance units	$D_3 = Dt - D_1$ absorbance units	K_3/mg^{-1} l
32	-	0.332	-	-
32	16	0.341	0.009	5.6×10^{-4}
32	32	0.349	0.017	5.3×10^{-4}
32	48	0.357	0.025	5.2×10^{-4}
32	64	0.365	0.033	5.2×10^{-4}
32	128	0.399	0.067	5.2×10^{-4}

From Soncek and Jelinkova[4]. Royal Society of Chemistry, London

METHOD 8 - DETERMINATION OF TOPANOL OC, BINOX M AND IONOX 330 PHENOLIC ANTIOXIDANTS IN POLYETHYLENE. ULTRAVIOLET SHIFT METHOD.[5]

8.1 SUMMARY

This is a low interference method for the determination of down to 0.01% Topanol OC, Binox M and Ionox 330 phenolic antioxidants in polyethylene

8.2 APPARATUS

A recording ultraviolet spectrophotometer capable of measurements over the range 200 to 700 nm.

8.3 REAGENTS

Ethanol - spectroscopic grade absolute ethanol.

Ethanolic potassium hydroxide solution. Dissolve 10 g of analytical reagent grade potassium hydroxide in 10 cm^3 of distilled water and dilute the solution to 100 cm^3 with absolute ethanol.

Nickel peroxide - B.D.H. laboratory reagent grade.

8.4 METHOD

Ultraviolet measurements (250 - 450 nm) are carried out on the following four solutions:

1. ethanol extract of polymer
ii. alkaline ethanol extract of polymer
iii. ethanol extract of polymer after treatment with nickel peroxide
iv. alkaline ethanol extract of polymer after treatement with nickel peroxide

These four spectra are compared with spectra obtained with synthetic standard solutions of the antioxidants, similarly treated, and the antioxidant contents of the polymer deduced.

8.5 EXPERIMENTAL PROCEDURE

Prepare the polymer compound containing the unknown stabiliser by cutting or grinding to small pieces and extract about 5 g by heating under reflux for 24 h with 50 ml of ethanol. Cool the extract to room temperature and filter off the polymer. Treat a solvent blank of 50 ml of ethanol identically.

Measure the ultraviolet absorption spectrum of the extract over the range 250 to 450 nm with the solvent blank in the comparison beam. Dilute the extract if necessary, so that the absorbance of the main band is about 0.6.

To 10 cm^3 of this solution add 1 cm^3 of distilled water and 2 cm^3 of ethanolic potassium hydroxide solution. Mix well and measure the ultraviolet absorption spectrum of this alkaline solution over the same range against a solvent blank similarly treated.

Transfer 20 cm^3 of the extract to a 50 cm^3 flask, add about 0.5 g of nickel peroxide, stopper the flask and shake, with a mechanical shaker for 5 m. Filter the solution and measure the absorption spectrum as before, over the range 250 to 450 nm with a solvent blank in the comparison beam. If necessary, because of the high absorption, dilute the filtered reaction products before measurement and note the dilution used. If the solutions are coloured, also measure the absorption spectrum over the range 450 to 700 nm.

Add 2 drops of ethanolic potassium hydroxide solution to the above cell solution, mix well and re-measure the absorption spectrum over the same range as before.

The above procedures give four spectra for each extract, which are then compared with the spectra obtained by carrying out the procedures with known stabiliser solutions.

8.6 DISCUSSION OF RESULTS

The three antioxidants can be individually determined in amounts down to 0.02% with an accuracy of $\pm$ 10%. There are no mutual interference effects.

METHOD 9 - DETERMINATION OF SANTONOX R PHENOLIC ANTIOXIDANT IN POLYOLEFINS. THIN LAYER CHROMATOGRAPHY.

9.1 SUMMARY

A thin layer chromatographic method for the determination of down to 0.002% of Santonox R in polyolefins.

9.2 APPARATUS

Thin layer chromatography tank.

Thin layer plate 20 cm x 20 cm, pre-coated with a 250 u layer of Merck.

CF254 silica gel, U.K. Distributors - Anderman & Co Ltd., Battlebridge House, Tooley Street, London SE1.

Hypodermic syringe - 25 ul capacity.

Ultraviolet lamp - wavelength 254 nm.

9.3 REAGENTS

Chloroform, analar, British Drug Houses.

Petroleum ether, 40/60 Analar, British Drug Houses.

Santonox R - (4,4'-thiobis-3-methyl 6-6-butylphenol).

Ethyl acetate, Analar, British Drug Houses.

2,5-dibromo-p-benzoquinone-4-chlorimine, British Drug Houses.

Sodium tetraborate (borax), Analar, British Drug Houses.

9.4 METHOD

A toluene extract of the polymer is spotted on to a thin layer plate which is then developed with a petroleum ether:ethyl acetate mixture (5:1). The developed plate is sprayed with 2,6-dibromo-p-benzoquinone-4-chlorimine solution to reveal coloured spots which are proportional in intensity to the Santonox R content of the original polymer.

9.5 EXPERIMENTAL PROCEDURE

9.5.1 Polyethylene Extraction Procedure.

Accurately weigh 5 g of polymer into a 500 ml round-bottomed flask. Add 65 ml of toluene and heat on a boiling water bath for 90 min using a reflux condenser.

Remove the flask from the water bath and immediately pour 85 ml of absolute ethanol down the condenser to precipitate the dissolved polyethylene from the hot toluene solution. Allow the flask to cool to room temperature, remove the condenser, stopper the flask and shake well. Filter the solution through a No. 42 Whatman filter paper into a 250 ml beaker. Wash the flask and residue with a further 100 ml of ethanol.

Evaporate the toluene/ethanol solution almost to dryness on a boiling water bath with the aid of a stream of nitrogen and finally to dryness using only the nitrogen stream. If more polymer is precipitated during the evaporation, refilter the solution into a smaller beaker. Wash the residue into a 2 ml volumetric flask with chloroform and make up to the mark.

Santonox R is extracted from polyethylene by precipitation of the polymer with ethanol from hot toluene solution as described above. The extract is evaporated to dryness, dissolved in chloroform and an aliquot applied to a thin-layer plate and the chromatogram developed. The concentration of Santonox R in the polyethyolene is estimated by visually comparing the intensity of the spot obtained with corresponding spots from known quantities of Santonox R.

Place 20 ul of the chloroform extract of the polyethylene sample as a spot on a thin layer plate. Also apply 20 ul aliquots of standard solutions of Santonox R in chloroform, 0.05%, 0.04%, 0.03%, w/v. Develop the chromatogram to a distance of 10 cm in a chromatography tank containing petroleum ether (40:60)/ethyl acetate (5:1 v/v mixture) as the eluent.

Inspect the plate under ultraviolet light (254 nm) and compare the intensity of the Santonox R spot from the polyethylene extract with the intensities of the standard spots. If the spot is of lower intensity than that of the 0.03% w/v standard than a new chromatogram should be developed using standards, 0.025%, 0.015% and 0.005% w/v of Santonox R in chloroform.

Visually compare under an ultraviolet lamp the intensity of the spot from the polyethylene extract with the standard spots and thus estimate the percentage level of Santonox R in the chloroform solution. Spray the thin-layer plate with a 2% w/v ethanol solution of 2,6-dibromo-p-benzoquinone-4-chlorimine. Allow the thin-layer plate to dry and re-spray using a 2% w/v aqueous solution of borax. Re-estimate the amount of Santonox R present in the chloroform solution by visually comparing the purple spots produced.

9.5.2 Calculation of Results

Santonox R from 5 g of polyethylene is concentrated into a 2 ml chloroform solution.

Thus the weight % of Santonox R in high density polyethylene = 2/5 x the level of Santonox R present in the 2 ml chloroform solution (as determined by thin-layer chromatography in % w/v).

9.6 DISCUSSION OF RESULTS

Santonox R can be determined in polyolefins in amounts down to 0.002% with an accuracy of ± 20% of the determined amount.

METHOD 10 - DETERMINATION OF PHENOLIC ANTIOXIDANTS IN POLYALKENES. THIN LAYER CHROMATOGRAPHY.[6]

10.1 SUMMARY

A general thin-layer chromatographic method for the determination of 0.02% of a range of phenolic antioxidants in polyalkenes.

10.2 APPARATUS

Soxhlet extraction apparatus, 250 ml.

Extraction thimble, F & S no. 603 size 33 x 94 mm.

Thin layer plates. Silica Gel G Uniplates (8 in x 8 in) and Silica Gel G Uniplates, reverse phase 5% Dow Silicone (8 in x 8 in) or equivalent. The plates are given a short equilibration time in the development solvent vapours.

Capillary pipet, 20 ml.

Chromatographic chamber. Inner diameter, $10^3/_8$ in x $2^3/_4$ in x $10^1/_4$ in with glass cover. The chamber contains a 8 in x 8 in sheet of Whatman no. E-17 filter paper to maintain a saturated solvent atmosphere during development.

Chromatography sprayer.

Densitometer. 0.0 - 1.0 optical density wedge with 620 nm filter.

10.3 REAGENTS

Extraction solvent. n-heptane:-n-octane 4:1 v/v.

Detection system. Phosphomolybdic acid (3.0g) was dissolved in a 100 cm^3 volumetric flask with ethanol and diluted to volume with water. Filter if necessary to remove any solids prior to use. The solution must be prepared fresh every two days.

Development solvent. System A (for identification of antioxidants). 240 volumes of ethanol were added to 80 volumes of distilled water. System B (for quantitative analysis). 300 volumes of practical grade cyclohexane were added to 6 volumes of reagent grade methanol.

10.4 METHOD

A n:hepane-n-octane (4:1 v.v) extract of the polymer is applied to a thin-layer plate which is then developed with ethanol, water or cyclohexane-methanol development solvents. Spraying the plate with phosphomolybdate revealed spots due to separated antioxidants. The method was calibrated against standard solutions of antioxidants.

10.5 EXPERIMENTAL PROCEDURE

10.5.1 Extraction Procedure.

The polyethylene or polypropylene film, (5.0 g precut into strips approximately 1 in x 1 in) is weighed into the extraction thimble, 130 ml of the extraction solvent and a few glass beads are added to the 250 cm^3 flat bottomed flask, the extraction apparatus is connected and heated to boiling. After 3 h the system is allowed to cool to room temperature and the solvent drained into the 250 cm^3 flat bottom flask. The extraction thimble and extraction apparatus are rinsed down with 25 - 30 cm^3 of the extraction solvent and drained into a 250 cm^3 flask. The contents of the 250 cm^3 flask are quantitatively transferred to a 250 cm^3 Pyrex glass beaker and concentrated to a 25 - 30 cm^3 volume by evaporation on a hot plate then quantitatively transferred to a 50 cm^3 beaker, concentrated to 6 - 7 ml, and transferred to a 10 cm^3 volumetric flask and diluted to volume with the extraction solvent.

10.5.2 Chromatographic Procedure.

Two 20 ul aliquots of both the sample and the synthetic standard are applied to the thin-layer plate, the spots are dried with a heat gun and the chromatograms eluted for 30 - 40 min in the development solvent. The resulting chromatogram is dried with a heat gun, sprayed with the detection reagent, redried, exposed to ammonium hydroxide vapors and scanned on the photodensitometer.

10.6 TYPICAL RESULTS

The areas of each zone were calculated by triangulation (area calculated by multiplying the height times the width at half height). The quantitative results are calculated from the following ratio:

$$\% \text{ antioxidant in the polyolefin} = \frac{A' \times P}{A}$$

where A = the average numerical area recorded for the synthetic standard;
A′ = the average numerical area recorded for the sample;
P = actual percentage of the antioxidant present in the synthetic standard based upon the weight of the sample.

The method was applied to the following six antioxidants:

4,4′-butylidene (2-tert-butyl-5-methyl) phenol; 4,4-thiobis (6-tert-butyl-m-cresol); pentaerythritol tetrakis (3,5-di-tert-butyl-4-hydroxyhydrocinnamate); 2,2′-methaylenebis

(4-methyl-6-tert-butylphenol); octadecyl (3,5-di-tert-butyl-4-hydroxyphenol) acetate; 2,6-di-tert-butyl-o-cresol.

In Table 7.4 are reproduced R_f values obtained by Dobies[6] when 20 ul of standard solutions (100 mg/200 ml) of the above mentioned antioxidants were applied to thin-layer plates and eluted with the development solvent.

In Table 7.5 are shown the results obtained by Dobies[6] in applying his method over a three day period to samples of polyethylene and polypropylene containing approximately 0.1%. 2.2′-methylenebis (4-methyl-6-tert-butylphenol), compound with results obtained by an ultraviolet spectrophotometric method. Although there are minor differences in the results of the methods, the thin-layer procedure shows the presence of an additional zone, attributed to the presence of 2,6-di-tert-butyl-p-cresol. The presence of this antioxidant, although not determined by the thin-layer chromatographic procedure, could interfere with the ultraviolet procedure thus giving higher results.

10.7 DISCUSSION OF RESULTS

The method has been used to determine down to 0.02% of six antioxidants in polyalkenes with an accuracy of $\pm$ 10%. As mentioned above, there exists a risk of interference effects by other antioxidants. Consequently, the method must be applied with caution.

METHOD 11 - DETERMINATION OF PHENOLIC AND AMINE ANTIOXIDANTS, ULTRAVIOLET ABSORBERS AND ORGANOTIN STABILIZERS IN POLYALKENES AND OTHER POLYMERS. THIN LAYER CHROMATOGRAPHY.[7]

11.1 SUMMARY

This thin-layer method is capable of determining a wide range of polymer additives, viz amine antioxidants, uv absorbers and organotin stabilizers in amounts down to 0.02% in polyalkenes.

11.2 APPARATUS

Thin layer plates, Merck.

Kieselgel GF254 (silica gel containing a fluorescent additive, thickness C.25 mm).

11.3 REAGENTS

Benzene-ethyl acetate - acetone (100 + 5 + 2) used for antioxidants.

Chloroform - hexane (2 + 1) - used for ultraviolet absorbers.

Butanol - glacial acetic acid (97 + 3) - used for organotin compounds.

2,6-Dichloro-*p*-benzoquinone-4-chlorimine, 0.2 per cent, w/v solution in absolute ethanol.

Table 7.4 - R_f Values of Antioxidants

System A: Silica Gel G, reverse phase 5% Dow Silicone; developing solvent, ethanol-water.

System B: Silica Gel G Uniplates; developing solvent, cyclohexane-methanol

Antioxidant	R_f Values System A	R_f Values System B
4,4'-Butylidenebis (2-tert-butyl-5-methyl) phenol	0.80	0.0
4,4'-Thiobis (6-tert-butyl-m-cresol)	0.84	0.0
Pentaerythritol tetrakis (3,5-di-tert-butyl-4-hydroxyhydrocinnamate)	0.60	0.29
2,2'-Methylenebis (4-methyl-6-tert-butyl-phenol)	0.76	0.34
Octadecyl (3,5-di-tert-butyl-4-hydroxy-phenyl) acetate	0.33	0.70
2,6-Di-tert-butyl-p-cresol	0.71	0.72

From Dobies[6]. Elsevier Science Publishers, Netherlands

Table 7.5 - Replicate Determination of Approximately 0,1% 2,2'-methylenebis (4-methyl-6-tert-butylphenol) in Polyethylene and Polypropylene

Sample	% Antioxidant 1st day	2nd day	3rd day	Average analysis	By U.V. analysis
Polyethylene	0.095, 0.094 0.096, 0.092	0.103, 0.100	0.087, 0.089 0.087	0.094	0.102, 0.105
Polypropylene	0.080, 0.084 0.073, 0.086	0.079, 0.078	0.078, 0.078 0.079	0.079	0.083, 0.083

From Dobies[6] Elsevier Science Publishers, Netherlands

Borax, 2 per cent, w/v solution, dissolve 2 g of sodium tetraborate decahydrate in 100 cm^3 of 5 per cent, v/v ethanol.

Catechol violet, 0.1 per cent, w/v solution in 95 per cent ethanol.

Diethyl ether (Analar).

Chloroform (Analar).

Ethanol (Analar).

11.4 METHOD

A diethyl ether extract of the polymer is applied to a thin-layer plate which is then developed with various solvent mixtures. The plate is then sprayed with a range of reagents specific for the compounds to be determined and the coloured spots compared visually with standards to determine the concentration of polymer additives.

11.5 EXPERIMENTAL PROCEDURE

Weigh two 2 - 5 g amounts of plastics material and extract each with diethyl ether for 8 h in a Soxhlet continuous extraction apparatus.

Remove the diethyl ether by distillation on a water-bath and add 5 cm^3 of ethanol to each flask. Allow the flasks to stand for 10 minutes. Evaporate the contents of one flask to dryness (sample A) and those of the other to very small bulk (sample B).

Dissolve sample A in a small amount of chloroform and make up to volume in a 5 cm^3 calibrated flask. Transfer sample B quantitatively to a 5 cm^3 calibrated flask and dilute to volume with ethanol. Use sample A solution for the determination of antioxidant and ultraviolet absorber, and sample B solution for the determination of organotin compounds.

Spot 2 ul of solution A on to each of two Merck Kieselgel GF254 plates and 2 ul of solution B on to a third plate. Place each plate into a chromatogram sandwich chamber and, by using the appropriate mobile phase, allow the solvent to travel a suitable distance. Locate the spots as described below:

11.5.1 Determination of Antioxidant.

Mobile phase - Benzene - ethyl acetate - acetone (100 + 5 + 2), Chromogenisis - View the plate under ultraviolet light and mark any spots located. Then spray the plate with 2,6-dichloro-p-benzoquinone-4-chlorimine solution, heat it with a drier for a few minutes and mark any spots coloured by the reagent. Finally, allow the plate to stand in a tank in contact with iodine vapour for 5 min.

11.5.2 Determination of Ultraviolet Absorbers.

Mobile phase - Chloroform - hexane (2 + 1), Chromogenesis - View the plate under ultraviolet light and mark any spots located. Spray the plate with borax solution followed by 2,6-dichloro-p-benzoquinone-4-chlorimine solution, heat it with a drier for a few minutes and mark any spots coloured by the reagents.

It was possible to detect down almost to 1 ug per spot applied in certain cases and at least 10 ug per spot applied.

11.6 TYPICAL RESULTS

In Table 7.6 are tabulated some R_f values obtained by this method for various types of polymer additives.

11.7 DISCUSSION OF RESULTS

The method is capable of determining down to 0.05% of a wide range of additives in polyalkenes with an accuracy of $\pm$ 10%.

Table 7.6 - Results Obtained for Each Additive in Each Mobile Phase

Substance	R_f value in Benzene-ethyl acetate-acetone-hexane	R_f value in chloroform-hexane
Antioxidants:		
Nonox SP	0.61, 0.93	0.50
Nonox TBC	0.98	0.62
Nonox WSP	0.85	0.52
Nonox EX	0.56	0.38
Nonox WSL	0.82	0.54
BHT	0.98	0.91
2246	0.75	0.56
Nonox CI	0.65, 0.84	0.49
Nonox DPPD	0.66	0.48
Nonox OD	0.56	0.40
Nonox ZA	0.43, 0.38	0.38
Santoflex	0.15, 0.61, 0.74	0.47, 0.54
Santoflex AW	0.80, 0.51, 0.55	0.54, 0.45
Santoflex R	0.26, 0.53	0.39
DLTP	0.61	0.41
Nonox NS	0.52	0.4
Superlite	0.64, 0.93	0.58
Polygard	0.52, 0.61	0.53
Nonox HO	0.46, 0.64, 0.60, 0.99	0.52
Nonox WSO	0.58	0.39
Irganox 1076	0.57, 0.75	0.56
Ultraviolet absorbers:		
Eastman DOBP	0.76	0.39
Uvinul 400	0.22	0.08
Permyl B100	0.24	0.08
Salol	0.73	0.45
Tinuvin HE	-	0.71
Tinuvin P	0.76	0.59
Eastman OPS	0.79	0.65
UV 318	0.26	0.14
Cyasorb 1988	0.26	0.12
Cyasorb 1084	0.24	0.08
Uvinul N 35	0.28	0.16

From Simpson and Currell[7]. Royal society of Chemistry, London

METHOD 12 - DETERMINATION OF IRGANOX 1076, IRGANOX 1010 AND BUTYLATED HYDROXYTOLUENE PHENOLIC ANTIOXIDANTS IN POLYETHYLENE. HIGH PERFORMANCE LIQUID CHROMATOGRAPHY.[8]

12.1 SUMMARY

This high performance liquid chromatographic method is capable of determining down to 0.001% of Irganox 1076, Irganox 1010 and butylated hydroxy toluene antioxidants in polyethylene.

12.2 APPARATUS

Waters Model 204 liquid chromatograph equipped with two Model 6000 A pumps, a Model 660 solvent programmer and a U6K injector. Elution was monitored with a Waters Model 450 variable wavelength detector set at 280 nm and a 10 mv strip chart recorder.

Column - 3.9 mm x 30 cm u-Porasil column packed with 10 um porous silica obtained from Waters Associates, Milford, Mass.

Thermolyne type 1000 stir plates (Sargent Welch).

Stir bars $^3/_8$ inch o.d. x $1^1/_2$ inch Teflon coated magnetic stir bars.

Sample filtering apparatus - a Waters 20 - 30 nm stainless steel solvent reservoir filter was connected to about a 2" length of 3 mm i.d. Teflon tubing. The other end of the Teflon tubing was connected to a $1^1/_2$ inch long blunt 16 gauge Luer lock needle with a $^1/_{16}$ inch stainless steel nut and ferrule at the end of the needle. The needle was connected to a Hamilton No. 1010 W gastight 10 ml syringe with Teflon plunger.

12.3 REAGENTS

Heptane, spectrograde.

Chloroform, Mallinckrodt AR grade, Scientific Products.

Methylene chloride, distilled in glass.

The above mobile phase solvents were all filtered through Millipore Type F-H 0.5 um filters prior to use.

12.4 METHOD

A decalin extract of the polymer was chromatographed on a program set to gradient elute with mixtures of heptane and methylene dichloride. Detection of eluted compounds was achieved by a uv detector. The method was calibrated against appropriate standard solutions.

12.5 PROCEDURE

Heated standard solution: A 50 ml portion of a standard solution containing about 0.02 mg/ml each of BHT, Irganox 1076 and Irganox 1010 is pipetted into a 100 ml beaker. A stirring bar is added and the solution is heated to 110°C with a gentle stirring for 30 min. The solution is transferred to a cool stirrer and cooled to room temperature. This heated and cooled standard solution is used to obtain quantitative data on the sample extract solutions.

About 2 g of polyethylene pellets are weighed into a 100 ml beaker. A stirring bar is added and 50 ml of decalin is pipetted into the beaker. The mixture is heated at 110°C on a hot plate with gentle stirring for about 30 minutes or until dissolution is complete. The beaker is then transferred to a cool stirrer and cooled to room temperature with stirring to precipitate the polyethylene.

The precipitated polyethylene from the above extraction is pushed aside with a microspatula. The porous metal filter portion of the filter apparatus is inserted into the solution and about 5-10 ml of solution is drawn into the syringe. The Teflon tube is removed from the ferrule on the needle and the filtered solution is dispensed into a small vial. The filter apparatus is rinsed with acetone and dried between samples. After extensive use, the metal filter became partially clogged and is regenerated by placing it in hot decalin and stirring.

The Model 660 solvent programmer is set at Program 6 (linear) going from 100% heptane to 100% methylene chloride in 5 min. The total flow rate is 3 ml/min. The Model 450 UV detector is set at 0.2 or 0.4 absorbance unit sensitivity and the recorder chart speed is 1 cm/min. Duplicate injections of 100 ul of each of the standard and sample solutions are made. The mobile phase gradient is started at the point of injection.

12.6 TYPICAL RESULTS

The retention volumes (VR) for BHT, Irganox 1076 and Irganox 1010 are 10.3, 14.8 and 22.2 ml respectively. The amount of each additive is determined from each sample injection by comparing peak heights for samples and standards. A blank decalin injection is made to determine from what points on the base line, peak heights should be measured. Gradient reset is instantaneous, from 100% methylene chloride to 100% heptane. Sample injection could be made any time after the appearance of a refractive index peak from the UV detector, signifying the emergence of heptane from the column.

Excellent antioxidant recoveries are obtained by this procedure, Table 7.7.

12.7 DISCUSSION

The method is very sensitive, being capable of determining down to 0.001% of the three antioxidants with an accuracy of ± 10%. As always with chromatographic methods the possible effects of interfering substances must be considered.

METHOD 13 - DETERMINATION OF NONOX CI AMINE ANTIOXIDANT IN POLYALKENES. ACID HYDROGEN PEROXIDE SPECTROPHOTOMETRIC METHOD.[9]

13.1 SUMMARY

This spectrophotometric method is capable of determining down to 0.01% of Nonox CI amine antioxidant in polyalkenes.

13.2 APPARATUS

(a) Miscellaneous

Conical flasks B24, 250 ml, T-piece with B19 cone and side arc.
Reflux condenser, Liebig type 16 in with B25 cone and B19 socket.
Volumetric flasks, 25 cm^3, 100 cm^3.
Pipettes, 2 ml, 20, 50 cm^3.
Rubber suction bulb.
Filter funnels, 3 in.
Glass rod $^1/_8$ in diameter.
Wash bottle (all glass).
Porous pot broken into approximately $^1/_{16}$ in pieces.

Hotplates.

(b) Spectrophotometer
Visible and or visible/uv instrument

13.3 REAGENTS

Toluene - redistilled B.D.H. Laboratory reagent grade (sulphur free).

Ethanol - redistilled industrial methylated spirit.

Hydrogen peroxide reagent. Dilute 25 cm^3 of 20% v sulphuric acid solution and 4 ml 100 vol. hydrogen peroxide (Analar) to 100 cm^3 with distilled water and shake. Allow to stand for 2 hours before use. Prepare this reagent fresh daily.

Natural polyethylene (powder or nibs), antioxidant free.

Pure Nonox CI for calibration purposes - add 50 cm^3 redistilled industrial methyl alcohol to approximately 5 g of commercial "Nonox" and warm to 60°C with constant stirring. Filter the mixture through a Whatman No. 4 filter paper. Thoroughly wash the solid remaining on the paper with 6 x 40 cm^3 portions of hot alcohol as quickly as possible. Reject the highly coloured filtrate. Allow the paper to drain for 2 to 3 min and transfer to a vacuum desiccator. Apply vacuum until paper and solid are absolutely dry. Store the pure "Nonox" under nitrogen in a brown glass bottle.

13.4 METHOD

A toluene extract of the polymer is reacted with ethanolic hydrogen peroxide to produce a coloured compound with an absorption maximum of 830 mm proportional to the Nonox CI content of the polymer. The method is calibrated against solutions of purified Nonox CI.

13.5 EXPERIMENTAL PROCEDURE

13.5.1 Extraction of Antioxidant from Polymer.

Weigh a suitable amount (eg 0.5 - 1 g if 0.1% Nonox present) of polyethylene powder, shavings or nibs, usually not exceeding 0.6 g into a 250 ml conical flask.

Pipette 50 cm^3 toluene into the flask and add 2 g of broken porous pot. Purge the interior of the flask with nitrogen and connect immediately to a condenser which has been previously purged with nitrogen. Connect a T-piece to the open end of the condenser and pass a gentle purge of nitrogen through the T-piece during the reflux period. Reflux gently for one hour. Allow the contents of the flask to cool to approximately 40°C and add 20 cm^3 of ethanol to reprecipitate any dissolved polymer. Filter the contents of the conical flask through a Whatman No. 54 filter paper into a 100 ml volumetric flask and make up to 100 ml with ethanol.

13.5.2 Determination of Antioxidant.

Pipette 20 cm^3 of the polyethylene extract into a 25 cm^3 volumetric flask. Pipette in 2 cm^3 ethanol and 2 cm^3 hydrogen peroxide reagent and adjust the solution temperature to 25 $\pm$ 3°C and make up to the 25 cm^3 mark with ethanol and mix well.

Table 7.7 - Recoveries of BHT, Irganox 1076 and Irganox 1010 from three Polyethylene Samples

Sample[a]	amount g	Amount Added, mg BHT	Irganox 1076	Irganox 1010	Amount Found, mg BHT	Irganox 1076	Irganox 1010	Percent Recovered BHT	Irganox 1076	Irganox 1010
A	1.96	1.02	1.00	1.01	1.04	1.08	1.03	130	108	102
A	2.01	1.02	1.00	1.01	1.06	1.10	1.13	104	110	112
B	2.02	1.01	1.00	1.01	1.02	1.07	0.99	100	107	98
B	1.97	1.02	1.00	1.01	0.99	1.09	0.94	98	109	93
C	1.97	1.02	1.00	1.01	1.04	1.05	0.98	102	105	97
C	2.04	1.02	1.00	1.01	1.07	1.06	1.05	105	106	104

From Schabon and Fenska[a]. American Chemical Society

After 25 to 30 min fill a spectrophotometer cell with the test solution. Record the optical density at 5 min intervals until the maximum reading is obtained. The maximum optical density is usually obtained between 40 min and 120 min after addition of the hydrogen peroxide reagent depending on the concentration of Nonox in the test solution.

Spectrophotometer conditions.

Cells: 4 cm glass.

Blank solution: 1:1 (v/v) toluene:ethanol.

Wavelength: 830 nm.

13.5.3 Preparation of Calibration Graph.

The procedure is calibrated against Nonox which has been purified by elution with ethanol. Weigh out 0.0250 $\pm$ 0.00002 g purified Nonox into a dry 100 cm^3 volumetric flask. Add 50 cm^3 toluene and swirl in a water bath until the solid is dissolved. Cool to room temperature, make up to 100 cm^3 with toluene and mix well. Use this stock solution of the day of preparation. Transfer 0.2, 0.5, 1.0, 2.0, 3.0, 4.0 and 5.0 cm^3 aliquots, (ie. containing 50 to 1250 micrograms Nonox) of this solution to dry 250 cm^3 conical flasks and add sufficient toluene to make the volume up to 50 cm^3. To the contents of each flask add approximately 0.3 g of natural antioxidant free polyethylene powder or nibs and make up to 100 ml with ethanol. Continue the colour development on 20 ml of this solution as described in Section 13.5. Plot a calibration graph relating optical density readings obtained and the number of micrograms of Nonox present in the 20 cm^3 aliquots of test solution used for colour development.

13.5.4 Calculation.

From the calibration graph read off the number of micrograms of Nonox corresponding to the maximum optical density reading obtained. The concentration of Nonox in the polyethylene is given by:

$$\text{Nonox \% (w/w)} = \frac{10^2 \times 10^2 \times 10^{-6} \times P}{20 \times W}$$

where P = weight of Nonox read off from the calibration graph (microgram)

W = weight of polyethylene taken for analysis (g).

13.6 DISCUSSION

This method is capable of determining down to 0.01% Nonox CI in both low and high density polyethylene with an accuracy of $\pm$ 10% of the determined Nonox CI content.

METHOD 14 - DETERMINATION OF AMINE ANTIOXIDANTS IN POLYALKENES AND OTHER POLYMERS. P-NITROANILINE COUPLING - SPECTROPHOTOMETRIC PROCEDURE.[10]

14.1 SUMMARY

This spectrophotometric method is capable of determining down to 0.01% of a wide range of amine antioxidants in polyalkenes.

14.2 APPARATUS

All of the absorption measurements were made on a visible spectrophotometer using quartz cells having a light path of 1 cm.

14.3 REAGENTS

Methanol - pure synthetic methanol (99.85%).
Hydrochloroic acid - Analar, concentrated.
p-Nitroaniline - micro analytical reagent.
Sodium nitrite - reagent grade.

Coupling agent - 2.800 g of *p*-nitroaniline is dissolved in 32 cm^3 of hot concentrated hydrochloric acid and the solution is diluted with distilled water to 250 cm^3. After cooling to room temperature the volume of liquid is adjusted to 250 cm^3. A second solution is made containing 1.44 g of sodium nitrite in 250 cm^3 of distilled water. Twenty-five cm^3 of each of these solutions are pipetted into separate 100 cm^3 beakers, mixed and chilled in ice to below 10°C. Pure nitrogen is bubbled through the mixture as it warms to room temperature. Finally, 10 mg of urea (or 1 cm^3 of 0.1 g/10cm^3 solution) is added to destroy any excess nitrous acid. Fresh reagent should be made every day.

Methanol -hydrochloric acid solvent. Three volumes of methanol plus one volume of concentrated hydrochloric acid.

14.4 METHOD

An ethanol or methanol extract of the polymer is reacted with diazotized *p*-nitroaniline to produce a colour in the 350 - 700 nm range proportional to the amine antioxidant content of the polymer. The colour is evaluated spectrophotometrically at its absorption maximum and compared with calibration standards.

14.5 EXPERIMENTAL PROCEDURE

The sample to be analysed must be very thinly sheeted or powdered. Weigh a 1.000 ± 0.0005 g sample. Wrap with extraction cloth which has been previously extracted to remove sizing, etc. Place in an Underwriter's extraction cup and extract for 16 hr with 95% ethanol or methanol. Transfer the alcohol extract to a 100 cm^3 volumetric flask. Cool to room temperature and bring to the mark with the extraction solvent. Transfer a 10 cm^3 aliquot to a 100 cm^3 volumetric flask. Add 15 cm^3 methanol-hydrochloric acid solution and 1 cm^3 of coupling agent. Place in dark for 1.5 h and then bring to the mark with methanol-hydrochloric acid. Determine the absorption spectrum from 350 - 700 nm. Measure the optical chemistry at the absorption maximum. Calibrate the methanol against amine antioxidant standards prepared in ethanol or methanol, treated as above and evaluated at the same wavelength.

14.6 TYPICAL RESULTS

Amine antioxidants which can be determined:

Agerite Excel	Flectol H
Agerite Gel	Flexamine
Agerite Resin	Neozone A
Agerite Resin D	Neozone C
Agerite Stalite	Octamine
Akroflex C	Polylite
Albasan	Stabilite
Aminox	Stabilite ALBA
Antox	Stabilite L
Aranox	Thermoflex

Amine antioxidants which cannot be determined include:

Agerite White	Santoflex 35
Antioxidant 4010	Santoflex 75
BXA	Tenamine 2
Eastozone 32	Tenamine 30
JZF (DPPD)	Tonox
OZO 88	Tonox D
Santoflex AW	U.O.P. 88
Santoflex BX	U.O.P. 288
Santoflex DD	

14.7 DISCUSSION

The method is capable of determining down to 0.02 to 0.05% of amine antioxidants in polyalkenes with an accuracy of ± 5%. The colour obtained with phenyl-ß-naphthylamine antioxidant fades too rapidly to enable it to be determined.

METHOD 15 - DETERMINATION OF DIORGANOSULPHIDE AND TERTIARY PHOSPHITE ANTIOXIDANT IN POLYETHYLENE. M-CHLOROPEROXYBENZOIC ACID OXIDATION METHOD.[11]

15.1 SUMMARY

This titration method is capable of determining down to 0.05% of diorganosulphide and tert phosphite type of antioxidants in polyethylene.

15.2 APPARATUS

Titration burette, 10 ml.

15.3 REAGENTS

Heptane, chloroform, sodium iodide, glacial acetic acid and isopropanol.

m-chloroperoxybenzoic acid, technical grade (85%) (available from Aldrich Chemical Co., main impurity being m-chlorobenzoic acid).

Oxidizing solution, dissolve 1.6 g of m-chloroperoxybenzoic acid in chloroform and dilute to 200 cm^3 with chloroform.

Sodium iodide solution, prepared by saturating iso-propanol with the solid salt at refluxing temperature.

Sodium thiosulphate 0.05N.

15.4 METHOD

The antioxidant is reacted with an excess of *m* chlorophenoxybenzoic acid.

$$(RO)_3P + ClC_6H_4COOOH = (RO)_3PO + ClC_6H_4COOH \qquad 7.1$$

$$(ROOCCH_2CH_2)_2S + 2\ ClC_6H_4COOOH = (ROOCCH_2CH_2O)_2S + 2\ ClC_6H_4COOH$$

Unused m-chloroperoxybenzoic acid is then reacted with potassium iodide in acidic medium to produce iodine which is estimated by sodium thiosulphate titration.

$$Cl\ C_6H_4COOOH + 2\ HI = Cl\ C_6H_4\ COOH + I_2 + H_2O$$

The amount of *m*-chloroperoxybenzoic acid consumed by a known weight of the sample can be used to calculate the concentration of antioxidant present.

15.5 EXPERIMENTAL PROCEDURE

The polyolefin sample was ground in a Wiley Mill to pass through a 60 mesh screen. Extract a 10 - 20 g portion of sample in a Soxhlet apparatus with heptane for a minimum of 6 h. Transfer the extract to a 300 cm^3 beaker, evaporate to approximately 20 cm^3 and then dilute to 40 cm^3 with chloroform. Pipette ten milliliters of the m-chloroperoxybenzoic acid solution into the beaker and allow to react for 45 min at room temperature. Pipette 2 cm^3 glacial acetic acid into the beaker followed by 100 cm^3 of dry isopropanol and 10 cm^3 of the sodium iodide-saturated isopropanol. Allow 60 min for complete decomposition of the excess peroxide. Add 25 cm^3 of water and titrate with 0.05 M sodium thiosulfate to the disappearance of iodine colour. A blank consisting of the oxidant and all reagents used was run along with the samples.

15.6 TYPICAL RESULTS

Calculations. For sulfide antioxidants, the calculations to obtain percentage present in the original polymer sample was:

$$\% \text{ sulfur-containing antioxidant} = \frac{(cm^3)\,(N)\,(MW)\,(0.1)}{(\text{sample wt})\,(4)}$$

Weight per cent phosphite compound present in the original polymer sample was calculated as:

$$\% \text{ phosphite antioxidant} = \frac{(cm^3)\,(N)\,(MW)\,(0.1)}{(\text{sample wt})\,(2)}$$

where cm^3 is the milliliters of thiosulfate used for the sample subtracted from the milliliters used for the blank, M is the molarity of the sodium thiosulfate, and MW is the molecular weight of the antioxidant being determined which divided by 4 for sulfides and by 2 for phosphites places the equation of a molar basis and allows the calculation of percentage. Recoveries of the following antioxidants were usually in excess of 99%:

Distearyl thiodipropionate
Dilauryl thiodipropionate
4,4′-Thiobis (6-tert-butyl-m cresol)
1,1′-Thiobis (2-naphthol)
Triphenyl phosphite
Triethyl phosphite
Tri-p-tolyl phosphite
Tris(dinonyl phenyl)-phosphite
2,2′-Thiobis (6-tert-butyl-p-cresol)
Tri isopropyl phosphite

15.7 DISCUSSION

The method is capable of determining down to 0.05% of diorganosulphide and tertiary phosphite type antioxidants in polyalkenes with an accuracy of $\pm$ 5%. Table 7.8 shows the results obtained by applying the oxidation method to blends of antioxidants in unstabilized polyethylene. The results are in excellent agreement with the theoretical antioxidant content of these polymers.

METHOD 16 - DETERMINATION OF DIORGANOSULPHIDE ANTIOXIDANTS IN POLYALKENES. GAS CHROMATOGRAPHY

16.1 SUMMARY

This gas chromatographic method is capable of determining diorganosulphide secondary antioxidants in polyalkenes in amounts down to 0.02%.

16.2 APPARATUS

Dual column chromatograph equipped with flame ionisation detectors and an isothermal column oven. The oven was operated isothermally at 165°C and the injection port at 300°C.

Column - 6 foot x 1 inch glass column, packed with 1.5 per cent of fluorisilicone oil FS-1265 on 80 to 100 mesh Chromosorb W AW-DMCS.

Argon carrier gas, flow rate of 30 ml min^{-1}.

16.3 REAGENTS

Dilauryl ßß'-thiodipropionate.

Dilauryl sulphenyl- ßß-dipropionate - prepared by oxidation of dilauryl ßß'-thiodipropionate with chromic acid in an acetic acid medium to 60 to 80°C (melting point 74^{oC}).

Dilauryl sulphonyl- ßß'dipropionate - prepared from dilauryl- ßß'-thiodipropionate by oxidation with hydrogen peroxide.

n-Octadecane - analytical reagent grade quality.

Chloroform - analytical reagent grade quality.

Methanol - analytical reagent grade quality.

Potassium hydroxide - analytical reagent grade quality.

Standard lauryl alcohol solution 0.07% in chloroform.

n-Octadecane solution - 0.2% in chloroform.

Standard potassium hydroxide, 5 M solution in methanol.

Diorganosulphide calibration solution. Place 5 cm^3 of solution containing 0.1 to 3 mg of dilauryl ßß'-thiodipropionate, dilauryl sulphenyl- ßß'-dipropionate or dilauryl sulphonyl- ßß'-dipropionate in a test tube provided with a B29 ground glass joint and evaporate off the solvent under a stream of nitrogen. Add 1 cm^3 of freshly prepared methanolic potassium hydroxide solution, fit a reflux condenser and immerse the tube in a heating bath at 80°C. After heating for 30 min, transfer the contents of the tube into a 25 cm^3 separating funnel with 10 cm^3 of water. Finally, rinse the tube with 2 ml of chloroform and add the rinsings to the water in the funnel. Shake the funnel well and when the layers have separated collect the bottom layer in a 10 cm^3 calibrated flask. Repeat the extraction a further three times, using 1.5 cm^3 of chloroform for each run. Add 1 cm^3 of n-octadecane solution to the combined extracts in the flask and make the solution up to 10 cm^3 with chloroform.

Table 7.8 - Analysis of typical Polypropylene Samples with Various Stabilizers Present

Additive	Per Cent Added	Per Cent Found
Distearyl thiodipropionate	0.20	0.19
Distearyl thiodipropionate	0.40	0.36
Distearyl thiodipropionate	0.10	0.095
4,4'-Thiobis (6-tert-butyl-m-cresol)	0.10	0.096
4,5'-Thiobis (6-tert-butyl-m-cresol)	0.20	0.19
Distearyl thiodipropionate	0.20	0.39 total
4,4'-Thiobis (6-tert-butyl-m-cresol)	0.20	0.30 total
Distearyl thiodipropionate	0.20	0.30 total

From Kellum[11]. American Chemical Society

16.4 METHOD

A chloroform extract of the polymer is treated with methanolic potassium hydroxide to hydrolyse the diorganosulphide antioxidant to lauryl alcohol. n-octadecene internal standard is added and the resulting solution gas chromatographed to determine lauryl alcohol which is proportional in amount to the diorganosulphide content of the original polymer. The procedure is calibrated against standard lauryl alcohol solutions that have been put through the whole test regime.

16.5 EXPERIMENTAL PROCEDURE

Pipette 50 cm^3 of chloroform into a 250 cm^3 conical flask and add 0.6 g of polyolefin sample. Reflux gently for 1 h. Allow the contents to cool to 40°C and add 20 cm^3 of ethanol to precipitate dissolved polymer. Carefully transfer the flask contents via Whatman No. 54 filter paper into a 100 cm^3 volumetric flask and make up to 100 cm^3 with ethanol.

To 5 cm^3 of this solution add 1 cm^3 of 5M methanolic potassium hydroxide and heat to 80°C for 30 min. Transfer this solution to a 25 cm^3 separating funnel using 10 cm^3 of distilled water. Run off the lower chloroform layer into a 10 cm^3 volumetric flask together with three 1.5 cm^3 chloroform washings of the aqueous phase. Add 1 cm^3 of n-octadecene in chloroform internal standard and make up to 10 cm^3 with chloroform.

Calibrate against the appropriate hydrolysed diorganosulphide calibration solution referred to under Reagents.

16.6 DISCUSSION OF RESULTS

This method is capable of determining down to 0.02% diorganosulphide antioxidants in polyalkenes with an accuracy of $\pm$ 5%. Because of the specificity of the gas chromatographic method there is little liklihood of interferences.

METHOD 17 - DETERMINATION OF CYASORB UV 531 ULTRAVIOLET ABSORBER IN POLYETHYLENE. INFRARED SPECTROSCOPY.

17.1 SUMMARY

Cyasorb UV 531 ultraviolet absorber (2-hydroxy-4-n octoxybenzophenone) is determined in polyethylene by infrared spectroscopy of a polymer extract in amounts down to 0.02%.

17.2 APPARATUS

Double beam spectrometer covering the 15 - 17 micron region (eg. Grubb Parsons GS2A).

Hydraulic press with heated and water-cooled platens.

Stainless steel moulding plates (6" x 6" x $^1/_{16}$").

Shims 0.06 cm thick (circular 1" diameter or rectangular 1" long).

Aluminium foil.

Clear plastic rule calibrated in millimetres.

Dial gauge calibrated in 0.01 mm divisions.

17.3 EXPERIMENTAL PROCEDURE

17.3.1 Preparation of Sample Film.

Cover two stainless steel moulding plates with aluminium foil and place up to six 0.06 cm thick shims on one of these. Place approximately 0.2 g of polymer sample in the centre of each shim and carefully place the second moulding plates on top. Position the two plates in the press and apply contact pressure. Switch on the heating supply and set the thermostat to 120°C. When the temperature reaches 120°C increase the pressure to 3000lb/in^2, switch off the heating supply and water cool to room temperature. Carefully strip off the aluminium foil from the polymer films and push out the films from the shims. Carefully strip off the aluminium foil from the polymer films and push out the films from the shims. Check the thickness of each film by means of the dial gauge. Six readings on each film should not vary by more than 0.03 mm. Reject any which has air bubbles, is uneven or is wedge-shaped. Shape the film to fit the spectroscopic sample holder and gently scrape one of the surfaces with a fine emery board to produce a series of fine parallel lines. This reduces the incidence of interference fringes.

17.3.2 Recording the Infrared Spectrum.

Place the film in the sample holder and position in the infrared instrument so that the beam passes through the film at right angles to the scratch marks.

Record the infrared spectrum from 15.5 to 16.5 micron using a scanning speed of $^1/_2$ micron per minute.

Before removing the film from the instrument, mark the position of the infrared beam. Remove the sample from the holder and measure the thickness to the nearest 0.01

mm by means of the dial gauge at six points within the marked area. Calculate the mean of these six measurements.

17.3.3 Measurement of Absorbance.

Remove the chart from the spectrometer and with a sharp pencil rule a base-line from approximately 15.8 micron to approximately 16.2 micron. With a ruler measure I and I_0 to the nearest 0.1 mm at the wavelength of the peak maximum. Calculate the absorbance at 15.94 micron and hence the absorbance per unit thickness by means of the following expression:

$$\text{Absorbance per unit thickness} = \frac{\log_{10} I_0}{\text{film of thickness (in cm)}}$$

Prepare duplicate films from the standard sheets containing 0.1, 0.3, 0.5 and 1.0 wt%. UV 531 as above. Record the infrared spectrum as described above and calculate the absorbance per unit thickness as described under measurement of absorbance. Construct a calibration curve by plotting absorbance per unit thickness against percentage weight UV 531 for each standard film. Use this calibration curve to obtain the UV 531 content of the polyethylene sample.

17.4 TYPICAL RESULTS

This method is capable of determining down to 0.02% Cyasorb UV 531 ultraviolet absorber in polyethylene with an accuracy of $\pm$ 5%.

METHOD 18 - DETERMINATION OF CYASORB UV 531 ULTRAVIOLET ABSORBER IN POLYOLEFINS. THIN LAYER CHROMATOGRAPHY.

18.1 SUMMARY

This thin-layer chromatographic method is capable of determining Cyasorb UV 531 ultra-violet stabilizer in polyolefins in amounts down to 0.02%.

18.2 APPARATUS

Thin layer chromatography tank.
Thin layer plate 20 cm x 20 cm pre-coated with a 250 u layer of GF_{245} silica gel.
Hamilton syringe (25 ul)
Ultraviolet lamp (wavelength 254 nm).
Sintered glass filter sticks, $^1/_2$" diameter.
10 ml standard flasks.
Ultraviolet spectrophotometer
Silica spectrophotometer cells, 3 cm.

18.3 REAGENTS

Methylene chloride.
Chloroform A.R.
Cyasorb UV 531 ex Cyanamid Ltd.
Standard UV 531 solution, 0.2% w/v in chloroform, for use as a "marker" to locate correct position of UV 531 on plate.
Standard UV 531 polymer blends - prepare by milling a series of blends of 0.05, 0.1, 0.2, 0.3, 0.4 and 0.5% w/w UV 531 in additive-free polyethylene powder.

18.4 EXPERIMENTAL PROCEDURE

UV 531 is extracted from the polymer by precipitation of the polymer with ethanol from a hot toluene solution. An aliquot of the extract is applied to a thin layer plate and chromatographed. The zone corresponding to UV 531 is removed from the plate and extracted with ethanol. The concentration of UV 531 is determined by measuring the ultraviolet absorption peak near 295 nm in ethanol solution and referring to a calibration graph.

To calibrate the procedure, weigh out 0.5 g of each of the standard polymer - UV 531 blends and 0.5 to 1.0 g of the unknown polyethylene sample and put each through the polyethylene solvent extraction procedure (toluene-ethanol) as described in Method 3.

Apply 25 ul portions of each of the standard polymer extracts and the sample extract to the base line of a thin layer plate in the form of a spot. Develop the chromatogram to a distance of 10 cm by using methylene chloride as eluant. Remove the plate from the tank and allow the solvent to evaporate.

Inspect the plate under ultraviolet light (254 nm) and locate UV 531 as a dark blue spot. Mark with a pointed marker the zones corresponding to the UV 531 in the standards and the sample and also an equivalent blank area of silica gel. Carefully scrape the silica gel zones from the plate and transfer to separate sintered glass filter columns. Extract the UV 531 from each with absolute ethanol into a 10 ml standard flask and finally make to volume.

Measure the optical density at 295 nm of the standard and sample extracts against the TLC plate blank extract in the reference cell in 3 cm silica cells.

Plot a calibration curve for net optical density at 295 nm against micrograms UV 531 in final solution (10 ml). Hence, deduce the concentration of UV 531 in the unknown polyethylene sample.

18.5 DISCUSSION OF RESULTS

This method is capable of determining Cyasorb UV 531 in polyethylene in amounts down to 0.02% with an accuracy of $\pm$ 10%.

METHOD 19 - DETERMINATION OF TINUVIN 326 ULTRAVIOLET ABSORBER IN POLYPROPYLENE. ULTRAVIOLET SPECTROSCOPY.[12]

19.1 SUMMARY

This ultraviolet spectroscopic procedure determines Tinuvin 326 ultraviolet absorber in polypropylene in amounts down to 0.02%.

19.2 APPARATUS

Ultraviolet spectrophotometer, recording type. Sample cells, quartz - 1 cm path length.

Water bath

Miscellaneous glassware.

19.3 REAGENTS

Chloroform AR

Tinuvin 326

19.3 METHOD

The Tinuvin 326 is extracted from the polymer with chloroform under reflux conditions. The concentration of Tinuvin 326 in the chloroform extract is determined by measuring the ultraviolet absorption peak at 355 nm and by reference to a prepared calibration curve.

19.5 EXPERIMENTAL PROCEDURE

19.5.1 Procedure

Weigh accurately into a 250 cm^3 round-bottomed flask a portion of the polypropylene sample, containing about 5 mg Tinuvin 325. Add 50 cm^3 of chloroform, place the flask under a reflux condenser and boil for 30 min. Decant the cooled solution into a 100 cm^3 volumetric flask after filtering through a No. 42 Whatman filter paper. Transfer any polymer collected in the filter paper to the round-bottomed flask and add 40 cm^3 of chloroform. Repeat the extraction procedure and combine with the first extract. Make up to the 100 ml mark with chloroform. Shake the solution well and then pipette 5 cm^3 into a 50 ml volumetric flask and make up to the mark with chloroform.

Record the ultraviolet spectrum of the diluted chloroform extract from 300 to 370 nm against a chloroform blank using 1 cm cells.

19.5.2 Calibration.

Dilute the 0.20% w/v chloroform solution of Tinuvin 326 to obtain solutions of 0.001, 0.0008, 0.0004, 0.0002% w/v.

Record the ultraviolet spectrum of each of these solutions against a chloroform blank from 300 mu to 380 nm using 1 cm cells.

Measure the absorbance of the Tinuvin 326 absorption peak at 355 nm. Determine the amount of Tinuvin 326 present in the diluted extract by reference to a prepared calibration curve.

$$\% \text{ w.w Tinuvin 326 in polymer} = \frac{(\% \text{ w/w Tinuvin 326 in diluted extract} \times 1000}{\text{weight of original polymer}}$$

19.6 DISCUSSION OF RESULTS

This method is capable of determining Tinuvin 326 ultraviolet absorber in amounts down to 0.02% with an accuracy of $\pm$ 5%.

METHOD 20 - DETERMINATION OF AROMATIC VOLATILES IN POLYSTYRENE. GAS CHROMATOGRAPHY PROCEDURE.

20.1 SUMMARY

This gas chromatographic method is capable of determining down to 0.001% of a wide range of saturated and unsaturated aromatic volatiles in polystyrene.

20.2 APPARATUS

Gas chromatograph with a hydrogen flame ionization detector and an injection port fitted with a glass liner, very loosely packed with glass fibre (Figure 7.5).

Column, copper tube (15 ft x $^1/_4$ in o.d. x $^3/_{16}$ in i.d.) packed with 10 per cent w/w Carbowax 15 - 20 m supported on 60 - 72 B.S. mesh Celite.

Gas pressures and flows:

Helium	30 lbf/in^2 gauge, 100 ml/min
Air	7 lbf/in^2 gauge, 650 ml/min
Hydrogen	12 lbf/in^2 gauge, 75 ml/min

Temperatures:

Injection	150°C
Detector	150 °C
Column	80 °C
Flame	200 °C

Recorder, 0.2 to + 1.0 mV range, 1 sec response, 15 in (381 mm)/hr chart speed.

20.3 REAGENTS

Carbowax 15 - 20 M and 60 - 72 B.S. mesh Celite

Styrene and propylene oxide, Ortho-, meta- and para-xylenes, cumene,n-propyl benzene, iso-butyl benzene, alpha-methyl styrene and mixed meta-/para-diethyl benzenes, tert-, sec- and n-butyl benzenes, ortho-methyl styrene, mixed meta-/para-methyl styrenes and n-undecane, meta- and para-ethyl toluenes, ortho-ethyl toluene.

20.4 EXPERIMENTAL PROCEDURE

20.4.1 Calibration.

Weigh in turn into a 1- cm^3 volumetric flask 1.0 cm^3 n-undecane, internal standard plus 1.0 cm^3 of any of the above aromatic hydrocarbons it is required to determine. Dilute to 10 cm^3 with propylene oxide and then further dilute 0.25 cm^3 of the solution to 25 ml with propylene oxide. Chromatograph 5 ul of the calibration blend. This calibration procedure should be carried out daily.

To analyse samples, weigh 1 g polymer into a 25 cm^3 stoppered measuring cylinder, add exactly 10 cm^3 of the propylene oxide containing 0.1% v/v n-undecane internal standard, seal with a serum cap and shake until the polymer has dissolved. Gel, pigment and filler may remain as an undissolved suspension without detriment to the analysis. Chromatograph 5 ul of the solution at a range setting of x 10 and suitable attenuation settings.

20.4.2 Relative Retention Distances.

Retention distances (injection point to peak centres) are corrected for the gas hold-up of the column and are expressed relative to styrene.

20.5 TYPICAL RESULTS

Calculation. In the calibration and analysis chromatograms, measure the peak heights of the n-undecane and the aromatic hydrocarbons allowing for any attenuation factors.

From the calibration chromatogram, determine the response factor for each component as follows:

A = weight of component in calibration blend
B = weight of n-undecane in calibration blend
C = peak height of component on calibration chromatogram
D = peak height of n-undecane on calibration chromatogram

$$\text{Component response factor } f = \frac{A \times D}{B \times C}$$

From the analysis chromatogram, determine the concentration of each component as follows:

F = component response factor
G = peak height of component on analysis chromatogram
H = peak height of n-undecane on analysis chromatogram

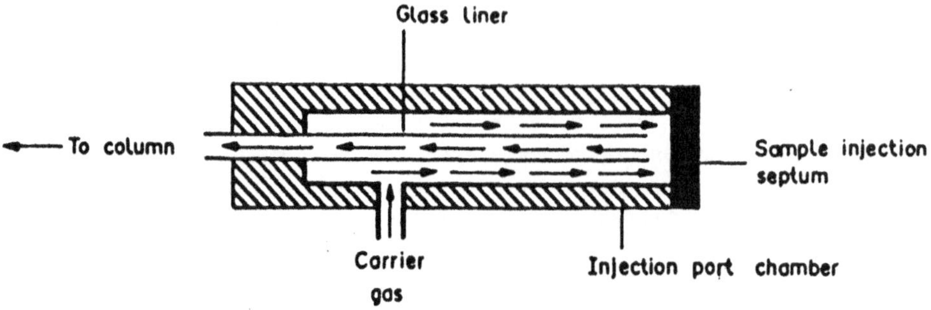

Figure 7.5 - Injection port glass liner fitted to F & M Model 1609 gas chromatograph. The glass liner measures 60 mm x 40 mm o.d. x 2 mm i.d. and is very loosely packed with glass fibre.

J = per cent w/v n-undecane in polymer solvent
K = weight of polymer taken

$$\text{Component (per cent w/w)} = \frac{10 \times G \times F \times J}{H \times K}$$

Report results to two significant figures.

The above method of calculation applies only to those components for which calibration factors have been determined. The remaining components are either determined coincidentally with styrene (sec-butyl benzene) or are likely to be present in such small amounts (o-ethyl toluene, iso-*tert*-butyl benzenes, m-diethyl benzene, n-butyl benzene/para-diethyl benzene and o/m/p methyl styrenes that they may be determined by comparing directly their peak areas with that of styrene or n-undecane.

Table 7.9 shows some results obtained for various types of polystyrene.

Using this method several polystyrenes from five different manufacturers were heated at 200°C for 15 min under helium, and the liberated volatiles were examined by gas chromatography. All the samples liberated the same range of aromatic hydrocarbons, these differing only in their relative concentrations. However, the non-aromatic hydrocarbon material, eluted from the gas chromatographic column prior to ethyl benzene, showed marked differences from sample to sample.

20.6 DISCUSSION

The method is capable of determining down to 0.001% of various aromatic volatiles in polystyrene with an accuracy of $\pm$ 5%. The quantitative aspects of this technique itself were investigated by analysing the volatiles liberated by heating polystyrene at 200°C and then re- analysing the polymer as a solution in propylene oxide. Several samples of polystyrene were examined in this way and in each case component peak areas were normalized and compared. The results (Table 7.10) show that the values obtained by head-space analysis technique differ by up to 20% from those obtained by the solution procedure which are known to be correct.

Table 7.9 - Analysis of some Different Grades of Polystyrene

Component	Weight Per Cent				
	Crystal grade	Expandable grade	Self extinguishing grade	High impact grade	Foodstuff packaging grade
Benzene	< 0.001	< 0.001	< 0.001	< 0.001	< 0.001
Toluene	0.001	0.003	< 0.001	< 0.001	< 0.001
Ethyl benzene	0.066	0.081	0.047	0.086	0.060
m-/p-Xylene	0.002	0.002	0.063	< 0.001	< 0.001
Cumene	0.046	0.049	0.039	0.011	0.030
o-Xylene	0.007	0.008	0.006	0.002	< 0.001
n-Propyl benzene	0.017	0.016	0.014	0.004	0.016
m/p-Ethyl toluenes	0.016	0.011	0.013	0.002	< 0.001
Styrene	0.033	0.32	0.080	0.18	0.04

Table 7.10 - Comparison of Normalized peak areas calculated from chromatograms of polystyrene solutions and polystyrene volatiles liberated by heating at 200°C

Component	Normalized Peak Area							
	Analysis of Polymer as a Solution in Propylene Oxide				Analysis of Volatiles Liberated by Heating Polymer at 200°C			
	A	B	C	D	A	B	C	D
Ethyl benzene	16.4	17.4	27.2	27.5	20.2	23.4	35.4	34.4
n-Propyl benzene	5.5	5.3	3.5	3.2	3.2	2.5	1.9	2.3
Cumene	17.7	18.1	10.4	10.8	15.4	12.7	10.2	10.7
m/p-Xylenes	4.1	2.5	2.7	2.0	1.1	1.3	0.6	0.7
m/p-Ethyl toluenes	1.7	1.7	2.0	1.5	0.8	0.7	0.5	0.6
Styrene	54.6	55.0	54.2	55.0	59.3	59.4	51.4	51.3

METHOD 21 - DETERMINATION OF AROMATIC VOLATILES IN POLYSTYRENE. HEAD SPACE ANALYSIS.[13]

21.1 SUMMARY

This head space method is capable of determining down to 0.001% of a wide range of aromatic volatiles in polystyrene.

21.1 APPARATUS

The apparatus is illustrated in Figure 7.6. A glass ignition tube is supported as shown in a Wade $^1/_4$ inch diameter brass coupling nut, covered with a silicone rubber septum and sealed with a Wade $^1/_4$ inch brass stop-end body. The stop-end body has two 1 mm diameter holes drilled through the cap. The whole unit is placed in a slot in a cylindrical copper block (3 in long x 2 in diam) which is heated by two Bray (240 V, 85

W) cartridge heaters and controlled at temperatures up to 300°C, from a variable transformer. The temperature is measured with a thermocouple capable of accurately measuring temperatures in the 100° - 300° C temperature range with a maximum error of ± 5 per cent. The thermocouple is inserted in the slot adjacent to the ignition tube, it has been shown that under these conditions the thermocouple records the true temperature of the contents of the tube. The provision of a slot in the copper block enables more than one ignition tube to be heated simultaneously if required.

Gas chromatography, see Method 20.

Pure standard aromatic compounds, see Method 20.

21.3 METHOD

In this method the polymer is heated to 300°C in a sealed tube in the absence of solvents. The head space gas in the tube is then swept into a gas chromatograph where the individual aromatics are determined. The method is calibrated against standard injections of pure aromatics into the gas chromatograph.

21.4 EXPERIMENTAL PROCEDURE

A sample of the polymer (0.25 - 0.50 g) is placed in an ignition tube and sealed with Wade fittings and a septum (Figure 7.6). If necessary, the tube is then purged with a suitable gas by inserting two hypodermic needles through the septum via the holes in the cap of the stop-end body and passing the gas into the tube through one hole and allowing it to vent through the other. After purging, the two needles are removed simultaneously and the tube is then heated in the copper block under the required conditions of time and temperature. A sample (1 - 2 ml) of the head space gas is withdrawn from the ignition tube into a Hamilton gas-tight hypodermic syringe via the septum and injected into a gas chromatograph. It is advisable to fill the syringe with the gas used initially in the ignition tube and to inject this into the tube before withdrawing the sample. This facilitates sampling by preventing the creation of a partial vacuum in the ignition tube or the syringe, or both. It also minimizes any undesirable entry of air into the ignition tube.

Calibrate the gas chromatograph as described in Method 20.

21.5 DISCUSSION OF RESULTS

With the apparatus, a polymer may be heated under any desired gas and, while this may frequently be the carrier gas used with the gas chromatograph, it is also possible to carry out studies in oxidising or reducing atmospheres. A polymer may also be heated to any known temperature and samples of the head space gas may be withdrawn at intermediate temperatures and times to determine under what conditions any particular volatile is liberated. By using gas chromatographic detectors of suitable sensitivities and selectivities, it is possible to examine polymer for the presence or formation of volatiles at both the percentage and parts-per-million levels. For example, traces of organic volatiles may be examined with an ionization detector and traces or organic halogen compounds lend themselves to analysis with an electron capture detector. Thermal conductivity cells of the hot-wire or thermistor type are suitable for the detection or inorganic volatiles and a helium ionization detector could be used for analysing trace amounts of permanent gases.

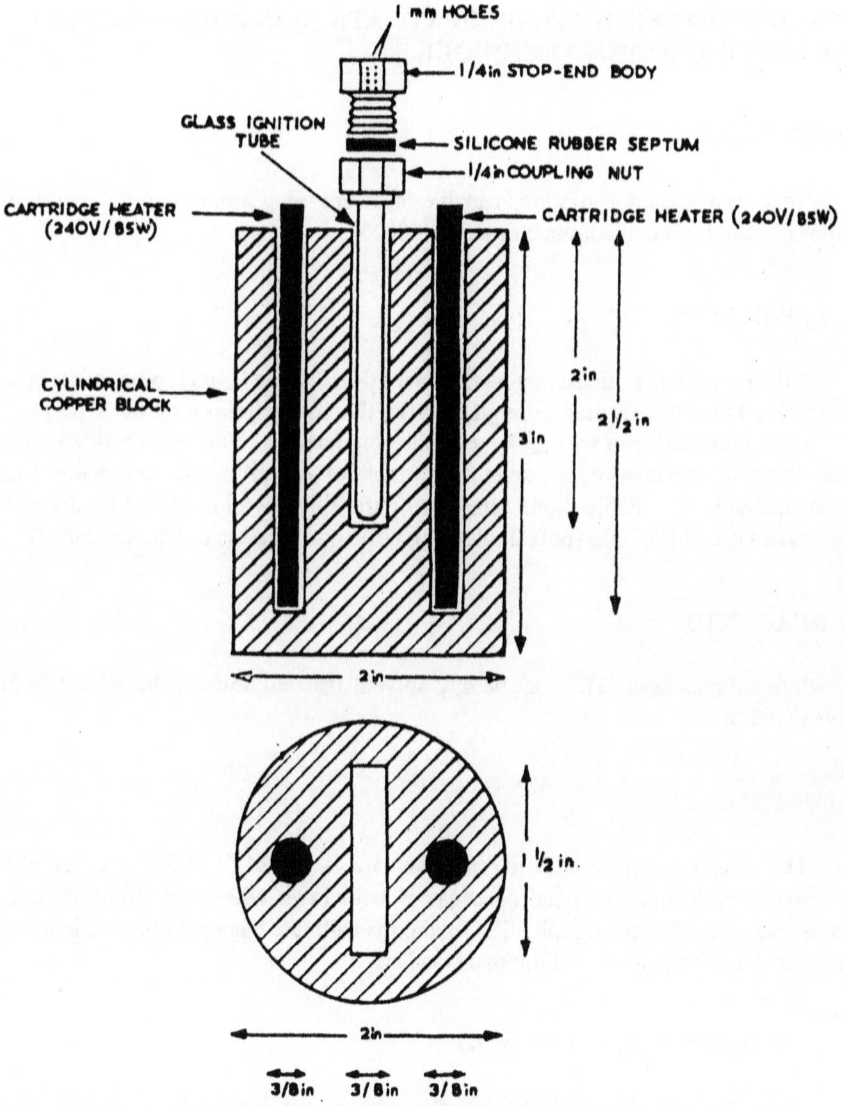

Figure 7.6 - Apparatus for liberating volatiles from polymers

The method is capable of determining down to 0.001% aromatic volatiles in polystyrene with an accuracy of ± 5%.

METHOD 22 - DETERMINATION OF VOLATILES IN POLYSTYRENE. HEAD-SPACE ANALYSIS TECHNIQUE.[14]

22.1 SUMMARY

This head space technique is capable of determining a range of different types of volatiles in polystyrene in amounts down to 0.002%.

22.2 APPARATUS

Kilner jars (1 lb) - these jars have metal lids. A brass $^{1}/_{4}$ inch pipe union is sealed with Araldite into a hole drilled through each lid, this firmly holds a rubber septum.

Gas chromatograph with a flame-ionisation detector. This can be fitted with any column that will satisfactorily separate the various solvents from each other and from the internal standard. A suitable liquid phase is 9 parts silicone oil modified by the addition of 2 parts of UCON HB2000 (polyalkylene glycol) and 100 parts of Chromosorb W.

22.3 REAGENTS

Internal standard. This can be any solvent that is known to be absent from the sample material.

22.4 METHOD

The sample together with internal standard is heated to 100°C until equilibrum then a measured volume of headspace gas is withdrawn with a gas tight syringe and injected into a gas chromatograph. The method is calibrated against known blends of the substances to be determined and internal standard.

22.5 EXPERIMENTAL PROCEDURE

About 250 cm^3 of sample is placed in a Kilner jar and the jar is closed. Internal standard (3.00 ul) is injected into the jar through the rubber septum. The jar is then placed in an oven at 100°C for the minimum specified period (see under discussion) at the end of which it is removed from the oven and about 2 ml of the atmosphere within are immediately extracted with an all-glass unlubricated syringe and injected into the gas chromatograph. The ratios of the peak heights of the solvents, (a, b, c, etc) to that of the internal standard are calculated. The same amount of internal standard as used with the samples, plus 2.00 ul of each of the solvents a, b, c, etc is introduced, for calibration purposes, into each of the Kilner jars. These jars are heated together with those containing samples.

22.6 TYPICAL RESULTS

The content of retained solvent, eg, a, is given in milligrams per square metre by the expression:

$$\frac{A \times B \times 20{,}000 \times D \times E}{C \times F \times G}$$

where
A = Peak height of solvent "a" in sample jar
B = Specific quantity of solvent, g cm^3
C = Peak height of internal standard in sample jar
D = Peak height of solvent "a" in calibration jar
E = "a" content of retained solvent mg square metre
F = Peak height of internal standard in calibration jar
G = Area of sample in square cm.

The method has been used to determine many solvents in polystyrene, polyethylene, polypropylene and cellophane including ethanol, ethyl acetate, methyl ethyl ketone, 2-ethoxy ethanol, propanol-1 and toluene.

METHOD 23 - DETERMINATION OF RESIDUAL ISO-BUTANE IN EXPANDABLE AND EXPANDED POLYSTYRENE. GAS CHROMATOGRAPHY.

23.1 SUMMARY

This gas chromatographic method determines iso-butane in amounts down to 10 ppm in expanded polystyrene.

23.2 APPARATUS

Gas chromatograph run under the following conditions:

Column - Copper tube (10 ft x $^1/_4$ in o.d. x $^3/_{16}$ in i.d.) packed with 25% w/w di-n-butyl phthalate on 44 - 60 Celite.

Gases -

Air	7 psig
Helium	30 psig
Hydrogen	22 psig

Temperatures -

Injection	150°C
Column	40 °C
Detector	150 °C
Flame	200 °C

Recorder - 1 mV full scale deflection, 6- in/hr chart speed.

23.3 REAGENTS

Solvent - Weigh a stoppered 100 cm^3 volumetric flask containing about 98 ml propylene oxide. Add 1 cm^3 2,2-dimethyl butane, internal standard, reweigh and then dilute to 100 cm^3 with propylene oxide. Shake thoroughly.

23.4 METHOD

A solution of the polystyrene sample and 2,2-dimethylbutane internal standard in propylene oxide is injected into the gas chromatograph. Calibration is achieved with known blends of internal standard and iso-butane in propylene oxide.

23.5 EXPERIMENTAL PROCEDURE

Weigh accurately about 1 g sample into a stoppered type 25 cm^3 measuring cylinder. Add exactly 10 cm^3 of the prepared solvent, seal the cylinder with a serum cap and shake to dissolve. Chromatograph 1 micro-litre at a range of x 1000 and attenuations suitable for the sample (usually x 4 for the iso-butane and x 2 for the 2,2-dimethyl butane). For each peak, draw in the escribed triangle and calculate the product of peak height and peak width at the base, allowing for any attenuation factors.

23.6 TYPICAL RESULTS

Calculation.

B = product of peak height and peak width at the base of the isobutane peak.

D = product of peak height and peak width at the base of the 2,2-dimethyl butane peak.

S = weight of 2,2-dimethyl butane in 10 cm^3 solvent.

W = weight of sample taken for analysis.

$$\text{Iso-butane (\% w/w)} = \frac{100 \times B \times S}{D \times W}$$

The following are the uncorrected retention distances (to peak centres) and the corresponding relative values (2,2-dimethyl butane = 1.00).

Component	DR	
	mm	Rel
Impurity ex propylene oxide	24.5	0.25
Iso-butane	32	0.35
2,2-dimethyl butane	98	1.00
Propylene oxide	150	1.53

The analysis is complete with the elution of propylene oxide. This takes 10 min.

23.7 DISCUSSION OF RESULTS

This method is capable of determining down to 10 ppm of iso-butane in polystyrene with an accuracy of $\pm$ 5%.

It is necessary to ensure that the injection port on the gas chromatograph is fitted with a glass liner (very loosely packed with glass fibre) and check that this is not blocked with polymer. Daily replacement is advisable.

METHOD 24 - DETERMINATION OF RESIDUAL ISOHEXANE IN EXPANDABLE AND EXPANDED POLYSTYRENE. GAS CHROMATOGRAPHY.

24.1 SUMMARY

This gas chromatographic method determines isohexane in amounts down to 10 ppm in expanded polystyrene.

24.2 APPARATUS

Gas chromatograph run under the following conditions:

Column - see Method 23.
Gases -

Helium	30 psig
Air	2.5 psig
Hydrogen	7.5 psig

Temperatures -

Injection	150°C
Column	50 °C
Detector	150 °C
Flame	140 °C

Recorder/chart speed 10 in/hr.

24.3 REAGENTS

Prepared solvent. Weigh 1 cm^3 2,4-dimethyl pentane into a 100 cm^3 volumetric flask and dilute to 100 ml with methylene dichloride.

24.4 METHOD

A solution of the polystyrene sample and 2,4 dimethyl-pentane internal standard in methylene dichloride is injected into the gas chromatograph. Calibration is achieved with known blends of internal standard and 2,4 dimethylbutane in methylene dichliride.

24.5 EXPERIMENTAL PROCEDURE

Weigh 1 g sample into a 25 cm^3 measuring cylinder, add 10 cm^3 prepared solvent, seal with a serum cap and shake to dissolve. Chromatograph 5 micro-litres at a range of x 100 and suitable attenuations.

24.5.1 Calculation.

Measure the retention distances (injection point to peak apex) and peak heights of all the components. For each, calculate the product of retention distance and peak height, allowing for any attenuation factor. Determine the concentration of each component as follows:

W = % w/v 2:4-dimethyl pentane in solvent
S = weight of sample taken
C = product of retention distance and peak height of component
D = product of retention distance and peak height of 2:4-dimethyl pentane

$$\text{Component (\% w/w)} = \frac{C}{D} \times \frac{10 \times W}{S}$$

Report results to the nearest 0.1% w/w.

24.5.2 Retention Data.

The following are uncorrected retention distances (injection point to peak apex) and the corresponding relative values (2:4,dimethyl pentane = 1.00).

Component	DR	
iso-Pentane	12	0.24
n-Pentane	18	0.35
2,2-dimethyl butane	23.5	0.46
2-methyl pentane + 2,3-dimethyl butane	31	0.61
3-methyl pentane + cyclo-pentane	36	0.71
n-Hexane	41	0.80
2:4-dimethyl pentane	51	1.00

24.6 DISCUSSION OF RESULTS

This method is capable of determining down to 10 ppm of isohexane in expanded polystyrene with an accuracy of ± 5%. It is necessary to ensure that the injection port on the gas chromatograph is fitted with a glass liner (very loosely packed with glass fibre) and check that this is not blocked with polymer. Daily replacement is advisable.

METHOD 25 - DETERMINATION OF RESIDUAL *N* AND *ISO*PENTANE IN EXPANDED AND EXPANDABLE POLYSTYRENE. GAS CHROMATOGRAPHY.

25.1 SUMMARY

This gas chromatographic method determines n- and iso-pentane in amounts down to 10 ppm in expanded polystyrene.

25.2 APPARATUS

Gas chromatograph run under the following conditions:

Column	see Method 23
Carrier gas	Helium at 0.6 kg/cm^2
Temperature	Injection 45°C, column 45°C, detector 45°C.
Bridge	250 mA
Chart speed	30 cm/hr

25.3 REAGENTS

Prepared solvent. Into a 100 cm^3 volumetric flask weigh 1.0 cm^3 2:2 dimethyl butane, sealing the flask with a serum cap during the weighing. Dilute to 100 cm^3 with propylene oxide and re-seal with the serum cap.

25.4 METHOD

A solution of the polystyrene sample and 2,2 dimethylbutane internal standard in propylene oxide is injected into the gas chromatograph. Calibration is achieved with known blends of internal standard and n- and iso-butane in propylene oxide.

25.5 EXPERIMENTAL PROCEDURE

Into a 25 cm^3 measuring cylinder (stoppered type) weigh 2 g sample and add exactly 25 cm^3 of the prepared solvent. Seal with a serum cap and shake to dissolve. Chromatograph 10 ul (from a 50 ul syringe) at an attenuation of XI. Immediately wash out the syringe with propylene oxide. Measure the peak heights or the integrated areas of the iso-pentane, n-pentane and 2:2-dimethyl butane.

25.5.1 Calibration.

Into a 100 cm^3 volumetric flask, weigh in turn 0.40 cm^3 iso-pentane, 0.50 cm^3 n-pentane and 0.80 cm^3 2:2-dimethyl butane. Seal the flask with a serum cap during each weighing. Dilute the mixture to 100 cm^3 with propylene oxide, seal with the serum cap and mix thoroughly. Chromatograph 10 ul (from a 50 ul syringe) at an attenuation of XI. Measure either the peak heights or the integrated areas of the iso-pentane, n-pentane and 2:2-dimethyl butane.

25.6 TYPICAL RESULTS

Calculation.
W1 = wt of iso-pentane in calibration blend.
W2 = wt of n-pentane in calibration blend
W3 = wt of 2:2 dimethyl butane in calibration blend.
W4 = wt of 2:2 dimethyl butane in 100 cm^3 solvent.
W5 = wt of sample
P1 = peak height or integrated area of iso-pentane in calibration blend.

P2 = peak height or integrated area of n-pentane in calibration blend
P3 = peak height or integrated area of 2:2 dimethyl butane in calibration blend.
P4 = peak height or integrated area of iso-pentane in analysis solution.
P5 = peak height or integrated area of n-pentane in analysis solution.
P6 = peak height or integrated area of 2:2 dimethyl butane in analysis solution.

$$\% \text{ w/w } iso\text{-pentane} = \frac{25 \times W1 \times W4 \times P3 \times P4}{W3 \times W5 \times P1 \times P6}$$

$$\% \text{ w/w } n\text{-pentane} = \frac{25 \times W2 \times W4 \times P3 \times P5}{W3 \times W5 \times P2 \times P6}$$

Relative retention values (corrected and to peak centre)

	DR
Air	= 0.00
Propylene oxide impurity	= 0.19
iso-pentane	= 0.77
n-pentane	= 1.00
2:2 dimethyl butane	= 1.41
Propylene oxide	Clear within 30 min.

25.7 DISCUSSION OF RESULTS

This method is capable of determining down to 10 ppm of n-and iso pentane in polystyrene with an accuracy of $\pm$ 5%.

It is necessary to ensure that the injection port on the gas chromatograph is fitted with a glass liner (very loosely packed with glass fibre) and check that this is not blocked with polymer. Daily replacement is advisable.

METHOD 26 - DETERMINATION OF RESIDUAL NEO-HEXANE IN EXPANDED POLYSTYRENE AND EXPANDABLE POLYSTYRENE. GAS CHROMATOGRAPHY.

26.1 SUMMARY

This gas chromatographic method determines neo-hexane in amounts down to 10 ppm in expanded polystyrene.

26.2 APPARATUS

Gas chromatograph run under the following conditions:

Column	see Method 23
Gases:	
helium	see Method 23
air	see Method 23
hydrogen	see Method 23

Temperatures:
injection see Method 24
column see Method 24
detector see Method 24
flame see Method 24
Recorder/chart speed 1mV, 10 in/hr.

26.3 REAGENTS

Prepared solvent. Weigh 1 cm^3 n-pentane internal standard into a 100 cm^3 volumetric flask and dilute to 100 cm^3 with propylene oxide.

26.4 METHOD

A solution of the polystyrene sample and n-pentane internal standard in propylene oxide is injected into the gas chromatograph. Calibration is achieved with known blends of internal standard and neo-hexane in propylene oxide.

26.5 EXPERIMENTAL PROCEDURE

Weigh 1 g sample into a 25 cm^3 measuring cylinder, add 10 ml solvent, seal with a serum cap and shake to dissolve. Chromatograph 5 micro litres at a range of x 100 and suitable attenuations. Measure the retention distances (injection point to peak apex) and peak heights of the n-pentane and neo-hexane. For each, calculate the product of retention distance and peak height, allowing for any attenuation factor.

26.6 TYPICAL RESULTS

Determine the concentration of neo-hexane as follows:
W - % w/v n-pentane in solvent
S = weight of sample taken.
C = product of retention distance and peak height of neo-hexane.
D - product of retention distance and peak height of n-pentane.

$$\text{neo-Hexane (\% w/w} = \frac{C \times 10 \times W}{D \times S}$$

Report results to the nearest 0.1% w/w.

26.7 DISCUSSION OF RESULTS

The method is capable of determining down to 10 ppm neo-hexane in polystyrene with an accuracy of $\pm$ 10%. It is necessary to ensure that the injection port on the gas chromatograph is fitted with a glass liner (very loosely packed with glass fibre) and check that this is not blocked. Daily replacement is advisable.

METHOD 27 - SOLVENTLESS PROCEDURE FOR THE DETERMINATION OF RESIDUAL *N*- AND ISO-PENTANE IN EXPANDABLE AND EXPANDED POLYSTYRENE. GAS CHROMATOGRAPHY.

27.1 SUMMARY

This head space method is capable of determining down to 0.001% of n- and iso pentane in polystyrene.

27.2 APPARATUS

Head space analysis equipment is discussed in Method 21 and illustrated in Figure 7.6. Gas chromatograph operated under the following conditions:

Column. Copper tube (15 ft x $^3/_{16}$ in i.d.) packed with 10% wt/wt Carbowax 15 - 20 M on 60 - 62 BS mesh acid washed Celite.

Gas flows.	Helium - 30 psig, rotameter = 10.0 (100 cm^3 /min)
	Hydrogen - 13 psig rotameter = 10.0 (75 cm^3 /min)
	Air - 7 psig, rotameter = 10.0 (650 cm^3 /min)
Temperatures.	Injection 155°C
	Column 80°C
	Detector 125°C
	Flame 200°C

Recorder. 1 mV full scale deflection, 1 second response, 10 in/hr chart speed.

27.3 METHOD

In this method the polymer is heated to 240°C for 5 min in a sealed tube in the absence of solvents. The head space gas in the tube is then swept into a gas chromatograph where the n- and iso-pentane are determined. The gas chromatograph is calibrated against known blends of n- and iso-pentane and n-undecane internal standard dissolved in propylene oxide.

27.4 EXPERIMENTAL PROCEDURE

See Method 21.

METHOD 28 - DETERMINATION OF HYDROCARBON IMPURITIES IN STYRENE MONOMER. GAS CHROMATOGRAPHY.

28.1 SUMMARY

This method is capable of determining a wide range of alkyl benzene impurities in styrene monomer in amounts down to 5 ppm.

28.2 APPARATUS

Gas chromatograph operating conditions are listed in Table 7.11.

28.3 REAGENTS

Internal standard, n-butyl benzene.
Pure aromatic hydrocarbons for calibration.

28.4 METHOD

A weighed amount of internal standard, usually n-butyl benzene, is added to the styrene monomer sample which is then gas chromatographed on two columns (Squalane and Carbowax). The method is calibrated against known blends of internal standard and the aromatics to be determined.

28.5 EXPERIMENTAL PROCEDURE

28.5.1 Analysis on Squalane Column.

Under the conditions given in Table 7.11 chromatograph 5 ul of the sample and from the chromatogram decide on a suitable internal standard, eg *n*-butyl benzene. Prepare an accurate w/w solution of the chosen internal standard in the sample and chromatograph 5 ul of the solution under the conditions given in Table 7.11. Determine the peak areas of the standard and of the relevent impurity peaks and, knowing the concentration of the standard and assuming that all components have the same response factor, calculate the results:

$$\% \text{ w/w component} = \frac{Ac}{As} \times \frac{100W}{(100-W)}$$

where Ac = peak area of impurity being determined
As = peak area of internal standard
W = concentration (%w/w) of internal standard in the sample

28.5.2 Analysis on Carbowax Column.

Under the conditions given in Table 7.11 chromatograph 1 ul of the sample. As soon as the styrene has been eluted, it is convenient to raise the column temperature to 180°C to hasten the elution of those high boiling components which it is not necessary to determine at this stage. Determine the peak areas of the relative impurity peaks, choosing one of them already determined (eg. n-propyl benzene) on the squalane column as a reference standard.

Assuming that all the components have the same response factor, calculate the results:

$$\% \text{ w/w component} = \frac{Ac \times Cp}{Ap}$$

Table 7.11 - Gas Chromatograph Operating Conditions

	Squaline Column	Carbowax Column
Instrument used	Perkin-Elmer Model F/11 fitted with a flame ionisation detector and a sample injection splitter	F. and M. Model 810, fitted with a flame ionisation detector
Column	Stainless steel tube (50m x 1/16 in O.D. ¼ mm I.D.) coated with squalane	Copper tube (15 ft x ¼ in O.D. x 3/16 in I.D.) packed with 10% w/w Carbowax 15-20M on 60 - 72 Celite
Gases		
Air	30 lbf/in^2 gauge	20 lbft/in^2 gauge
Nitrogen	20 lbf/in^2 gauge	---
Helium	---	50 lbft/in^2 gauge contolled at 100 ml/min
Hydrogen	20 lbf/in^2 gauge	30 lbf/in^2 gauge
Temperatures		
Injection	185°C	200°C
Column	80°C	80°C
Detector	80°C	200°C
Recorder	Honeywell-Brown, 1mV full scale deflection, 30 in/h chart speed	Honeywell-Brown, 1mV full scale deflection, 15 in/h chart speed
Sample split ratio	160:1	

Each chromatograph was operated in conjunction with an electronic integrator

where Ac = peak area of impurity being determined
Ap = peak area of n-propyl benzene
Cp = concentration (%w/w) of n-propyl benzene as determined on the squalane column.

For the purpose of calculating results it can be assumed that each trace impurity in the mixture has the same response factor.

28.6 TYPICAL RESULTS

Retention data for some of the impurities found in styrene monomer are given in Table 7.12.

Table 7.13 gives the results for the analysis of some typical crude and polymerisation grade styrene monomer samples. It is seen that in the crude material approximately 40 impurities were detected, their concentration being 2.6% w/w (unsaturated) 3.3% w/w (saturated and unsaturated). In comparison, the purified monomer contained far fewer impurities at a total concentration of less than 0.5% w/w.

28.7 DISCUSSION OF RESULTS

The method is capable of determining up to 40 aromatic impurities in styrene monomer in amounts down to 5 ppm with an accuracy of $\pm$ 5%.

Table 7.12 - Relative Corrected Retention Time of Likely Styrene Impurities on Squalane and Carbowax Columns

Pure reference compounds were not available for components marked with asterisks. Tentative identities have been ascribed from boiling points

		Relative corrected retention times	
	B.Pt (°C)	Squalane	Carbowax
Benzene	80.1	0.042	0.238
Toluene	110.6	0.102	0.394
Ethyl benzene	136.2	0.212	0.628
n-propyl benzene	159.3	0.439	1.000
n-butyl benzene	183.3	1.000	
Cumene	152.5	0.346	0.818
iso-butyl benzene	172.5	0.683	
sec-butyl benzene	173.5	0.683	
tert-butyl benzene	169.0	0.597	
o-xylene	144.4	0.286	0.874
m-xylene	139.1	0.243	0.654
p-xylene	138.4	0.238	0.680
o-ethyl toluene*	165.2	0.553	
m-ethyl toluene	161.0	0.490	1.104
p-ethyl toluene	162.0	0.499	1.104
o-diethyl benzene*	183.4	1.000	
m-diethyl benzene	181.1	0.952	
p-diethyl benzene	183.8	1.037	
Styrene	145.2	0.264	1.268
-methyl styrene	165.4	0.541	
ß-methyl styrene*	176.0	0.777	
o-methyl styrene	171.6	0.623	
m-methyl styrene	170.9	0.631	
p-methyl styrene	170.6	0.641	
o-ethyl styrene*		1.131	
m-ethyl styrene		1.253	
p-ethyl styrene		1.345	
o-divinyl benzene*		1.414	
m-divinyl benzene		1.558	
p-divinyl benzene		1.705	

METHOD 29 - DETERMINATION OF HYDROCARBON IMPURITIES IN STYRENE MONOMER, TEMPERATURE-PROGRAMMED GAS CHROMATOGRAPHY.

29.1 SUMMARY

This method is capable of determining a wide range of alkyl benzene impurities in styrene monomer in one hour and in amounts down to 5 ppm.

Table 7.13 - Analysis of Crude and Pure Styrene Monomer Samples

Components	Crude Styrene Monomer % w/w		Pure Styrene Monomer % w/w	
	Sample 1	Sample 2	Sample 1	Sample 2
Benzene	0.0004	0.004	< 0.0005	0.0003
Toluene	0.0018	0.0019	0.0005	0.0007
Ethyl benzene	0.0097	0.0175	0.025	0.051
m-xylene/p-xylene	0.0013	0.0023	0.0003	0.0017
Cumene	0.094	0.093	0.021	0.021
o-xylene	0.0054	0.0069	0.0032	0.0041
n-propyl benzene	0.066	0.067	0.0067	0.016
m-ethyl toluene	0.039	0.046	0.0022	0.013
p-ethyl toluene	0.023	0.023	0.0013	0.0071
-methyl styrene	2.0	2.1	0.083	0.14
o-ethyl toluene	0.092	0.10	0.0013	0.0027
tert-butyl benzene	< 0.0005	< 0.0005	< 0.0005	< 0.0005
o-methyl styrene	< 0.0005	< 0.0005	< 0.0005	< 0.0005
m-methyl styrene	0.064	0.067	< 0.0005	0.0049
p-methyl styrene	0.029	0.039	< 0.0005	0.0022
iso-butyl benzene)				
sec-butyl benzene)	0.093	0.096	0.0009	0.0066
ß-methyl styrene	0.24	0.25	< 0.0005	0.0007
m-diethyl benzene	0.19	0.15	< 0.0005	0.0005
n-butyl benzene)				
o-diethyl benzene)	0.0043	0.0031	< 0.0005	< 0.0005
p-diethyl benzene	0.059	0.040	< 0.0005	< 0.0005
o-ethyl styrene	0.0006	0.0010	< 0.0005	< 0.0005
m-ethyl styrene	0.098	0.10	< 0.0005	< 0.0005
p-ethyl styrene	0.020	0.019	< 0.0005	< 0.0005
o-divinyl benzene	< 0.0005	< 0.0005	< 0.0005	< 0.0005
m-divinyl benzene	0.041	0.039	< 0.0005	< 0.0005
p-divinyl benzene	0.0097	0.0075	< 0.0005	< 0.0005
Unidentified A	0.033	0.016	< 0.0020	0.0063
Unidentified B	0.057	0.057	0.0005	0.0057
Total number of impurities exceeding 0.0005% concentration	36 - 41	36 - 41	20	24 - 29

Tentative identities ascribed from boiling point data.
'Unidentified A' includes 10 components of unknown identities eluted from the Carbowax column.
'Unidentified B' includes 5 components of unknown identities eluted from the squalane column.

29.2 APPARATUS

Gas chromatograph, see Table 7.14.
Gas chromatograph operating conditions are listed in Table 7.14.

29.3 REAGENTS

Internal standard, n-undecane.
Pure aromatic hydrocarbons for calibration.

29.4 METHOD

A weighed amount of n-undecane internal standard is added to the styrene monomer sample which is then gas chromatographed. The method is calibrated against known blends of internal standard and the aromatics to be determined.

29.5 EXPERIMENTAL PROCEDURE

Using the conditions given in Table 7.14 chromatograph 5 micro litres of the styrene sample at a range of x 10 and an attenuation appropriate to the concentration of impurities present. Prepare an accurate 1% w/w solution of n-undecane internal standard in the styrene sample. Chromatograph 5 microlitres of this solution attenuating the peaks as necessary. Measure the peak areas of the internal standard and of the components to be determined.

29.6 TYPICAL RESULTS

From the known concentration of the internal standard obtain weight per cent values, assuming that styrene impurity peak area is directly proportional to concentrations. The concentrations of impurities in the styrene monomer sample are obtained as follows:

$$\text{\% w/w impurity in styrene monomer} = \frac{A_C \times 100W}{A_S\ (100-W)}$$

where W = concentration (% w/s) of internal standard in styrene sample
A_S = peak area of internal standard
A_C = peak area of impurity

In Table 7.15 are tabulated relative corrected retention distance data obtained using isothermal and temperature programmed Carbowax columns for a range of impurities in styrene monomer. It is seen that different separations are obtained by the two methods of analysis. The selection of method is dependent on the separations it is required to achieve.

29.7 DISCUSSION OF RESULTS

This method is capable of determining up to 40 aromatic impurities in styrene monomer with an accuracy of $\pm$ 5%.

Table 7.14 - Instrumental Conditions

	Temperature Programmed Method
Instrument used	Carbowax column
	F & M Model 810 with flame ionization detector
Column	Copper tube (15 ft x 0.25" in o d x $^3/_{16}$" i d) packed with 10% w/w Carbowax 15-20 M on 60-72 mesh Celite.
Gases	
Air	20 lb f/in^2 gauge
Nitrogen	50 lb f/in^2 gauge (controlled at 100 ml/min)
Helium	30 lb f/in^2 gauge
Temperature, °C	
Injection	200
Column	Isothermally at 100°C for 12 minutes, then programmed at 4° per minute to 140°C then held at 140°C for 20 minutes.
Detector	200
Recorder	Honeywell Brown 1MV F.S.D. 15 in/hr chart speed
Internal standard	1% w/w n-undecane in sample

Table 7.15 - Impurities in Styrene Monomer, Relative Corrected Retention Distances

	Earlier Method Isothermal Column		Temperature Programmed Column
	Column: Squalane (Relative to n-butyl benzene)	Carbowox (Relative to n-propyl benzene)	Carbowax Relative to n-undecane for compounds appearing below styrene (b) and relative to alpha methyl styrene for compounds appearing above styrene (c)
n-undecane	-	-	1.000 (b) used as internal standard
Benzene	a	0.283	0.580 (b)
Toluene	a	0.394	0.850 (b)
Ethyl benzene	a	0.628	1.230 (b)
o-xylene	a	0.874	1.600 (b)
m-xylene	a	0.654	1.310 (b)
p-xylene	a	0.680	1.310 (b)
Cumene	a	0.818	1.510 (b)
n-propyl benzene	0.439	1.000	1.780 (b)
m-ethyl toluene	0.490	1.104	1.920 (b)
p-ethyl toluene	0.499	1.104	1.920 (b)
-methyl styrene	0.541		1.000 (c)
o-ethyl toluene	0.553		a
tert-butyl benzene	0.597	a	a
o-methyl styrene	0.623		1.090 (c)
m-methyl styrene	0.631		1.090 (c)
p-methyl styrene	0.641		1.090 (c)
iso-butyl benzene	0.683	a	a
sec-butyl benzene	0.683	a	a
ß-methyl styrene	0.777		not examined
m-diethyl benzene	0.952		0.910 (c)

Table 7.15 continued

n-butyl benzene	1.000 (used as internal standard)		0.950 (c)
o-diethyl benzene	1.000		not examined
p-diethyl benzene	1.037		0.950 (c)
o-ethyl styrene	1.131		not examined
m-ethyl styrene	1.253		1.350 (c)
p-ethyl styrene	1.345		1.380 (c)
o-divinyl benzene	1.414		not examined
m-divinyl benzene	1.558		1.730 (c)
p-divinyl benzene	1.705		1.800 (c)
benzaldehyde	not examined		1.600 (c)
Number of unidentified peaks:	5	13	14
	1 prior to 0.439	1 - 6 prior to 0.283	1. 0.230, 8. 0.650
	2 between 0.641 & 0.683	7 - 12 between 0.283 & 0.394	2. 0.260, 9. 0.720
	3 between 0.683 & 0.677	13 between 0.934 & 0.628	3. 0.290, 10. 0.790
	4 between 0.777 & 0.952		4. 0.310, 11. 0.860
	5 between 1.414 & 1.558		5. 0.360, 12. 1.80
			6. 0.490, 13. 1.24
			7. 0.560, 14. 1.93

a Masked by styrene (major constituent)

METHOD 30 - DETERMINATION OF BENZALDEHYDE IN STYRENE MONOMER. GAS CHROMATOGRAPHY.

30.1 SUMMARY

This gas chromatographic method is capable of determining benzaldehyde in amounts down to 5 ppm in styrene monomer.

30.2 APPARATUS

Gas chromatograph, see Table 7.14
Gas chromatograph operating conditions, see Table 7.14

30.3 REAGENTS

Internal standard, n-hexadecane
Benzaldehyde
Propylene oxide

30.4 METHOD

This gas chromatographic method determines benzaldehyde in amounts down to 5 ppm in styrene monomer.

30.5 EXPERIMENTAL PROCEDURE

Into a 100 cm^3 stoppered volumetric flask, weigh 10 ml of styrene sample and exactly 1 ml of a 0.5% v/v solution of n-hexadecane internal standard in propylene oxide. Chromatograph 2 ul of this solution on a Carbowax Celite column at a range of x 10 and suitable attenuations using the conditions indicated in Table 7.14 with the exception that the injection and column temperatures are set at 150°C and 140°C respectively.

Calibrate the procedure as follows. Weigh into a 100 cm^3 volumetric flask exactly 0.5 cm^3 of pure benzaldehyde and 1.0 cm^3 n-hexadecane and dilute to 100 cm^3 with propylene oxide. Chromatograph 2 ul of this solution at a range of x 10 and suitable attenuations. Run the chromatogram to the peak following that of benzaldehyde.

30.6 TYPICAL RESULTS

Retention distances uncorrected injection point to peak apex

	mm	relative
n hexadecane	145	1.00
Benzaldehyde	173.5	1.20

On the calibration chromatogram determine the response factor for benzaldehyde as follows:

where A = weight of benzaldehyde in calibration blend
B = weight of n-hexadecane in calibration blend
C = peak height of benzaldehyde on calibration chromatogram
D = peak height of n-hexadecane on calibration chromatogram

$$\text{Benzaldehyde factor, } F = \frac{A \times D}{B \times C}$$

From the analysis chromatogram, determine the concentration of benzaldehyde as follows:

where A = weight of benzaldehyde in calibration blend
B = weight of n-hexadecane in calibration blend
C = peak height of benzaldehyde on calibration chromatogram
D = peak height of n-hexadecane on calibration chromatogram

$$\text{Benzaldehyde factor } F = \frac{A \times D}{B \times C}$$

From the analysis chromatogram, determine the concentration of benzaldehyde as follows:

where G = peak height of benzaldehyde on analysis chromatogram
H = peak height of n-hexadecane on analysis chromatogram
J = % w/v n-hexadecane in solvent
K = weight of sample taken
F = benzaldehyde response factor

$$\text{Benzaldehyde (\% w/w)} = \frac{10 \times G \times F \times J}{H \times K}$$

It is seen in Table 7.15 that in the temperature programmed gas chromatographic method benzaldehyde with a retention distance relative to methyl styrene of 1.60 is well resolved from all other impurities in styrene monomer.

30.7 DISCUSSION OF RESULTS

This method is capable of determining benzaldehyde in amounts down to 5 ppm in styrene monomer with an accuracy of $\pm$ 5%.

METHOD 31 - DETERMINATION OF PHENOLIC ANTIOXIDANTS IN POLYSTYRENE. *P*-NITROANILINE COUPLING - SPECTROPHOTOMETRIC METHOD.[3]

31.1 SUMMARY

This spectrophotometric procedure determines phenolic antioxidants in amounts down to 0.02% in polystyrene.

31.2 APPARATUS

Visible spectrophotometer, 1 cm quartz cells.

31.3 REAGENTS

Ethyl alcohol 95%
Sodium hydroxide, 4M, Analar.
p-Nitroaniline.
Sodium nitrite, analytical reagent grade.

Coupling agent, 2.800 grams of *p*-nitroaniline dissolved in 32 cm^3 of hot concentrated hydrochloric acid and diluted with water to 250 cm^3. After cooling to room temperature, the volume of liquid is adjusted to exactly 250 cm^3. A second solution is made containing 1.44 (0.0209 mole) of sodium nitrite in 250 cm^3 of distilled water. Both the above solutions are stable indefinitely. Pipette 25 cm^3 of each of these solutions into separate 100 cm^3 beakers and chill to below 10°C. Bubble pure nitrogen through the mixture as it is allowed to warm to room temperature. Finally, add 10 mg of urea to destroy any excess nitrous acid. Fresh reagent should be made every day.

31.4 METHOD

A 95% alcohol extract of the polymer is reacted with diazotisation reagent to produce a colour which is evaluated spectrophotometrically. The method is calibrated against ethanol solutions of the phenolic antioxidant being determined.

31.5 EXPERIMENTAL PROCEDURE

The sample to be analysed must be very thinly sheeted (or powdered by passing through a Wiley Mill). Weigh a 2.000 $\pm$ 0.020 gram sample and wrap with extraction cloth which has been previously extracted to remove sizing. Place the sample in an Underwriter's extraction cup and extract for 16 hours with 95% ethanol or methanol. Transfer the extract to a 100 cm^3 volumetric flask, cool and adjust to 100 ml with the extraction solvent. Transfer 10 ml of this solution to a 100 ml volumetric flask and add 2 cm^3 of coupling reagent and 3 cm^3 of 4 M sodium hydroxide solution then adjust to 100 ml with 95% ethanol or methanol.

Determine the absorption spectrum from 700 to 400 nm. The colour is stable for at least 2 hours. Ethyl alcohol is used in the reference cell unless the alcohol extract is strongly coloured. In this case, use a 10 cm^3 aliquot of the ethyl alcohol extract diluted to 100 cm^3 with ethyl alcohol as reference. Plot absorbance against concentration on semilogarithmic graph paper. The per cent antioxidant is calculated using the equation developed for the antioxidant concerned.

31.6 DISCUSSION OF RESULTS

This method is capable of determining down to 0.02% of phenolic antioxidants in polymers with an accuracy of $\pm$ 5%. To calibrate the method it is necessary to know the particular antioxidant being determined, ie. the method is only applicable to known antioxidants.

METHOD 32 - DETERMINATION OF UVITEX OB ULTRAVIOLET ABSORBER IN POLYSTYRENE. FLUOROMETRIC PROCEDURE.

32.1 SUMMARY

This fluorometric method determines Uvitex OB ultraviolet absorber in polystyrene in amounts down to 0.02%.

32.2 APPARATUS

Spectrometer with fluorimetric attachment.
1 cm silica fluorimetric cells.
25 cm^3 and 100 cm^3 graduated flasks.
Ultraviolet lamp (wavelength 350 mu).
5 cm^3 and 1 cm^3 graduated pipettes.
Mechanical shaker.

32.3 REAGENTS

Chloroform, spectroscopic or analar grade.
Polystyrene, free of Uvitex OB.
Uvitex OB standard, 0.01% in chloroform.

32.4 METHOD

A chloroform extract of the polystyrene is subject to excitation by ultraviolet radiation of wavelength 370 nm from a mercury vapour lamp and the fluorescence spectrum of the sample recorded over the range 400 - 440 nm. The reading from the fluorimeter is noted and the Uvitex OB concentration in the polystyrene determined by reference to a prepared calibration graph.

32.5 EXPERIMENTAL PROCEDURE

32.5.1 Procedure

Run 0.8% w/v solutions of polystyrene in chloroform as below and record peak height at 435 nm. Obtain the Uvitex OB content of the polymer by reference to the calibration curve.

32.5.2 Calibration.

Into each of seven 100 cm^3 volumetric flasks weigh 0.800 g of additive free polystyrene then make up to 100 ml with additions of 0.1 to 5 cm^3 of 0.01% w/v Uvitex standard. Record the fluorescence spectrum of each solution, prepared as above from 400 to 440 nm at a suitable slit width. If the 435 nm peak is less than 15 divisions, or greater than 80 divisions in height, re-run the trace on a higher or lower slit width, respectively. Measure the height of the peaks at 435 nm for the set of calibration solutions at each slit width setting used. Plot a calibration curve of peak height versus the corresponding concentrations of Uvitex OB in the standard solutions.

32.6 DISCUSSION OF RESULTS

Antioxidants such a Ionol CP (2,6,di-tert-butyl *p*-cresol), Ionox 330 (1,3,5-tri-methyl-2,4-6-tri(3,5-di-tert-butyl-4-hydroxybenzyl)benzene), Polygard (tris (nonylated phenyl) phosphite), Wingstay T (described as a butylated cresol), and Wingstay W and many others, do not interfere in this procedure.

This method is capable of determining down to 0.02% Uvitex OB in polystyrene with an accuracy of $\pm$ 5%.

METHOD 33 - DETERMINATION OF *P*-TERT-BUTYL PERBENZOATE IN POLYSTYRENE. CATHODE-RAY POLAROGRAPHY.

33.1 SUMMARY

This polarographic method determines down to 0.02% *p*-tert-butyl perbenzoate in polystyrene.

33.2 APPARATUS

Cathode-ray polarograph with dropping mercury electrode, 10 cm^3 polarograph cells and thermostatted (25°C) water bath.

Micrometer syringe capable of delivery 0.01 cm^3 with an accuracy of 0.0002 cm^3.

Volumetric glassware, pipettes, 50 cm^3, 25 cm^3, 5 cm^3, volumetric flasks 100 cm^3.

Centrifuge to take 250 cm^3 centrifuge bottles.

33.3 REAGENTS

Toluene, redistilled pure grade, rendered peroxide-free by shaking 1 L of the solvent with 10 cm^3 of iron sulphate solution, prepared by mixing 60 g iron II sulphate, 6 cm^3 concentrated sulphuric acid and 110 cm^3 distilled water, followed by redistillation. Store in a stoppered brown glass bottle.

Base electrolyte (0.6 M) remove peroxides from 1 L of methyl alcohol as described above. Weigh out 2.544 g lithium chloride (Analar) and make up to 100 ml with methyl alcohol.

p-tert-butyl perbenzoate.

Standard addition solution. Prepared by diluting a suitable weight of *p-tert*-butyl perbenzoate with toluene.

Nitrogen, oxygen content less than 25 ppm Mercury, pure for polarographic analysis, use trebly distilled mercury, the manufacturers should be requested to supply this mercury in stone containers as over a period of time mercury picks up an impurity from polyethylene storage bottles which interferes in polarography.

33.4 METHOD

To a toluene extract of the polymer is added methanolic lithium chloride base electrolyte. Precipitated polymer is removed and the resulting clear layer polarographed. Calibration is achieved by polarography of known toluene-methanol solutions of p-tert butyl perbenzoate.

33.5 EXPERIMENTAL PROCEDURE

33.5.1 Procedure

Transfer 5.0 ± 0.01 g of sample to a 250 cm^3 Pyrex glass centrifuge bottle and add 50 cm^3 of toluene. Drop a polythene coated magnetic stirrer rotor into the bottle and stopper with a cork (avoid rubber bungs). Stand the bottle on a magnetic stirrer and leave for several hours until the sample has completely dissolved or dispersed in the solvent. Accurately pipette in 50 cm^3 of 0.6 M lithium chloride reagent.

Centrifuge the bottle at 700 - 900 g until insolubles have completely settled to the bottom of the bottle leaving an absolutely clear upper phase containing the peroxide.

Dilute 50 cm^3 toluene with 50 cm^3 of lithium chloride solution to serve as a blank. Pipette 5 cm^3 of the sample solution and 5 cm^3 of the blank solution into two polarographic cells and immerse these in the constant temperature tank of the cathode-ray polarograph (thermostatted at 25°C). Carry out the degassing operations with oxygen free nitrogen on the sample and on the blank solutions immediately before carrying out polarographic measurements.

The analytical condition with the cathode-ray polarograph (with single cell operation) are as follows:

Cathode	dropping mercury
Reference anode	mercury pool
Circuit	forward sweep (with derivatives units control switch off)
Sensitivity contol	select a suitable current scale factor and keep this constant throughout the analysis, adjust the

	instrument sensitivity by means of the shunt scale factor control.
Start potential	- 0.7 V

By means of the "Y" shift contol, adjust the light spot to the origin of axes on the left-hand side of the graticule on the cathode-ray tube. Repeat this operation at different shunt scale-factor settings, until the polarographic wave is visible on the graticule.

Take the readings on the freshly degassed solutions as follows. Adjust the polarograph to the *p-tert*-butyl perbenzoate start potential (-0.7 V) and obtain the wave as described above for the polystyrene sample solution. Read off from the graticule the maximum height of the peak (at above -0.9 V). Raise the dropping mercury electrode from the cell and deliver into the sample solution a suitable "standard addition" of a toluene solution of *para-tert*-butyl perbenzoate.

Limit the volume of "standard addition" solution to less than 0.05 cm^3 in order to avoid dilution errors. Lower the electrode into the sample cell and again pass oxygen free nitrogen for 2 min. Immediately read the new peak height at - 0.9 V. Similarly, determine the peak height at - 0.9 V on the degassed polystyrene-free reagent blank solution.

33.5.2 Calculations

p-tert-butyl perbenzoate (ppm w/w)

$$= \frac{100 \times M \times 10^6}{5W} \times \frac{(h_1S_1 - h_3S_3)}{(h_2S_2 - h_1S_1)}$$

where

W = weight (g) of polystyrene sample taken for analysis (assuming 5 cm^3 portion taken for polarography)

h_1 = peak height (graticule divisions) of sample solution before standard addition

h_2 = peak height (graticule divisions) of sample solution after standard addition

h_3 = peak height (graticule divisions) of polymer-free reagent blank solution.

S_1, S_2 and S_3 are the corresponding instrument sensitivity settings (the product of h and S are known as peak currents in microamps)

M = weight (g) of *p-tert*-butyl perbenzoate in volume of "standard addition" solution added to cell solution.

33.6 DISCUSSION OF RESULTS

This method is capable of determining down to 0.02% of *p-tert*-butyl perbenzoate in polystyrene with an accuracy of $\pm$ 5%. Other organic peroxides which reduce at or near - 0.9 V will interfere.

METHOD 34 - DETECTION AND DETERMINATION OF *P-TERT* BUTYL PERBENZOATE AND BENZOYL PEROXIDE RESIDUES IN EXPANDABLE POLYSTYRENE. THIN LAYER CHORMATOGRAPHY.

34.1 SUMMARY

This thin layer chromatographic method determines *p-tert*-butyl perbenzoate and/or benzoyl peroxide residues in expandable polystyrene in amounts down to 0.002%.

34.2 APPARATUS

Thin layer chromatography tank to take 8 in x 8 in plates.

Thin layer chromatography plates (8 in x 8 in) precoated with a 1 micron layer Merck GF 254 silica gel.

Hypodermic syringes - 25 ul capacity.

Filter paper - Whaman No. 42

Spray for chromogenic reagents.

Volumetric flasks, 2 ml capacity.

Beakers 50 ml.

Ultraviolet lamp, 254 mu

34.3 REAGENTS

Potassium iodide starch, mix 1 g potassium iodide Analar and 0.1 g zinc dust with 100 ml 4:1 (v/v) glacial acetic acid water.

Starch, 1%.

Ferric rhodamide reagent, dissolve 0.2 g ammonium rhodamide in 15 cm^3 acetone. Shortly before use, dilute this solution with 10 cm^3 4% aqueous ferrous sulphate, use the solution immediately.

Diethyl ether, redistilled pure grade. Render peroxide free by shaking with 1 litre of solvent with 10 - 20 cm^3 of iron II sulphate solution, prepare by mixing 60 g iron II sulphate, 6 cm^3 concentrated sulphuric acid and 110 cm^3 distilled water followed by redistillation. Store in a well stoppered brown glass bottle.

Benzene, redistilled pure grade, peroxide free (see above).

Benzoyl peroxide.

Standard benzoyl peroxide stock, 0.5% w/v in peroxide free benzene, prepare 0.005% to 0.15% dilutions of this by diluting with benzene.

p-tert butyl perbenzoate.

Standard *p-tert*-butyl perbenzoate, 0.5% w/v in peroxide free benzene, prepare 0.005% to 0.15% dilutions of this by diluting with benzene.

34.4 METHOD

An ether extract of the polystyrene is applied to a thin-layer chromatographic plate and developed with benzene . Expose to ultraviolet light and the application of specific spray reagents reveals spots due to the peroxides which are compared visually with standard solutions of pure peroxides to quantify the procedure.

34.5 EXPERIMENTAL PROCEDURE

34.5.1 Extraction

Weigh out 5 $\pm$ 0.1 g of sample and add 25 cm^3 diethyl ether. Cover with a watch glass and leave overnight. Decant the ether and wash the beads with two further 10 cm^3 portions of ether, decanting each washing into the main extract. By means of an air line and warm water bath carefully evaporate off the ether until about 1 ml of solution remains. Transfer this to a 2 cm^3 volumetric flask. Rinse round the walls of the beaker with 5 - 10 cm^3 of ether and concentrate to about 0.5 cm^3. Transfer this solution to the 2 cm^3 volumetric flask and make the flask contents up to 2 cm^3 with diethyl ether.

34.5.2 Chromatography.

Activate the silica gel coated plates by heating in a ventilated oven for 30 min at 110°C. Store the plates in a dessicator until use, they should be used on the same day that they are activated.

Fill a chromatograph tank to a depth of about 0.5 in with benzene and place sheets of filter paper against the internal walls of the tank and dipping in the benzene. Replace the lid and leave the tank for 30 min to equilibriate in a constant temperature area free from draughts.

Transfer by microsyringe two 20 ul portions of the diethyl ether extract of the sample as separate spots on the starting line on the plate. Also apply 20 ul portions of the standard solutions of the peroxides in benzene (ie. 0.005 to 0.15% equivalent to 1 to 30 ug peroxide) to the plate. Develop the chromatogram to a distance of 15 cm from the starting line. Remove the plate from the tank and allow to air dry.

Inspect the plate under a 254 nm ultraviolet lamp and compare the intensities of the sample and standard spots which show as dark areas on a blue fluorescent background. Spray the plate with either of the spray reagent systems (ie. potassium iodide/starch or ferric rhodamide). Estimate the concentration of peroxides in the sample spots by comparison with the standards.

34.5.3 Calculation.

Peroxide from 6 g polyethylene is concentrated into a 2 ml ether solution and 0.02 ml (20 ul) taken for thin-layer chromatography.

$$\text{Thus \% peroxide in polymer} = \frac{2 \times 5 \times P}{0.02}$$

Where

P = weight (g) of peroxide present in standard spot which compares in intensity with the intensity of the sample spot.

34.6 DISCUSSION OF RESULTS

This method is capable of determining down to 0.002% p-tert-butyl peroxide or benzoyl peroxide in polystyrene with an accuracy of $\pm$ 10%.

METHOD 35 - TESTS FOR INHIBITORS IN STYRENE MONOMER. IRON III CHLORIDE/POTASSIUM FERRICYANIDE PROCEDURE.

35.1 SUMMARY

This spectrophotometric method determines inhibitors such as hydroquinone, *p-tert* butyl catechol or p-methoxy phenol in amounts down to 10 ppm in styrene monomer.

35.2 APPARATUS

Ultraviolet-visible spectrophotometer

35.3 REAGENTS

Iron III chloride, 0.05% w/v aqueous
Potassium ferricyanide, 0,01% w/v aqueous.

35.4 METHOD

Toluene solutions of the styrene monomer sample are extracted with the reagents to provide aqueous extracts which are evaluated spectrophotometrically. The procedure is calibrated against standard solutions of the inhibitors.

35.5 EXPERIMENTAL PROCEDURE

35.5.1 Procedure

If the monomer sample contains up to 0.005% of inhibitor then pipette 0.1 to 2 cm^3 of the sample directly into a 100 cm^3 separatory funnel, make the volume up to 2 ml with toluene. Into a second (reagent blank) separatory funnel pipette 2 cm^3 of pure toluene. To each funnel add 1 cm^3 of freshly prepared iron III chloride solution (0.5%) and shake for one min. Add 10 cm^3 water and allow the phases to separate. Run the aqueous phases into two 100 cm^3 volumetric flasks. Into each volumetric flask pipette 1 cm^3 of potassium ferricyanide solution and dilute to 100 cm^3 .

Evaluate the sample and blank solutions on a spectrophotometer under the following conditions:

Cells	1 cm glass
Blank solution	Fill comparison cell with the inhibitor free reagent blank solution referred to above.
Wavelength	Determine the wavelength at which maximum absorption occurs.
Temperature	$25 \pm 2°C$
Colour development time	15 to 20 min from addition of first reagent

35.5.2 Calibration.

Accurately weigh out on a watch glass 0.006g of the stabilizer that is being determined in the monomer sample. Transfer this solid to a 100 cm^3 volumetric flask with toluene and make up to the mark. Into each of seven 100 cm^3 separatory funnels pipette the volumes of 0.006% monomer solution and pure toluene shown below. Continue as described under procedure. Construct a calibration graph by plotting the determined optical densities of the solution against the corresponding number of grams of inhibitor in each 100 ml separatory funnel.

Volumes of 0.006% inhibitor solution - required for preparation of calibration graph.

Volumes added to 100 cm^3 separatory funnel		Weight of inhibitor added g
0.006% w/v inhibitor in toluene cm^3.	Pure toluene	
0.0 (blank)	3.0	0.00000
0.5 (sample)	2.5	0.00003
1.0 (sample)	2.0	0.00006
1.5 (sample)	1.5	0.00009
2.0 (sample)	1.0	0.00012
2.5 (sample)	0.5	0.00015
3.0 (sample)	0.0	0.00018

35.5.3 Calculation

Convert the optical density obtained for the sample solution to grams of inhibitor by means of the appropriate calibration graph obtained for a pure specimen of the inhibitor. Calculate the inhibitor content of the monomer as follows:

Inhibitor content of monomer

$$\text{Inhibitor (\% 2/v)} = \frac{N\text{x } 100}{V}$$

where

N = weight (g) of inhibitor obtained by refering the sample optical density to the calibration graph.

V = volume (cm^3) of monomer sample taken for analysis (allow for any preliminary dilution of the monomer that has been made before analysis)

35.6 DISCUSSION OF RESULTS

This method is capable of determining styrene monomer inhibitors in styrene in amounts down to 10 ppm with an accuracy of $\pm$ 5%. To obtain quantitative results it is necessary to know the identity of the particular inhibitor present in the styrene monomer sample.

METHOD 36 - TESTS FOR INHIBITORS IN STYRENE MONOMERS. SODIUM PHOSPHOTUNGSTATE PROCEDURE NESSLERIZATION METHOD.

36.1 SUMMARY

This visual colour comparison method determines inhibitors such as hydroquinone, p-tertiary butyl catechol and o-methoxy phenol in amounts down to 1 ppm in styrene monomer.

36.2 APPARATUS

Nesslerizer tubes, 50 ml.
Separatory funnels, 50 ml

36.3 REAGENTS

Toluene
Sodium phosphotungstate, 2% aqueous.
Sodium carbonate, 10% aqueous.

36.4 METHOD

Toluene solutions of the styrene monomer are extracted with distilled water and to the extract are added sodium phosphotungstate and sodium carbonate solutions. The intensity of the blue colour produced is compared visually with that of standards comprising solutions of the inhibitor in toluene which have been similarly treated.

36.5 EXPERIMENTAL PROCEDURE

Into a 50 cm^3 separatory funnel pipette 1 cm^3 (or for greater sensitivity use 10 cm^3) of the monomer sample.

Into seven further 5 cm^3 separatory funnels pipette 0.1, 0.5, 1.0, 2.0, 5.0, 7.5 and 10.0 cm^3 of a 10 ppm toluene solution of the inhibitor which is being determined in the monomer sample (ie. additions of between 5 and 100 u g of inhibitor. Extract the sample funnel and the seven standard funnels with 20 cm^3 distilled water then two 10 cm^3 portions of distilled water and collect the 40 cm^3 aqueous extracts in 50 cm^3 Nessler cylinders (filtering if necessary). Dilute to 50 cm^3 with water. To the sample and standard solutions add 2 cm^3 sodium phosphotungstate reagent and 4 cm^3 of sodium carbonate (10%) in this order and mix well after the addition of each reagent. Leave the solutions for 15 min and compare the depth of the blue colour obtained for the sample with that obtained from the various standard solutions. If necessary, repeat the run this time more closely bracketing the sample solution with the standards to obtain a better colour match between sample and standards.

36.6 DISCUSSION OF RESULTS

This method is capable of determining styrene monomer inhibitors in styrene in amounts down to 1 ppm with an accuracy of ± 10%. To obtain quantitative results it is necessary to know the identity of the particular inhibitor present in the styrene monomer sample.

METHOD 37 - DETERMINATION OF HYDROQUINONE INHIBITOR IN MONOMERS. CATHODE-RAY POLAROGRAPHY.

37.1 SUMMARY

A cathode-ray polarographic procedure is described which determines hydroquinone, substituted benzoquinone and anthraquinone inhibitors in styrene monomer in amounts down to 5 ppm.

37.2 APPARATUS

Cathode ray polarograph complete with dropping mercury electrode polarographic cells (10 cm^3) and thermostated (at 75°C) water bath. The sensitivity of the procedure for determining hydroquinone depends on the cleanness of the capillary used in the dropping mercury electrode. A new capillary should be used when any decrease in sensitivity in the polarographic procedure is noted.

Agla micrometer syringe capable of delivery 0.01 cm^3 with an accuracy of 0.002 cm^3.

Volumetric glassware. Pipettes 25 cm^3, 5 cm^3, volumetric flasks, 10 cm^3.

37.3 REAGENTS

For polarography of aqueous samples.

Aqueous lithium phosphate base electrolyte PH 7.0. Dissolve 3.60 ± 0.03 g of lithium hydroxide ANALAR in distilled water and make up to 100 cm^3. Dilute 11.5 ± 0.1 g of 85% phosphoric acid to 100 cm^3 with distilled water. Pipette 10 ml of the lithium hydroxide solution, 500 cm^3 distilled water and 10 cm^3 of the phosphoric acid solution into a 1 litre flask and dilute to the one litre mark with distilled water.

Hydroquinone, for calibration.

Aqueous solution of hydroquinone, 0.1%. Prepare freshly each day.

For polarography of alcohol soluble samples.

Alcoholic acetate base electrolyte PH 7.4. Dissolve 8.2 ± 0.1 g of anhydrous sodium acetate ANALAR in distilled water and dilute to 100 cm^3. Dilute 6.05 ± 0.05 g glacial acetic acid ANALAR to 100 cm^3 with distilled water. Pipette 5 cm^3 of each of these solutions and 2.5 cm^3 distilled water into a 500 cm^3 volumetric flask. Dilute to the 500 cm^3 mark with absolute ethanol and mix.

Standard hydroquinone for calibration, alcoholic solution in absolute ethanol 0.1%, prepare freshly each day.

Mercury, pure for polarographic analysis, use trebly distilled mercury supplied in stone containers. Over a period of time mercury picks up an impurity from polyethylene storage bottles which might interfere in polarography (see note 1).

Nitrogen, oxygen content, 25 ppm.

37.4 METHOD

The monomer sample is diluted with a suitable aqueous or alcoholic polarographic base electrolyte solution.

The deoxygenated base electrolyte solution of monomer is then polarographed in the range - 0.2 volts to + 0.3 V. The wave occuring at 0.0 V due to hydroquinone is then evaluated. The concentration of hydroquinone in the cell solution is then determined by making "standard additions" of a solution of hydroquinone.

37.5 EXPERIMENTAL PROCEDURE

37.5.1 Procedure

Depending on the expected hydroquinone content of the sample, weigh out a suitable amount of monomer (up to 1 g) into a clean 10 cm^3 volumetric flask. Make the sample up to the 10 cm^3 mark with the alcoholic or the aqueous base electrolytes.

Pipette 5 cm^3 of the sample solution and 5 cm^3 sample-free blank base electrolyte solution into two polarographic cells and immerse these in the constant temperature tank of the cathode ray polarograph (thermostated at 25°C)

The degrassing operations described in this section must be performed on the sample and on the blank solutions immediately before carrying out all polarographic measurements. Connect a supply of oxygen-free nitrogen to the polarographic cell. Lower the dropping mercury electrode system over this cell, so that the outer glass sleeve of the electrode dips 1 to 2 mm into the water tank (providing a water seal to prevent the ingress of atmospheric oxygen - note 2). Immerse the anode connection in the side arm of the polarographic cell. Pass nitrogen for 3 min to completely displace oxygen from the cell solution, then switch off the nitrogen. Leave the glass sleeve in position to prevent re-entry of atmospheric oxygen into the cell solution and carry out the polarographic measurements described below within 1 to 3 min of stopping the nitrogen purge.

The analysis is carried out using a dropping mercury electrode and a mercury pool reference electrode (note 3). Operate the polarograph using the anodic direct circuit. First adjust the instrument to -0.25 V start potential. Move the X-shift control until the light spot on the cathode ray tube graticule commences its horizontal sweep exactly at the left hand vertical axis of the graticule. Now move the Y shift control until the light spot commences its sweep at the origin of axes of the graticule (left hand side).

If the polarographic wave is not now visible on the graticule then the instrument is being operated at too low a sensitivity setting. Switch to higher sensitivity settings and repeat the operations described in the previous paragraph until the 0.0 V polarographic wave for hydroquinone is visible. Read off from the graticule the maximum height of the hydroquinone peak.

Similarly, in the case of the reagent blank solution obtain the height of the wave occuring at 0.0 V (use the same instrument sensitivity setting that is employed when evaluating the sample solution).

Raise the electrode from the cell and deliver into the sample solution a suitable "standard addition" (note 4) of a 0.1% solution of hydroquinone in ethanol (or in water) using a micrometer syringe for delivery (see note 5). In order to avoid sample dilution errors "standard additions" should be kept below 0.05 cm^3 at this stage of the analysis. Lower the electrode into sample cell and pass nitrogen for 3 min before continuing the determination. If necessary, reset the polarograph at a suitable sensitivity setting and note the new height of the hydroquinone wave at 0.0 V (equivalent to reagent blank plus hydroquinone in sample plus hydroquinone in "standard addition" solution.

37.5.2 Calculations

$$\text{Hydroquinone (ppm/w/v) in monomer} = \frac{10^6 \times M\,(H_1S_1 - H_2S_2)}{W\,(H_3S_3 - H_1S_1)}$$

where

W = weight (g) of monomer sample taken for analysis.

M = weight (g) of hydroquinone present in "standard addition" of hydroquinone solution made to polarographic cell.

H_1 = peak height (cms) at 0.0 V of sample solution before standard addition.

H_2 = peak height (cms) at 0.0 V of reagent blank solution.

H_3 = peak height (cms) at 0.0 V of sample solution after "standard addition".

S_1, S_2 and S_3 are the corresponding instrument sensitivity settings (the product of H and S is known as peak current in u amps)

37.5.3 Note 1. Cleanness of Mercury

Blocking or contamination of the dropping mercury electrode capillary is prevented only by using mercury of the highest degree of purity. Mercury should be transferred to the electrode reservoir through a pin hole in the cone of a No. 40 Whatman filter paper. The filter paper should be cleaned and unmarked at the end of the mercury filtration.

37.5.4 Note 2. Removal of Dissolved Oxygen from Cell Solution.

Prior to polarography oxygen must be completely removed from the cell solution in order to prevent it interfering in the determination of hydroquinone.

37.5.5 Note 3. Operation of Polarograph.

After the solution has been placed in the cell and de-oxygenated by bubbling with nitrogen, the start potential control is set to a voltage about 0.1 V more positive than the half-wave potential of the ion under examination. The X and Y shift controls are then employed to set the spot to the zero point on the graticule. The sensitivity control is set to give a convenient step height and the instrument is allowed to sweep several times in order to come into synchronism with the mercury drops. This usually takes place in about two sweeps, but may be hastened by detaching the mercury drop from the capillary by a sharp tap on the stand base as the spot is passing between 0.4 and 0.5 V on the horizontal scale. As soon as synchronism has been achieved, the height of the peak may be read of the graticule. Further peaks, if present, may then be examined by simply re-setting the start potential control and bringing the spot back to the zero point.

37.5.6 Note 4. Calibration of Polarograph.

The "standard addition" calibration technique is used. The peak current (C_1) (corrected for reagent blank) due to hydroquinone in the sample solution is noted. "Standard addition" of a known weight of hydroquinone (Mgrams) is then made to the cell solution and the peak current increase (C_2) corresponding to this addition is noted.

Then: weight of monomer in sample = $\frac{M \times C_1}{C_2}$

Note 5. Use of Microsyringe.

More accurate deliveries result when the barrel of the microsyringe are held in a horizontal position when delivering a "standard addition" into the cell solution.

37.6 DISCUSSION

A cathode-ray polarographic method is described for the determination of hydroquinone in amounts down to five parts per million in the following organic monomers with an accuracy of $\pm$ 5%.

Methyl methacrylate.
Methyl alphachloroacrylate.
Alpha methyl styrene.
Styrene.
Vinylidene chloride.
Methacrylanitrile.
Acrylonitrile.
Dimethyl itaconate.

The procedure is applicable to samples which are soluble in alcohol or in water. Benzoquinone is also included in the determination of hydroquinone. (Benzoquinone is the oxidation product of hydroquinone and might also be present in monomer samples).

Copper, iron and other materials which reduce near 0.0 V will interfere in the polarographic determination of hydroquinone.

Substituted benzoquinones and anthraquinones can also be determined by the described procedure. They are reduced in the range 0.0 to 1.0 V and some of these will interfere in the determination of hydroquinone.

Hindered phenolic types of monomer stabilizers (eg. p-methoxy phenol and *p-tert* butyl catechol) do not interfere in the described procedure.

METHOD 38 - DETERMINATION OF *P-TERT* BUTYL CATECHOL INHIBITOR IN STYRENE MONOMER. SPECTROPHOTOMETRIC METHOD.

38.1 SUMMARY

This spectrophotometric method determines *p-tert* butyl catachol inhibitor in styrene monomer in amounts down to 0.1 ppm.

38.2 APPARATUS

Ultraviolet visible spectrometer.

Cells, 1 cm path length, fused glass with lids.

Miscellaneous glassware - pipettes, separator funnels, $1 dm^3$ capacity, filter funnels.

38.3 REAGENTS

Sodium hydroxide, 1 M, aqueous, carbonate free.

p-tert-butyl catechol – solid styrene monomer - purified. Wash several litres of styrene monomer with separate portions of aqueous 1 normal sodium hydroxide solution until the aqueous layer shows no red colouration due to p-tert butyl catechol.

38.4 METHOD

The sample of styrene monomer is shaken with sodium hydroxide solution in the presence of air. The inhibitor is oxidised to form a red coloured quinoid compound which extracts into the sodium hydroxide phase. The colour intensity of the complex is measured using a spectrophotometer at a wavelength of 480 nm and hence the amount of *p-tert* butyl catechol present in the styrene monomer may be read off directly from a prepared calibration graph.

38.5 EXPERIMENTAL PROCEDURE

Pipette 10 cm^3 of 1 M sodium hydroxide solution into a 1 cm^3 separatory funnel and add 500 cm^3 of the styrene monomer sample. Shake the mixture for exactly three min and allow to settle for 8 $\pm$ 1 min (see note 1). Run off the aqueous layer into a 1 cm cell (see note 2) and measure the optical density of the solution versus 1 M sodium hydroxide solution in the reference cell at a wavelength of 480 nm and at 12 min after the initial addition of monomer sample to the sodium hydroxide reagent.

38.5.1 Calibration.

Pipette 1, 2, 3, 4, 5 and 10 cm^3 of a 100 ppm solution of *p-tert* butyl catechol dissolved in purified styrene monomer, into a separate 500 cm^3 volumetric flasks and dilute to the mark with purified monomer. Mix the solutions thoroughly and proceed as described under Method. Plot the optical density values obtained against the concentration of p-tert butyl catechol (ppm) present in the 500 cm^3 dilutions.

38.5.2 Calculation.

From the optical density obtained read off the *p-tert* butyl catechol concentration present in the 500 cm^3 styrene monomer sample, directly from the calibration graph.

38.5.3 Note 1.

The colour of the oxidised *p-tert* butyl catachol fades gradually with time and the optical density of the solution must be read a stipulated time after shaking.

38.5.4 Note 2.

If the aqueous layer shows any turbidity it must be filtered through glass wool into the 1 cm^3 cell, to remove turbidity.

38.6 DISCUSSION OF RESULTS

This method is capable of determining down to 0.1 ppm *p-tert* butyl catechol inhibitor in styrene monomer with an accuracy of $\pm$ 5%.

METHOD 39 - IDENTIFICATION OF TIN CONTAINING DIALKYLTIN STABILIZERS IN PVC.

39.1 SUMMARY

This procedure involving the use of infrared and NMR spectroscopy is used for the identification of sulphur containing organotin stabilizers in PVC. Compounds belonging to this group include:

dialkyl tindialkylthioglycollates,
R_2S_n (SCH_2COOR') $(S\ CH_2COOR'')$
dialkytin dilauryl mercaptides,
R_2S_n $(S\ CH_2(CH_2)_{10}\ CH_3)_2$

and stabilizers in which the dialkyl tin group is combined with both thiol and carboxyl groups. R, R′ and R″ denote alkyl groups - they can be the same or different.

39.2 APPARATUS

Infrared spectrometer.
NMR spectrometer.
X-Ray fluorescence spectrometer (for checking for the presence of sulphur).
Gas liquid chromatograph (for identification of alcohols).
Filter paper, Whatman No. 541.
Thin-layer chromatography apparatus.

39.3 REAGENTS

Acetone, Analar.
Silver nitrate, aqueous 50%.
Diethyl ether, Analar.
Hydrochloric acid, concentrated.

Sodium hydroxide, 5M.

39.4 METHOD

A variety of techniques as reviewed below are used to identify different sulphur containing stabilizers in PVC.

esters and metal carboxylates	infrared spectroscopy
alcohols	infrared spectroscopy and gas chromatography
thio acids and thiols	infrared spectroscopy and NMR spectroscopy
alkyl groups attached to tin	infrared spectroscopy and thin-layer chromatography

39.5 EXPERIMENTAL PROCEDURE

The scheme of analysis in Figure 7.7 illustrates the chemical procedures involved in the characterisation of the structural groups present.

39.5.1 Infrared Spectroscopy in Presence of Ester and Carboxylate.

The infrared spectrum of the sample will show the presence of ester groups, metal carboxylates, etc., and will indicate whether or not the sample is a mixture and in need of separation (eg. the detection of impurities and of stabiliser in the presence of excess of plasticiser).

Place about 1g of the sample into a 150 cm^3 flat-bottomed stoppered flask and add 100 cm^3 of acetone. Swirl to dissolve, then add dropwise, with swirling 1 cm^3 of a 50 per cent solution of silver nitrate, in distilled water. Shake the flask well and allow it to stand until the precipitate has settled out. Filter the precipitate through a Whatman No. 541 filter paper, wash it with a small amount of distilled water, then with acetone and finally with a small amount of diethyl ether. Retain the filtrate and washings (filtrate A, Figure 7.7). Dry the precipitate (precipitate B) in a desiccator and determine its infrared spectrum of using the nujol emulsion technique.

39.5.2 Infrared Spectroscopy and Gas Chromatography. Identification of Alcohols.

If the infrared spectrum of precipitate B (Figure 7.7) shows the presence of ester bands transfer the precipitate to a 150 cm^3 flat bottomed flask and shake it with 50 cm^3 of diethyl ether. Add 1 cm^3 of concentrated hydrochloric acid and shake the flask vigorously for 1 min to precipitate the silver as silver chloride. Allow it to stand for 30 min then decant into a 150 cm^3 flat bottomed extraction flask through a Whatman No. 541 filter paper, washing the flask and the precipitate with 20 cm^3 of diethyl ether. Reject the precipitate but evaporate the ether extract to dryness on a water-bath. This yields the residue C (see Figure 7.7). Determine the infrared spectrum of the residue. This spectrum should show the presence of ester bands and may already suggest the nature of the alcohol involved.

Next, add 20 cm^3 of 10 per cent aqueous sodium hydroxide solution and saponify the ester by refluxing on a hot-plate. Normally a refluxing time of 1.5 h is sufficient. Wash the condenser with a small amount of distilled water and allow to cool, then transfer the reaction mixture to a distillation apparatus. Distil, separate and identify the alcohols by gas - liquid chormatographic methods. If the gas - liquid chromatography

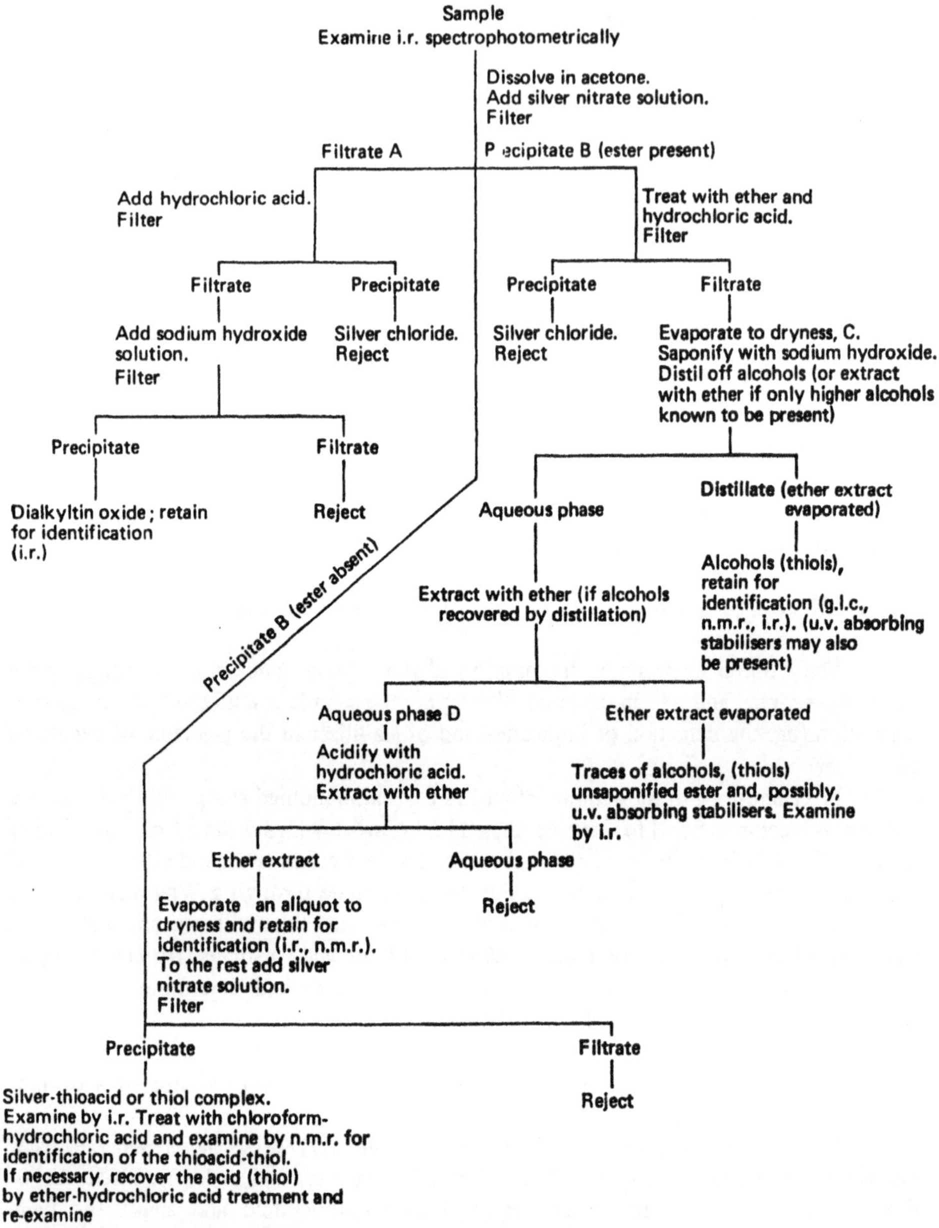

Figure 7.7. Scheme for analysis of dialkytin thiocompounds. From Udris with permission.[15] Royal Society of Chemistry, London.

retention times alone are considered unsatisfactory for complete identification, use the separated fractions for infrared and for nuclear magnetic resonance spectroscopic examination. Sometimes the gas - liquid chromatographic examination can provide information as to the approximate ratio of the alcohols present.

Extract the aqueous phase by shaking it with diethyl ether. Evaporate the ether extract to dryness and examine by infrared spectrophotometry. Retain the aqueous phase (Figure 7.7) for examination for thioacids.

If the infrared spectrum of precipitate B suggests the presence of only a mercaptide or carboxylate or mecapto-carboxylate with no ester bands proceed as outlined under the identification of thioacids and thiols as described below, ie. omitting the saponification step.

In the case of the less volatile alcohols, such as 2-ethylhexanol and higher alcohols, the alcohol can be recovered by extraction of the alkaline saponification mixture with diethyl ether, followed by careful evaporation of the solvent. Subsequent infrared (or nuclear magnetic resonance) examination of the residue usually suffices for the identification of the alcohol. Of course, if more than one alcohol is present, the gas - liquid chromatographic examination is called for.

39.5.3 Infrared Spectroscopy and NMR Spectroscopy. Identification of Thioacids and Thiols.

Acidify the aqueous phase D, remaining after the removal of alcohols (see above) with concentrated hydrochloric acid and extract the liberated thioacid with diethyl ether. Liquid - liquid extraction for 4 hours produces more material, but shaking with 50 cm^3 of diethyl ether for about 1 min normally gives enough extract for identification. Because of partial decomposition of the thioacid during the foregoing saponification, sulphides are produced and the above acidification and extraction should be carried out in a fume cupboard. The ether extract can be evaporated to dryness on a water bath and examined by infrared or nuclear magnetic resonance spectroscopy or by both techniques, but, because of partial decomposition a better way of identification at least for the thioacid, is as follows:

Evaporate to dryness and examine only an aliquot of the ether extract. To the rest, prior to evaporation, add a few drops of 50 per cent silver nitrate solution, dropwise, with shaking. Shake well and allow the precipitate to settle, then filter through a Whatman No. 541 filter paper washing well with ether. Dry in a desiccator and examine by infrared spectrophotometry and the nujol emulsion technique. Also, suspend an aliquot of the precipitate in chloroform, acidify with a little hydrochloric acid, shake and carry out nuclear magnetic resonance spectroscopic examination of the chloroform extract. This should suffice for the identification of the thioacid. The dark colour of the silver complex is probably due to the presence of silver sulphide.

If precipitate B contains no ester groups (see Figure 7.7) the hydrolysis with sodium hydroxide should be omitted. In this event, examine the precipitate by the above infrared and nuclear magnetic resonance technique for identification of the thioacid or the thiol.. Although this particular nuclear magnetic resonance technique was not used for the identification of thiols, it could prove equally effective.

If required, the free acid, usually thioglycollic, or thiol can be obtained in a purer state by acidifying the silver complex with hydrochloric acid and extracting the liberated acid and thiol with diethyl ether. The infrared and nuclear magnetic resonance examinations can then be repeated on the thioacid or thiol itself.

Mass spectrometric examination is also a useful test for the identification of thioacids, for example, thioglycollic acid gives the characteristic ion at m/e = 91.

39.5.4 Infrared Spectroscopy and Thin-Layer Chromatography. Identification of the Alkyl Groups Attached Directly to Tin. Precipitation with Sodium Hydroxide and Infrared Examination.

To filtrate A add a slight excess of concentrated hydrochloric acid (about 1 cm^3) to precipitate the excess of silver ions as silver chloride. Shake the mixture well, allow to settle and filter through a Whatman No. 541 filter paper. Make the filtrate alkaline by the addition of 25 cm^3 of 10 per cent sodium hydroxide solution and again shake and filter through a Whatman No. 541 filter paper, this time washing well with acetone. Dry at 105°C for 30 mins. This gives a crude sample of the dialkyltin oxide but this is sufficiently pure for identification by infrared spectrophotometric examination (potassium bromide).

Good samples of dibutyl and dioctyltin oxides have also been obtained by dispersing about 0.8 g of the stabiliser in 25 cm^3 of methanol and refluxing for 3 h with 20 ml of M sodium hydroxide solution. The white powder, isolated after filtering through a G3 sintered glass crucible, washing with distilled water followed by methanol and drying at 105°C, usually gives a characteristic infrared spectrum of the dialkyltin oxide, believed to be in polymeric form.

In addition to the infrared spectroscopy techniques described above thin-layer chromatography with hexane: glacial acetic acid (12 + 1) as eluting agent is satisfactory for the identification of thioacids and thiols. A small amount of the original tin stabiliser is dissolved in the elution solvent, applied to a thin-layer chromatographic plate coated with 0.25 mm thick layer of Kieselger G as the stationary phase, and eluted with the hexane: glacial acetic acid mixture. After drying, the stationary phase is sprayed with 0.1 per cent solution of catechol violet in 95 per cent ethanol, blue spots appearing where tin compounds are present.

If mixed tin stabilisers are believed to be present, for example, a thiocompound and a dialkyltin dialkylcarboxylate, it is advantageous to acidify the acetone filtrate after removal of the dialkyltin oxides with hydrochloric acid, evaporate most of the acetone on a water-bath under an air-jet, cool and extract by shaking with diethyl ether. The ether extract then contains hemiesters, acids, etc., liberated from the other type of tin stabilisers originally present and can be examined further. Although reaction products produced by the action of sodium hydroxide or acetone are often present, esters of maleic and benzoic acids have been identified in this way.

39.6 DISCUSSION OF RESULTS

These methods are basically qualitative although in some instances the methods can be put on a quantitative basis, eg. the gas chromatographic determination of alcohols produced by the hydrolysis of ester groups. Figure 7.8 shows an infrared spectrum of dibutyl tin dichloride showing that this compound can be easily identified by this technique. Good samples of dialkyltin dichlorides have been isolated by the above method from mixtures of tin stabilisers with diisooctyl phthalate and tritolyl phosphate; the dialkyltin groups have been identified in dibutyltin dinonylmaleate, dioctyltin dilaurate, dibutyltin dinonylthioglycollate and dioctyltin thioglycollate.

METHOD 40 - IDENTIFICATION OF DIALKYLTIN CARBOXATED AND HEMIESTERS STABILIZERS IN PVC.[15]

40.1 SUMMARY

This procedure involving the use of infrared spectroscopy, gas chromatography and thin-layer chromatography is used for the identification of sulphur-free type organotin stabilizers in PVC.

To this group belong stabilisers like dialkltin maleates, ($R_2Sn(OOCCH = CHCOOR^1)_2$, dialkyl tin dilaurates ($R_2Sn(OOC(CH_2)_{10}CH_3)_2$, where R and R′ may be the same or different.

40.2 APPARATUS

Infrared spectrometer.
Gas chromatograph.
Thin-layer chromatography apparatus.
Separating funnel: 150 ml.
Filter paper Whatman No. 541.
Water bath.

40.3 REAGENTS

Sodium hydroxide, aqueous, 10%.
Diethyl ether, Analar.
Hydrochloric acid, concentrated

40.4 METHOD

A variety of techniques, as discussed below, are used to identify different types of non-sulphur containing stabilizers in PVC.

ester and carboxylate	infrared spectroscopy
alcohols	gas chromatography
alkyl groups attached to tin	infrared spectroscopy and thin-layer chromatography

40.5 EXPERIMENTAL PROCEDURE

The scheme in Figure 7.9 illustrates the analytical procedures involved in the characterisation of the structural groups present.

40.5.1 Infrared Spectroscopy. Ester and Carboxy Groups.

Examine the infrared spectrum of the sample and deduce its general nature from the absence or presence of ester groups, free carboxyl groups, plasticizers or significant

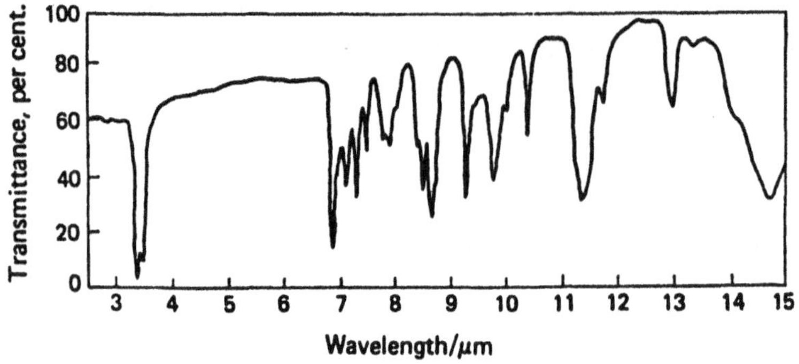

Figure 7.8. Infrared spectrum of dibutyltin-dichloride. From Udris with permission.[15] Royal Society of Chemistry, London.

amounts of other constituents. More detailed methods of examination for the presence of impurities and additivies, or both, and techniques of separation are described below.

40.5.2 Gas Chromatography. Identification of Alcohols.

Place about 1 g of the sample containing ester groups into a 150 cm^3 extraction flask and saponify by refluxing with 50 cm^3 of 10 per cent aqueous sodium hydroxide solution for 4 hours. Wash the condenser with distilled water, cool the flask, transfer it to a distillation apparatus and distill off the alcohols. Identify the alcohols and, if possible, determine their approximate ratios by gas - liquid chromatographic methods. Retain the alkaline solution (see aqueous phase A, Figure 7.9).

40.5.3 Infrared Spectroscopy and Thin-Layer Chromatography. Identification of the Alkyl Groups Attached to Tin.

Transfer the alkaline solution, aqueous phase A, to a 150 cm^3 separating funnel, adding more water if necessary and extract by carefully shaking with 50 cm^3 of diethyl ether.

Filter the aqueous phase, diluting with distilled water or warming if necessary, through a Whatman No. 541 filter paper and wash the precipitate with hot distilled water, especially if a long-chain acid salt is present. Dry the precipitate at 105°C and examine by infrared spectrophotometry to identify the dialkyltin group. Retain the filtrate, filtrate B (Figure 7.9) but evaporate the ether to dryness and examine the residue by infrared spectrophotometry. This may contain traces of alcohols, unsaponified esters, etc.

If the only concern is the identification of the dialkyl groups attached directly to tin, then a method using methanolic sodium hydroxide solution is recommended (Figure 7.9). The thin-layer chromatographic method described is also suitable. For pure samples of dialkyltin carboxylates with no ester present, the time of reflux with alkali can usually be much reduced, the recovery and identification of the dialkyltin oxide, believed to be in polymeric form , still being quite satisfactory. In addition, maleic acid can be recovered from a sample of dibutyltin maleate after decomposition by this simplified method.

40.5.4 Infrared Spectroscopy. Identification of Acids.

Reduce the volume of filtrate B (Figure 7.9) to about 150 cm^3, by boiling, acidify it with hydrochloric acid and extract by shaking it with 50 cm^3 of diethyl ether. Evaporate the solvent on a water-bath, dry the residue at 105°C for 20 minutes and determine its infrared spectrum to identify the acid readily soluble in ether.

Extract the aqueous phase with diethyl ether for 4 h by liquid - liquid extraction, evaporate the solvent, dry the residue at 105°C and again examine it by infrared spectrophotometry, this time to identify the acid less soluble in ether, usually maleic acid.

If the infrared spectrum of the first ether extract, ie. the one obtained by shaking, indicates a mixture of a long-chain and another acid, triturate the residue several times with small amounts of distilled water. Recover the fraction soluble in water by evaporation and dry both fractions at 105°C. Examine both the water-soluble and the water-insoluble fractions by infrared spectrophotometry. If necessary, prepare and examine the barium salt of the long-chain acids. Udris[15] separated and identified a mixture of phthalic and lauric acids by this method.

The long-chain acids obtained by this method are usually contaminated by metal salts and a dialkyltin dichloride, the latter originating from the presence of residual dialkyltin groups. For positive identification of the acid by infrared spectroscopic examination, the preparation of the barium salt has been found helpful. Alternatively, the acid can be identified by mass spectrometric examination.

40.6 DISCUSSION OF RESULTS

These methods are basically qualitative although in some instances the methods can be put on a quantitative basis, eg. the determination of ester groups and alcohols by gas chromatography.

METHOD 41 - IDENTIFICATION OF ACCELERATORS AND ANTIOXIDANTS IN NON-VULCANIZED RUBBER COMPOUNDS. THIN LAYER CHROMATOGRAPHY.[16]

41.1 SUMMARY

This thin-layer chromatographic method determines a range of accelerators and antioxidants in non-vulcanized rubber compounds in amounts down to 5 ppm.

41.2 APPARATUS

Sil G UV_{254} pre-coated plastic sheets of silica gel.
Laboratory mill with freeze drying attachment.
Equipment for thin-layer chromatography.

41.3 REAGENTS

Liquid nitrogen.
Developing solvents:

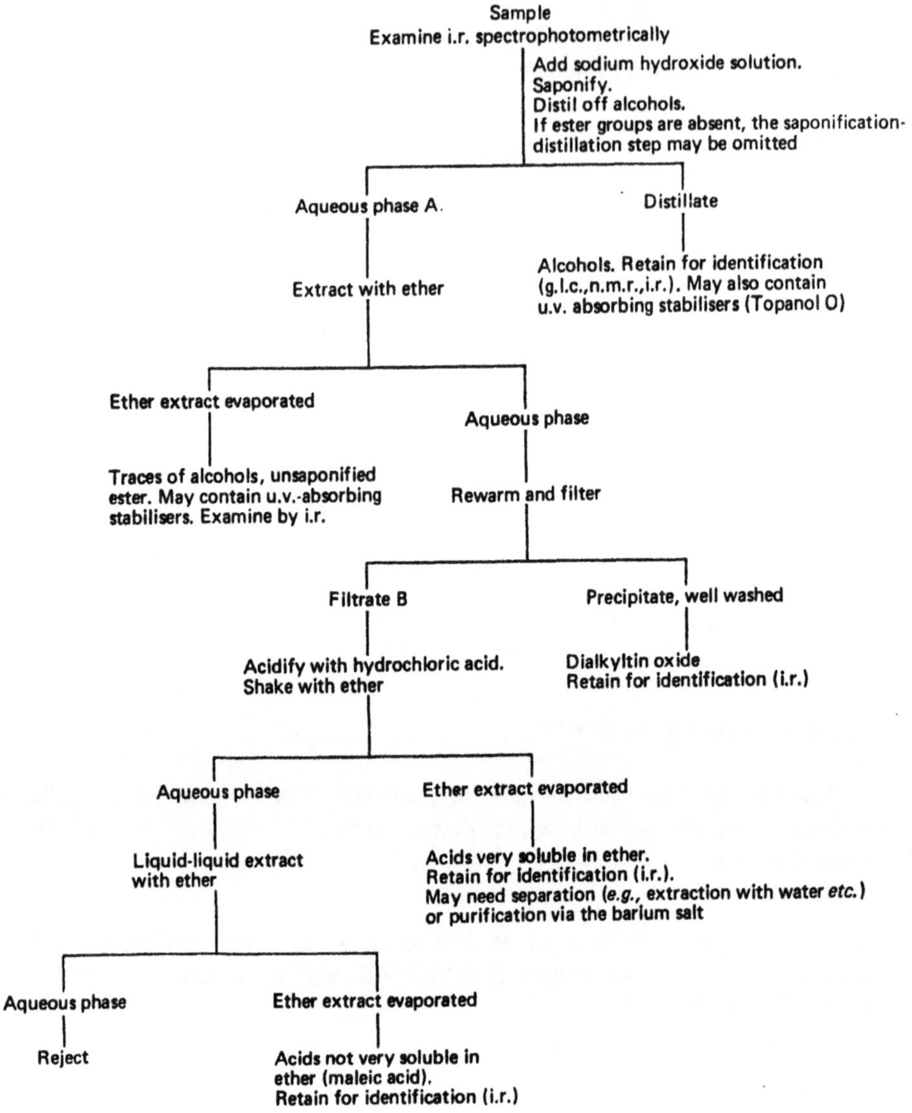

Figure 7.9. Scheme of analysis of dialkytin carboxylates and hemiesters. From Udris with permission.[15] Royal Society of Chemistry, London

(a) Petroleum ether (30° - 40°C) diethyl ether 110:20 (pre-developing solvent),
(b) Benzene: ethyl acetate, acetone, 100:7:2 v/v (primary solvent mixture)
(c) Cyclohexane
(d) Toluene:ethyl acetate:ammonium hydroxide 100:5:0.1
(e) Cyclohexane:diethylamine, 75:25 v/v
(f) Chloroform: Benzene, 100:90 v/v
(g) Acetone: ammonium hydroxide, 100:1 v/v (confirmation guanidines)
(h) Isopropanol, A.R

Visualizing reagents:

(a) Iodoplatinate solution, 3 cm^3 of 10% platinum chloride solution mixed with 97 cm^3 of water to which are added 100 cm^3 of a 6% aqueous solution of potassium iodide.

(b) Dibromo benzoquinone chlorimide, 1% solution in methanol
Sodium hypochlorite, 4% aqueous
Sodium bicarbonate, 1%.

41.4 METHOD

The unvulcanized rubber sample is extracted with isopropanol to remove accelerators and antioxidants. Isopropanol is removed and the residue dissolved in chloroform and applied to a thin-layer plate. Development with a range of solvent mixtures separates the components which are then detected with a range of spray reagents. Calibration is achieved by applying the procedure to prepared pure standard solutions.

41.5 EXPERIMENTAL PROCEDURE

Immerse 3.0 grams of non-vulcanized sample, cut into small pieces, in liquid nitrogen until hard and grind to a fine state using a laboratory mill with freeze drying attachment. Transfer the ground sample to a glass stoppered 30 cm^3 test tube, add 5 cm^3 of isopropanol and extract by infusion for 1 hour at room temperature.

Decant off extract and leave overnight in a deep freezer, filter and evaporate off solvent from the filtrate at room temperature. Redissolve the extract in 1 cm^3 of chloroform and apply amounts of 1 - 10 ul to a thin-layer chromatography plate evaporating off solvent with cold air between applications. After development examine plate under short and long UV and record observations.

Spray the sheet with the iodoplatinate solution and dry in an air oven at 100°C. The iodoplatinate reagent visualizes accelerators as white, yellow, orange and brown spots on a brick-red background.

When using the combination of sprays, spray with the chlorimide and dry at 65°C. Examine and then spray with iodoplatinate and dry at 100°C. A second plate is sprayed with chlorimide reagent. Both accelerators and antioxidants produce coloured spots with this reagent. On overspraying with the iodoplatinate, the antioxidant spots retain their color, while the accelerator spots fade or become a darkened centre with an intense white halo.

Calibrate the procedure against chloroform solutions of pure specimens of the compounds being determined.

Accelerators studied were representative of a wide range of chemical types including guanidines, thiazoles, thiurams, sulfenamides, diethiocarbamides and morpholine disulfides. All the accelerators tested gave a positive response to the iodoplatinate reagent and a list of colour reactions is shown in Table 7.16. In Table 7.16 are shown the colour reactions of a wide range of antioxidants tested involving both phenolic and amine types. Only those antioxidants based on phenylene diamine gave a colour reaction with the iodoplatinate reagent. Colours produced are distinctively different - cyclamens, purples and greens.

An overspray of sodium bicarbonate immediately following the chlorimide spray and then overspraying with iodoplatinate can be useful for confirming the presence of the following accelerators: tetramethyl thiuram monosulfide (TMTM), tetramethyl thiuram disulfide (TMTD), zinc diethyl dithiocarbamate (ZDC), 2-(morpholinothio) benzothiazole (Santocure MOR) and mercaptobenzothiazole (MBT).

Chromatograms should be observed at intervals through a period of 48 hours. Colours tend to become more definite or fade, leaving a pale spot with a discoloured centre. This would appear to be characteristic of accelerators treated with the iodoplatinate reagent, and could serve as a distinguishing feature when analyzing accelerators in the presence of antioxidants, (Table 7.16).

41.6 DISCUSSION OF RESULTS

This method is capable of determining accelerators and antioxidants in unvulcanized rubber compounds with an accuracy of $\pm$ 10%.

For quantitative determinations it is of course necessary to know the identity of the compound being determined and to calibrate the procedure accordingly.

METHOD 42 - IDENTIFICATION AND DETERMINATION OF ANTIOZONANTS AND ANTIOXIDANTS IN RUBBER VULCANIZATES - MASS SPECTROMETRY.[17]

42.1 SUMMARY

This mass spectrometric method determines down to 0.01% antiozonants and antioxidants in rubber vulcanizates.

42.2 APPARATUS

Mass spectrometer/data system Finnigan MAT 311/A

Incos 2400 or equivalent with combined EI/CI (electron inpace/chemical ionization ion source)

EI, FD, FAB analysis

Computer mass range up to m/z 1450. Ion source temperature, 250°C, accelerating voltage 3 kV, resolution M/dM 1000 (10% valley). The electron energy for EI was 70 eV; the CI (chemical ionization) reagent gas was generally isobutane, but methane was also used.

Spectra were acquired as the direct probe was heated over the range 50 - 300°C. Accurate mass measurements made by peak matching in either EI or CI mode (resolution M/dM about 10 000).

FD and FAB (fast atom bombardment) spectra acquired with the same instrument using a combined EI/FD/FAB ion source. For FD analysis, the residues were taken up in the extraction solvent (either acetone, acetonitrile or dichloromethane) and deposited using the standard dipping technique onto the emitter wire. Standard high temperature activated carbon emitters were used. For FD the ion source temperature was 90°C, the accelerating voltage 3 kV, the extraction plate voltage 6kV and the resolution M/dM about 600. Spectra were acquired as the emitter was heated manually with 0 - 30 mA heating current. For FAB the viscous extraction residues were deposited directly on a stainless steel FAB probe, either neat or else mixed with a little thioglycerol solvent. For

Table 7.16 - Reactions[a] of Antioxidants and Accelerators

Chemical	Trade name or abbreviation	Spray 1 iodoplatinate	Spray 2 chlorimide	Spray 2 plus Spray 1	Primary solvent system R_f
(a) Reactions of Antioxidants					
Acetone diphenylamine condensation product	Nonox BL		Pale green	Green blue	0.53
Butraldehyde aniline condensation product	Antox		Brown yellow edge	Grey purple	0.40
4,4′-Butylidene-bis(6-tert-butyl-3-methylphenol)	Santowhite powder		Purple	Grey	0.37
4-Phenyl-N″cyclohexyl-p-phenylene diamine	Antioxidant 4010	Grey blue	Purple	Purple	
2,2′-Methylene-bis(4-ethyl-6-tert-butylphenol)	Antioxidant 425		Green white centre	Rust yellow	0.58
Diphenylamine acetone reaction product	Ble	Green blue (weak)	Yellow	Rust	0.48
N-Phenyl-N′-1,3-dimethyl butyl p-phenylenediamine	Santoflex 13	Blue brown	Brown yellow edge	Purple	0.48
6-Dodecyl-1,2-dihydro-2,2,4-trimethylquinoline	Santoflex DD		Yellow	Rust yellow centre	0.38
Diphenyl-p-phenylene-diamine	DPPD	Green	Purple	Purple	0.47
Mixture of diaryl p-phenylenediamines	Wingstay 100	Brown green	Brown yellow edge	Purple	0.50
6-Ethoxy-1,2-dihydro-2-2,4-trimethylquinoline	Santoflex AW		Cyclamen	Purple	0.40
N,N′-Bis(1-ethyl-3-methyl-pentyl)-p-phenylenediamine	UOP88	Cyclamen	Cyclamen	Cyclamen	
N,N′-Bis(1-methylheptyl)-p-phenylenediamine	Santoflex 217	Purple	Pink grey edge	Purple	0.25
N,N′-Bis(1,4-dimethyl pentyl)-p-phenylenediamine	Santoflex 77	Purple	Brown	Purple	0.23
2,2′-Methylene-bis(6 alpha-methylcyclohexyl-4-methyl-phenol)	Nonox WSP		Yellow	Yellow orange edge	
4-Isopropyl-aminodiphenylamine	Nonox ZA	Grey green	Brown yellow edge	Purple	0.33
A substituted phenol	Nonox WSL		Rust	Yellow purple edge	0.55
beta-Naphylamine			Brown	Purple	0.22
Tri-(nonylated phenyl)-phosphite	Polygard	White	Purple, orange	White purple edge	0.57, 0.33
Substituted styrenated phenol	Wingstay S		Purple, blue	Purple	0.60, 0.37
2,6-di-tert-butyl-4-methylphenol	Ionol		Rust	Yellow brown edge	0.90
2,2′-Methylene-bis(4-methyl 6-tert-butylphenol)	Antox 2246		Yellow	Orange brown edge	0.57
Modified phenyl beta-naphthylamine	Nonox DN		Pink	Purple	0.53

Table 7.16 continued

A phenol condensation product	Nonox EX		Yellow	Yellow purple streak	0.38
A blend of arylamines	Nonox HFN		Pink	Purple	0.53
Sym. di-beta-naphthyl-p-phenylenediamine	Nonox CI		Brown streak	Lavender streak	0.00
Phenyl-beta-naphthylamine + diphenyl-p-phenylenediamine	Nonox HP	Weak pale green	Yellow, purple	Purple	0.47, 0.53
A phenol condensation product	Nonox WSP		Yellow	Rust	0.37
Octylated diphenylamines	Octamine		Pale Pink	Brown green	0.99
2,2,4-Trimethyl-1,2-dihydroquinoline	Flectol Flakes	(Weak response)	Turquoise	Green	0.45, 0.35, 0.20, 0.13, 0.05
A phenol condensation product	A.F.D.		Grey		0.60
N,N′-Bis(1-methyl heptyl)-p-phenylenediamine	UOP288	Cyclamen	Pink, blue	Purple	0.50, 0.08
<u>(b) Reactions of Accelerators (Included are a few Peptizers)</u>					
Diphenylguanidine	DPG	Brown, white edge	Blue	Green, white edge	0.00
Di-o-tolyguanidine	DOTG	Brown, white edge	Blue	Green, white edge	0.00
2-Mercaptobenzothiazole	MBT	Yellow	Orange	Rust, white edge	0.30
Benzothiazyl disulfide	MBTS	Cream streak	Pale Yellow	Orange streak, white edge	0.50
Tetramethylthiuram disulfide	TMTD	Yellow, brown white edge	Dark red pink edge	Red, white edge	0.45
Tetramethylthiuram monosulfide	TMIM	Yellow, brown white edge	Dark red pink edge	Yellow, red edge	0.45
N,N-Dicylcylohexylbenzothiazole-2-sulfenamide	Vucazit DZ	Yellow, white edge	Yellow orange	Orange, white edge	0.70
Phosphorus-containing polysulfide	Vulcadone 3 SN	White, grey edge	Yellow	Rust, white edge	0.85
N-Cyclohexyl-2-benzothiazole sulfenamide	Santocure	Yellow, pale edge	Yellow orange	Orange, white edge	0.63
N-tert-butyl-2-2benzothiazole sulfenamide	Santocure NS	Yellow, pale edge	Yellow orange	Orange, white edge	0.55
Zinc diethyl-dithiocarbamate	ZDC	Yellow	Brown, red edge	Rust, white edge	0.73

Table 7.16 continued

Zinc salt of 2 mercaptobenzo-thiazole	ZMBT	Yellow white	Red	Rust	0.25
2-Benzothiazyl-N,N-diethylthio-carbamyl sulfide	Ethylene	Yellow, white edge	Yellow orange	Rust, white edge	0.39
4-Morpholine disulfide	Morfax	White	Yellow	Brown, white edge	0.35
Copper dimethyl-dithiocarbamate	Cumate	Pale brown streak	Red streak	Brown streak	0.20-0.70
Diphenylmethylene thiuram tetrasulfide	Tetrone A	Brown streak cream centre	Orange yellow	Red streak	0.30-0.80
2-(Morpholinthio)-benzothiazole	Sanocure MOR	Yellow, cream	Yellow	Orange, white edge	0.30, 0.63
Hexamethylene tetramine	Vulcacit H30	Red, white edge	Yellow	Red, white edge	0.00
Morpholine disulfide	Sulfasan R	Grey, white edge	Yellow	Rust, white edge	0.38, 0.20
2-Mercapto imidazoline	Na 22	White	Brown	Brown orange	0.00
2,2'-Dibenzamido-diphenyl disulfide	Pepton 22	Cream, brown edge	Pale orange	Red, white edge	0.35

[a] Colour reactions given in the tables refer to pure compounds. At the dilution level in a rubber stock, many of the positive responses given by antioxidants to the iodoplatinate reagent are weak and in many instances, could be regarded as negative. From Milligen[16] American Chemical Society

FAB the ion source temperature was 70°C and the resolution M/dM about 1000. The fast atom gun (IonTech Model FAB-11-GG) provided xenon atoms at an energy of 8.0 keV.

FI analysis. For FI analysis a Varian MAT 731/SS200 mass spectrometer/data system (Wiesbaden) was used. Standard high temperature activated carbon emitters were used with a combined EI/FI/FD ion source. The ion source temperature was 50°C, the accelerating voltage 8 kV, the extraction plate voltage 3 kV, and the resolution M/dM about 2000. A heated direct probe was used for sample introduction.

42.3 METHOD

The technique can be applied either to solvent extracts of rubbers or directly to the rubber sample. Electron impact or chemical ionization mass spectra are prepared and the individual additives identified and, if pure calibration standards are available, determined.

42.4 EXPERIMENAL PROCEDURE

42.4.1 Rubber Extract Analysis.

The cured rubbers are cut into small pieces (about 2 mm square) and extracted for 24 h with either acetone, acetonitrile or dichloromethane in a Soxhlet apparatus. The extraction solvents are evaporated under a stream of nitrogen at room temperature. The extract residues are placed in aluminium crucibles for EI or CI mass spectral analysis.

42.4.2 Direct Rubber Analysis.

Small pieces of rubber are cut from the ASTM sheets using a razor blade and placed in aluminium crucibles for direct mass spectral analysis. EI and CI spectra were acquired by using the same MAT 311A instrument with the conditions as described above. Spectra were acquired as the probe was heated from 50 to 300°C. For direct FAB analysis, the vulcanizates were cut into strips (8 x 3 x 2 mm) which were attached to the stainless steel FAB probe with scotch 924 transfer tape. The experimental conditions for FAB were the same as described above (no FAB solvent was used with the samples).

42.5 TYPICAL RESULTS

The various classes of additives are discussed individually below in terms of their ability to be detected in the vulcanizates by mass spectrometry.

Antioxidants and antiozonants examined include aromatic amines (HPPD, DOPPD, DODPA and poly-TMDQ) and a hindered bisphenol (AO 425). These compounds could be identified quite readily by either extraction or direct analysis and by use of any vaporization/ionization method.

Stearic acid could be identified only by direct analysis of the rubber, using any vaporization/ionization method. Stearic acid was not observed in any of the extracts; this indicates that it remains in the rubber bulk during the extraction process.

METHOD 43 - DETERMINATION OF ACID AND GLYCOL UNITS IN TERYLENE AND OTHER POLYESTERS. HYDROLYSIS - GAS CHROMATOGRAPHY.

43.1 SUMMARY

This gas chromatographic method determines acid and glycol units in polyesters in amounts down to 0.05%.

43.2 APPARATUS

Hewlett-Packard model 5711A gas chromatograph equipped with a flame ionization detector and a Hewlett-Packard model 3352B laboratory data systems were used.

43.3 REAGENTS

Redistilled Eastman Kodak 2-ethoxyethanol (practical grade). Internal standard solutions:

Dodecane, 1 g diluted to 100 ml with 2-ethoxyethanol.
Nonyl alcohol, 5 g diluted to 100 ml with 2-ethoxyethanol.
Propyl alcohol, 0.4 g diluted to 100 ml with 2-ethoxyethanol.
Heptadecane, 1 g diluted to 100 ml with 2-ethoxyethanol.
N,O-bis (trimethylsilyl) trifluoroactamide.
Potassium hydroxide (1 M) in 2-ethoxyethanol.
Hydrochloric acid, concentrated.
Pyridine, Analar.

43.4 METHOD

The polyester is refluxed with a 2-ethoxyethanol solution of potassium hydroxide to break down ester linkages to alcohols and carboxylic acids, eg.

7.2

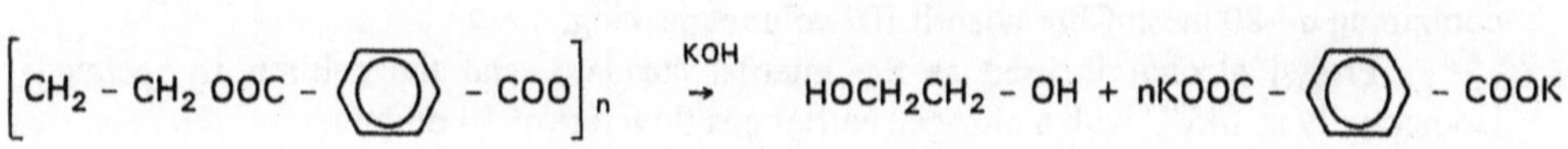

After acidification and the addition of an internal standard the extract is reacted with N,O-bis (trimethylsilyl) trifluoroacetamide to produce the silyl ethers of glycols and the silyl esters of carboxylic acids. Gas chromatography of this reaction mixture enables the numbers of acid and glycol units in the polymer to be estimated.

43.5 EXPERIMENTAL PROCEDURE

The amount of sample used for the alkaline hydrolysis is controlled by the type of determination. For example, a 1 g sample is sufficient for the determination of the glycols and acids comprising the major portion of the repeat units. A 4 g sample is used when determining trace components such as the methyl ester end-groups.

The sample is weighed into a 100 ml flask and 50 ml of 1 N potassium hydroxide in 2-ethoxyethanol is added to the flask. A condenser cooled by chilled water is attached to the flask and the reaction mixture is protected from carbon dioxide by means of a tube packed with Ascarite absorbent and Drierite desicant. The contents of the flask are heated and maintained at reflux temperature for 10 min with constant stirring. The flask is allowed to cool to room temperature and the hydrolysate is adjusted to a pH of 1 using concentrated hydrochloric acid (5 ml). An internal standard is added to the flask. Pyridine (25 ml) is added to dissolve the acids present and an aliquot sample is centrifuged to remove the potassium chloride. Approximately 50 ul of hydrolysed sample is allowed to react at room temperature for 5 to 10 min with 500 ul of N,O-bis(trimethylsilyl)trifluoracetamide to form silyl ethers and esters of the glycols and acids, respectively. The silyated hydrolysate is chromatographed by injecting 0.1 ul of sample into the gas chromatograph.

The silyl derivative of diethylene glycol is separated from the silyl derivative of ethylene glycol using a 10 ft x 1/8 inch stainless steel column packed with 100-200 mesh Chromosorb G-HP solid support containing 3% by weight of Versilube F-50 liquid phase, Versilube F-50 is the trademark for a mixture of trichlorophenyl silicone (10%) and methyl silicone manufactured by General Electric.

Figure 7.10 is a chromatogram obtained for the silyl derivatives of the hydrolysate of an experimental polyethylene terephthalate. Dodecane is used as an internal standard and the column is operated isothermally at 127°C with a nitrogen carrier gas flow rate of 10 ml/min.

High boiling acids, such a terephthalic and isophthalic acids are separated using a 6 ft x 1/8 in stainless steel column packed with 100-200 mesh Chromosorb W-HP solid support containing 10% by weight Versilube F-50 liquid phase.

Another important use of the alkaline hydrolysis/gas chromatographic procedure is the determination of methyl ester end groups. If the polyester is terminated with methyl ester end groups, then methyl alcohol is produced when the polyester is hydrolysed. By determining the concentration of methyl alcohol, one can calculate the number of methyl ester end groups in polyethylene terephthalate. The hydrolysis procedure for determining methyl ester end-groups is the same as discussed for the glycols and acids. The gas chromatographic procedure is different in that no silylation reagent is used. The hydrolysed sample is separated using a 6 ft x 1/4 inch glass column containing 60-80 mesh Chromosorb 102 column packing.

Propyl alcohol is used as the internal standard, and the column is operated isothermally at 145°C with a nitrogen carrier gas flow rate of 10 ml/min.

43.6 DISCUSSION OF RESULTS

The data given in Table 7.17 shows that the precision of the overall method is good. This method is relatively simple and fast. The method has been used to analyse experimental polyester for the amount of monomers present in the repeat units and for determining the composition of copolymers and polymer blends. The relative standard deviation varies from 0.8% for the determination of 1,4-butanediol to approximately 2% for 1,4-cyclohexanedimethanol. For the determination of different concentrations of isophthalic acid, the standard deviation varies from 0.7 to 2.4%.

METHOD 44 - DETERMINATION OF WATER IN POLYESTERS. DERIVATIVIZATION - GAS CHROMATOGRAPHY.[19]

44.1 SUMMARY

This gas chromatographic procedure determines water in polyesters in amounts down to 50 ppm.

44.2 APPARATUS

Auto injection vial, 1.9 ml (Wheaton Glass) with cape lined with Teflon resin are sealed with a crimping tool

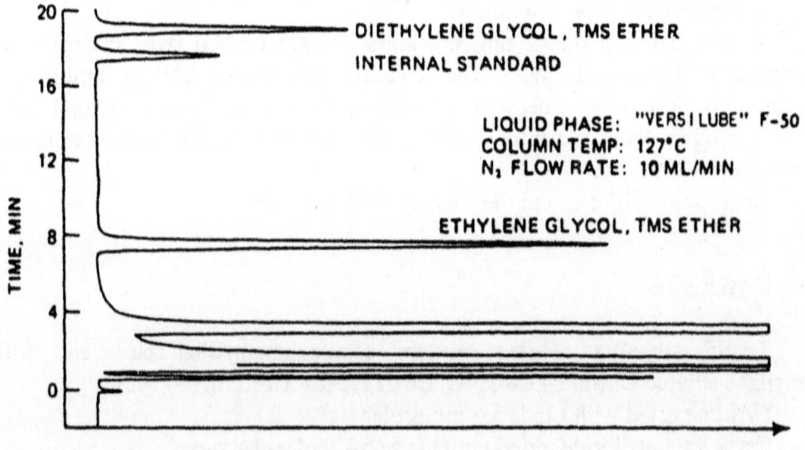

Figure 7.10. Chromatogram of TMS derivatives of the glycols from the hydrolysis of poly(ethylene terephthalate). From Allen et al with permission.[18] American Chemical Society.

Table 7.17 - Precision of Hydrolysis - Gas Chromatographic Method

Component	Concn. %	SD	No.
Diethylene glycol, TMS ether	1.781	0.019	40
1,4-Butanediol, TMS ether	6.59	0.050	20
1,4-Cyclohexanedimethanol TMS ether	20.45	0.40	20
Isophthalic acid, TMS ether	5.00	0.12	20
Isophthalic acid, TMS ether	60.00	0.44	20
Methyl alcohol	0.088	0.002	20

From Allen[18]. American Chemical Society

Serum vials, 10 cm^3 Neutraglass with Teflon-lined caps are sealed with a larger crimping tool. The latter vials contain a maximum of 14 cm^3.

Hamilton syringes, 1 and 10 ul.

A dry box flushed with dry nitrogen from a liquid air supply.

Gas chromatograph used with a computer for peak integration.

Gas chromatograph conditions:

A 4 m 6.4 mm o.d. 2 mm i.d. glass column packed with 1% OV-1 on high performance Chromosorb W. The column temperature set to ambient, (actual temperature is close to 50°C because of heat from the manifold and injection block).

Hydrogen flame detector, 150°C, injection port 100°C and manifold 100°C. Helium flow, 33 cm^3/min.

Attenuation 10^3 x 2, and the chart speed 5 min/inch.

44.3 REAGENTS

Tri-Sil silyation reagent mixture of hexamethyldisilazane and trimethylchlorosilane (Pierce Chemical Co., P.O. Box 117, Rockford, Ill. 61105).

Pyridine dried with Linde 5A molecular sieve.

Carbon tetrachloride, dried with Linde 5A molecular sieve.

Molecular sieves.

44.4 METHOD

The water in the sample is reacted with a mixture of hexamethyldisilazane and trimethylchlorosilane to form hexamethyldisiloxane

$$2(Me_3Si)_2NH + 3\,H_2O + 2Me_3SiCl = 3\,Me_3SiOSiOMe_3 + 2NH_4Cl \qquad 7.3$$

The hexamethyldisiloxane equivalent to the water content of the sample is separated from other silanized active hydrogen constituents and determined by gas chromatography.

44.5 EXPERIMENTAL PROCEDURE

Weigh 0.5 - 1.0 g of polymer in ground or pellet form into a dry 2 cm^3 vial and add 1 cm^3 silylation reagent, 1 cm^3 dry carbon tetrachloride internal standard and 10 cm^3 dry pyridine by syringe then crimp the cap. Pick up of water may be avoided by filling the vial in a dry box. Place the samples in the oven at 105°C for 40 min. After cooling to ambient temperature, inject 1 ul samples into the gas chromatograph. The peak area ratios between sample and standard are converted to micrograms of water by means of the calibration curve.

44.5.1 Calibration.

Fill a large dry serum vial with 4 ml Tri-Sil, 1 cm^3 carbon tetrachloride, as internal standard and 10 cm^3 dried pyridine. Seal immediately with a Teflon lined cap using a crimping tool. Transfer 1 cm^3 of this reagent to a 2 cm^3 auto-injector vial by means of a syringe. From this smaller vial, inject 1 ul amounts into the gas chromatograph. Having established a stable blank value, inject 200 ul increments of water into the vial and after each increment has reacted for a few minutes, inject into the gas chromatograph. The ratio of the areas of the hexamethyldisiloxane peaks to the carbon tetrachloride peaks is plotted vs. the ug water/ cm^3. A calibration curve is thus constructed which is linear and intersects the ordinate at the reagent blank ratio value. It is recommended that the analyses is carried out after standardization without delay, in order to avoid a blank value due to adsorbed water on the column.

44.6 DISCUSSION OF RESULTS

This procedure is capable of determining down to 50 ppm of water in polyesters with a relative standard deviation of 13% at the 200 ppm water level.

METHOD 45 - DETERMINATION OF ACRYLIC ACID MONOMER IN POLYACRYLATES. HIGH PERFORMANCE LIQUID CHROMATOGRAPHY.[20]

45.1 SUMMARY

This high performance liquid chromatographic procedure determines acrylic acid monomer in polyacrylate resins in amounts down to 0.005%.

45.2 APPARATUS

Liquid chromatograph fitted with a Whatman PXS 10/25 um PAC column (250 x 4 mm i.d.) a variable wavelength spectrophotometer and a recorder.

A Rheodyne, model 7105 (Perkins Elmer) injection valve with a 175 ul sample loop is used to inject the sample on to the analytical column.

Syringe, 100 ul.

Ultrasonic bath, to deairate the solvents used as the eluents.

Orbital shaker.

45.3 REAGENTS

All standards and solvents are of Analar or Nanograde quality unless otherwise stated.

Solvents. - Isopropanol, hexane and methanol (HPLC grade).

Concentrated orthophosphoric acid, sp. gr. 1.75.

Eluent, 0.01% v/v orthophosphoric acid in distilled water.

Acrylic acid, 99% pure. 0.25% aqueous.

Anhydrous sodium sulphate heated at 600°C for 24 h, dried and stored in a grease-free desiccator.

Distilled water. Clean up by passing through a Whatman SAX (strong anion exchange) column stored in glass under nitrogen.

Glassware. - All glassware is soaked in a chromic acid bath, washed in distilled water and oven dried at 110°C.

45.4 METHOD

A methanol extract of the polymer is injected into a high performance liquid chromatograph and the acrylic acid peak evaluated. The method is calibrated against standard methanolic solutions of acrylic acid.

45.5 EXPERIMENTAL PROCEDURE

Add a known mass of polymer (500 mg) to a methanol distilled water mixture (1 + 1, 50 cm^3) and allow to stand overnight.

Inject an aliquot (12 ul) of the sample into the liquid chromatographs Rheodyne valve and chromatograph under the following conditions: column Whatman PXS 10/25 um PAC (250 x 4 mm i.d.); mobile phase, 0.01% v/v orthophosphoric acid in distilled water; flow rate, 4 cm^3 min^{-1}; pressure, 2000 lb in^{-2}; detector wavelength 195 nm; chart speed, 0.5 cm min^{-1}; and absorbance scale 0.02.

45.5.1 Calibration.

Treat standard solutions of acrylic acid (0, 0.05, 0.117, 0.5, 1.17, 3.5, 4.0, 5.0, 7.0, 8.0, 10.0, 11.7 and 14.0 mg l^{-1}) in the same manner as the samples described above. The slope of the calibration graph of absorbance (peak height) versus concentration does not change by what was considered an appreciable amount for an eluent flow rate of 4 cm^3 min^{-1} over the period studied.

45.6 TYPICAL RESULTS

Calculations. The concentration of acrylic acid in a sample calculated from the absorbance of a 100 ul injection on an absorbance scale of 0.01 a.u.f.s., ie:

$$\text{Total absorbance} = \frac{\text{mean absorbance of concentrate} \times 100}{\text{volume of concentrate injected}} \times \frac{\text{absorbance scale}}{0.01}$$

The concentration of acrylic acid is found by comparison with a previously prepared calibration graph of total absorbance versus original acrylic acid concentration.

45.7 DISCUSSION OF RESULTS

The method is capable of determining down to 50 ppm acrylic acid monomer in polyacrylates with an accuracy of $\pm$ 5%.

METHOD 46 - DETERMINATION OF HYDROXYL GROUPS IN POLY(ETHYLENE GLYCOL). SILATION - SPECTROPHOTOMETRY.[21]

46.1 SUMMARY

This silation spectrophotometric method is capable of determining down to 0.1% of hydroxy groups in a molecular weight 3000 poly(ethylene glycol).

46.2 APPARATUS

Gas chromatograph equipped with flame ionization detectors.[21] The carrier gas was helium.

Spectrographs: IR spectrometer, UV spectrophotometer.

46.3 REAGENTS

Pentane, spectroscopic grade.

Petroleum ether, 40:60.

Ethanol, spectroscopic grade.

1-napthyl dimethyl (dimethylamino) silane (preparation). A 10% solution of 1 equiv monochlorosilane (napthyl dimethyl chlorosilane) in hexane (in benzene if necessary for solubility reasons) is stirred in a flask equipped with a reflux condenser cooled at - 20°C. From a communicating flask 2.2 equiv of dimethylamine vapor is slowly introduced over the solution and the mixture kept at room temperature overnight. After filtration and evaporation of the solvent, the residue is distilled.

7.4

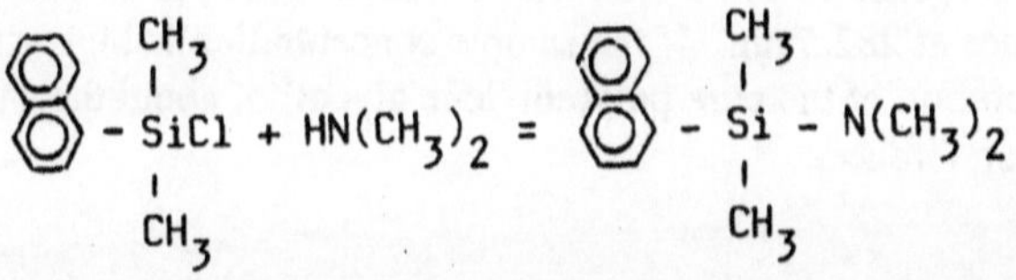

Purification of the poly(ethylene glycols) (Hydroxyl terminated) H-PEG.

A solution of 100 g of commercial practical grade polymer in 400 ml of 0.01 M sodium hydroxide is digested at 40°C for 24 h, then passed through two ion-exchange columns, the first containing 50 cm^3 of Amberlite IR 120 (H+ form) and the second 50 cm^3 of Lewatite MP 7080 (base form). After evaporation of the water, the residue is dissolved in 1 cm^3 of toluene and filtered on an aluminium oxide column (50 g). The agitated solution is then thermostated at the temperature indicated below and 5.0 cm^3 of of petroleum ether added dropwise. The slurry is cooled to 0°C, allowed to stand for 15 min and the precipitate filtered. The temperature of precipitation is critical; it was 0°C for H-PEG of molecular weight 600 (H-PEG-600), 10°C for H-PEG-1000), 20°C for H-PEG-2000 and 30°C for polymers of higher molecular weight. The purified product is dried at room temperature and 0.1 Torr pressure.

46.4 METHOD

A 1,2 dimethoxyethane solution of the polyglycol is silylated with napthyl dimethyl dimethylamino-silane. After some working up the extract is dissolved in ethanol and the chromophore concentration evaluated spectrophotometrically. The method is calibrated against polyglycols of known equivalent weight.

46.5 EXPERIMENTAL PROCEDURE

Determination of the hydroxyl content of poly(ethylene glycols). For polymers with expected equivalent weights higher than 1000 a sample of 1 g (0.5 g for lower equivalent weights) is dissolved in 5 cm^3 of dry 1,2-dimethyoxyethane. The dissolution is omitted if the melting point of the polymer is less than 60°C. The flask is purged with dry argon (or nitrogen) and three equivalents for each expected equivalent hydroxyl (but not less than 200 mg) of 1-napthyl dimethylaminosilane is added. The homogenized mixture is kept at 60°C for 30 min. The sample is then diluted at room temperature to 10 cm^3 with 1,2-dimethyloxyethane. After cooling it to an appropriate temperature, the silyated polymer is precipitated by adding dropwise 50 cm^3 of low boiling petroleum ether to the stirred solution (the temperature of the precipitation is critical: see under section " Purification of H-PEG"). After precipitation, the mixture is stirred at 0°C for 15 min, then filtered on a sintered glass filter (G3) and washed with 5 cm^3 portions of petroleum ether. The precipitate is redissolved on the filter in 5 cm^3 of warm 1,2-dimethyloxyethane and precipitated a second time as above. If the expected hydroxyl content of the polymer is lower than 10^{-1} mol kg^{-1}, the precipitation should be repeated a third time. After the last precipitation the polymer is washed on the filter with 20 cm^3 of spectroscopic grade pentane, dried for 5 h at room temperature/0.1 Torr or in a stream of dry nitrogen. A portion of the dry sample is dissolved in spectroscopic grade ethanol so as to yield an absorbance of 0.2 - 0.8. The absorbance of the solution is measured vs. ethanol as reference at 282.5 nm. If the sample concentration is higher than 10 g dm^{-1} (= c) an ethanolic solution of the pure poly(ethylene glycol) of about the same concentration must be used as a reference.

46.6 TYPICAL RESULTS

The molarity of the silyl groups in the solution M_{sol} is given by:

$$M_{sol} = A/a \qquad = c/EW_{PSil} \quad \text{mol L}^{-1}$$

where A is the absorbance of the solution, c is the concentration of the silylated polymer in dm^3, a is the molar absorptivity of the silylating agent and EW_{PSil} is the equivalent weight of the silylated polymer which can be calculated by:

$$EW_{PSil} = \frac{Ca}{A} \quad \text{g mol}^{-1}$$

In order to obtain the equivalent weight of the hydroxy terminated polymer, EW_{POH} the molecular weight of the silyl group MW_{Sil} and that of the substituted proton must be taken into account

$$EW_{POH} = \frac{Ca}{A} - (MW_{Sil} - 1) \quad \text{g mol}^{-1}$$

The molarity of the hydroxyl groups in 1 kg of original polymer, m_{OH} is then given by:

$$m_{OH} = 1000/EW_{POH} = 1000\ A/.(Ca - A\ (MW_{Sil} - 1)) \quad \text{mol kg}^{-1}$$

For the 1-naphthyldimethylsilyl substituent, $MW_{Sil} = 185$

The values of the molar absorptivities are given in Table 7.18. 1-naphthyldimethyl(dimethylamino) silane quantitatively silated primary and secondary groups at 60°C but little of tertiary hydroxyl groups.

The above procedure with 1-naphthyldimethyl(dimethylamino)silane was applied to four polymers, two of them hydroxyl terminated, the other two with methoxy end groups of unknown, but very low, hydroxyl content. Two nominal molecular weights were chosen for both types of polymers, 600 and 20,000 representing extreme types regarding the difficulties involved in the purification of the silylated polymers by precipitation and filtration. For this operation, the best pair of solvents proved to be 1,2-dimethyloxyethane and low boiling petroleum ether. The samples were redissolved and precipitated repeatedly and each time a portion of the precipitate was subjected to photometric analysis. Apparent hydroxyl contents expressed as a function of the number of precipitations showed that in the case of the hydroxyl terminated polymers a constant silyl concentration is attained after the first precipitation. Three precipitations were necessary however, for the methylated polymers in order to obtain consistent results.

46.7 DISCUSSION OF RESULTS

This method is capable of determining down to 0.01% of hydroxy groups in a 3000 molecular weight poly(ethylene glycol), (or 1% of hydroxy groups in a 30,000 molecular weight poly(ethylene glycol)) with a precision of ± 45%.

METHOD 47 - DETERMINATION OF AMINE ANTIOXIDANTS AND ANTIDEGRADANTS IN RUBBER. GEL PERMEATION, CHROMATOGRAPHY.[22]

47.1 SUMMARY

This gel permeation chromatographic method determines a range of amine antioxidants and antidegredants in rubbers in amounts down to 0.02%.

47.2 APPARATUS

Chromatographic system comprising a series of six stainless steel columns (1200 x 8 mm) connected to a flow refractometer (R 4oz Waters Ass., Framingham, Mass., U.S.A.) or equivalent and to a 254 flow ultraviolet analyzer. The data provided by both detectors and the photocell which was used for checking the volume of the elution agent were recorded. The measurements were carried out at 30°C, the flow rates of the elution agent through the columns were 35 - 40 cm^3/h and the pressure in the columns was about 5 atm. The columns were packed with Copolymer ST-DVS, No. IX polystyrene gel (United Chemical and Metallurgical Works, Usti., Czechoslovakia), 0.040 - 0.056 mesh.

47.3 REAGENTS

Tetrahydrofuran, distilled prior to use, was used as the elution agent.

Table 7.18 - Physical Data of R′, R″ and R‴ (2′ methoxyethoxy) Silanes

$\overline{R}$′, R″, R‴	bp(°C)/Torr	mp(°C)	d20g cm^{-3}	20_D	Molar Absorbtivity mol^{-1}(nm)
1-napthyl-dimethyl	111°/0.03	---	1.051	1.5582	7.33 x 10^3 at 282.5 nm

47.4 EXPERIMENTAL PROCEDURE

The samples were injected as 2 - 5% solutions in tetrahydrofuran in amounts of about 0.5 $\pm$ 0.1 cm^3 into the columns. The elution volumes, V, characteristic of each compound in the given system and corresponding to the size of the molecules of the given compound in solution, were determined.

Since the gel permeation method does not allow a direct determination of the molecular weights or molar volumes of the samples under investigation, calibrations are made by using standard compounds in the form of a graphic dependence of their molar volume on the elution volume.[23-26] Normal hydrocarbons (pentane, hexane, heptane, dodecane, hexadecane, octadecane) and aliphatic esters (octyl adipate and octyl sebacate) are used as standards. The´molar volumes (ml/mole) were plotted against elution volumes, V (ml) in the calibration curves, as shown in Table 7.19. The molar volumes were calculated from the atomic volumes and structural coefficients.[24]

47.5 TYPICAL RESULTS

The results of the gel permeation chromatography measurements by Protivova and Pospisil[22] of the elution volumes of aromatic amines, their molecular weights, calculated molar volumes and the effective molar volumes observed and read from the calibration curves are given in Table 7.20. A comparison of the calculated and effective molar volumes revealed deviations in the behaviour of all the amines they investigated, compared to similar aliphatic hydrocarbons.

47.6 DISCUSSION OF RESULTS

This method is capable of determining down to 0.02% of amine antioxidants and antidegradants in rubbers with an accuracy of $\pm$ 5%.

METHOD 48 - DETERMINATION OF WATER IN CELLULOSE - KARL FISCHER TITRATION.

48.1 SUMMARY

This titration method is capable of determining down to 0.2% water in cellulose.

Table 7.19 - The Behaviour of Standard Compounds in Gel Permcation Chromatography Molar Volumes Calculated

Compound	Molecular weight	Molar Volume (ml/mole)	V_e (ml)
n-Pentane	72.15	118.4	246
n-Hexane	86.18	140.6	239
n-Heptane	100.20	162.8	233
n-Dodecane	170.33	273.8	220
n-Hexadecane	226.43	362.6	212
n-Octadecane	254.48	414.4	206
Octyl adipate	270.14	495.4	20
Octyl sebacate	326.24	613.8	192

From Protovova and Posposal[22]. Elsevier Science Publishers, Netherlands

Table 7.20 - The Behaviour of Amine Antioxidants, Antiozonants and Model Compounds in Gel Permeation Chromatography

Chemical Structure	Molecular Weight	V_e (ml)	Molar Volume (ml/mole) Calculated	Effective	Deviation
Aniline	93.12	238	110.2	150.3	+ 40.1
4-methylaniline	107.15	238	132.4	130.3	+ 17.9
2,3-Dimethylaniline	121.18	247	154.6	115.6	- 39.0
2,4,6-Trimethylaniline	135.2	232	168.8	183.2	+ 14.4
2,3,5,6-Tetramethylaniline	149.24	247	199.0	115.6	- 83.4
N.Methylaniline	107.15	253	133.9	95.9	- 38.0
N,N-Dimethylaniline	121.18	278	156.1	44.2	- 111.9
1-Naphthylamine	143.18	242	161.8	134.9	- 26.9
2-Naphthylamine	143.18	241	161.8	139.0	- 22.8
Diphenylamine	169.22	229	200.3	201.0	+ 0.7
Phenyl-2-naphthylmine[a]	219.27	235	251.9	166.7	- 85.2
o-Phenylenediamine	108.14	238	124.4	150.3	+ 25.9
m-Phenylenediamine	108.14	220	124.4	266.1	+ 141.7
p-Phenylenediamine	108.14	248	121.4	111.4	- 10.0
4-Aminodiphenylamine	184.23	221	214.5	257.6	+ 43.1
4,4′-Bis-(dimethylamino) diphenylamine	255.41	228	320.5	212.3	- 108.2
Benzidine	184.23	217	213.0	291.7	+ 78.7
o-Tolidine	212.28	224	257.4	234.4	- 23.0
N,N′-Dimethyl-p-phenylene-diamine	136.22	247	171.8	115.6	- 56.2
N,N′-Diethyl-p-phenylene-diamine	164.14	222	216.2	249.5	+ 33.3
N.N′-Di-sec-butyl-p-phenylenediamine[b]	220.38	236	305.0	162.2	- 142.8
N,N′-Di-iso-hepyl-p-phenylenediamine[c]	305.4	202	438.2	462.4	+ 24.2
N,N′-Di-iso-octyl-p-penylenediamine[d]	332.58	200	482.6	495.5	+ 12.9
N,N,N′-Trimethyl-p-phenylenediamine	150.28	252	186.0	98.9	- 87.1
N,N′-Dimethyl-2-methyl-p phenylenediamine	150.28	250	186.0	105.2	- 80.8

Table 7.20 continued

N,N'-Diphenyl-p phenylenediamine[e]	260.36	208	304.6	384.5	+ 79.9
N,N'-Dinaphthyl-p phenylendiamine[f]	360.46	205	415.2	421.7	+ 6.5
N-iso-Propyl-N'-phenyl-p-phenylenediamine[g]	226.34	222	282.6	249.5	- 33.1
N-iso-Butyl-N'-phenyl-p-phenylenediamine[h]	240.36	214	304.8	319.9	+ 15.1
N-Cyclohexyl-N'-phenyl-p-phenylenediamine[i]	266.41	206	326.8	410.2	+ 83.4
N-octyl-N'-phenyl-p-phenylenediamine[k]	296.47	206	393.6	410.2	+ 16.6
N,N'-Bis-4-(beta,N-dimethyl-amino)-phenyl-p-phenylene-diamine	346.55	218	424.8	283.1	- 141.7

[a] Agent Rite Powder (Anchor Chemical Company Ltd., Manchester, Great Britain)
[b] Topanol M (ICI, Macclesfield, Great Britain)
[c] Santoflex 77 (Monsanto, St. Louis, Mo., USA)
[d] UOP 88, UOP 288 (UOP Chemical Company, East Rutherford, N.J., USA)
[e] DPPD (Monsanto), Altofane DIP (Etablissements Kuhlmann, Paris, France) JZF
[f] Santowhite CI (Monsanto), Antioxidant 123 (Anchor), DNPD (Chemical works of J. Dimitrov, Bratislava, Czechoslovakia), ASM DNP (Bayer, Leverkusen, Germany)
[g] ASM 4010 NA (Bayer), Nonox ZA (Arnold, Hoffman & Co., Providence, R.I., USA)
[h] Santoflex 13 (Monsanto)
[i] Flexone 6-H (U.S. Rubber Co., Naugatuck Division)
[k] UOP 688 (UOP)

From Protivova and Posposil[22]. Elsevier Science Publishers, Netherlands

48.2 APPARATUS

Burette assembly. An all glass enclosed buret system. The connection glass tubes fitted with spherical joints afford flexibility and ease of assembly. Teflon stopcocks are used in the buret and connecting lines.

Protect all vents against atmospheric moisture by adequate drying tubes, the use if "Indicating Drierite" is recommended. Lubricate all joints with "High Vacuum Silicone Grease" (Dow-Corning Corp., Midland, Michigan).

Glass stoppered flasks, 250 cm^3.

48.3 REAGENTS

Fischer reagent. Dissolve 133 g cp iodine in 425 cm^3 of cp pyridine in a dry, glass stoppered bottle, add 425 cm^3 of anhydrous methyl cellosolve (see note 2) and cool in an ice water bath, add with constant swirling, 70 cm^3 of anhydrous liquid sulfur dioxide. The water equivalent to this reagent is approximately 6 mg of water per ml of reagent. Standardize the reagent as follows: introduce 10 cm^3 of anhydrous methanol into a volumetric flask and titrate with Fischer reagent until the straw-yellow colour changes to an orange-red which remains after vigorous swirling of the stoppered flask, view the colour by transmitted light.

To calibrate, weigh from a Lunge weighing bottle between 0.15 and 0.18 g of water. Mix and titrate with Fischer reagent to the original orange-red end point colour.

Calculate the equivalency factor, F, as milligrams of water equivalent to 1 cm^3 of the Fischer reagent.

48.4 EXPERIMENTAL PROCEDURE

Introduce 25 cm^3 anhydrous methanol into each of two 250 cm^3 glass-stoppered volumetric flasks and titrate with Fischer reagent to the end-point colour. Retain one flask and, to the other flask, add a weighed quantity of sample containing between 0.1 and 0.2 g of water. Small particle size is necessary when the sample is insoluble in methanol. Allow the reaction mixture to stand 1 hour and titrate to the end-point colour of the standard. An end-point stable for fifteen minutes usually indicates complete extraction, (notes 1 and 2).

48.5 METHOD

In this Karl Fischer titration procedure a methanol extract of the sample is titrated with standardized Karl Fischer reagent to a visual end-point.

Calculation. From the standardization, calculate the equivalency factor, F, in milligrams of water equivalent to 1 ml of Fischer reagent, by means of the following equation:

$$F = \frac{(W_s)(1000)}{V_s}$$

where
W_s = weight of water grams, used for the standardization if Fischer reagent.
V_s = volume of Fischer reagent, millilitres, consumed in the standardization.

Calculate the water content of the sample as follows:

$$\text{Water, \%o} = \frac{(S - B)(F)}{(W)(10)}$$

where
S = volume of Fischer reagent, millilitres, required by samples.
B = volume of Fischer reagent, millilitres, required by blank determination (if necessary)
F = equivalency factor for Fischer reagent.
W = weight of sample, grams.

Note 1. View the colour by transmitted daylight or by transmitted light.

Note 2. If the water content of methyl Cellosolve or ethylene glycol is appreciably greater than 0.1% they may be dried by distilling off about 5% and using the remaining 95%.

Note 3. The analysis should be completed within four hours, since longer standing encourages side reactions. Other inert suspending media besides methanol are sometimes useful. Ethylene glycol, for example, exerts a greater solvent action upon some salts.

48.6 DISCUSSION OF RESULTS

This method is intended for the determination of water in substances that are inert to Fischer reagent and from which all the water can be released. It may be applied, by proper choice of technique, to a wide variety of polymers. The method is not applicable in the presence of hydrogen sulfide, mercaptans or appreciable amounts of hydroperoxides and it is not applicable to any compound or mixture which partially reacts under the conditions of the test, to produce water.

METHOD 49 - DETERMINATION OF AMINO GROUPS IN AROMATIC POLYAMIDES, POLYIMIDES AND POLY(AMIDES-IMIDES). POTASSIUM HYDROXIDE FUSION-GAS CHROMATOGRAPHY.[27]

49.1 SUMMARY

This alkali fusion gas chromatographic method determines down to 0.02% of amino groups in aromatic polyamides, polyimides and poly(amides-imides).

49.2 APPARATUS

The total analysis (reaction, trapping, separation and quantitation) is performed in a single piece of apparatus, Figure 7.11. The reaction portion consists of a pyrolysis tube, combustion furnace assembly and a mounting frame taken from a commercial unit (Perkin Elmer Pyrolysis Accessory 154-0825, Norwalk, Conn. 06856). Samples, standards and reagent are contained in miniature platinum boats (10 mm x 4 mm x 4 mm, Fisher Scientific Co., Pittsburgh, Pa. 15219) and are magnetically manipulated within the confines of the glass tubing with metal cylinders and a hoe-shaped retriever. The combusion furnace assembly monitors and controls the temperature within the reaction zone.

A loop-shaped piece of 316 grade stainless steel tubing (0.125 inch o.d. 0.093 inch i.d.) connected between the outlet of the reaction furnace and a special low dead volume GC injector assembly (H) (Perkin Elmer Part No. 009-1276) serves as a trap for efficiently concentrating most volatile reaction products. All of the transfer tubing from the reaction zone to the injection port, except the lower section of the loop, is heated to 300°C by a clam-shell type combustion furnace clamped around the tubing. The exposed piece of tubing, which is loosely packed with quartz wool, is cooled in a liquid nitrogen-filled Dewar flask when concentrating the evolved products and heated with a 400°C nichrome wire heater when "injecting" the products into the gas chromatograph. Helium carrier gas flows (60,l/min) through the reaction unit and cold trap before entering the chromatographic column.

49.3 REAGENTS

Fusion reagent. The alkali fusion is a prefused mixture of potassium hydroxide (J.T. Baker, Analytical reagent grade) and 0.5% sodium acetate.

Typically, the mixture melts around 110°C and contains 13-14% water. When preparing this reagent, it is important to avoid excessive heating. If too much water is lost, the molten reaction mass may solidify before the hydrolysis reaction is complete. On the other hand, too much water will make the reagent sticky and difficult to handle. Additionally, it is possible to plug the cold trap if large amounts of water are released. To prevent adsorption of moisture from the atmosphere, the powdered reagent is stored in a small desiccator within a nitrogen-filled glove bag.

49.4 METHOD

Samples are hydrolyzed with a molten potassium hydroxide reagent at elevated temperatures in a flowing inert atmosphere. Volatile reaction products are concentrated in a cold trap before separation by gas chromatography. The identity of amine and/or diamine products aids in the characterization of the monomer or polymer; the amount of each compound generated is used as the basis for quantitative analysis. The average per cent relative standard deviation of the method is $\pm$ 1.0%.

49.5 EXPERIMENTAL PROCEDURE

One to five milligrams ($\pm$ 1 ug) of powdered sample or standard is weighed into the tared platinum micro-boats. The boats are then filled with the powdered caustic reagent in the nitrogen-purged glove bag and quickly loaded into the storage arm of the reaction tube with a metal cylinder behind each. When making quantitative analysis, the first sample reacted in a series of runs is used to condition the apparatus. Boats are usually loaded at the end of a work day, thus permitting entrapped air to be purged from the system overnight. When samples are loaded in the morning, the tube is purged for 1 hour before turning on the thermal conductivity detector current.

Once a stable baseline is obtained, the liquid nitrogen-filled Dewar flask (-196°C) is positioned around the unheated section of the trap. The first metal cylinder is magnetically move forward, pushing the boat in front of it into the heated reaction zone. The cylinder is removed to storage area. The optimum reaction temperature profile for a fusion reaction will depend on the reactivity of the samples. The furnace temperature was programmed from 100 to 300°C over a period of 10 min. As the reaction occurs, the volatile reaction products and water, liberated from the molten mixture, are carried by the flowing helium carrier gas and concentrated in the trap. At the end of the reaction period, ie. 10 minutes, the boat is withdrawn from the furnace with the magnetic retriever and deposited with its cylinder. After adjusting the furnace control to its initial temperature setting, the trapped compounds are revolatilized and directed into the gas chromatograph by replacing the Dewar flask with the heater (400°C). When the separtion and integration are completed and the initial conditions re-established, the procedure is repeated for each sample and standard. Matched 6 foot by 1/4 inch o.d. stainless steel columns packed with 60/80 mesh Chromosorb 103 (Johns-Manville) were used to separate the amine reaction products. Diamines were chromatographed on 4 foot by 1/8 inch o.d. stainless steel columns packed with 10% FFAP on 60/70 mesh Anakrom AB.

Calibration curves were prepared daily for each compound determined. Anhydrous ammonia was injected with a calibrated 1000 ul gas tight syringe through the septum inlet of the reaction tube. Ambient temperature and pressure corrections were made to determine the actual amount of gas injected. Benzylamine was generated from reaction of the hydrochloride salt with potassium hydroxide. The diamine standards were weighed into boats, covered with caustic reagent and volatilized by moving the boat into the heated furnace. In each case, the standard compounds were trapped and

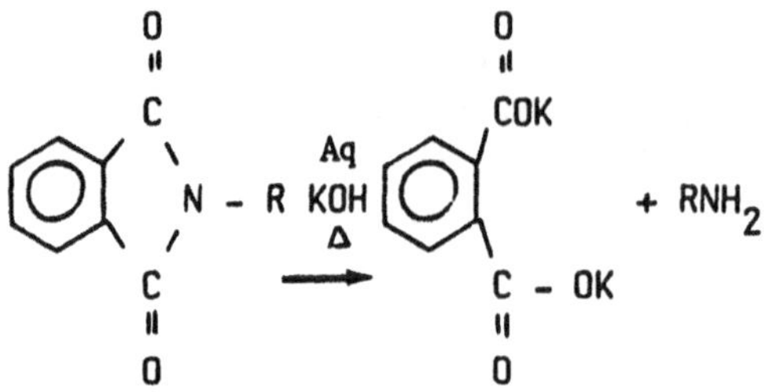

7.5

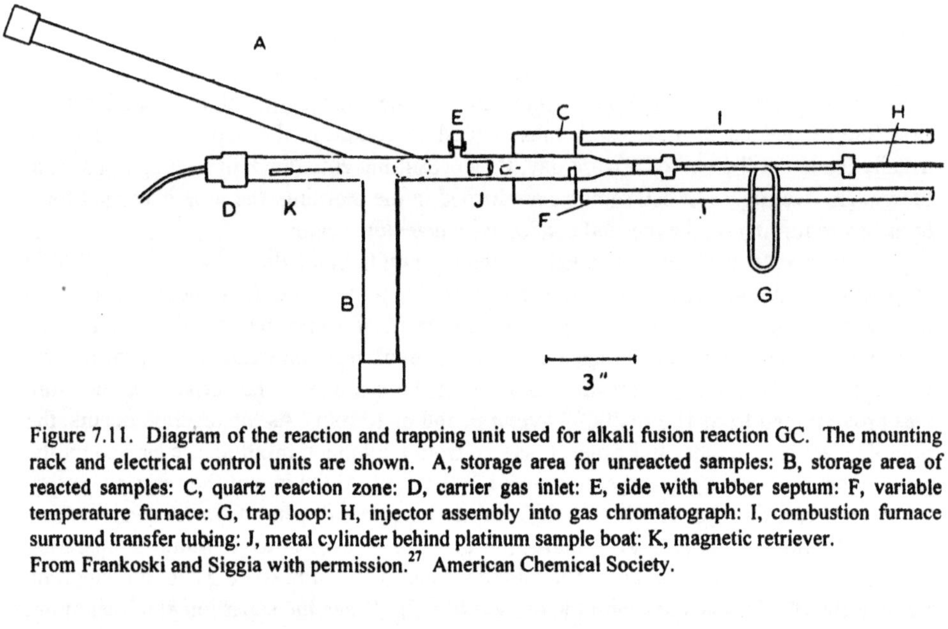

Figure 7.11. Diagram of the reaction and trapping unit used for alkali fusion reaction GC. The mounting rack and electrical control units are shown. A, storage area for unreacted samples: B, storage area of reacted samples: C, quartz reaction zone: D, carrier gas inlet: E, side with rubber septum: F, variable temperature furnace: G, trap loop: H, injector assembly into gas chromatograph: I, combustion furnace surround transfer tubing: J, metal cylinder behind platinum sample boat: K, magnetic retriever.
From Frankoski and Siggia with permission.[27] American Chemical Society.

chromatographed in the same manner as the volatile reaction products. The best straight line calibration curves were determined by a least-square regression curve fitting computer programme.

49.6 TYPICAL RESULTS

Table 7.21 lists the results obtained from the analysis of mixtures of polymers. The amount of m-phenylenediamine produced was, in all cases within 1.2% relative of the theoretical value.

Table 7.21 - Analysis of Polymer Mixtures by Alkali Fusion Reaction Gas Chromatography

Polymer Mixture	Micromoles of *m*-phenylenediamine			Recovery %
	Taken[a]		Found	
PI-2	2.24)			
PAI-2	9.26)	11.50	11.36	98.8
PA-1	7.23)			
PA-2	7.43)	14.66	14.71	100.3
PAI-1	4.99)			
PI-2	7.67)	12.66	12.58	99.4
PAI-1	5.28)			
PAI-2	7.09)	12.37	12.48	100.9
PA-1	5.51)			
PAI-2	5.81)	14.21	14.33	100.8
PI-2	2.89)			
PA-2	8.79)			
PAI-1	3.25)	14.80	14.98	101.2
PA-1	2.76)			

[a] These values were calculated from the weight of polymer taken and were corrected for the weight of adsorbed water (Table 7.22) and the average experimental recovery (Table 7.23).

From Frankoski and Siggia[27]. American Chemical Society

49.7 DISCUSSION OF RESULTS

Table 7.22 summarizes the chemical structures and sample designations of the polymers studied by Fronkoski and Siggia.[27] The weight percent of adsorbed water and the decomposition temperature, both determined by thermogravimetric analysis, are also included. The water content was needed to calculate the dry weight of polymer in each sample. It was not possible to simply dry and weigh the polymers because they tended to readsorb moisture from the atmosphere too rapidly. Table 7.23 summarizes the diamine recoveries obtained with isothermal (5 min at 380°C) or programmed (100 to 390°C over 30 min) reaction temperatures. These numbers represent the mole per cent of theoretical diamine based on the dry weight of sample and the idealized linear polymer repeat units depicted in Table 7.22. Recoveries were, in all cases, between 90 and 101% of theory. The amount of diamine recovered from a polymer reacted isothermally, was within $\pm$ 3% of the amount produced by the programmed temperature approach. Average percent relative standard deviations were similar in both cases ($\pm$ 0.9% vs. 1.1%).

The reaction gas chromatograph obtained from poly(amide-imide) PAI-2 showed only one diamine was produced whilst that from the fusion reaction of the polyimide terpolymer PI-3, contains peaks for each diamine. The quantitative analysis of both products revealed that the weight ratio of 2,4 toluenediamine to 4,4′-methylenedianiline in the polymer was 2 to 1.

In an attempt to correlate the alkali fusion recoveries (expressed as weight percent nitrogen) with the elemental nitrogen analyses of the polymers, it was noted that the elemental nitrogen analyses of the polymers were higher. Analysis of the fusion residue indicated that no detectable amounts of nitrogen were present. This led to the discovery that ammonia was also produced from the samples. When the nitrogen contribution from

Table 7.22 - Structure, Water Content and Decomposition Temperatre of the Polymers Studied

Designation	Structure of Repeat Unit	Wt% Water	Decomposition Temp., °C
PI-1	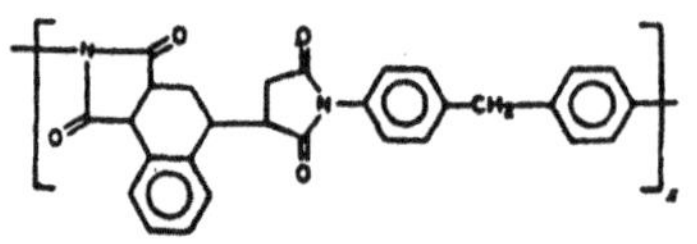	6.5	385
PI-2	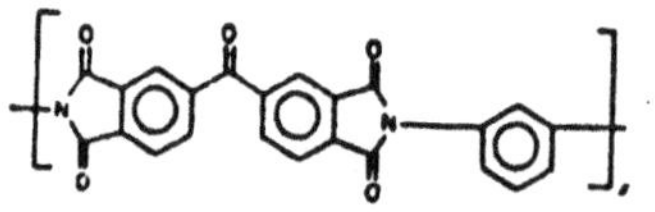	2.2	410
PI-3	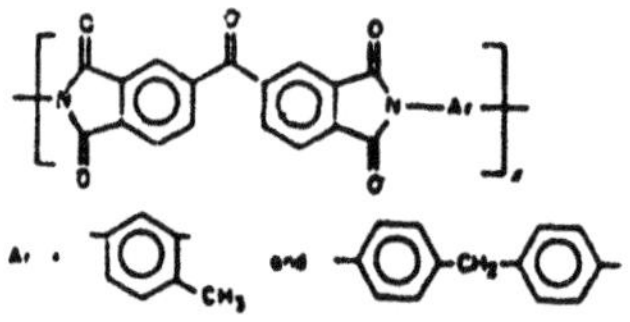	3.4	410
PI-4	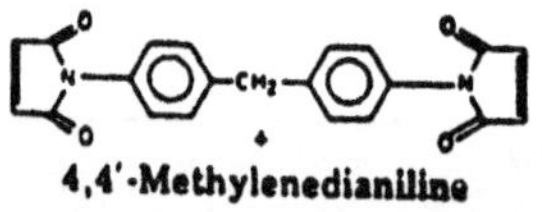 4,4'-Methylenedianiline	0.6	310
PA-1	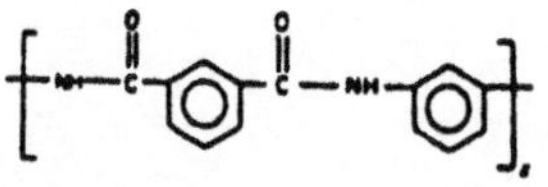	5.8	315
PA-2	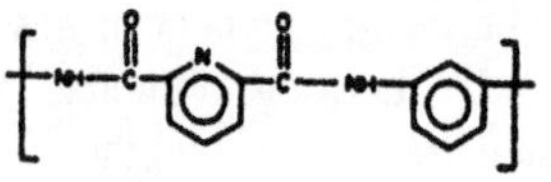	9.7	330
PAI-1	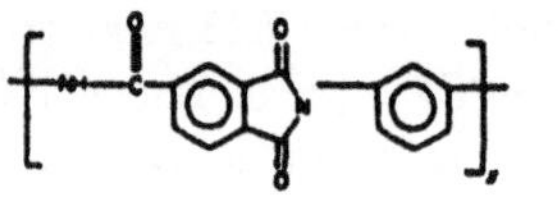	6.2	340
PAI-2	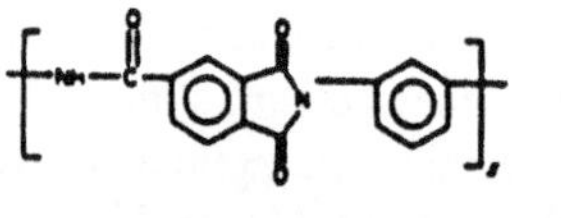	12.6	395
PAI-3	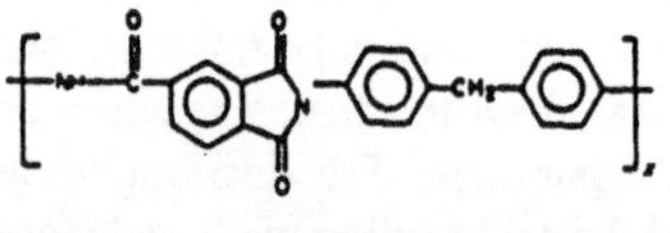	8.9	340

From Frankoski and Siggia[27]. American Chemical Society

ammonia was added to that of the diamine, the total was in good agreement with the elemental value.

Table 7.23 - Analysis of Polyimides, Polyamides and Poly(amide-imides) by Alkali Fusion Reaction Gas Chromatography

Sample	Diamine Produced	Mol % of theoretical[a] ± RSD[b]	
		5 min at 380°C	30 min at 100-390°C
PI-1	4,4'-Methylenedianiline	98.3 ± 0.8	97.5 ± 0.9
PI-2	m-Phenylenediamine	89.1 ± 0.7	
		91.2 ± 0.9	91.0 ± 0.8
		91.6 ± 1.3	91.0 ± 0.7
PI-3	2,4-Toluenediamine	73.1 ± 0.2	73.2 ± 1.0
		97.8 ± 0.5	99.1 ± 1.2
	4,4-Methylenedianiline	24.8 ± 0.3	25.9 ± 0.2
PA-1	m-Phenylenediamine	100.7 ± 0.6	97.0 ± 1.7
			97.7 ± 1.7
			97.9 ± 1.0
			99.6 ± 2.6
PA-2	m-Phenylenediamine	95.9 ± 0.6	92.9 ± 1.0
		96.4 ± 1.5	93.3 ± 0.6
PAI-1	m-Phenylenediamine	93.5 ± 0.9	93.1 ± 0.7
			95.3 ± 0.5
PAI-2	m-Phenylenediamine	93.5 ± 1.1	97.4 ± 0.9
		94.1 ± 0.6	
		95.1 ± 1.2	
PAI-3	4,4'Methylenedianiline	98.0 ± 1.1	98.8 ± 1.1

From Frankoski and Siggia[27]. American Chemical Society
[a] These recovery values are based on the structures shown in Table 7.22, which assume that one mole of diamine is produced for each mole of repeat unit. [b] The relative standard deviation is based on five or more determinations.

METHOD 50 - DETERMINATION OF ACRYLAMIDE MONOMER IN POLYACRYLAMIDE. DIFFERENTIAL PULSE POLAROGRAPHY.[28]

50.1 SUMMARY

This differential pulse polarographic method enables down to 0.02% acrylamide monomer to be determined in polyacrylamide.

50.2 APPARATUS

Differential pulse polarographic analyzer equipped with a PARC Droptimer.

X - Y recorder.

The polarographic cell assembly comprising cell base and cover with a dropping mercury electrode as the cathode, a platinum wire auxiliary electrode as the anode and a saturated calomel reference electrode making solution contact via a liquid junction bridge. The mercury column is kept at a constant height, to have a natural drop time of about 4 s. The instrumental conditions for the differential pulse technique were: 0.5 s drop time, -10 mV/S scan rate, 50 mV modulation amplitude.

50.3 REAGENTS

All reagents were reagent grade and were used without further purification unless specified.

Tetra-*n*-butylammonium hydroxide and chloride solutions polarographic grade.

Resin: Analytical grade mixed bed resin, AG 501-X8, 20 - 50 mesh fully regenerated (Bio-Rad Laboratories, Richmond, Calif.). Prepared by washing with copious quantities of pure methanol for several hours, washing with the 80 - 20 methanol/water solvent and then air-drying.

Methanol, distilled-in-glass.

Acrylamide.

50.4 METHOD

The polyacrylamide sample is extracted with aqueous methanol and tetra n-butyl ammonium hydroxide added as base electrolyte. This solution is polarographed and the peak height at -2V of the polyacrylamide peak evaluated using a platinum wire anode electrode and a saturated calomel reference electrode. The method is calibrated against standard aqueous methanolic solutions of acrylamide.

50.5 EXPERIMENTAL PROCEDURE

Add 1 to 10 g of the polymer to 50 cm^3 of an 80/20 volume: volume methanol /water solution in a glass bottle. A magnetic stirring bar coated with Teflon resin is added, the solution stirred and the acrylamide monomer allowed to extract for at least 3 h, preferably overnight.

After an appropriate length of time, 25 cm^3 of the methanolic extraction solution are removed and added to a bottle containing 5 g of mixed resin and a magnetic stirring bar coated with Teflon resin and stirred for 20 min. This mixed resin treatment removes interfering cationic and anionic species. However, impurities extracted from the resin do increase the polarographic residual current and, therefore, it is necessary that a blank be run using the same resin treatment time (viz about 20 min).

After 20 min, 10 cm^3 of the resin treated analyte are added to the polarographic cell, 0.5 cm^3 of 1 M tetra-*n*-butylammonium hydroxide in water is also added as a supporting electrolyte.

The dropping mercury electrode, a platinum wire auxiliary electrode and a saturated calomel reference electrode assembly are inserted into the cell. Solution contact, from the methanolic solution to the saturated calomel reference electrode, is made via a salt bridge containing 0.1 M tetra-*n*-butylammonium chloride in water. The cell solution is deaerated for 5 min by purging with pre-purified nitrogen which has been passed through a scrubbing tower containing an 80;20 methanol/water (v/v) solution. After deaeration, a differential pulse polarogram is obtained on the solution, the solution is spiked with an appropriate amount of an acrylamide standard (such that the total solution concentration does not exceed ca, 100 ppm acrylamide), deaerated and a second differential pulse polarogram obtained.

50.6 TYPICAL RESULTS

The differential pulse polarogram of a 10 ug/ml methanolic solution of acrylamide showed a well-defined and well-resolved peak. The differential pulse polarographic acrylamide reduction current is directly proportional to concentration as shown in Table 7.24. The polarographic detection limit for acrylamide is less than 1 pg acrylamide/cm^3.

The reduction of the acrylamide monomer occurs at ca. - 2.0 V vs SCE. In this region of the polarogram, reduction of alkali cations occurs. The mixed resin treatment is used to remove polarographically interfering ionic species, such as sodium and potassium cations, which are major interferences.

50.7 DISCUSSION OF RESULTS

Acrylo-type compounds, such as acrylic acid, acrylonitrile and acrolein can be present in polyacrylamides. Their unsaturated characteristics could cause them to interfere with the acrylamide analysis. Acrylic acid is electroreducible, but its anionic form, acrylate is not. A polarogram of acrylic acid and acrylamide in an 80:20 (v/v) methanol/water solution with tetra-*n*-butylammonium chloride as the supporting electrolyte showed that the reduction of the associated acid occurs at ca. - 1.7 V vs SCE, 0.3 V more positive than the acrylamide reduction. The reduction peak of acrylic acid is easily resolved from acrylamide. Esters of acrylic acid are electroreducible and some, such as the alkyl esters (eg. ethyl acrylate) reduce in the same potential region as acrylamide and do constitute an interference. Acrylonitrile is also electroreducible at the same potential as acrylamide. However, the high volatility of acrylonitrile allows it to be readily purged from the analyte solution. The detection limit of acrylamide monomer by this technique is less than 1 ppm.

METHOD 51 - DETERMINATION OF ACRYLONITRILE AND METHACRYLONITRILE MONOMERS IN POLYACRYLAMIDE. HIGH PERFORMANCE LIQUID CHROMATOGRAPHY.[29]

51.1 SUMMARY

This high performance liquid chromatographic method enables acrylonitrile and methacrylonitrile to be determined in polyacrylamide in amounts down to 0.1 ppm.

51.2 APPARATUS

Liquid chromatograph with a variable wavelength detector, operated at 208 nm; (Perkin Elmer LC-55 or equivalent instrument).

A mini pump (model 396-31 Milton Roy instrument or equivalent)

Injection valve with 20 ul loop (Rheodyne model 70-10 or equivalent),

Recorder, having 1 - 100 mV output.

Whatman Inc. Partisil-10 ODS-2 4.6 x 250 mm reverse phase column. The analytical column was protected with a guard column 2.1 x 60 mm in size containing pellicular reverse phase packing.

8 ul 6 mm path length flow through cell.

a 3000 psig pressure gauge was located between the pump and the injection valve.

51.3 REAGENTS

Water eluent. Laboratory deionized water was circulated through a Milliport Corp. Milli-Q water system. It was then degassed by laboratory vacuum.

Acrylamide (Eastman Kodak electrophoresis grade).

Methanol (distilled in glass grade).

51.4 METHOD

Water soluble compounds such as acrylamide and methacrylamide have sufficient lipophilic character such that they can be retained and separated on HPLC reverse-phase columns using water as the eluent. By employing a low-wavelength ultraviolet detector, these compounds can be measured with high sensitivity.

An aqueous methanol extract of the polyacrylamide sample was injected into the high performance liquid chromatograph equipped with an ultraviolet detector and eluted with methanol-methylene chloride 15:85 v/v. The acrylonitrile and methacrylonitrile peaks were evaluated. The method was calibrated against standard aqueous methanol solutions of the two monomers.

51.5 EXPERIMENTAL PROCEDURE

Five g of polyacrylamide are placed into a 4 ox bottle. Fifty cm^3 of 80:20 methanol/water (v/v) are added and the mixture placed on a mechanical shaker for 3 - 4 h. Twenty ul of the clear phase are injected under the following chromatographic conditions and the peak height recorded.

Liquid chromatographic conditions. The water eluent is pumped at 2 cm^3/min which gave a pressure of about 1600 psig. The variable wavelength ultraviolet detector is set at 208 nm. Recorder response is generally at 0.04 aufs for acrylamide monomer in polymer. Chart speed is 0.2 in/min and injection volume 20 ul.

The above extracts can also be examined by ion exclusion liquid chromatography under the following conditions.[30]

Column	Partial 10 PAC (250 x 4.6 mm) (% octadecyl loading = 5 acrylonitrile retention for 5 min.
Mobile phase	15% methanol, 85% v/v methylene chloride.
Flow rate	1 cm^3/min
Pressure	500 psi
Detector	240 mm

The acrylamide response is linear from 1 - 500 ppm monomer in solution using a 20 ul injection. This equates to 0.02 - 10 ug acrylamide injected. Area response obtained from a computing integrator is also found to be linear.

Table 7.24 - Differential Pulse Polarographic Response to Acrylamide Monomer

Acrylamide concn[a], ppm	Peak current[b] A	Peak potential, V vs. SCE	i_{peak}/concn, uA/ppm
0	0.00	...	...
1	0.35	-2.01	0.35
2	0.68	-2.01	0.34
3	1.07	-2.01	0.36
4	1.37	-2.02	0.34
5	1.67	-2.02	0.33
11	3.61	-2.02	0.33
15	5.51	-2.02	0.37
21	7.44	-2.02	0.35
31	11.3	-2.03	0.36
41	15.0	-2.03	0.37
51	19.2	-2.03	0.38
96	37.3	-2.04	0.39

[a] 9.5ml 80/20 methanol/water + 0.5ml 1 N tetra-*n*-butylammonium hydroxide. [b] Peak current is calculated as current above baseline

From Betso and McLean[28]. American Chemical Society

51.5.1 Calibration.

A 10 ppm acrylamide standard is prepared in 80:20 v/v methanol/water. 20 ul are injected and the peak height recorded.

51.6 TYPICAL RESULTS

The retention times for acrylamide and related compounds are given in Table 7.25.

51.7 DISCUSSION OF RESULTS

The sensitivity of acrylamide detection is about 0.1 ppm in solution based on a 20 ul injection. Relative precision of the 95% confidence level for acrylamide is ± 7.5%.

METHOD 52 - DETERMINATION OF TRACES OF 2,4- AND 2,6-DIAMINOTOLUENES IN FLEXIBLE POLYURETHANE FOAMS. THIN-LAYER CHROMATOGRAPHY-FLUORIMETRIC METHOD.[31]

52.1 SUMMARY

This fluorimetric method determines 2,4 and 2,6 diaminotoluene in amounts down to 1 ppm in flexible polyurethane foams.

52.2 APPARATUS

Spectrophoto fluorimeter with a thin film chromatographic scanner and photomultiplier microphotometer (Amino Bouman or equivalent).

Strip chart recorder, 10 V output fsd.

52.3 REAGENTS

Methanol and acetone (reagent grade)

2,4 and 2,6 diaminotoluenes (Eastman and Aldrich).

Fluram reagent (American Instrument Company, 8030 Georgia Avenue, Silver Spring, Md.) 20910.

2,4 and 2,6 diaminotoluene standards, 2-20 mg cm^3 in methanol.

52.4 METHOD

A methanol extract of the polyurethane is applied to a thin-layer chromatography plate and developed with a mixture of chloroform, ethylacetate, ethanol, glacial acetic acid (120:33:20:7 v.v). Fluran spray reagent reveals the two diaminotoluene isomers. The intensity of the spots are evaluated (500 nm excitation 340 nm emission) using a spectrofluorimeter coupled to a thin film scanner and photomultiplier microphotometer. The method is calibrated against standard solutions of the two diaminotoluene isomers.

52.5 EXPERIMENTAL PROCEDURE

Weigh 1 - 2 g of foam and cover with 75 cm^3 of methanol and soak for 5 min with occasional compression. Decant the methanol and squeeze the foam to express the methanol. Repeat twice with fresh methanol, concentrate the combined extracts and make up to 25 cm^3 with methanol. On the TLC plate spot the standard solutions (20 uL) at six positions 3 cm from the bottom of the plate, spot the foam extract at the seventh position.

After the spots have dried, place the plate in a developing tank which contains 120 cm^3 of chloroform, 33 cm^3 of ethyl acetate, 20 ml of ethanol and 7 cm^3 of glacial acetic acid. Development takes 1 h and is complete when the developing solution reaches a line 15 cm above the bottom of the plate.

Dry the plate in a horizontal position for 5 - 10 min, then spray uniformly with a 0.015% solution of Fluram in acetone. The spots, which appear approximately 6 cm (2,4-isomer) and 8 cm (2,6-isomer) above the bottom of the plate, can be located with an ultraviolet light. The sides of the plate are marked to indicate the line through which the plate is to be scanned.

Place the plate in the scanner and set the excitation wave-length at 500 nm to produce a visible light spot. Adjust the plate position so that the visible spot is in position to scan across the line of diaminotoluene spots. Close the cover of the scanner and set the excitation and emission wavelengths at 390 and 500 nm respectively. Scan the plate starting with the strongest standard spot to generate a calibration. The plate must be read within 1 h after development because the Fluram adducts are unstable.

Peak heights are measured from the baseline and are plotted against concentration in ug ml^{-1}. The concentration of diaminotoluene in the extract is then calculated from the

Table 7.25 - HPLC Retention Times[a] for Acrylamide and Related Compounds

Compound	Min.
Acrylic acid	1.4
ß-hydroxypropanamide	2.1
Acetamide	3.0
Acrylamide	5.4
Propanamide	7.3
Acrylonitrile	11.8
Methacrylamide	18.0
Butanamide	20.8
Methacrylonitrile	46.0

[a] Partisil-1 0 ODS-2, water, 2.0mL/min, 208 nm, 0.04 aufs

From Skelly and Husser[29]. American Chemical Society

known concentrations of the standard solutions. To express this value as parts per million in the original foam, the following equation is used:

$$\text{Diaminotoluene, ppm in foam} = \frac{(\text{Diaminotoluene in extract, ug/mL})\ (\text{mL extract})}{\text{foam weight, g}}$$

52.6 DISCUSSION OF RESULTS

This method is capable of determining down to 1 ppm of 2,4 and 2,6 diaminotoluenes in polyurethanes with a precision of $\pm$ 30% at the 20 ppm level.

CHAPTER 7.2

COPOLYMERS AND CONDENSATION PRODUCTS

Methods 53 - 64

METHOD 53 - DETERMINATION OF STYRENE MONOMER AND OTHER VOLATILES IN CONVENTIONAL POLYSTYRENE AND STYRENE - BUTADIENE COPOLYMERS. GAS CHROMATOGRAPHY.[32]

53.1 SUMMARY

This gas chromatographic method is capable of determining styrene monomer and a wide range of aromatic volatiles in polystyrene and styrene-butadiene copolymers in amounts down to 50 ppm.

53.2 APPARATUS

Gas chromatograph with a hydrogen flame ionization detector and an injection port fitted with a glass liner, very loosely packed with glass fibre (Figure 7.5).

Column, copper tube (15 ft x 1.4 o.d. x 3.16 in i.d.) packed with 10 per cent w/v Carbowax 15 - 20 M supported on 60 - 72 B.S. mesh Celite.

Gas pressures and flows.	
Helium	30 lbf/in^2 gauge, 100 cm^3/min
Air	7 lbf/in^2 gauge, 650 cm^3/min.
Hydrogen	12 lbf/in^2 gauge, 75 cm^3/min.
Temperatures	
Injection	150°
Detector	150°
Column	80°
Flame	200°

Recorder, -0.2 to + 1.O mV range, 1 sec response, 15 in (381 mm).hr chart speed.

53.3 REAGENTS

Celite, 60 - 72 BS mesh.
Carbowax 15 - 20 M.
Styrene.
Propylene oxide.
o, m- and p- xylenes.

Cumene.
n-propyl benzene.
iso-butyl benzene.
alpha-methyl styrene.
m and p diethylbenzenes
tert, sec- and n- butyl benzenes
o,m and p methyl styrene.
o, m and ethyl toluenes
n-undecane (internal standard).

53.4 METHOD

The polymer sample and a known weight of n-undecane internal standard are dissolved in propylene oxide and injected into the gas chromatograph. Calibration is achieved against propylene oxide standard solutions of internal standard and the substances to be determined.

53.5 EXPERIMENTAL PROCEDURE

53.5.1 Solution of Polymer.

Prepare the polymer solvent by weighing 0.1 cm^3 n-undecane into a 100 cm^3 volumetric flask and then diluting to 100 cm^3 with propylene oxide. Weigh accurately about 1 g polymer into a 25 cm^3 stoppered measuring cylinder, add exactly 10 cm^3 of the prepared polymer solvent from a pipette, seal with a serum cap and shake until the polymer has dissolved. Gel, pigment and filler may remain as an undissolved suspension without detriment to the analysis. Chromatograph 5 ul of the solution at a range setting of x 10 and suitable attenuation settings.

53.5.2 Calibration.

Weigh in turn into a 10 cm^3 volumetric flask 1.0 cm^3 n-undecane, 1.0 cm^3 alpha-methyl styrene, 1.0 cm^3 styrene, 0.5 cm^3 ethyl benzene, 0.5 cm^3 cumene, 0.5 cm^3 of *n*-propyl benzene, 0.25 cm^3 *m* or *p* xylene, 0.25 cm^3 *o*-xylene, 0.25 cm^3 *m* or *p*-ethyl toluene, 0.25 cm^3 toluene and 0.1 cm^3 benzene. Dilute to 10 cm^3 with propylene oxide and then further dilute 0.25 cm^3 of the solution to 25 cm^3 with propylene oxide. Chromatograph 5 ul of the calibration blend at a range setting of x 10 and suitable attenuation settings. This calibration procedure should be carried out daily.

53.5.3 Calculation.

On the calibration and analysis chromatograms, measure the peak heights of the n-undecane and the aromatic hydrocarbons allowing for any attenuation factors.

From the calibration chromatogram, determine the response factor for each component as follows:

A = weight of component in calibration blend.
B = weight of n-undecane in calibration blend.
C = peak height of component on calibration chromatogram.

D = peak height of n-undecane on calibration chromatogram.

Component response factor, $f = \frac{A}{B} \times \frac{D}{C}$

From the analysis chromatogram, determine the concentration of each component as follows:

F - component response factor.
G = peak height of component on analysis chromatogram.
H = peak height of *n*-undecane on analysis chromatogram.
J = per cent w/v *n*-undecane in polymer solvent.
K = weight (g) of polymer taken.

$$\text{Component (per cent w/w)} = \frac{10 \times G \times F \times J}{H \times K}$$

Relative retention distances. Retention distances (injection point to peak centres) are corrected for the gas hold-up of the columns and are expressed relative to styrene:

Benzene	0.17
Toluene	0.29
m-undecane	0.40
Ethyl benzene	0.47
m-xylene	0.50
p-xylene	0.50
Cumene	0.62
o-xylene	0.66
n-propyl benzene	0.77
p-ethyl toluene	0.84
m-ethyl toluene	0.84
iso-butyl benzene	0.92
tert-butyl benzene	0.92
sec-butyl benzene	1.00
Styrene	1.00
o-ethyl toluene	1.05
m-diethyl benzene	1.35
p-diethyl benzene	1.44
n-butyl benzene	1.44
alpha-methyl styrene	1.60
o-methyl styrene	1.86
m-methyl styrene	1.86
p-methyl styrene	1.86

53.6 TYPICAL RESULTS

Some results obtained by this method for different types of polystyrene and copolymers are shown in Table 7.26

53.7 DISCUSSION OF RESULTS

In this procedure down to 50 ppm of styrene monomer and other aromatic volatiles can be determined in polystyrene and styrene-butadiene latices with an accuracy of ± 5%.

METHOD 54 - DETERMINATION OF STEARIC ACID AND SODIUM STEARATE IN STYRENE-BUTADIENE RUBBER SOLUTIONS. TITRIMETRIC PROCEDURE.

54.1 SUMMARY

This titrimetric procedure is capable of determining down to 0.01% of stearic acid and sodium stearate in styrene-butadiene rubber solutions.

54.2 APPARATUS

Burettes, 2 cm^3, 5 ml., 10 ml.
Conical flasks, 500 cm^3.
Graduated cylinders, 100 cm^3.
Pipettes, 25 cm^3.

54.3 REAGENTS

Toluene/absolute ethanol mixed solvent 5:1, five volumes of redistilled toluene and one volume absolute ethanol mixed.

Ethanolic sodium hydroxide 0.1 M, standardize against potassium hydrogen phthalate to the m-cresol purple end-point.

0.05 M hydrochloric acid/alcohol, standardized, dissolve 4.2 ml concentrated hydrochloric acid in 1 cm^3 absolute ethanol.

m-cresol purple, dissolve 0.3 g m-cresol purple in 95% ethanol and add 7.9 cm^3 0.1 M sodium hydroxide. Make up to 100 cm^3. with 95% ethanol.

54.4 METHOD

A suitable amount (50 to 100 g) of sample is weighed into a flask and diluted with toluene/absolute ethanol mixed solvent. The stearic acid is determined by titration with 0.1 M ethanolic sodium hydroxide solution to the *m*-cresol purple end-point. To determine sodium stearate a separate 100 g quantity of diluted sample is titrated with 0.05 M ethanolic hydrochloric acid solution to the m-cresol purple end-point.

Table 7.26 - Analysis of Some Different Grades of Polystyrene

Component	Weight Per Cent				
	Crystal Grade	Expandable Grade	Self Extinguishing Grade	High Impact Grade	Foodstuff Packaging Grade
Benzene	<0.001	<0.001	<0.001	<0.001	<0.001
Toluene	0.001	0.003	<0.001	<0.001	<0.001
Ethyl benzene	0.066	0.081	0.047	0.086	0.060
m/p-Xylene	0.002	0.002	0.063	<0.001	<0.001
Cumene	0.046	0.049	0.039	0.011	0.030
o-Xylene	0.007	0.008	0.006	0.002	<0.001
n-Propyl benzene	0.017	0.016	0.014	0.004	0.016
m/p-Ethyl toluenes	0.016	0.011	0.013	0.002	<0.001
Styrene	0.033	0.32	0.080	0.18	0.04

From Crompton and Myers[32]

54.5 EXPERIMENTAL PROCEDURE

54.5.1 Determination of Stearic Acid.

Weigh out a suitable quantity of sample (50 g if 0.3%) to within ± 0.5 g into a 500 cm^3 conical flask and add 35 cm^3 toluene/absolute ethanol dilution solvent and 7 drops m-cresol purple indicator. Titrate with 0.1 M ethanolic sodium hydroxide contained in a 10 cm^3 burette from the yellow colour to the first permanent blue colour.

54.5.2 Determination of Sodium Stearate.

Weigh out a suitable quantity of sample (100 g if 0.02%) to within ± 0.5 g into a 500 cm^3 conical flask and add 70 ml of 5.1 toluene: absolute ethanol dilution solvent and 15 drops of m-cresol purple indicator. Titrate with 0.05 M ethanolic hydrochloric acid from the yellow colour to the first permanent pink colour.

54.5.3 Calculations.

Stearic acid in rubber solution,

$$= \frac{(T_1 - T_B) \times f_1 \times 284}{10 \times W_1}$$

where

T_1 = Titration of 0.1 M alcoholic sodium hydroxide solution (sample) cm^3.
T_B = Titration of 0.1 M alcoholic sodium hydroxide solution (styrene monomer blank) cm^3.
f_1 = Molarity of alcoholic sodium hydroxide solution.
W_1 = Weight of sample taken, g.

Sodium stearate in rubber solution, %w,

$$= \frac{T_2 \times f_2 \times 306}{10 \times W_2}$$

where

T_2 = titration of 0.05 M alcoholic hydrochloric acid, cm^3.
f_2 = molarity of alcoholic hydrochloric acid, cm^3.
W_2 = weight of sample taken, g.

54.6 DISCUSSION OF RESULTS

This method is capable of determining down to 0.01% of stearic acid and sodium stearate in styrene-butadiene latices with an accuracy of ± 5%.

METHOD 55 - DETERMINATION OF STYRENE AND ACRYLONITRILE MONOMERS IN STYRENE - ACRYLONITRILE COPOLYMERS. CATHODE-RAY POLAROGRAPHY.[33]

55.1 SUMMARY

This polarographic method is capable of determining down to 20 ppm styrene and 2 ppm acrylonitrile monomers in styrene-acrylonitrile copolymers.

55.2 APPARATUS

Cathode ray polarograph. Complete with stand for dropping mercury electrode, polarographic cells (10 ml) and thermostatted (at 25°C) waterbath.

Micrometer syringe delivering 0.01 cm^3 with an accuracy of 0.0002 cm^3.

Calibrated glassware - pipettes, 25 and 5 cm^3 capacities; 50 cm^3 calibrated flasks.

55.3 REAGENTS

NN'dimethylformamide, if necessary, purify as follows, dry 1 litre N.N'dimethyl-formamide by standing over 50 g of anhydrous potassium carbonate for 3 d. Distill the product through an 80 cm Fenske column. Discard the first 200 cm^3 of distillate and collect in a dry flask the fraction boiling between 151.5° and 153°C.

Tetrabutylammonium iodide base-electrolyte, 0.2 M in dimethylformamide:water. Purge with nitrogen and use within 1 d.

Acrylonitrile and styrene monomers. Store in a refrigerator. Re-distill the monomers immediately before use.

Mercury - for polarographic analysis. Use trebly distilled mercury, supplied in stone containers. Mercury picks up an impurity that might interfere in polarography from polyethylene bottles.

Hydrogen or nitrogen - extremely low oxygen content.

55.4 METHOD

A dimethyl formamide solution of the sample containing 0.2 M tetrabutyl ammonium iodide base electrolyte is polarographed at start potentials of -2.0 V (styrene) and -1.7 V (acrylonitrile). The styrene and acrylonitrile waves are evaluated. The method is calibrated by standard additions of the two monomers.

55.5 EXPERIMENTAL PROCEDURE

55.5.1 Sample Preparation

Purge the stock bottle of dimethylformamide with oxygen-free nitrogen or hydrogen for 30 to 40 min to sweep out dissolved oxygen. Keep the bottle well stoppered and re-purge each time it is opened. Use freshly purged dimethylformamide solvent throughout the analysis.

Weigh 1.25 ± 0.01 g of the styrene-acrylonitrile sample into a 50 cm^3 calibrated flask. Transfer 25 cm^3 of anhydrous dimethylformamide to dissolve the polymer. Dilute the solution to the mark with the 0.2 M tetrabutylammonium iodide reagent. For the blank solution, transfer 25 cm^3 of dimethylformamide and 25 cm^3 of the 0.2 M tetrabutylammonium iodide reagent to a 50 cm^3 calibrated flask. Transfer 5 cm^3 of the sample solution and 5 cm^3 of the blank solution into two polarographic cells and immerse these in the constant temperature bath of the polarograph.

Purging the cell with hydrogen or nitrogen must be performed immediately before any polarographic measurements are made. Lower the dropping mercury electrode system over this cell, so that the outer sleeve of the electrode dips 1 to 2 mm into the water tank (providing a water seal to prevent the ingress of atmospheric oxygen) and immerse the anode connection in the side arm of the polarographic cell. Pass hydrogen or nitrogen for 15 min, then switch off the gas supply. Carry out the polarographic measurements described below within 1 to 3 min of stopping the purge with gas.

The analytical conditions are as follows:

Cathode	dropping mercury
Reference anode	mercury pool
Circuit	cathodic direct
Start potential	styrene - 2.0 V
	acrylonitrile - 1.7 V.

Adjust the poalrograph to its minimum sensitivity setting (ie. x 25). Adjust the X-shift and the Y-shift contols until the light spot commences its sweep at the origin of axes at the left-hand side of the graticule on the cathode-ray tube. Repeat this operation at decreasing sensitivity settings until the polarographic wave is visible on the graticule.

55.5.2 Determination of Styrene.

Take the readings of the freshly de-gassed sample solutions as indicated below.

Adjust the polarograph to the styrene start potential (-2.0 V) and obtain the styrene wave as described above. Read off from the graticule the maximum height of the styrene peak and note the voltage. V_{STY}, at which this maximum polarographic reading occurs. Raise the electrode from the cell and deliver into the sample solution by means of a horizontally held Agla syringe a suitable "standard addition" of a solution of styrene in dimethylformamide: water (95:5 v/v) mixture. Limit the volume of standard addition solution to less than 0.05 cm^3 in order to avoid dilution errors. Lower the electrode into the sample cell and again pass hydrogen or nitrogen for 5 min. Note the new height of the styrene wave corresponding to V_{STY} volts.

55.5.3 Determination of Acrylonitrile.

Take the readings indicated below on the same sample solution as used to determine styrene.

Adjust the polarograph to the acrylonitrile start potential (-1.7 V). To determine acrylonitrile, repeat the operations described above for styrene by using a solution of acrylonitrile in dimethylformamide - water (95:5, v/v) mixture to make the "standard addition" of acrylonitrile. Note the voltage V_{ACN}, at which the maximum polarographic reading occurs.

Measurement of reagent blank solution. Take the readings indicated above on the freshly de-gassed reagent blank solution. Set the instrument at the styrene start potential. Adjust the instrument to obtain the reagent blank wave on the graticule. Measure the blank peak height corresponding to V_{STY} volts. Similarly adjust the instrument to the acrylonitrile start potential and measure the blank peak height corresponding to V_{ACN} volts.

55.5.4 Calculations

Acrylonitrile monomer content of copolymer, ppm =

$$\frac{50 \times 10^6 \times A_{ACN}}{5 \times W} \frac{(h_1S_1 - h_3S_3)}{(h_2S_2 - h_1S_1)}$$

Styrene monomer content of copolymer, ppm =

$$\frac{50 \times 10^6 \times A_{STY}}{5 \times W} \frac{(h_4S_4 - h_6S_6)}{(h_5S_5 - h_4S_4)}$$

where:

W = weight in g. of original styrene - acrylonitrile copolymer sample made up to 50 cm^3 in base electrolyte solution (5 cm^3 portion used for polarography)

For acrylonitrile determinations:

h_1 = peak height in cm at V_{ACN} V, of sample solution before the standard addition.
h_2 = peak height in cm at V_{ACN} V, of sample solution after the standard addition.
h_3 = peak height in cm at V_{ACN} V, of the polymer-free reagent blank solution.
S_1, S_2, S_3 are the corresponding instrument sensitivity settings (the products of h and S are known as peak currents in uA) and

A_{ACN} = weight in g of acrylonitrile present in volume of "standard addition" solution added to the cell solution.

For styrene determination:

h_4 = peak height in cm at V_{STY} V, of sample solution before the standard addition.
h_5 = peak height in cm at V_{STY} V, of sample solution after the standard addition.
$h_{=}$ = peak height in cm, at V_{STY} V, of the polymer free reagent blank solution.
S_4, S_5, S_6 are the corresponding instrument sensitivity settings and
A_{STY} = weight in g of styrene present in volume of "standard addition" solution added to the cell solution.

55.6 TYPICAL RESULTS

Table 7.27 shows the results obtained in determinations of acrylonitrile monomer in some copolymers by the polarographic procedure. Comparison of these results with those obtained by the dodecyl mercaptan procedure[34] show that acrylonitrile contents some 30 per cent higher are obtained by the latter method.

The acrylonitrile and styrene monomer contents of several copolymers were determined by the polarographic procedure. The styrene monomer content of these samples was also determined by a procedure involving evaluation by ultraviolet spectroscopy at 292 nm of solutions of the samples in carbon tetrachloride. The results obtained (see Table 7.28) show that styrene contents determined by the two procedures are in good agreement with each other.

55.7 DISCUSSION OF RESULTS

The above method is capable of determining down to 2 ppm of acrylonitrile and 20 ppm of styrene monomer in styrene-acrylonitrile copolymers with an accuracy of $\pm$ 5% when a 2.5% solution of polymer in base electrolyte is used.

METHOD 56 - DETERMINATION OF STYRENE AND ACRYLONITRILE MONOMERS IN STYRENE-ACRYLONITRILE COPOLYMER. TITRATION PROCEDURES.

56.1 SUMMARY

These titration procedures are capable of determining, respectively, down to 50 ppm and 100 ppm of styrene and acrylonitrile monomers in styrene-acrylonitrile copolymers.

56.2 APPARATUS

pH meter.
Read stop titrimeter.

56.3 REAGENTS

Styrene determination.
Benzene, Analar.

Table 7.27 - Determination of Acrylonitrile: Comparison of Polarographic and Chemical Methods

Acrylonitrile content, percent w/w, determined by: Polarograph	Titration with Dodecyl Mercaptan
0.06 0.07	0.09 0.09
0.08	0.12
0.11	0.15
0.11	0.19
0.12	0.15
0.12	0.17
0.14	0.18
0.21	0.31 0.29

From Crompton and Buckley[33]. Royal Society of Chemistry, London

Table 7.28 - Determination of Acrylonitrile and Styrene Monomers in Copolymers

Acrylonitrile Determined by Polarography, Per Cent, w/w	Styrene, Per Cent w/w Determined by:	
	Polarography	Spectroscopy
0.015 0.014	0.20	0.21
0.015 0.015	0.15	0.13 0.15
0.035 0.036	1.60	1.53

From Crompton and Buckley[33]. Royal Society of Chemistry, London

Potassium bromide-bromate solution 0.083M (0.5N). Dissolve 13.92 g of potassium bromate and 51 g of potassium bromide in 1 cm^3 of water.

Potassium bromide - bromate solution, 0.003M (0.02N). Dissolve 0.557 g of potassium bromate and 2.04 g of potassium bromide in 1 L of water.

Titration solvent. Prepare 1 L of titration solvent by mixing the following: 714 cm^3 of glacial acetic acid, 134 cm^3 of carbon tetrachloride, 116 cm^3 of methanol, 18 cm^3 of sulphuric acid (1 + 5) and 18 cm^3 of a methanolic solution of mercury (II) chloride ($100g^{-1}$).

Acrylonitrile determination.

Propan-2-ol.

Ethanolic potassium hydroxide solution. Dissolve 25 g of potassium hydroxide in 1 L of ethanol.

Standard silver nitrate solution, 0.05 and 0.01 M.

Nitric acid. Mix 3 volumes of concentrated nitric acid (sp. gr. 1.42) with 5 volumes of water.

Standard ammonium thiocyanate solution, 0.05M. Standardise the solution by titration against standard silver nitrate solution.

Indicator solution. Dissolve 40 g of ammonium iron(III) sulphate dodecahydrate (ferric alum) in a mixture of 20 cm^3 of nitric acid:water (3:5 v.v) and 80 cm^3 of water. Heat to boiling and dilute with 3 volumes of water.

Dodecanethiol solutions. Solution A: dissolve 16 cm^3 of dodecanethiol in 1 L of propan-2-ol. Solution B: dissolve 4 cm^3 of dodecanethiol in 1 L of propan-2-ol.

56.4 METHOD

Styrene monomer. A benzene solution of the sample is acidified and titrated with standard potassium bromide bromate solution. The styrene consumes a stoichioimetric amount of the liberated free bromine. The end-point of the titration is detected using a dead-stop titrimeter equipped with glass and calomel electrodes. The styrene content is calculated from the amount of bromine consumed.

Acrylonitrile monomer. An isopropanol solution of the polymer is treated with an excess of standard dodecanethiol solution and standard ethanolic potassium hydroxide. Excess dodecanethiol is determined by titration with standard silver nitrate. The acrylonitrile content of the sample is calculated from the amount of dodecanethiol consumed.

56.5 EXPERIMENTAL PROCEDURE

56.5.1 Determination of Residual Styrene Monomer in Styrene Acrylonitrile Copolymers.

A 2.5 g sample of polymer (accurately weighed to the nearest milligram) is taken in a 150 cm^3 wide-mouthed stoppered bottle and 25 cm^3 of benzene added. The bottle and contents are kept for 50 h with occasional stirring and the resulting extract is added to 100 cm^3 of titration solvent. Styrene is then determined by titration with 0.083M or 0.003M bromide-bromate solution in a dead stop titrimeter using glass and calomel electrodes.

56.5.2 Determination of Residual Acrylonitrile.

The amount of acrylonitrile is determined by using the following procedure. A known amount of the synthetic mixture is dissolved in 100 cm^3 of propan-2-ol and to this solution 10 cm^3 of dodecanethiol solution A is added with continuous stirring by means of a magnetic stirrer. A 2 cm^3 volume of ethanolic potassium hydroxide solution is added and after 2 min the excess of alkali is neutralised by adding 2 cm^3 of nitric acid:water (3:5 v/v). Then 25 cm^3 of 0.05M silver nitrate solution and 25 ml of water are added. The excess of silver nitrate is titrated with 0.05M ammonium thiocyanate solution, using ammonium iron III sulphate as indicator.

56.5.3 Calculations.

$$\text{Styrene, \%} = \frac{(A - B) \times M \times 5.2}{W \times 6}$$

where A ml is the volume of bromide-bromate solution required for the sample. B ml is the volumes of bromide-bromate solution required for the blank. M is the molarity of the bromide-bromate solution and W is the mass of sample solution.

$$\text{Acrylonitrile, \%} = \frac{(A - B) \times M \times 5.3}{W}$$

where A ml is the volume of ammonium thiocyanate solution required for the sample, B ml is the volume of ammonium thiocyanate solution required for the blank, M is the molarity of the ammonium thiocyanate solution and W g is the mass of sample solution.

56.6 DISCUSSION OF RESULTS

These procedures are capable of determining styrene and acrylonitrile monomers, respectively, in amounts down to 50 and 100 ppm in styrene-acrylonitrile copolymers with an accuracy of $\pm$ 10%.

METHOD 57 - DETERMINATION OF VINYL CHLORIDE, BUTADIENE, ACRYLONITRILE, STYRENE AND 2-ETHYLHEXYL ACRYLATE MONOMERS IN POLYMERS. HEAD SPACE ANALYSIS.[36]

57.1 SUMMARY

This head space method is capable of determining vinyl chloride, butadiene, acrylonitrile styrene and 2-ethylhexyl acrylate monomers in a range of polymers in amounts down to 0.05 - 0.5ppm.

57.2 APPARATUS

Gas chromatograph.
Automated head space analyzer equipped with a flame ionization detector.
Operating conditions are listed in Table 7.29.
Glass vials (24 cm^3 of capacity) equipped with septa and aluminium sealing rings.
Manual crimper for sealing vials.
6 ft x 1/8 in 0.4% Carbowax 1500 on Carbopak A, GC column.

57.3 REAGENTS

N,N'-dimethylacetamide.
o-dichlorobenzene
Vinyl chloride monomer and 1,3-butadiene in pressurized cylinders.

The polymerization inhibitors in styrene, acrylonitrile and 2-ethyl hexyl acrylate were removed by passing the monomer over Amberlyst resins (Rohm and Haas) - A-27 resin for styrene and A-21 resin for acrylonitrile and 2-ethylhexylacrylate.

Preparation of standard monomer solutions, (see Table 7.30). Vinyl chloride and butadiene. Pour about 1 cm^3 of monomer into a tared vial containing 20.0 cm^3 of N,N-dimethylacetamide. Seal the vial immediately with a septum and weigh. This standard is diluted to obtain an appropriate concentration of monomer.

Preparation of standard monomer solutions, acrylonitrile, styrene and 2-ethyl acrylate. Weigh 0.2 g portion of monomer into a tared 25 cm^3 volumetric flask and dilute with solvent. Dilute 1 cm^3 of this solution to 25 cm^3 and store in a septum sealed vial.

The above standard solutions are prepared fresh each week and refrigerated when not in use.

Preparation of monomer free calibration standards (see Table 7.30). Vinyl chloride, butadiene and acrylonitrile. Dissolve four 0.500 g portions of monomer free polymer (ie. < 8 ppm monomer) at 90°C in septum sealed vials containing 5.0 cm^3 of N,N′ dimethylacetamide. Cool the vials to room temperature. Inject aliquots of the standard monomer solution (5, 10, 20 and 40 ul) into the polymer solutions using a 50 ul syringe. These solutions are equilibrated at 90°C for a minimum of 30 min prior to head-space sampling.

Preparation of calibration standards, styrene and 2-ethylhexyl acrylate. Dissolve four 0.5 g portions of monomer free polymer (ie. < 8 ppm monomer) in septum sealed vials containing measured aliquots of N,N′-dimethylacetamide. Heat vials to 90°C to aid dissolution of the polymer then cool to room temperature. Inject aliquots of the standard monomer solution (5, 10, 20 and 40 ul) into the polymer solutions using a 50 ul syringe. Swirl the solutions to mix and inject an aliquot of distilled water into each polymer solution. Shake the vials briefly to assure complete mixing of the water with the organic phase and to prevent the precipitated polymer from forming a film on top of the solution. Equilibriate the vials at 90° C for 60 min prior to head space sampling.

57.4 METHOD

A N,N′ dimethyl acetamide solution of the polymer is subject to head space analysis. Measurement of the concentrations of monomers in the head space by gas chromatography enables their concentrations in the original polymers to be calculated using appropriate calibration data.

The more volatile monomers (vinyl chloride, butadiene, and acrylonitrile) are determined by dissolution of the polymer. The injections of water into polymer solutions containing styrene and 2-ethyl hexyl acrylate monomers prior to head-space analysis greatly enhanced the detection capability for these monomers.

57.5 EXPERIMENTAL PROCEDURE

Dissolve a weighed portion of polymer sample with the appropriate solvent in a septum sealed vial. (Table 7.30). Dissolution of the polymer is assisted by heating at 90°C. After the entire sample has dissolved, one of the procedures listed below is followed.

Polymers containing vinyl chloride, butadiene or acrylonitrile. Equilibrate the vials containing polymer solution for a minimum of 30 min at 90°C prior to head space sampling.

Polymers containing styrene and ethylhexylacrylate. Inject a measured volume of water into each vial and shake the vial vigorously. The vials and their contents are equilibrated for 60 min at 90°C prior to head space sampling.

57.6 TYPICAL RESULTS

Peak heights are used for quantitation. A response factor for each monomer is determined by dividing the weight (ug) of monomer added to each standard solution by the adjusted peak height. The adjusted peak height is the product of the measured peak height, the attenuation and the range settings.

Table 7.29 - Operating Conditions for the GC Multifract F-40

GC columns	VCM[a], BD and AN	6 ft x in S.S. tubing packed with 0.4% Carbowax 1500 on Carbopak A[b]
	Styrene and 2-EHA	8 ft x in S.S. tubing packed with 10% poly(phenyl ether) (6R) on Chromosorb W (60/80 mesh)
Column temperatures	VCM, BD, AN and	100°C
	Styrene, and EHA	130°
Injection times	VCM, BD and AN	$I = 2_s$
	Styrene	$I = 10_s$
	2-EHA	$I = 20_s$
Thermostated sample tray		90°C
Injector/detector		180°C
Dosing Line		180°C
Carrier Gas		Helium, 30 cm^3/min

[a] The GC conditions used for VCM were suggested by N. A. Sebestyen of Perkin-Elmer Corporation. [b] Supelco, Inc. recently announced (Chromatography Lipids, IX, July (1975)) that this column packing is being replaced by an equivalent packing, 0.2% Carbowax 1500 on Carbopak C.

From Steichen[36]. American Chemical Society

Table 7.30 - Details for Preparation of Calibration Standards[a] and Sample Solutions

Monomer	Polymer Weight, g	Volume and Solvent[b]		H_2O injection ml
VCM	0.500	5 cm^3	DMA	None
Butadiene	0.500	5 cm^3	DMA	None
Acrylonitrile	0.500	5 cm^3	DCB	None
Styrene	0.200	4 cm^3	DMA	3
2-EHA	0.100	2 cm^3	DMA	5

[a] Calibration standards also contain added known amounts of monomer. [b] N,N' Dimethylactamide (DMA) and orthodichlorobenzene

From Steichen[36]. American Chemical Society

The monomer concentration in the polymer is calculated from the response factor (Rf) the adjusted peak height of the sample (H), and the weight of the sample in grams (W).

$$\text{Ppm} = \frac{\text{Rf H}}{\text{W}}$$

57.7 DISCUSSION OF RESULTS

The presence of polymers in solution affected the equilibrium head-space concentration of monomer. To compensate for this effect, polymer containing a low level of residual monomer are included in the calibration standards to simulate the polymer matrix of the sample solution. At the polymer concentrations recommended, failure to incorporate polymer into the calibration standards resulted in a -8% error in the determination of vinylchloride and a -30% error for 2-ethylhexyl acrylate.

The sensitivity achieved by the head space method can be improved by decreasing the solubility of the 2-ethylhexylacrylate monomer in the solution of polymer in N,N′-dimethylacetamide through the introduction of a second solvent. Water is the most effective solvent for this purpose (see Table 7.31). A greater than 200 fold increase in the 2-ethylhexylacrylate equilibrium head space concentration resulted when water is injected into its polymer solutions. A significant increase in the styrene head space concentration is also obtained when polystyrene solutions are treated with water. In the region where styrene and 2-ethylhexylacrylate monomers elute, no increase in baseline noise results from the injections of water.

The response for styrene and 2-ethylhexylacrylate was affected by the relative amount of water injected into the polymer solution. Increasing the relative amounts of water caused a continuing increase in monomer peak height with no plateau.

It is possible to determine vinyl chloride and butadiene at the 0.05 ppm level and acrylonitrile down to 0.5 ppm. The injection of water in the case of styrene and 2-ethylhexyl acrylate makes it possible to determine styrene down to 1 ppm and 2-ethyl hexyl acrylate at 5 ppm. The relative precision and error in the determination of these monomers near the quantitation limit is less than 7%.

METHOD 58 - DETERMINATION OF HYDROXYL NUMBER OF GLYCEROL-ALKYLENE OXIDE POLYETHERS AND BUTANE, 1,4-DIOL ADIPIC ACID POLYESTERS. DIRECT INJECTION ENTHALPIMETRY.[37]

58.1 SUMMARY

This direct injection enthalpimetric method is capable of determining hydroxy values of polyesters and polyethers in less than 10 minutes.

58.2 APPARATUS

The basic electrical circuit[38] is a simple d.c. Wheatstone bridge incorporating in one arm a thermistor of nominal resistance 10 k ohm at 25°C. The off-balance voltage is recorded on a recording potentiometer (10 mV full-scale deflection) with a chart speed of 4 cm min^{-1}.

A glass bottle (of capacity about 25 ml) is used for titrations, it was thermally insulted in a block of polystyrene. The titrand is stirred at a constant rate by a PTFE covered magnet.

58.3 REAGENTS

Titrant for polyethers. Polyethers 2 % v/v sulphuric acid (98 % v/v); 23% v/v acetic anhydride, 25% v/v carbon tetrachloride and 50% v/v glacial acetic acid.

Titrant for polyesters. 2% v/v sulphuric acid (98% v/v); 23% v/v acetic anhydride, 25% v/v glacial acetic acid and 50% v/v methylene chloride.

Carbon tetrachloride.

Methylene chloride.

Table 7.31 - Investigation of Solvents for Displacing 2-EHA Monomer from DMA-Polymer Solutions into Headspace[a]

Displacing Solvent	GC Headspace Response[b] for 2-EHA	Effect of Displacing Solvent upon Polymer Solution
None added	0	Clear solution
Water	652%	Polymer precipitates
Heptane	0	Turbid solution
Ethyl-Cellosolve	0	Clear solution
Ethylene glycol	54%	Polymer precipitates

From Steichen[36]. American Chemical Society

[a] A 0.10 g portion of resin was dissolved in a septum-sealed vial with 2 ml of DMA. An aliquot of a standard solution containing 76 ug of 2-EHA was added. To this, 5.0 ml of the displacing solvent was injected. [b] Percent of full scale response at lowest instrument attenuation.

58.4 EXPERIMENTAL PROCEDURE

The titrant consists of 100 to 200 ul of a solution of a sample dissolved in either carbon tetrachloride (for polyethers) or methylene chloride (for polyesters) delivered from a micro litre syringe. These organic solvents are dried over molecular sieve (grade 5A). When the hydroxyl value is less than 100 mg g^{-1} of potassium hydroxide, the titrant is used at a 33.3 per cent m/V solution; for hydroxy values exceeding 100 mg g^{-1} it is made up to 20 per cent m/V.

Six portions, each 200 ul in volume of a titrant added separately to one sample of titrand yields the same recorder deflection after each addition. The deflections produced by two samples and one standard can thus be determined in duplicate without renewing the titrand. All samples are examined in duplicate or triplicate and deflections on the recorder are usually reproducible to $\pm$ 0.025 per cent of full-scale deflection.

58.5 TYPICAL RESULTS

Plots of dT (defined as the height of the recorder deflection) against hydroxyl value (determined by the conventional phthalic anhydride method) for five gylcerol-propylene oxide condensates ranging in molecular weight between 300 and 4000 pass very near to the origin.

58.6 DISCUSSION OF RESULTS

This method is capable of determining hydroxy values of polyethers and polyesters down to 10 mg kg^{-1} with an accuracy of $\pm$ 2%.

METHOD 59 - DETERMINATION OF PRIMARY AND SECONDARY HYDROXYL GROUPS IN ETHYLENE OXIDE TIPPED GLYCEROL-PROPYLENE OXIDE CONDENSATES. PHENYL ISOCYANATE KINETIC METHOD.

59.1 SUMMARY

In many situations it is necessary to be able to determine both primary and secondary hydroxy groups in polymers. This kinetic method, based on the difference in reaction rate of primary and secondary hydroxy groups with phenyl isocyanate enables both groups to be determined.

The reaction is carried out under standard conditions in which a calculated weight of the polyol (depending on its hydroxyl number) is reacted under standard conditions with an excess of a standard toluene solution of phenyl isocyanate in the presence of a basic catalyst.

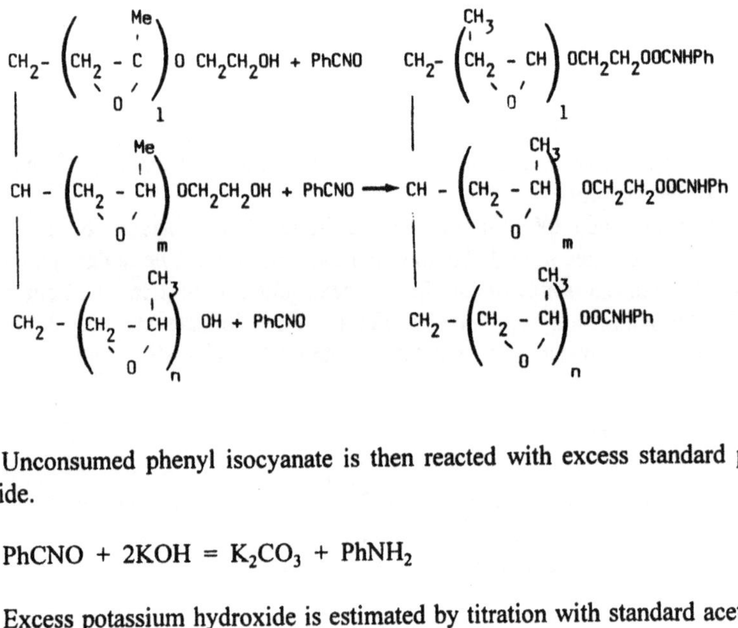

7.6

Unconsumed phenyl isocyanate is then reacted with excess standard potassium hydroxide.

$$PhCNO + 2KOH = K_2CO_3 + PhNH_2 \qquad 7.7$$

Excess potassium hydroxide is estimated by titration with standard acetic acid to the phenol phthalein end-point. A blank run is performed in which the sample is omitted. From the difference between the sample and the blank titrations it is possible to calculate the hydroxyl content of the original polymer:

1 mole hydroxyl groups $\equiv$ 1 mole phenyl isocyanate $\equiv$ 2000 ml N KOH

Reactivity is calculated by a kinetic procedure in which the percentage of the original phenyl isocyanate addition, which reacts with the sample in a given time, it taken as an index of its reactivity. A calibration graph can be prepared in which the determined phenyl isocyanate reactivity is plotted against the ethylene oxide content, and this enables a determination to be made of the ethylene oxide content of unknown samples from the calibration graph by interpolation.

59.2 APPARATUS

(a) apparatus for preparation, storage and dispensing of dry toluene (containing less than 15 ppm water) A description of the apparatus required is given with the method of drying toluene described later. Figure 7.12 shows the apparatus required for storing and dispensing the dried toluene.

(b) apparatus for reducing water content of polyol sample to 100 - 150 ppm (Figure 7.13). The water content of the neat polyol is reduced to 100 - 150 ppm by subjecting it to vacuum treatment for 2 hours at 70 - 80°C. Connect a vacuum pump of approximately 50 cu ft/min capacity (capable of attaining a vacuum of 0.005 mm mercury) to a 700 ml Quickfit flange type flask assembly and a Manostat gauge, (via a cardice/isopropanol filled cold trap, see Figure 7.13). Provision is made for feeding dry nitrogen into the flask when the vacuum treatment is completed. During vacuum treatment the temperature of the reaction flask is maintained at 70° to 80° by immersion in a Simmerstat controlled water bath.

(c) apparatus for reaction of polyol with phenyl isocyanate (Figure 7.14). The reaction of the polyol with phenyl isocyanate is carried out in a 700 cm^3 Quickfit flange type vessel (as described above). This apparatus is mounted in a rectangular aluminium tank dimensions 24 in x 7 in high, thermostatted at 40 $\pm$ 0.1°C. The tank is fitted with a Perspex top having holes cut to accommodate two reaction flasks and the thermostatted heater/stirrer.

(d) miscellaneous apparatus. Precision torsion balance, capacity 4000 g, scale divisions 1 g.

Stopwatches.

Long stemmed sampling pipette, capacity 10 - 12 cm^3 from tip to graduation on top of bulk.

Miscellaneous volumetric glassware.

59.3 REAGENTS

(a) preparation and storage of dry toluene (containing less than 15 ppm water). Reflux 4 cm^3 of toluene with 30 g calcium hydride for 6 to 8 h in a 5 cm^3 B24 neck flask with 24 in vertical condenser attached. Connect a horizontal condenser to the 5 cm^3 flask, (excluding moist air). To the end of this condenser connect a dry three-neck B25 5 dm^3 flask containing 100 g freshly ignited molecular sieves, (6 in x 1 in molecular sieve trap also connected to 5 dm^3 flask - molecular sieves dried 3 to 4 h at 120°C). Now distil 3 to 3.5 dm^3 of the toluene from the calcium hydride into the dry receiver containing molecular sieves. Deactivate the calcium hydride remaining in the distillation flask by slowly adding a 10:1 hexane:ethanol solution and dispose of this solution in an open drain.

Connect the toluene receiver to a dry dispensing system as shown in Figure 7.12. In this apparatus the solvent is kept dry (ie. less than 15 ppm water) by standing over molecular sieves under an atmosphere of dry nitrogen. To the outlet end of this dispensing apparatus connect a dry 500 cm^3 graduated cylinder. By suitable adjustment of stopcocks keep this cylinder dry by continuously purging with dry nitrogen when the apparatus is not in use.

Immediately before dry toluene is required for an analysis syphon a suitable volume into the dry 500 cm^3 graduated cylinder by adjustment of the stopcocks. The

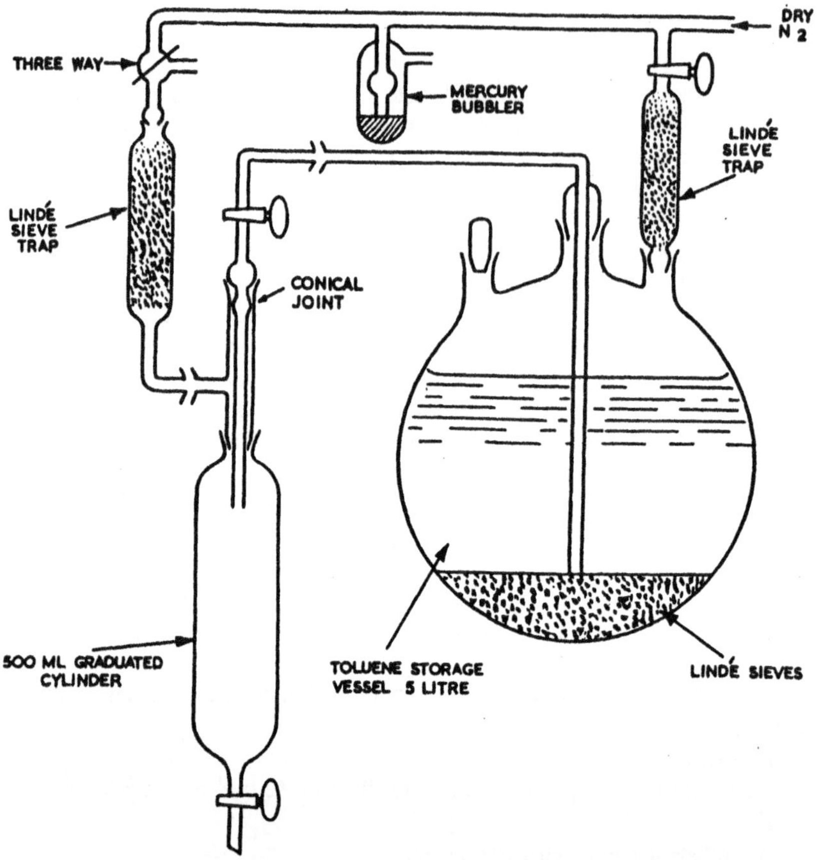

Figure 7.12 Apparatus for preparation, storage and dispensing of dry toluene.

water content of the toluene stock should be checked periodically by a Karl Fischer procedure to ensure that it is below the required limit of 15 ppm.

(b) Toluene solution of phenyl isocyanate, 1.00 ± 0.005 M. Weigh 120 ± .2 g phenyl isocyanate into an oven dried, stoppered, 1 dm^3 volumetric flask. Run dry toluene (less than 15 ppm water) into this flask from the dispenser (see section a and Figure 7.12) until approximately 950 cm^3 of liquid is present in the volumetric flask. Immerse the stoppered volumetric flask in a water bath, thermostatted at 25 ± 1°C and leave for 15 min with occasional mixing. Finally make the volume of liquid in the flask up to 1 dm^3 at 25°C with dry toluene. Stopper and mix well. This solution is to be maintained in a water bath thermostatted at 25 ± 1°C when portions of it are withdrawn for standardization or for use in the determination of the reactivity of polyols. A measured volume of the solution then contains a constant weight of phenyl isocyanate.

Standardisation of phenyl isocyanate solution.

Hydrochloric acid, 0.1 M aqueous.

Bromophenol blue indicator, 0.1%.

Dissolve 0.1 g bromophenol blue in 100 cm^3 distilled water, add 1.5 ml of 0.1 M sodium hydroxide solution.

Methanol redistilled.

Di-n-butylamine 0.2 M (approx) in dry toluene, (less than 15 ppm water). Transfer 9.3 ml di-*n*-butylamine ANALAR to a dry 250 cm^3 volumetric flask. Make up

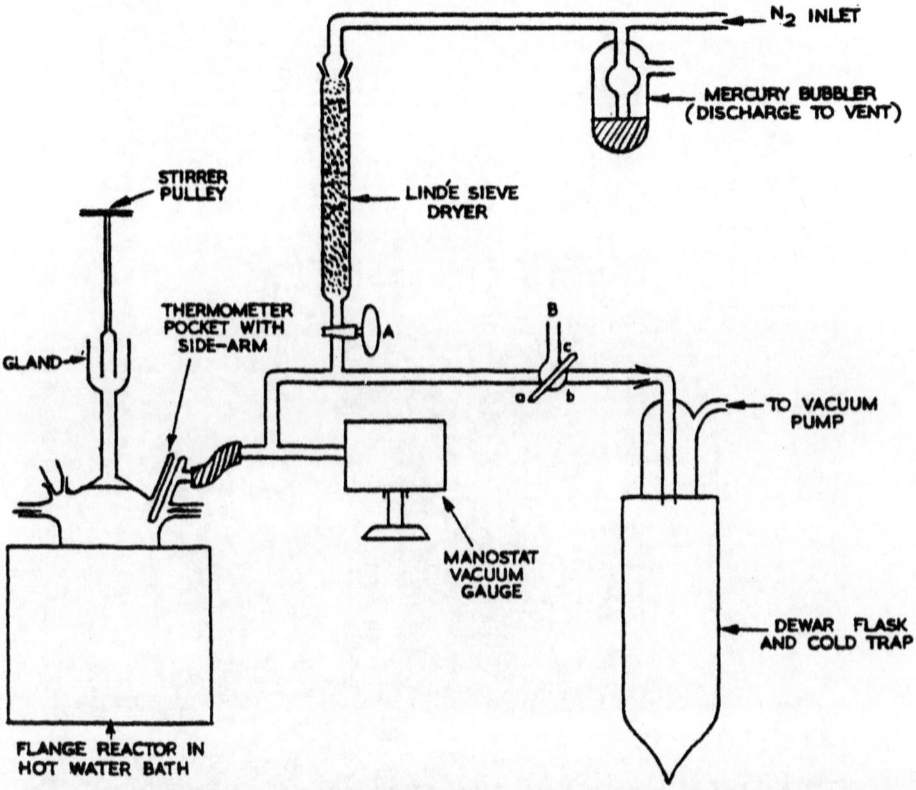

Figure 7.13 Apparatus for reducing water content of polyol sample.

to the mark with dry toluene and thoroughly mix the solution. Keep this reagent well stoppered when not in use and renew weekly.

Adjust the temperature of the 1 cm^3 flask of 1 M phenyl isocyanate solution by immersing it in a water bath at 25 ± 1°C with dry toluene (less than 15 ppm water) and thoroughly mix the contents of the 250 ml flask.

Pipette 25 ml 0.2 M di-*n*-butylamine solution into two dry stoppered 250 cm^3 conical flasks (ie. a sample and a blank flask). Allow a 15 s pipette draining time during these operations. Into one of the flasks pipette 25 cm^3 of the ten-fold diluted phenyl isocyanate solution (sampled at 25°C using a 15 s pipette draining time) and into the blank flask pipette 25 cm^3 of toluene. Leave both solutions for 15 min to react, then add to each 100 cm^3 methanol and 5 drops bromophenol blue indicator. Using 0.1 M hydrochloric acid titrate the sample (T_S ml of f normal hydrochloric acid) and the blank solutions (T_B ml of f normal hydrochloric acid) to the yellow-green end point.

The normality (F) of the 1 cm^3 undiluted stock phenyl isocyanate solution of 25 ± 1°C is then given by:

$$F = \frac{10 \times f \times (T_B - T_S)}{25}$$

The molarity of the phenyl isocyanate solution at 25 ± 1°C, prepared as described above, should be slightly greater than 1.000M. Finally, adjust the strength of the 1 cm^3 of

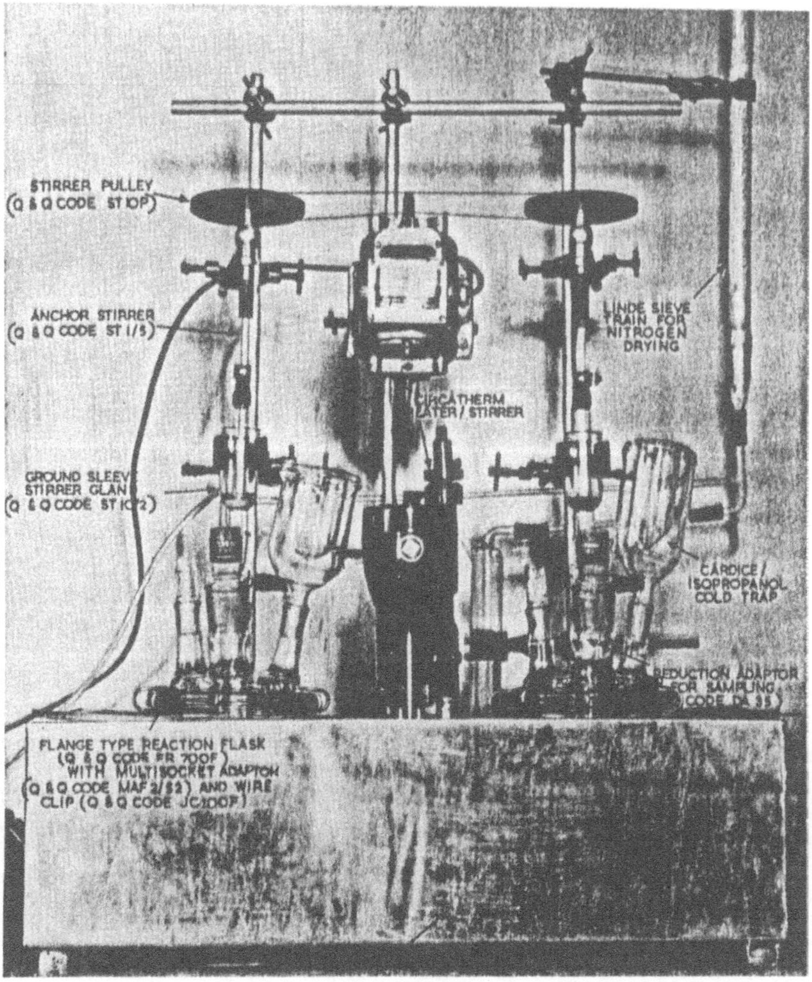

Figure 7.14. Apparatus for reaction of polyol with phenyl isocyanate.

stock phenol isocyanate solution to within the range 1.000 ± 0.005 M by addition of a calculated volume of dry toluene. Mix the 1 cm^3 stock solution well in the thermostatted 25 ± 1°C water bath and recheck the normality of a ten-fold dilution as described above.

Due to the high coefficient of cubical expansion of toluene, the normality of a toluene solution of phenyl isocyanate varies as its temperature is changed. It is for this reason that the solution is prepared at 25 ± 1°C and subsequently standardized at the same temperature.

For the purpose of calculating results, it is necessary to know the weight of 50 cm^3 of the phenyl isocyanate at 25 ± 1°C. This may be obtained by weighing 50 cm^3 of the reagent from the thermostatted stock flask in a dry 50 cm^3 stoppered volumetric flask.

The normal phenyl isocyanate stock solution should be well stoppered when not in use and should be left open for a minimum period when sampling (to exclude atmospheric moisture). Recheck the normality of the solution at 25 ± 1°C at frequent intervals.

(c) Dabco catalyst solution (0.56% w/v)

Dabco is the trade name for triethylene diamine (Jacobson and Van den Burg and Co. (U.K.) Ltd, 3 - 5 Crutched Friars, London, E.C.3). Weigh out 6 ± 0.1 g Dabco and transfer to an over-dried 1 litre volumetric flask. Add dry toluene (less than 15 ppm water) from the dispenser (see Figure 7.12) to make the volume up to approximately 950 cm^3. Dissolve the Dabco by shaking and stand the flask in a water bath thermostatted at 25 ± 1°C. When the solution has reached 25°C make up to the 1 dm^3 mark with dry toluene and mix thoroughly. Stand the solution in a water bath, thermostatted at 25 ± 1°C. A measured volume of this solution then contains a constant weight of Dabco catalyst.

Standardisation of the Dabco solution.

Immerse the 1 dm^3 flask of stock solution of Dabco in a water bath thermostatted at 25 ± 1°C and leave until the solution reaches this temperature. Pipette 25 cm^3 of this solution (15 s draining time) into a 250 cm^3 conical flask containing 100 cm^3 methanol and 5 drops 1% aqueous bromophenol blue indicator. Using 0.1 M hydrochloric acid, titrate the sample solution (T cm^3 of f M hydrochloric acid) to the yellow-green end point. The reaction which has occured at the bromophenol blue indicator end point is:

$$N\,(C_2H_4)_3N + HCl = N\,(C_2H_4)_3\,N.HCl \qquad 7.8$$

The strength (% w/v) of the 1 dm^3 of Dabco solution at 25 ± 1°C is given by:

$$\text{Dabco (\% w/v)} = \frac{4 \times 112 \times T \times f}{1000}$$

The strength of the Dabco solution at 25 ± 1°C must be in the range 0.56 ± 0.01 % w/v.

For the purpose of calculating results it is necessary to know the weight of 50 cm^3 of the Dabco catalyst solution at 25 ± 1°C. This may be obtained simply by weighing 50 ml of the reagent from the thermostatted stock flask into a dry 250 ml stoppered conical flask.

This solution should be well stoppered when not in use and should be left open for a minimum time when sampling (to exclude atmospheric moisture).

59.4 METHOD

The described kinetic procedure, based on the difference in the reaction rates with phenyl isocyanate of the terminal primary and secondary hydroxyl groups present in ethylene oxide tipped glycerol/propylene oxide condensates must be carried out under rigidly standardized conditions. Primary hydroxyl groups react with phenyl isocyanate to form a urethane faster than do secondary hydroxyl groups. The greater the amount of terminal ethylene oxide units in a glycerol/propylene oxide condensate, therefore, the greater its reactivity (ie. rate of reaction) with phenyl isocyanate. The method may be used, therefore, for evaluating the reactivity of ethylene oxide tipped polyols.

59.5 EXPERIMENTAL PROCEDURE

(a) Calculation of the weight of polyol required for analysis.

It is explained in note 1 that the analysis is carried out using a toluene solution of the polyol which is exactly decimolar with respect to total primary and secondary hydroxyl groups in the sample. First determine the hydroxyl value (H, mg KOH/g) of the polyol sample by a modified catalyzed acetylation procedure (see Method 60). Calculate the weight W (g) of polyol required to produce 500 cm^3 of a decinormal toluene solution (at 25°C ± 1°C) from the following equation:

$$W\ (g) = \frac{56.1 \times 50}{H}$$

Transfer approximately 95 g of polyol sample to a clean 100 cm^3 conical flask and weigh to the nearest 0.01 g. Transfer from this flask exactly the calculated weight (W) of polyol into a 700 cm^3 Quickfit flanged reaction vessel (see Figure 7.14). Reweigh the conical flask to check that the weight of polyol added to the reaction vessel is correct to within 0.01 g.

(b) Removal of water from polyol.

Assemble the reaction apparatus with multineck adaptor, stirrer and gland, thermometer pocket and vacuum assembly as shown in Figure 7.13.

Mount the flask in a suitable water bath and connect the stirrer motor and vacuum line. Maintain the cardice-isopropanol cold trap at -60°C and operate the stirrer at approximately 200 rpm.

Close stopcock A (Figure 7.13) and connect (a) to (c) with stopcock B. Heat the water bath up to 70 to 80°C and allow the stirred reactor contents to reach this temperature. Switch on the vacuum pump. Reduce the pressure in the reaction flask to 2 mm over a period of 5 min by momentarily turning tap B to connect (a) to (b) several times (if the pressure is reduced too rapidly to 2 mm then violent de-gassing of the polyol may take place due to the presence of dissolved propylene oxide in the sample). When de-gassing is complete turn tap B to connect (a) to (b) and continue the vacuum treatment for 2 hours. After half an hour the pressure in the flask should be less than 0.1 mm.

When the vacuum treatment is completed switch off the bath heater and the stirrer motor. Isolate the flask from the vacuum line by turning tap B to connect (c) to (b). Now apply a rapid purge of dry nitrogen at the nitrogen inlet and cautiously open stopcock A to allow nitrogen to fill the reaction vessel. Replace the thermometer pocket by a stopper, remove the flask from the water bath and dry the outside of the flask. The moisture content of the polyol should now be in the range 80 - 150 ppm.

(c) Toluene dilution of polyol.

Stand the stoppered reaction flask on a cork ring placed on a balance. Fill the 500 cm^3 graduated cylinder of the dry toluene dispensing unit (Figure 7.12) with toluene and attach a molecular sieve filled guard tube to the top of this cylinder. Run a suitable weight (Wg) of toluene (to the nearest 1 g) from the graduated cylinder into the reaction flask and immediately stopper the flask to exclude atmospheric moisture. The weight (Wg) of toluene required is given by the following expression.

Weight of toluene required to make the volume of test solution up to 400 cm^3 at 25°C.

$$W(g) = 0.861 \times \left(400 - \frac{W_1}{S}\right)$$

where:

W_1 = weight (g) of polyol used in experiment.

S = specific gravity of neat polyol at 25°C (usually assumed to be 1.00).

0.861 = specific gravity of pure toluene at 25°C.

(d) Commencement of the reaction.

Immerse the reaction flask with multineck adaptor in an aluminium water tank of the type shown in Figure 7.14. This tank is thermostatted at 40 ± 1°C. Insert the cold trap and sampling fitting with Gaco seal into the multineck adaptor. Maintain the cold trap at -60°C with cardice/isopropanol during the whole run.

Connect to the top of the cold trap a supply of dry nitrogen and maintain a slight pressure of nitrogen in the reaction flask. The nitrogen stream should be very gentle, otherwise some loss of toluene vapour, with consequential analytical errors may occur when the reaction flask is opened for sampling. Adjust the stirrer speed to 200 rpm and leave the contents of the flask for 30 min to reach thermal equilibrium with the bath at 40°C.

Whilst the reaction flask is reaching thermal equilibrium carry out the following two operations:

(1) Place the two 1 litre reagent flasks containing the stocks of 1.000 ± 0.005 M phenyl isocyanate solution and of 0.56% w/v Dabco catalyst solution in a separate water bath thermostatted at 25 ± 1°C, and allow these flasks to reach this temperature over a period of 30 min.

(2) Pipette 10 ml 0.2 N di-*n*-butylamine solution into each of twelve dry 100 cm^3 conical flasks and quickly replace the stoppers (15 s pipette draining time). Ensure that the same amount of di-*n*-butylamine is introduced into each flask by keeping the temperature of the stock di-*n*-butylamine solution constant during the pipetting. These flasks will be used later for the determination of the phenylisocyanate content of samples withdrawn from the polyol/phenyl isocyanate reaction at various times.

When the reaction flask containing the toluene solution of polyol has reached 40°C, pipette into this flask 50 cm^3 of normal phenyl isocyanate stock solution (at 25°C), allowing a 15 s pipette draining time. No reaction occurs between polyol and phenyl isocyanate in the absence of Dabco catalyst. Leave the mixture 15 min to reach thermal equilibrium. Transfer, by pipette, 50 cm^3 of stock 0.56% w/v Dabco solution from the storage vessel (thermostatted at 25°C) into a dry 50 cm^3 volumetric flask. Remove the B34/B24 sampling adaptor from the reaction flask assembly and insert the neck of the 50 cm^3 volumetric flask into the B34 socket to introduce the Dabco catalyst into the reaction mixture. Allow a 15 s draining time for the Dabco flask and replace the B34/B24 sampling adaptor. Start the stopwatch at the same moment that the Dabco solution is first tipped into the reaction flask, (ie. at the exact moment that the Dabco catalyzed reaction between polyol and phenyl isocyanate commences).

(e) Sampling the polyol/phenyl isocyanate reactor.

Samples are now withdrawn periodically from the reactor for determination of residual phenyl isocyanate. Immediately before starting the reaction between polyol and phenyl isocyanate by the addition of Dabco catalyst, (ie. time zero) weigh to the nearest 1 mg, one of the twelve 100 cm^3 conical flasks containing 10 cm^3 0.2 M di-n-butylamine (referred to in section (2)). Withdraw approximately 10 cm^3 of the polyol/phenyl isocyanate reaction mixture by means of a long stemmed 10 cm^3 pipette via the Gaco seal on the sampling point and exactly one minute after starting the reaction (ie. addition of Dabco) run this solution into the di-n-butylamine solution contained in the weighed flask. Stopper and reweigh the conical flask to obtain the weight of polyol/phenyl isocyanate reaction mixture withdrawn for analysis (this weight is needed in the calculation of results). Leave the mixture for 15 min, or longer, to react. Transfer the contents of this flask quantitatively to a 250 cm^3 conical flask with 100 cm^3 redistilled methanol and add 5 drops 0.1% bromo-phenol blue indicator. Using 0.1 M hydrochloric acid, titrate the sample to the yellow-green end-point and record this titration.

To one of the blank flasks referred to in section (2) add 10 cm^3 toluene and transfer the contents of a 250 cm^3 conical flask with 100 cm^3 methanol. Add 5 drops bromophenol blue indicator and titrate with 0.1 normal hydrochloric acid to the yellow-green end-point.

Similarly, withdraw samples from the polyol/phenyl isocyanate reaction solution at various other times intervals (see table below) after commencement of the reaction and immediately introduce the sample into 10 ml di-n-butylamine in order to analyze for phenyl isocyanate content as just described. Tabulate the titrations obtained. Titrations of the 10 cm^3 0.2 M di-n-butylamine blank solution carried out at the beginning and at the end of the experiment usually agree within 0.05 ml.

di-n-butylamine flask no.	Test	Time interval between adding Dabco catalyst (ie. start of reaction) and sampling for phenyl isocyanate determination (min).
1	Blank run	-
2	Sample run	1
3	Sample run	15
4	Sample run	30
5	Sample run	55
6	Sample run	60
7	Sample run	65
8	Sample run	95
9	Sample run	100
10	Sample run	105
11	Sample run	120
12	Blank run	-

The small amount of Dabco catalyst present in samples taken from the reaction flask interferes slightly in these determinations of residual phenyl isocyanate. This interference is allowed for in the method of calculation.

(f) Graphical plotting of experimental data.

The method of calculating % of the original phenyl isocyanate addition consumed under standard conditions (denoted by P%) is given below. On graph paper plot P% versus the corresponding time (in minutes) from the commencement of reaction of each of the ten samples withdrawn from the reaction flask. Draw a smooth line through the 10 points. From this graph read (P%) corresponding to 60 min and P% corresponding to 100 min., ie. $(P\%)_{60\ min}$ and $(P\%)_{100\ min.}$

(g) Correlation of reactivities with ethylene oxide content of sample.

As $(P\%)_{60\ min}$ and $(P\%)_{100\ min}$ are determined under rigidly standardized test conditions these quantities are dependent upon the reactivity of the polyol sample analyzed and therefore, constitute a parameter by which it is possible to compare one batch of polyol with another. In ethylene oxide tipped condensates prepared under specified conditions, the reactivity of the polyol increases with its ethylene oxide content.

A series of molecular weight 5000 glycerol/propylene oxide condensates containing between nil and 9 moles ethylene oxide per mole of glycerol can be used to prepare a calibration graph. The procedure under calculations was then used to determine $(P\%)_{60\ min}$ and $(P\%)_{100\ min}$for the standard samples.

Prepare plots of $(P\%)_{60\ min}$ and $(P\%)_{100\ min}$ respectively versus the weight addition ethylene oxide content (expressed in moles ethylene oxide per mole glycerol). The ethylene oxide contents of ethylene oxide tipped glycerol/propylene oxide condensates of unknown ethylene oxide content may be determined by referring their determined $(P\%)_{60\ min}$ and $(P\%)_{100\ min}$ reactivity figures to this calibration graph. The two determinations usually agree within $\pm$ 0.1 ethylene oxide units.

It is necessary to ensure that the samples used to prepare the calibration curve were manufactured in the same way as the unknown samples being analyzed, thus the standard samples should be manufactured using the same polymerization catalyst that is used to manufacture the unknown samples.

59.6 TYPICAL RESULTS

Calculations.

(a) To calculate total weight of initial reaction flask charge let;

W_1(g) = weight of neat polyol taken for analysis

W_2(g) = weight of toluene required to make the volume of polyol/toluene test solution up to 400 ml at 25 $\pm$ 1°C

W_3(g) = weight of 50 cm^3 1.000 $\pm$ 0.005 normal phenyl isocyanate (in toluene) reagent at 25 $\pm$ 1°C.

W_4(g) = weight of 50 cm^3 0.56% Dabco (in toluene) reagent at 25 $\pm$ 1°C

W_A (g) = total weight of initial reaction flask charge.

$= W_1 + W_2 + W_3 + W_4$

(b) to calculate weight of phenyl isocyanate in total initial weight of reaction flask charge before the polyol/phenyl isocyanate reaction is started

Let F = normality of phenyl isocyanate stock solution at 25 $\pm$ 1°C.

Then the original weight of phenyl isocyanate present W_B(g) in the total reaction flask charge W_A(g) (ie. before reaction with polyol commences) is given by:

$$W_B(g) = \frac{50 \times F \times 119.13}{1000}$$ g phenyl isocyanate present in W_A(g) of reaction flask charge.

(c) To calculate initial weight of phenyl isocyanate (ie. before reaction with polyol is started) in the portion of reaction solution withdrawn for phenyl isocyanate determination.

Let W_C = weight (g) of an approximately 10 cm^3 sample taken from the reactor during the run for determination of phenyl isocyanate.

Then calculated weight of phenyl isocyanate W_D(g) present in the W_C(g) portion of reaction solution withdrawn for phenyl isocyanate determination (ie. before reaction between polyol and phenyl isocyanate is started by addition of Dabco) is given by

$$W_D(g) = \frac{W_B \times W_C}{W_A}$$

(d) To calculate the titration correction (T_D cm^3) required to allow for the presence of Dabco catalyst in the portion of reaction solution W_C(g) withdrawn for phenyl isocyanate determination

112g Dabco = 1000 cm^3 1.0000 M hydrochloric acid (Dabco titrated to the bromophenol blue end-point)

In the reactivity determination there is present 50 cm^3 0.56% w/v Dabco solution (ie. 0.28g pure Dabco) in W_A(g) of initial reaction flask charge.

Now 0.28g pure Dabco consumes $\frac{0.56 \times 50 \times 1000}{100 \times 112}$ cm^3

= 2.5 cm^3 1.0000 M hydrochloric acid.ie. the Dabco present in W_A(g) of initial reaction flask charge consumes 2.5 cm^3 1.000 M hydrochloric acid.

The hydrochloric acid titration correction (T_D cm^3) required to allow for the presence of Dabco catalyst in the portion of reaction solutions W_C(g) withdrawn for phenyl isocyanate determination is given by:

$$T_D (cm^3) = \frac{2.5 \times W_C}{W_A} \quad cm^3 \text{ 1.000 M hydrochloric acid}$$

To obtain the actual weight of phenyl isocyanate, W_E(g) (corrected for Dabco present) remaining in W_C(g) portion of reaction solution withdrawn for phenyl isocyanate determination after time T min (ie. T min after the Dabco catalyzed reaction of polyol and phenyl isocyanate has been started).

If, in the phenyl isocyanate determination by the di-n-butylamine method:

T_B = volume of hydrochloric acid (cm^3) required in blank titration of 10 cm^3 of 0.2 M di-n-butylamine solution

T_A = volume of hydrochloric acid (cm^3) required to titrate a mixture of 10 cm^3 0.2 M di-n-butylamine and the W_C(g) of reactor sample which was withdrawn for phenyl isocyanate determination after time T min.

f = normality of hydrochloric acid used in phenyl isocyanate determination

T_D = Dabco titration correction, see Section 59.6.

119.13 = molecular weight of phenyl isocyanate

Then the actual weight W_E(g) of phenyl isocyanate (corrected for Dabco present) in a W_C(g) portion of reaction solution withdrawn for phenyl isocyanate determination after time T min is given by:

$$W_E(g) = (T_B \times f - (T_A \times f - T_D)) \frac{119.13}{1000}$$

g phenyl isocyanate

or

$$W_E(g) = \left(T_B \times f - \left(T_A \times f - \frac{2.5 \times W_C}{W_A}\right)\right) \frac{119.13}{1000}$$

g phenyl isocyanate

To calculate the percentage (p) of the original addition of phenyl isocyanate consumed in the reaction solution after time T min denoted by (P_T min)

$$= \frac{W_D - W_E) 100}{W_D}$$

This method has been applied to a range of glycerol/propylene oxide adducts containing various accurately known amounts of ethylene oxide tipping, up to 5.3 moles

(Table 7.32). Increasing the ethylene oxide tip content of a polyol leads to a distinct increase in the reactivity of the polyol with phenyl isocyanate. As expected, "untipped" polyols which are relatively free from primary hydroxyl groups react comparatively slowly with phenyl isocyanate. Decinormal solutions of "untipped" glycerol/propylene oxide condensates of molecular weight 3000 and 5000 has an identical rate of reaction with phenyl isocyanate. Thus, the rate of reaction with phenyl isocyanate of the terminal isopropanol end-groups in polyols is independent of molecular weight in the molecular weight range 3000 to 5000 and depends only on proportions of primary and secondary end groups present.

A calibration curve is prepared by plotting moles ethylene oxide per mole of glycerol for the range of standard tipped polyols of known ethylene oxide content against % of original phenyl isocyanate addition consumed after 60 and 100 min, ie. P60% and P100%. This curve can be used to obtain from P60% and P100% data obtained for tipped glycerol propylene oxide polyols of unknown composition their tipped ethylene oxide contents (in moles ethylene oxide per mole glycerol).

59.7 DISCUSSION OF RESULTS

For polyols which contain less than 4 moles ethylene oxide per mole of glycerol the reactivity method has an accuracy of better than $\pm$ 5% of the determined value. For polyols which contain between 5 and 15 moles ethylene oxide per mole of glycerol, the accuracy is in the range $\pm$ 10%.

METHOD 60 - DETERMINATION OF OXYALKYLENE GROUPS IN POLYETHER BASES AND POLYURETHANE POLYMERS. MIXED ANHYDRIDE CLEAVAGE - GAS CHROMATOGRAPHY.[39]

60.1 SUMMARY

This acetic anhydride - toluene-p-sulphonic acid cleavage - gas chromatography procedure is capable of determining down to 0.05% of oxyalkylene and oxypropylene groups in polyethers and polyurethanes.

60.2 APPARATUS

Gas chromatograph, equipped with a thermal conductivity detector or equivalent. The column consists of a 1 m length of 1 mm i.d. stainless steel tubing packed with 15% m/m FFAP (free fatty acid phase) or 10% m/m SE-30 silicone) coated on 60-80 mesh Uniport B. Helium is used as a carrier gas.

60.3 REAGENTS

Cleavage reagent. Acetic anhydride (80g) is added dropwise to 120 g of toluene-p-sulphonic acid contained in a 300 ml round-bottomed flask at room temperature and the mixture is refluxed at 120°C for 30 minutes. The product obtained is used as the reagent without removal of the acetic acid produced and the excess of acetic anhydride.

Sodium carbonate, 50% aqueous.

Table 7.32 - Application of Reactivity Method to Standard Ethylene Oxide Tipped Polyols

Sample Identific-ation	Approximate Molecular Weight	Hydroxyl Number mg KOH/g Polyol	Ethylene Oxide Tipp-ing, Moles Ethylene Oxide/Mole Glycerol (by Weight Addition)	Reactivity, ie. percentage of Original Phenyl Isocyanate Addition Consumed in:-	
				60 min. Reaction	100 min Reaction
A	3000	59.0	0.0	25.2	35.2
B	5000	34.9	0.0	27.1	36.8
C	5000	34.3	3.0	41.5	46.6
D	5000	35.9	3.5	44.4	51.0
E	5000	36.0	4.3	47.5	54.1
F	5000	33.3	5.3	51.9	57.4

Diethyl ether.

60.4 METHOD

The sample, for example, polyethylene glycol, is reacted with acetic anhydride in the presence of toluene-p-sulphonic acid catalyst to produce a glycol diacetate:

$$(CH_2CH_2O)_n + (CH_3CO)_2O = CH_3COOCH_2CH_2OOCCH_3 \qquad 7.9$$

The diacetate content, proportional to the oxyethylene content of the original polymer is determined by gas chromatography. The procedure is calibrated against polyols of known molecular weight and oxyalkylene content.

60.5 EXPERIMENTAL PROCEDURE

A 100 mg amount of sample and 2 ml of reagent are mixed in a 20 ml round-bottomed flask and the mixture refluxed on an oil-bath at 120°C for 2 hours. The contents of the flask are cooled to room temperature and then neutralised with 50% aqueous sodium carbonate solution, followed by extraction with about 20 ml of diethyl ether. The ether layer is washed with de-ionised water and concentrated to a small volume on a steam bath: the concentrate is then injected into the gas chromatograph. The FFAP column is operated isothermally at the appropriate temperature for the purpose of both identifying the base compounds and determining their oxyethylene groups contents. For the analysis of the polyether based on 1,2-diaminoethane or 2,2'-diaminodiethylamine the SE-30 column is used.

60.5.1 Identification of Polyol Base Compounds.

The acetate peaks are satisfactorily separated from each other and the peak of propylene glycol diacetate produced by the cleavage of the polyoxypropylene groups did not overlap those of the derivatives from polyol base compounds except that for the polyether based on propylene glycol. These base compounds could therefore be easily distinguished and identified. By decreasing the temperature of the gas chromatographic column, the peak for propylene glycol diacetate can be accurately identified.

Polyethers based on sorbitol yielded complex products that consist of the acetates of sorbitans and sorbides produced by dehydration. However, the gas chromatograms always showed similar patterns, so that the base compound (sorbitol) can be identified from the chromatogram.

In order to obtain the peaks of the reaction products of polyethers based on 1,2-diaminomethane and 2,2′-diaminodiethylamine, the SE-30 column is operated isothermally at 230°C and the flow-rate of the carrier gas was maintained at 60ml/min^{-1}. These base compounds can also be distinguished and identified.

60.5.2 Determination of Oxyethylene and Oxypropylene Groups.

The ethylene oxide propylene oxide copolymers and polyols that were oxypropylated and then oxyethylated were decomposed as described above for the determination of the polyol base compounds. The FFAP column is operated isothermally at 65°C and the flow-rate of the carrier gas is regulated at 60 ml/min^{-1}.

A typical chromatogram for the reaction of the 1,2-diaminoethane ethylene oxide-propylene oxide adduct showed ethylene glycol diacetate and propylene glycol diacetate peaks, produced from polyoxyethylene and polyoxypropylene groups respectively. The proportions of ethylene and propylene oxide can be determined by measuring the two peak areas (ie. of the ethylene glycol and propylene glycol diacetates) and applying the appropriate calculations. The derivative from the base compound itself (1,2-diamino-ethane) do not appear in the chromatogram and so do not interfere in the determination of the ratio of oxyethylene to oxypropylene groups.

60.6 DISCUSSION OF RESULTS

This procedure is capable of determining down to 0.005% of oxyalkylene and oxypropylene groups in polyethers and polyurethanes with an accuracy of $\pm$ 3%. The method is also capable of identifying and determining the base compounds in polyurethanes with similar accuracy.

METHOD 61 - IDENTIFICATION OF ACRYLIC ACID AND METHACRYLIC ACID IN ACRYLIC COPOLYMERS. PROPYLATION - PYROLYSIS GAS CHROMATOGRAPHY.[40]

61.1 SUMMARY

This pyrolysis - gas chromatographic method enables bound acrylic acid and methacrylic acid units to be determined in acrylic copolymers in amounts down to 0.1%.

61.2 APPARATUS

Glass vials of capacity 15-30 ml with Teflon-lines septa (Pierce Hypo-vials) are suitable.

A thermostatically controlled oven at 60°C capable of being evacuated in less than 2kPA, was used.

Pyrolysis - gas chromatographic - mass spectrometric equipment, Perkin Elmer filament pyrolysis unit fitted in a Perkin Elmer F11 gas chromatograph (pyrolysis temperature control 250-550°C) or equivalent. The gas chromatographic column used is a 1.7 m x 3 mm o.d. stainless steel column packed with 30% m/m silicone oil (Embaphase) on acid washed Celite, operated at 80°C with a helium flow rate of 30 ml/min^{-1}. The column effluent is split in the ratio 2:1 between a flame ionization detector and an AFI MS12 mass spectrometer equipped with a glass fit type of molecular separator at 150°C. Mass spectra are scanned from m/e 200 to 20 at 8 seconds per decade under standard electron bombardment conditions, electron energy 70 eV, emission current 500 uA, accelerating voltage 8 kV and source temperature 200°C.

61.3 REAGENTS

Dimethylformamide dipropyl acetal (2 mequiv ml^{-1} in pyridine) (Propyl-8, Pierce Chemical Co.).

61.4 METHOD

In this procedure the acrylic acid and methacrylic acid groups in acrylic copolymers are first propylated using dimethyl formamide dipropyl acetal then this product pyrolysed according to the following scheme:

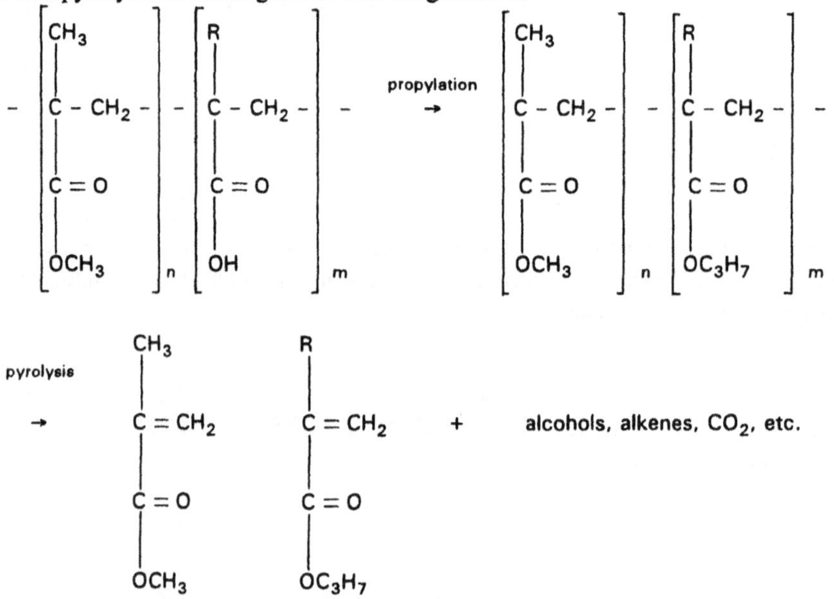

7.10

The resulting pyrolysis products (propylacrylate and methacrylate) are separated on a gas chromatograph and analysed by mass spectrometry.

61.5 EXPERIMENTAL PROCEDURE

61.5.1 Direct Pyrolysis.

Transfer 1 mg of polymer or an amount of latex or solution containing this mass of polymer on to the pyrolysis filament. Remove any water or solvent by blowing with hot air from a hairdryer. Insert the filament into the pyrolysis chamber and connect the carrier gas and gas chromatographic apparatus. Heat the filament at 250°C for 15 seconds to ensure the removal of water, solvent or volatile additives. Pyrolyse the polymer by heating the filament at 550°C for 15 seconds.

61.5.2 Propylation and Pyrolysis.

Transfer about 0.25 g of polymer or an amount of the acrylic latex or solution containing this mass of polymer into a glass vial and dry by evacuation at 60°C for 15 hours. Seal the vial and inject by means of a hypodermic syringe 1 ml of the Propyl-8 reagent on the dried polymer film. Heat at 60°C for 16 hours then let the vial and contents cool to ambient temperature. Transfer a smear of the gel or viscous solution produced on to the pyrolysis filament and remove the bulk of the propylation reagent by blowing with warm air from a hairdryer. Heat the filament for 15 second periods at 250°C in the stream of carrier gas until the chromatogram indicates the complete removal of reagent residues and finally pyrolyse at 550°C. Identify by mass spectrometry any eluted compounds that were not observed in the original direct pyrogram or that occur to a significantly greater extent in the pyrogram of the propylated polymer.

The presence of propyl acrylate or propyl methacrylate in the pyrogram of the propylated polymer indicates that the original polymer contained acrylic or methacrylic acid respectively.

The mass to charge (m/e) ratio and relative abundance of the molecular and fragment ions observed are listed below:

Propyl acrylate ($C_6H_{10}O_2$),M_n = 114

m/e	55	73	29	42	41	43	27	85	31	39	59
relative abundance	100	46	15	11		9	7	6	55	5	4
m/e	15	99	114								
ralative abundance	1	1	0.2								

Propyl methacrylate ($C_7H_{12}O_2$), M_n = 128

m/e	41	69	43	87	39	42	27	59	29	86	70
relative abundance	100	91	66	66	38	34	25	9	6	6	5
m/e	31	38	88	26	99	13	128				
relative abundance	4	3	3	2	2	2	2				

61.6 DISCUSSION OF RESULTS

By this procedure, copolymerised acrylic or methacrylic acid has been identified in terpolymers with (a) butyl acrylate and styrene (b) methyl methacrylate and ethyl acrylate and (c) ethylene and propylene. A methyl methacrylate - methylstyrene - maleic acid terpolymer, when examined by this propylation - pyrolysis procedure, yielded dipropyl fumerate and a smaller amount of dipropyl maleate.

This method enables acrylic acid and methacrylic acid units to be determined in acrylic copolymers in amounts down to 0.1% with an accuracy of ± 5%.

METHOD 62 - DETERMINATION OF ACRYLIC ESTER GROUPS IN COPOLYMERS. HYDRIODIC ACID REDUCTION - GAS CHROMATOGRAPHY.[41]

62.1 SUMMARY

This hydriodic acid reduction - gas chromatography procedure enables acrylic ester groups in copolymers to be determined in amounts down to 0.2%.

62.2 APPARATUS

Hewlett Packard 5756 gas chromatograph equipped with flame ionization detector or equivalent. A stainless steel column, 10 ft DO-401 on 60-80 mesh Gas Pak WAB. helium flow rate 10 ml/min with an inlet pressure of 65 psi. Injection port and detector temperatures: 210 and 250°C respectively.

Column temperature 150°C.

62.3 REAGENTS

Spectrograde acetone, methyl ethyl ketone, hexane, 1,1,2-trichloroethane and isopropylbenzene (Eastmen Organic Chemicals).

Tetrahydrofuran (E.I. du Pont de Nemours) is distilled over triphenyl phosphite and nitrogen to remove any water and butylated hydroxy toluene inhibitor. After distillation, tetrahydrofuran is stored in amber bottles over thin strips of copper metal to prevent the formation of peroxides.

Hydriodic acid (Fisher Scientific Company) is freshly distilled over hypophosphorus acid and nitrogen. Only the 57% hydriodic acid azeotrope boiling at 127°C is retained. The distillate is stabilised with hypophosphorus acid.

62.4 METHOD

A weighed portion of the polymer is dissolved or swollen in a mixture of glacial acetic acid, molten phenol and acetic anhydride.

7.11

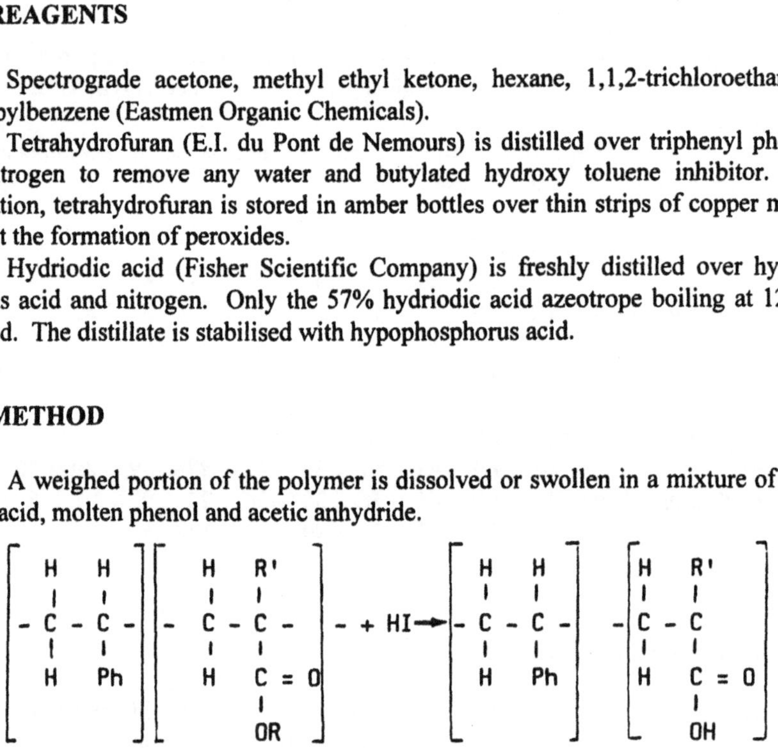

$R = H, CH_3, C_2H_5, C_4H_9, R' = H, OH_3$

The mixture in then digested with hydriodic acid to produce alkyl iodides corresponding to the acrylic ester groups present in the polymer.

The alkyl iodides up to butyl iodide is analysed by gas chromatography. The method is calibrated against standard solutions of alkyl iodides and internal standard.

62.5 EXPERIMENTAL PROCEDURE

62.5.1 Zeisel Cleavage Reaction.

Approximately 200 mg of the dried polymer is placed in the reaction flask. Figure 7.15. To that is added 10 ml of molten phenol, 5 ml of glacial acetic acid and 5 ml of acetic anhydride to swell and dissolve the polymer and scavenge any water that might be present. The reaction flask is then placed in a heating mantle held at 125°C until the polymer is completely dissolved or sufficiently swollen to allow attack by the hydriodic acid. While the sample mixture is heating, 5 ml of a 1% w/v solution of 1,1,2-trichloroethane in isopropylbenzene is pipetted into the receiver which is then placed in a dry ice-acetone bath. After the solution is obtained, the contents of the reaction flask are cooled and 25 ml of hydriodic acid is added. The nitrogen flow rate is set at 10 ml/min and the reaction flask placed in an oil bath at 132 ± 1°C. The position of the reaction flask placed in the oil bath is extremely important. The reaction mixture in the flask was level with the oil in the bath. Care is taken to prevent the refluxing mixture from reaching the desiccant.

62.5.2 Analysis of Alkyl Iodides.

Periodically, 2 ul samples of the reaction products are removed from the receiver and analysed by gas chromatography. Prior to taking a sample, the receiver is loosened and the trapping solution shaken around the glass spiral to form a homogenous solution of the alkyl iodides and the internal standard. Care is taken not to raise the delivery tube above the liquid level in the receiver. The reaction is allowed to proceed until the ratio of the longest chain alkyl iodide to the internal standard became constant. Details of the determination of relative response factors and calculation of copolymer composition are given below.

62.5.3 Determination of Response Factors.

An accurately known blend of the alkyl iodides and the internal standard (1,1,2-trichloroethane) is prepared and analysed using gas chromatography. The areas of the peaks in the chromatograms are determined and the response factors of the alkyl iodides calculated using the equation:

$$K_1 = \frac{(W_1)\ (A_x)\ (K_x)}{(W_x)\ A_1)}$$

where
K_1 = response factor of the respective alkyl iodide
K_x = response factor of 1,1,2-trichloroethane (arbitarily set equal to 1.00)

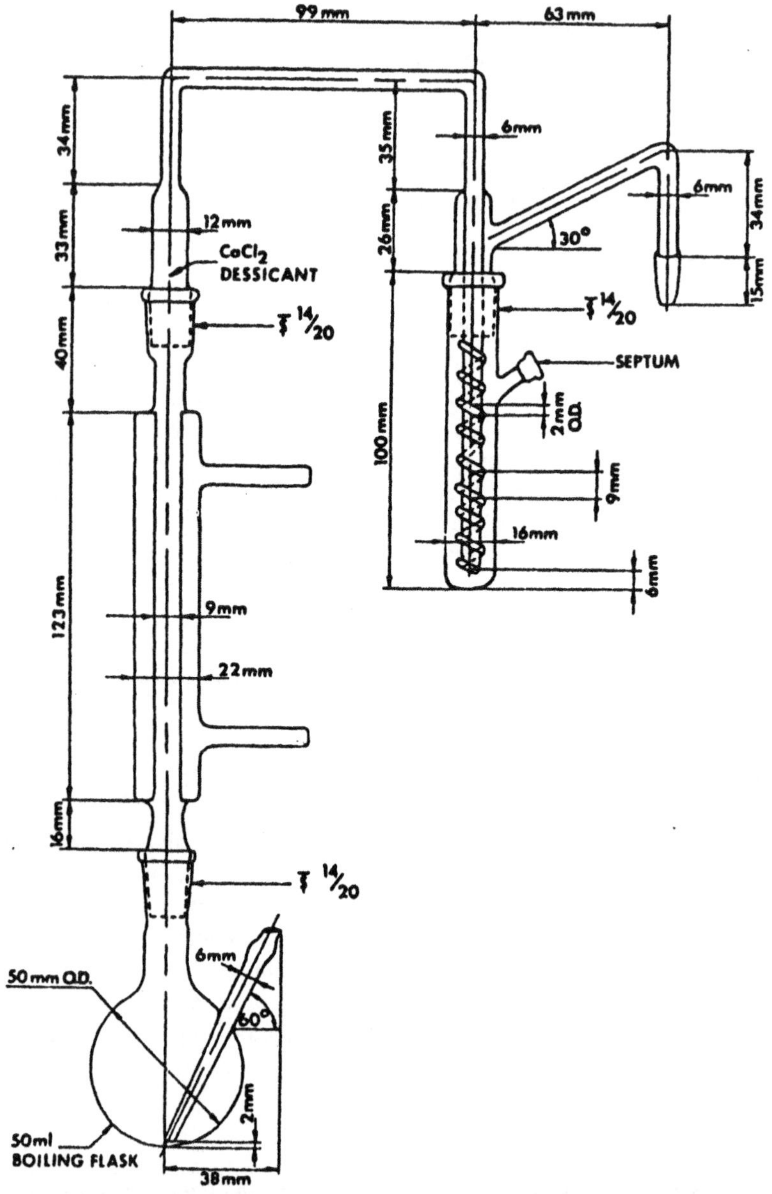

Figure 7.15. Apparatus for Zeisel cleavage of acrylic esters. From Anderson with permission.[41] American Chemical Society.

W_1 = weight of the respective alkyl iodide in the calibration blend
W_x = weight of 1,1,2-trichloroethane in the calibration blend
A_1 = area of the respective alkyl iodide peak
A_x = area of 1,1,2-trichloroethane peak

Response factors determined at two-week intervals were found to deviate. Best results are obtained when response factors were determined just prior to each analysis.

62.5.4 Determination of Copolymer Composition.

Samples from the receiver are periodically removed through the rubber septum and analysed using gas chromatography. The areas of the alkyl iodide and internal standard peaks are determined and the copolymer composition calculated using the following equation:

$$\% \text{ acrylate} = \frac{(K_I)\ (A_I)\ (W_x)\ (MW_A\,(100))}{(A_x)\ (W_{rs})\ (MW_I)}$$

where

K_I = response factor of the respective alkyl iodide
A_I = area of the respective alkyl iodide peak in the chromatogram
A_x = area of 1,1,2-trichloroethane peak in the chromatogram
W_x = weight of 1,1,2-trichloroethane added to the receiver
W_{rs} = weight of the sample used
Mw_a = weight of acrylate or methylacrylate being analysed for
MW_I - molecular weight of the respective alkyl iodide

Duplicate determinations should agree within $\pm$ 3% relative.

In Table 7.33 is shown some results obtained by applying this method to a range of acrylic polymers. The calculated recoveries are greater than 95%, for polymers containing between 10% and 100% acrylic monomer. The method has 99% confidence interval of 0.8. The presence of comonomers such as styrene, acrylonitrile, vinyl acetate, acrylamide or acrylic acid does not change the recovery of acrylate or methacrylate esters. Non-quantitative results are obtained, however, for polymers containing hydroxy-propyl methacrylate.

62.6 DISCUSSION OF RESULTS

This method enables methyl to butyl acrylate ester groups to be determined in acrylate copolymers in amounts down to 0.2% with an accuracy of $\pm$ 3%.

METHOD 63 - DETERMINATION OF COMBINED VINYL ACETATE IN VINYL-ACETATE-VINYLCHLORIDE COPOLYMERS. ISOTOPE DILUTION - DERIVATIVE METHOD.[42]

63.1 SUMMARY

This isotope dilution method enables vinylacetate units to be determined in vinyl acetate-vinyl chloride copolymers in amounts down to 0.2%.

Table 7.33 - Recovery of Alkyl Iodides from the Zeisel Cleavage of Acrylic Polymers[a]

Polymer	Methyl Acrylate %	Methyl Methacrylate %	Ethyl Acrylate %	Ethyl Methacrylate %	Butyl Acrylate %	Butyl Methacrylate %
1	33.7 (33.3)		33.0 (33.3)		32.7 (33.3)	
2		33.6 (33.3)		33.4 (33.3)		31.9 (33.3)
3			19.5 (20.0)			
4		50.1 (50.0)	30.5 (30.0)			
5		59.9 (60.0)			39.8 (40.0)	
6		29.7 (30.0)			29.8 (30.0)	
7		59.8 (60.0)				9.8 (10.0)
8	19.9 (20.0)		19.8 (20.0)			19.8 (20.0)
9		30.1 (30.0)		30.0 (30.0)	29.8 (30.0)	
10		29.9 (30.0)	29.6 (30.0)		39.7 (40.0)	
11		60.1 (60.0)	30.1 (30.0)			9.5 (10.0)

From Anderson[43]. American Chemical Society

[a] All values are the average of at least three determinations and are reported as % monomer found (% monomer in polymer).

63.2 APPARATUS

A Packard Model 3303 liquid scintillation spectrometer for radioassays.
Erlenmeyer flasks, 125 ml.

63.3 REAGENTS

An aqueous solution of acetic $2^{14}C$ acid containing 0.1-1.0 millimole per gram is prepared from acetic anhydride. Its specific activity of acetic anhydride was determined by treating a portion with an excess of aniline to form acetanilide ^{14}C which was crystallized from water and assayed in dioxane-naphthalene scintillator solution.

Methyl ethyl ketone reagent grade, is stirred with calcium hydride and distilled.

Sodium hydroxide, 1 molar, is prepared in 9:1 methanol: water.

o-Phenylene diamine is crystallized twice from water with the aid of activated charcoal.

The scintillator solution is prepared by dissolving naphthalene (200 grams) PPO (2,5-diphenyloxazole) (14.0 grams) and POPOD (1,4 bis-2-(5-phenyloxazolyl)benzene) (0.06 gram) in scintillation grade dioxane, diluting to 2 litres and adding 200 ml of methanol.

63.4 METHOD

In this method a known amount of methyl labelled acetic 2-^{14}C acid is added to a solution or suspension of the resin in methyl ethyl ketone and the ester is hydrolyzed with sodium hydroxide. The major portion of the sodium acetate ^{14}C is isolated and converted to 20methylbenzimidazole-methyl ^{14}C by means of the Phillips reaction[78] with *o*-phenylene diamine.

7.12

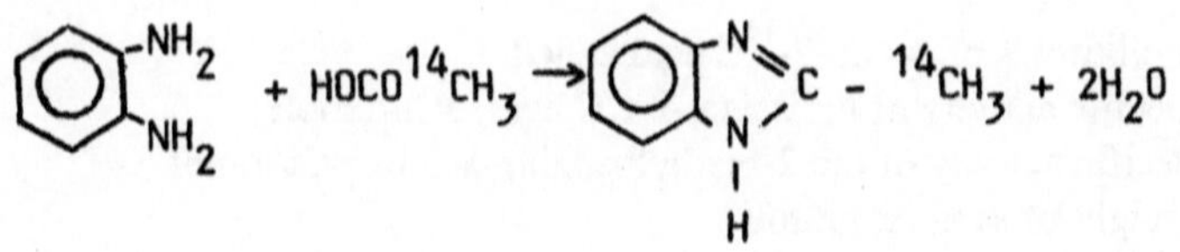

The product is radio assayed following addition of a scintillator and the vinyl acetate content calculated from the determined specific activities.

63.5 EXPERIMENTAL PROCEDURE

A specimen containing approximately one millimole of vinyl acetate, but not exceeding 4 grams, is weighed into a 125 ml Erlenmeyer flask and 40-50 ml of distilled methyl ethyl ketone is added. The solvent is added carefully so as to prevent agglomeration of the resin. A weighed aliquot of the solution of labelled acetic acid in water, containing 0.10-1.15 millimole of the acid is then added, after which 5 ml of the sodium hydroxide solution is introduced. The stoppered flasks are swirled with periodic immersion in an oil bath at 60°C to redisperse any precipitated resin and then placed in the bath for 16 hours.

The contents of the cooled flasks are transferred to a 125 ml separatory funnel and the flasks are rinsed with two 15 ml portions of water. The funnel is shaken vigorously and the lower phase is removed to a 250 ml beaker. After evaporation of the contents of the beaker to dryness, the sodium acetate ^{14}C is extracted from the residue with 10-12 ml of hot methanol and the solution is decanted into a 50 ml beaker. The methanol is then carefully evaporated on a water bath.

The derivative is prepared in a thick-walled glass (polymerization) tube to which an excess of approximately 10% (130 mg) of purified o-phenylene diamine is added initially. The residue in the 50 ml beaker is dissolved in 3-5 ml of water and the solution is transferred to the tube, to which 1.5 ml of 10 N hydrochloric acid is added. The tube is then sealed, enclosed in a protective sleeve of glass cloth and heated one hour in an oven at 180°C. After cooling, the contents of the tube are transferred to a 30 ml beaker and a slight excess of 1:1 ammonium hydroxide is added with vigorous stirring to precipitate the 2-methylbenzimidazole. The beaker is then placed in an ice bath for 30-60 minutes.

The crude derivative is collected on qualitative filter paper cut to fit a micro (1 cm) Buchner funnel and washed with 2-3 ml of ice water. The compound is then transferred to a 30 ml beaker, dissolved in 10-12 ml of water by heating to incipient boiling, and the solution is decolorized with 30-35 mg of activated charcoal. The suspension is stirred briefly then filtered as before to remove the charcoal. The filtrate is transferred to a 30 ml beaker, which is placed in an ice bath and the solution is stirred vigorously with a glass rod to initiate crystallization. After 30-60 minutes, the purified derivative is collected by filtration, washed with 2-3 ml of ice water and dried in vacuum over phosphorus pentoxide. The product is assayed by weighing 4-6 mg into a counting vial and adding 15 ml of scintillator solution, or by weighing a larger amount into a 50 ml volumetric flask and removing aliquots.

$$\text{Vinyl acetate, \% by weight} = (W_1 \left(\frac{S_o}{S_t} - 1\right) \frac{8.609}{W_s}$$

where
W_1 = millimoles of acetic-2-^{14}C acid added
S_o = specific activity of the acetic-2-^{14}C acid, uCi/mmole
S_t = specific activity of the 2-methylbenzimidazole, uCi/mmol
W_s = weight of sample, grams

63.6 DISCUSSION OF RESULTS

This method is capable of determining down to 0.2% vinyl acetate units in vinyl acetate-vinyl chloride copolymers with an accuracy of $\pm$ 5%. A relative standard deviation of approximately 0.48% was obtained for a copolymer containing 11.1% of vinyl acetate.

METHOD 64 - DETERMINATION OF ETHERIFICATION LEVELS IN ACRYLAMIDE INTERPOLYMERS. ALCOHOL EXCHANGE - GAS CHROMATOGRAPHY.[43]

64.1 SUMMARY

This alcohol exchange gas chromatographic method is applicable to the identification and determination of etherifying alcohols present in thermosecting acrylamide interpolymers of the type:

7.13

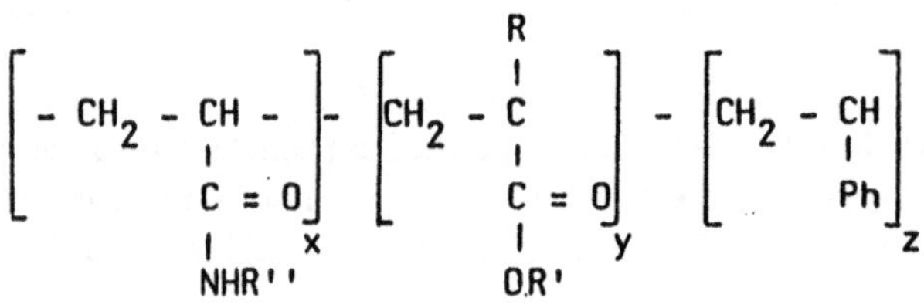

where
R = H or CH_3
R^1 = H, CH_3, C_2H_5 C_4H_9 or C_8H_{17}
R^{11} = H,CH2OC4H9 or CH_2OH

64.2 APPARATUS

Gas chromatographic separations are performed using a Hewlett-Packard 5760 gas chromatograph or equivalent equipped with flame ionization detectors. A 12 ft x 0.125 in o.d. stainless steel column is packed with 20% Carbowax 20 M on 60-80 mesh Gas Pak WAB. Helium flow rate: 20 ml/min with an inlet pressure of 65 psi. Injection port and detector temperatures: 220 and 240°C respectively. The column temperature: held at 105°C until the elution of the pentanol peak, then programmed from 105 to 205°C at 10°C/min and held until the 2-ethylhexanol eluted.

The glass assembly used to carry out the alcohol exchange reaction consists of a 50 ml erlenmeyer flask equipped with a 14/20 joint topped with a Liebig condenser with a 14/20 joint.

64.3 REAGENTS

Reagent grade 2-ethylhexanol, n-butanol, n-pentanol, butyl cellosolve and acetone (Eastman Organic Chemicals).

Paraformaldehyde, p-toluenesulfonic acid and propionamide (Eastman Organic Chemicals) used without further purification.

64.4 METHOD

Cast films of the acrylamide interpolymer are reacted with an alcohol, eg. octyl alcohol, to exchange with etherifying alcohols present in the polymer backbone as follows:

7.14

The etherifying alcohol content of the digest, in this case butanol, is determined by gas chromatography. The method is calibrated against standard solutions of the etherifying alcohol and an internal standard.

64.5 EXPERIMENTAL PROCEDURE

64.5.1 Drying of Polymer.

All acrylamide interpolymer samples are initially diluted to 10% nonvolatile resin with acetone. Thin films of the resin solutions are uniformly cast on clean glass plates using a smooth glass rod and immediately dried at 50°C under 5 cm Hg for 2.5 hours to remove all solvents. The film thickness of the dried polymers is kept below 0.5 ml to ensure complete solvent removal. The films are removed from the glass plates using oil free razor blades and stored in vials.

64.5.2 Alcohol Exchange Reaction.

The etherifying alcohols present in the polymer backbone are exchanged according to the above reaction. Approximately 200 mg of dried polymer is placed in a reaction flask along with 20 ml of 2-ethylhexanol. To this solution 5 ml of a 0.4% w/v solution of p-toluenesulfonic acid in 2-ethylhexanol and 5 ml of a 1% w/v solution of n-pentanol in 2-ethylhexanol are added. A magnetic stirring bar is placed in the flask, the condenser is fitted to the flask and placed on a magnetic stirrer hot plate. Heating and stirring are employed until a vigorous reflux begins. The reaction is allowed to continue for four hours with the reaction flask being checked periodically to ensure constant refluxing. After the reaction period is complete, the reaction mixture is allowed to cool to room temperature. The condenser is thoroughly washed with fresh 2-ethylhexanol to wash all products into the reaction flask. After allowing total drainage from the condenser, the reaction flask is stoppered until gas chromatographic analysis.

64.5.3 Analysis of Exchanged Alcohols.

Determination of response factors - an accurately known blend of the etherifying alcohols and internal standard (n-pentanol) is prepared and analysed by gas chromatography. The areas of the peaks in the chromatograms are determined by triangulation and the response factors of the etherifying alcohols calculated using the following equation:

$$K_1 = \frac{(W_1)\ (A_x)\ (K_x)}{(W_x)\ (A_1)}$$

where
K_1 = response factor of etherifying alcohol,
K_x = response factor of n-pentanol (arbitrarily set equal to 1.00)
W_1 = weight of respective etherifying alcohol in the calibration blend
W_x = weight of n-pentanol in the calibration blend
A_1 = area of respective etherifying alcohol peak
A_x = area of n-pentanol peak

Response factors determined at two-week intervals were found to deviate. Best results were obtained when response factors were determined just prior to analysis.

64.6 TYPICAL RESULTS

A 2 ul sample of exchanged alcohols is removed from the reaction flask and is analysed by gas chromatography. Prior to sampling, the reaction mixture is thoroughly shaken to ensure uniform composition. The areas of the etherifying alcohols and internal standard peaks are determined by triangulation and the moles of etherifying alcohol calculated using the following equation:

$$M_A = \frac{(A_A)\ (K_1)\ (W_x)}{(A_x)\ (MW_A)\ (W_A)}$$

where
M_A = moles of etherifying alcohol per gram of dried polymer
A_A = area of n-pentanol peak
W_x = weight of n-pentanol added to flask
W_A= weight of dried polymer added to flask
MW_A = molecular weight of etherifying alcohol

Each reaction mixture is analysed in duplicate and the average value used for further calculations. Duplicate determinations do not vary by more than $\pm$ 2% relative. 99.7% recovery was obtained with this procedure.

The moles of acrylamide present in the polymer are determined from the nitrogen content of the dried polymer. The acrylamide content is calculated using the following equation:

$$M_a = \frac{(\%N)}{(14.01)(100)}$$

where
M_a = moles of acrylamide per gram of dried polymer
14.01 = molecular weight of nitrogen
%N = percent nitrogen in dried polymer.

The percent etherification of the acrylamide in the polymer may be calculated as follows:

$$\%\ \text{etherification} = \frac{(M_A)\ (100)}{(M_a)}$$

where
M_A = moles of etherifying alcohol per gram of dried polymer
M_a = moles of acrylamide per gram of dried polymer

The total etherification is calculated by adding the % etherification due to all etherifying alcohols.

The level of alcohol obtained during the exchange reaction of acrylamide is correlated with data obtained using Zeisel cleavage of the alkoxy groups. These data are summarised in Table 7.34. Comparable results are obtained using both procedures, however, the Zeisel cleavage reaction will also cleave ester linkages as well as ether functionalities. This presents no problem with polymers which do not contain ester groups. In systems employing both butyl esters and butyl ethers the Zeisel cleavage reaction gives a total $O\text{-}C_4H_9$ content in the sample. Alcohol exchange, on the other hand, will cleave only the butyl ether groups in the sample. By subtracting the alcohol exchange data from that obtained from Zeisel cleavage, one can assess the relative amounts of alkyl ester and alkyl ether in the sample.

Table 7.34 - Correlation of Alcohol Exchange and Zeisel Cleavage

Sample	Percent Butylated Acrylamide[a]	
	Alcohol Exchange	Zeisel Cleavage
14% Acrylamide	100	101
23% Acrylamide	101	100

From Anderson[43]. American Chemical Society

[a] Based on experimentally determined butoxy and nitrogen content

CHAPTER 7.3

DETERMINATION OF ELEMENTS

Methods 65 - 89

METHOD 65 - DETERMINATION OF IRON, TITANIUM AND ALUMINIUM CATALYST RESIDUES IN POLYALKENES AND POLYALKENE COPOLYMERS. ASHING - SPECTROPHOTOMETRIC PROCEDURE.

65.1 SUMMARY

Most of the published work on the determination of functional groups, not unexpectedly, has been carried out on copolymers. This is because the determination of a functional group that is specific to one of the copolymer constituents is the key to the determination of the monomer ratios in the copolymer.

The method describes procedures for the determination of trace amounts of titanium, aluminium and iron in polyolefins. Analysis of a 5 g sample enables aluminium, iron and titanium to be determined in amounts down to 1 ppm with an accuracy of ± 1 ppm.

65.2 APPARATUS

Muffle furnace - maximum operating temperature 900°C.
Silica crucibles - 50 cm^3 capacity
Visible spectrophotometer
Standard volumetric glassware.

65.3 REAGENTS

Potassium hydrogen sulphate, AR grade.
Sulphuric acid, 0.5% and 50% v/v.
Tiron solution - 4 g of 1.2 dihydroxybenzene, 3:5 sulphonic acid sodium salt in 100 ml distilled water.
Buffer solution - pH 4.7 acetic acid/sodium acetate. Dissolve 136 g of $CH_3COONa3H_2O$ in 1 cm^3 of water. Dilute 57 cm^3 of glacial acetic acid to 1 cm^3 with water. The pH 4.7 buffer solution is prepared by mixing equal volumes of these two solutions.
Sodium hydrosulphite - solid AR grade.
Ammonia solution - mix equal volumes of 0.880 ammonia and water.
Potassium titanyl oxalate - pure grade.

Aluminon solution - dissolve each of the following reagents in a minimum volume of distilled water; 7 g ammonium acetate AR grade, 2.5 cm^3 concentrated hydrochloric acid, AR grade, 0.2 g aluminon reagent, 0.5 g gum acacia. Transfer the solutions in the above order to a 500 cm^3 standard volumetric flask and dilute to the mark with distilled water.

Aluminium powder - AR grade
Thioglycollic acid - 98% AR grade.
Ammonia - specific gravity 0.830, AR grade.
Iron wire - AR grade.

65.4 METHOD

Polymer samples are ashed to remove organic material. The ash residues are fused with potassium hydrogen sulphate to effect solution of the metals. The resulting ash in dissolved in hot water containing sulphuric acid and diluted to a standard volume. Suitable aliquots of the aqeuous solutions are reacted with organic reagents under controlled conditions, ie. titanium is determined spectrophotometrically with "Tiron", aluminium is determined spectrophotometrically by the "Aluminon" procedure and iron is determined spectrophotometrically by a procedure using thioglycollic acid.

65.5 EXPERIMENTAL PROCEDURE

Weigh accurately about 5 g of polymer into a silica crucible and ignite the sample in a muffle furnace. Allow the temperature to increase gradually to 800°C over a period of 4 hours, maintain this temperature for 1 h.

Remove the crucibles from the muffle furnace when cold and add 5 g of fused potassium hydrogen sulphate. Heat the mixture over a bunsen burner until a clear melt is obtained, cool and dissolve in 25 cm^3 of hot water containing 1 cm^3 of 50% w/v sulphuric acid. Make the solution up to 100 cm^3. Carry out a blank under identical conditions omitting the sample.

65.5.1 Titanium.

Pipette an aliquot of the test and blank solutions (maximum 35 cm^3) into two separate 50 cm^3 standard volumetric flasks. If the volume added is less than 35 cm^3 then adjust the volume to 35 cm^3. Add 5 cm^3 of Tiron to each flask and neutralize to Congo red paper with 50% ammonia solution. Add 5 cm^3 of sodium acetate-acetic acid buffer solution (pH 4.7) and dilute to 50 cm^3 with distilled water.

Transfer the coloured solutions to 1 cm cells and add 5 mg of sodium hydrosulphite, mix with a glass rod. Immediately measure the optical density of the sample solution against the blank solution in the comparison cell at a wavelength of 410 nm. Read off the titanium content of the solution from the calibration graph and calculate the titanium content of the original sample.

65.5.2 Aluminium.

Pipette an aliquot of the test and blank solutions (maximum 20 cm^3) into two separate 50 cm^3 standard volumetric flasks. If the volume taken is less than 20 cm^3 adjust the volume to 20 cm^3 with distilled water. Add 20 cm^3 of Aluminon reagent and immerse the flasks in a boiling water bath for 10 min to allow the colour to develop. Cool the flasks and dilute to 50 cm^3 with distilled water.

Measure the optical density of the sample solution against the blank solution in the comparison cell, using the spectrophotometer at a wavelength of 535 mu in 1 cm cells. Read off the concentration of aluminium from the appropriate calibration graph. Calculate the aluminium content of the original sample. The Aluminon reagent solution must be discarded when the optical density of the blank solution exceeds 0.09 compared with distilled water.

65.5.3 Iron.

Transfer an aliquot of the sample and blank solutions (maximum 40 ml) into two separate 50 cm^3 standard volumetric flasks and add 0.2 cm^3 of thioglycollic acid.

Adjust the volume of the solutions to 40 cm^3 with distilled water and add 5 cm^3 of 0.880 ammonia. Check that the solution is alkaline to litmus and dilute to 50 cm^3 with distilled water. Mix thoroughly and measure the optical density of the sample solution versus the blank solution in the comparison cell at a wavelength of 535 mu using 4 cm glass cells.

Read off the iron content of the sample solution from the calibration graph and calculate the iron content of the original sample.

65.5.4 Calculations.

$$\text{Ti or Al or Fe ppm} = \frac{M \times 100}{W \times V}$$

where

M = weight (microgram) of titanium or aluminium or iron found from calibration graph.

W = weight (g) of sample taken for analysis

V = volume (cm^3) of 100 cm^3 of ash solution taken from colour development.

65.5.5 Calibration Procedure for Determination of Titanium.

Into six 50 cm^3 standard volumetric flasks pipette the following volumes of standard 5 ug/ cm^3 titanium solution.

0 ml	2 cm^3	5 cm^3	10 cm^3	20 cm^3	25 cm^3
0	10	25	50	100	125 ug Ti

Add the required volume of distilled water to take the volume of each solution up to 35 cm^3. Add 5 cm^3 of Tiron reagent and neutralize to Congo red paper with 50% ammonia solution, add 5 cm^3 of pH 4.7 acetic acid/sodium acetate buffer solution and dilute to the mark with distilled water.

Transfer the solutions to 1 cm cells and read off the optical density at a wavelength of 410 nm using the solution containing no titanium as the reference solution. Plot a graph of optical density against ug of titanium contained in the 50 cm^3 standard volumetric flasks.

65.5.5 Calibration Procedure for Determination of Aluminium.

Into seven sets of 50 cm^3 volumetric flasks pipette the volumes of 7.5 ug/ cm^3 aluminium solution shown below and also 2 cm^3 of 5 ug 1 cm^3 titanium solution (ie. 10 mg Ti). Take the seven flasks comprising set 1. Add to each flask sufficient distilled water to make the volume up to 20 cm^3. Add 20 cm^3 Aluminon reagent and immerse the seven flasks in a boiling water bath for 10 min to allow the colour to develop. Cool the flasks under running water and dilute to 50 cm^3 with distilled water and mix well.

0	1	2	5	10	15	20	30 cm^3
0	7.5	15	22.5	30	67.5	90	135 ug ml

Measure the optical density at 535 nm of the solution in 1 cm cells against the aluminium and titanium free reagent blank solution in the comparison cell. Plot a graph of ug aluminium present in the 50 cm^3 solution versus determined optical density. Repeat these operations six times using the same volumes of aluminium standards but using 0.5, 10, 20 and 25 cm^3 of the 5 ug/ cm^3 titanium standard. Plot the seven aluminium calibration graphs and indicate on each graph the constant amount of titanium (micrograms) in each calibration solution (ie. between 0 and 125 ug of titanium). Recalibrate the procedure each time a new batch of Aluminon reagent is used in the analysis.

65.5.6 Calibration Procedure for Determination of Iron.

Pipette 10 cm^3 of the 100 ug/ cm^3 iron solution into a 100 cm^3 standard volumetric flask and dilute to the mark with 0.2 nitric acid solution, mix thoroughly. This solution contains 10 ug/Fe/ cm^3.

Into six 50 cm^3 standard volumetric flasks pipette the following volumes of 10 ug/ml iron solution.

0	1	2	5	10	15 cm^3
0	10	20	50	100	150 ug Fe

Add the required volume of 0.2 M nitric acid solution to make the volume of each solution 40 cm^3. Add 5 cm^3 of 0.880 ammonia solution and check that the solution is alkaline to litmus paper. Dilute to 50 cm^3 with distilled water and mix thoroughly. Measure the optical density of the solutions. Transfer the solutions to 4 cm glass cells and read off the optical density at a wavelength of 535 nm using the solution containing 0 ug iron as the reference solution. Plot a graph of optical density against ug of iron present in the 50 cm^3 standard volumetric flask.

65.6 DISCUSSION OF RESULTS

This method is capable of determining down to 1 ppm of aluminium, iron and titanium in polyolefins with an accuracy of $\pm$ 1 ppm.

METHOD 66 - DETERMINATION OF ALUMINIUM AND VANADIUM CATALYST RESIDUES IN POLYALKENES AND POLYALKENE COPOLYMERS. ASHING - SPECTROPHOTOMETRIC PROCEDURE.[44]

66.1 SUMMARY

This spectrophotometric procedure determines aluminium and vanadium catalyst residues in polyolefins and their copolymers in amounts down to 20 ppm (aluminium) and 1 ppm (vanadium).

66.2 APPARATUS

Glass chromatographic column - 10 in x 1 in.
Muffle furnace - capable of maintaining a temperature of 600°C.
Platinum dish, 75 mm in diameter.

66.3 REAGENTS

Toluene, AR.
Dowex 50 W x 8 cation exchange resin, 50 - 100 mesh.
Digest 5 g Dewex in 100 cm3 of 5M sulphuric acid transfer the slurry to the column to give a resin-bed depth of 50 mm. Run off the sulfuric acid until the level of the liquid falls to 1 to 2 mm from the top of the resin bed. Wash the resin bed with 2.5 M sulfuric acid until the effluent from the bottom of the column is colourless leaving 1 to 2 mm depth of liquid on top of the resin bed. Add 20 cm^3 of a 0.25 M sulfuric acid - 1% hydrogen peroxide solution as described above. When the last of the sulfuric acid-hydrogen peroxide solutions has been added, turn off the column tap.
Hydrochloric acid, AR, 38% sp. gr 1.19.
Hydrofluoric acid, AR, 40% sp. gr. 1.13.
Hydrogen peroxide, AR, 30%, 100 vol.
Nitric acid, AR, 65% sp. gr. 1.4.
Sulfur, cp crystals.
Sulfuric acid, AR, 98% sp. gr. 1.84.
Sulfuric acid, 5 M and 2.5 M.
Sulfuric acid, 0.25 M hydrogen peroxide 1% solution. Add 3.3 cm^3 of hydrogen peroxide (30%) to 25 cm^3 of 1M sulphuric acid and dilute to 100 ml.

66.4 METHOD

The sample is ignited and the ash dissolved in acids. Hydrogen peroxide is added and the solution passed down a cation exchange column. Vanadium, present as a pentavalent peroxide complex, passes unadsorbed down the column and is determined photo-absorptiometrically. Aluminium is removed from the column by elution with a more concentrated acid and determined by a complexometric titration.

66.5 EXPERIMENTAL PROCEDURE

66.5.1 Determination of Vanadium.

Weigh 10 g of sample into a platinum dish. Dissolve 1 g of sulfur in 20 cm^3 of toluene and add to the contents of the dish. Evaporate off the toluene.

Ignite the sample over a bunsen flame until fumes are no longer evolved, transfer the dish to a muffle furnace at 600°C until all the carbonaceous material has been removed.

Treat the ash with 5 cm^3 of water, 5 cm^3 hydrofluoric acid, 5 drops of nitric acid and 12 drops of sulfuric acid (98%) taking care to moisten the inner wall of the dish completely. Evaporate first on a water bath and then on a hot plate to dryness. Redissolve the sample in 1 cm^3 of hydrochloric acid, 5 cm^3 of water and 2.4 cm^3 of 2.5 M sulfuric acid, taking care to moisten the inner wall of the dish completely. Warm on the water bath until all the hydrochloric acid has been removed. Transfer the sample to a 25 cm^3 volumetric flask containing 1 cm^3 of 30% hydrogen peroxide and make up to 25cm^3. Transfer 10 cm^3 of this solution to the ion exchange column. Allow the sample to pass down the column at a flow rate of not more than 1 cm^3/min. Collect the effluent when the 10 cm^3 aliquot has fallen to 1 to 2 mm from the top of the resin bed elute with 15 cm^3 of sulfuric acid-hydrogen peroxide solution added at a flow rate of not more than 1 cm^3/min maintaining 1 to 2 mm depth of liquid on top of the resin bed. When the 15 cm^3 of sulfuric acid-hydrogen peroxide solution have been added transfer the effluent, which contains the vanadium, to the water bath and complete the vanadium determination as described below.

Evaporate the contents of the vanadium extract first on a water bath and then on a hot plate to dryness. Dissolve the vanadium in 1 cm^3 of hydrochloric acid, 5 cm^3 of water and 1 cm^3 of 1 M sulfuric acid. Heat the dish on the water bath until all the hydrochloric acid is removed. Transfer the vanadium solution to a conical flask using 70 cm^3 of water.

Complete the vanadium determination as described in the spectrophotometric procedure described in Method 67.

66.5.2 Determination of Aluminium.

Elute the resin column with 25 cm^3 of 2.5 M sulfuric acid at a flow rate of not more than 1 cm^3/min again always maintaining a 1 to 2 mm depth of liquid on top of the resin bed. Collect the effluent, which contains the aluminium in a 250 cm^3 beaker. Evaporate first on a water bath and subsequently on a hot plate to dryness.

Dissolve the aluminium in about 50 cm^3 of water and 3 drops of nitric acid by boiling. Complete the determination of aluminium by the spectrophotometric procedure as described in Method 65. Run a blank, taking care to use the same quantities of all reagents as used in the determination.

66.6 DISCUSSION OF RESULTS

This method describes a procedure for the determination of 20 to 250 ppm of aluminium and more than 1 ppm of vanadium in polymer samples. Tin and zinc and large amount of cobalt, nickel and lead interfere with the aluminium determination. These interfering metals can be removed by a suitable modification of the basic procedure.

METHOD 67 - DETERMINATION OF VANADIUM CATALYST RESIDUES IN ETHYLENE - PROPYLENE RUBBER. ASHING - SPECTROPHOTOMETRIC PROCEDURE.[44]

67.1 SUMMARY

This spectrophotometric procedure determines vanadium catalyst residues in amounts down to 2 ppm in ethylene-propylene rubbers.

67.2 APPARATUS

Visible spectrophotometer with 1 and 5 cm cells.

67.3 REAGENTS

Standard vanadium solution. Prepare 5 ug per cm^3 or 50 ug per cm^3 from ammonium metavanadate.

3,3′-diaminobenzidine tetrachloride solution. 0.5% w/v store at 4°C under nitrogen.

Phosphoric acid, 85%.

Nitric acid concentrated.

Potassium, pyrosulphate, AR solid.

67.4 METHOD

The polymer is ashed then dissolved in nitric acid - phosphoric acid. 3,3′diaminobenzidine tetrachloride is added and the colour produced evaluated spectrophotometrically at 470 nm. Calibration is achieved against similarly prepared solutions of pure vanadium.

67.5 EXPERIMENTAL PROCEDURE

Ash a 10 gram sample of elastomer solution in a platinum crucible by charring on a hot plate followed by heating over a meker burner. Add 10 cm^3 of water and 2 cm^3 of nitric acid then transfer to a 100 cm^3 beaker. To ensure complete removal of vanadium, melt a gram of potassium pyrosulfate in the crucible, cool, then wash the salt into the beaker with hot water. Evaporate to about 10 cm^3 and transfer the contents to a 25 cm^3 volumetric flask. Before diluting to the mark, add 1 cm^3 of 85% phosphoric acid and 1 cm^3 of 3,3′diaminobenzidine tetrachloride solution. Measure the absorbance at 470 nm in a suitable cell, using a reagent blank, after standing 15 min in the dark.

67.5.1 Calibration.

Add suitable aliquots of standard vanadium solution to 100 cm^3 beakers which contain 50 cm^3 of water, 2 cm^3 of nitric acid and 1 gram of potassium pyrosulfate. Reduce the volume to about 10 cm^3 by boiling and transfer the contents to a 25 cm^3 volumetric flask. React with phosphoric acid and 3,3′-diaminobenzidine tetrachloride solution as described for sample analysis. For vanadium contents from 5 to 50 ug measure the absorbance in a 4 cm cell using a reagent blank.

67.6 TYPICAL RESULTS

Data in Table 7.35 shows good agreement between data obtained by this method and neutron activation analysis for a range of ethylene-propylene copolymer samples.

67.7 DISCUSSION OF RESULTS

This method is capable of determining down to 2 ppm vanadium in ethylene propylene elastomers with an accuracy of $\pm$ 5%.

METHOD 68 - DETERMINATION OF SILICA CATALYST SUPPORT IN HIGH DENSITY POLYETHYLENE. INFRARED SPECTROSCOPY AND NEUTRON ACTIVATION ANALYSIS.[45]

68.1 SUMMARY

Three methods are described for the determination of down to 50 ppm of silica catalyst supports in high density polyethylene. These methods involve direct ashing and weighing, infrared spectroscopy and neutron activation analysis.

68.2 APPARATUS

Infrared spectrometer.
Quartz bowl.
Meker burner.
Buehler mold assembly.

68.3 METHODS

Ashing. The polymer is ignited at 650°C and the residual silica weighed.

Infrared spectroscopy. Films of the polymer are examined at 1118 and 470 cm^{-1}. The method is calibrated against known blends of silica in high density polyethylene.

Neutron activation analysis. Polymer powder is irradiated with neutrons. Silicon is activated by the reaction $^{28}Si(N,p)^{28}Al$. The concentration of silicon is measured by counting the 1.78 MeV gamma - ray emission from the decay of ^{28}Al. A silicon standard is used for calibration.

Table 7.35 - Vanadium (ppm)

Sample	Dry Ashing	Neutron Activation
A	10.2	9.9 ± 0.2
B	14.0	14.1 ± 0.1
C	14.6	15.6 ± 0.3
D	0.5	0.14 ± 0.01
E	13.0	14.8 ± 0.2
F	0.9	0.27 ± 0.01
G	15.2	18.8 ± 0.3
H	18.2	17.9 ± 0.3

68.4 EXPERIMENTAL PROCEDURES

68.4.1 Ashing Technique.

Productivity by the ashing technique is determined by weighing approximately 100 g of polymer into a dry quartz bowl, which is then placed on a small burner, and the sample gently ignited. After most of the polymer is burned away, the bowl is set in a muffle furnace at 650°C to oxidize residual carbon. Finally, the bowl is allowed to cool in a desiccator and then weighed.

68.4.2 Infrared Method.

Polymer films are prepared on a Buehler mold assembly by placing 0.52 g of the sample between aluminium foil disks in a 1 mm spacer. The mold is heated to 160°C, the pressure set to 6000 psig and the sample then cooled to 35°C by means of an air stream. The film is mounted on a holder and the thickness is measured to the nearest 0.01 mm with a micrometer.

All spectra obtained on the infrared spectrometer were taken under the following conditions: 100 scans, 4 cm^{-1} resolution, boxcar apodization TGS detector.

The absorbance bands at 1118 and 470 cm^{-1} are used to determine the quantity of silica in polyethylene.

68.4.3 Neutron Activation Analysis.

Approximately 1 - 2 g of polyethylene powder is irradiated with neutrons obtained with a 500 ke V neutron activator according to the following reaction: $^{3}H(D,n)^{4}He$. Silicon is activated by the reaction $^{28}Si(N,p)^{28}Al$, and the concentration of silicon is then measured by counting the 1.78 MeV gamma-ray emission from the decay of ^{28}Al. A Conostan 5000 ppm Si standard is used for instrument calibration. Two 3 in sodium iodide detectors are used to measure the 1.78 MeV gamma-rays.

68.5 TYPICAL RESULTS

Infrared method. Thicker films (500 - 1000 um) of high density polyethylene yield spectra in which the presence of down to 0.01% silica can be detected. The bands at about 1118, 800, and 470 cm^{-1} indicate silica. The absorbance bands at approximately 1118 and 470 cm^{-1} may both be used to determine the quantity of silica in polyethylene.

The standard base line technique of determining absorbance was used in both cases. A further infrared method involves the direct measurement of the absorbance of 470 cm^{-1} band of silica. The 470 cm^{-1} absorbance band of silica is in the region of the infrared spectrum which is relatively free of polyethylene absorbance bands. The absorbance value of the 470 cm^{-1} band is calculated for each standard by use of the peak height determined by means of the base line technique between minima near 350 and 580 cm^{-1}.

Results obtained for polyethylene samples containing residual silica support of three different catalysts by infrared (470 cm^{-1} method), neutron activation analysis, ashing and weight are shown in Table 7.36. Infrared results differ from neutron activation analysis results by 0 - 5% while ashing and weighing techniques differ from neutron activation analysis results by 5 - 21% and 5 - 28% respectively.

68.6 DISCUSSION OF RESULTS

These methods are capable of determining down to 50 ppm of silica in high density polyethylene with an accuracy of $\pm$ 2 to 6%.

METHOD 69 - DETERMINATION OF SODIUM NEUTRALIZING AID IN HIGH DENSITY POLYETHYLENE AND POLYPROPYLENE. EMISSION AND FLAME PHOTOMETRIC PROCEDURES.

69.1 SUMMARY

This emission and flame photometric procedure determines sodium in amounts down to 0.2 ppm.

69.2 APPARATUS

Flame photometry.
Flame photometer.
Platinum dish, 100 ml.
Muffle furnace.
Emission spectrography.
Emission spectrograph.
Platinum dish.
Muffle furnace.

69.3 REAGENTS

Flame photometry.
Standard sodium solution, 10 ug/cm^3 sodium in deionized water.
Sulphur, (Specpure).
Magnesium AC dope (magnesium in salt of long chain fatty acids) in Xylene (36 gl^{-1})
Nitric acid, N/1 Analytical reagent grade.
Filter paper, Whatman No. 42 or equivalent.

Table 7.36 - Catalyst Productivity for Some Test Samples

Sample	IR at 21.27 Micron	NNA	Ashing at 650°C	Weight
1[a]	1893 (3.3)[d]	1923 (4.9)	1923 (4.9)	
2[a]	4165 (0)	4165	4545 (9.1)	3571 (14.3)
3[b]	1759 (2.9)	1709	2222 (30.0)	2007 (17.4)
4[c]	1727 (5.4)	1825	2000 (9.6)	2000 (9.6)
5[c]	2207 (3.5)	1288	2778 (21.4)	2941 (28.5)

From Battiste et al[45]. American Chemical Society
[a] Catalyst A. [b] Catalyst B. [c] Cabosil S17. [d] Values in parentheses are percent deviation from NAA.

69.4 METHOD

Flame photometry. The polymer is ashed in platinum with sulphur and a xylene solution of magnesium AC dope. A nitric acid solution of the residual ash is analysed by flame photometry. Calibration is achieved against standard solutions of sodium in nitric acid.

Emission spectrography. The polymer is ashed as above and the ash blended with carbon containing 0.1% palladium prior to emission spectrographic analysis for sodium. Calibration is achieved against known blends of sodium carbonate in magnesium sulphate.

69.5 EXPERIMENTAL PROCEDURE

69.5.1 Sample Ashing.

Ten grams of sample are weighed into a clean, dry 100 ml platinum basin, and 1 g sulphur (Specpure and 5 ml of a solution in xylene of magnesium AC dope magnesium salt of long chain fatty acids) (35 g/litre) added.

The dishes are heated gently over a bunsen until the xylene has evaporated and the temperature then raised until the polymer starts to decompose. Careful heating is maintained such that the vapours never ignite and heating is continued until all light volatile matter has been removed. The dishes are transferred to a cold electric muffle and ashed overnight at 500°C then weighed.

69.5.2 Flame Photometric Procedure.

Nitric acid solution (25 cm^3 of 1 M) is added to the ash in the platinum crucible which is then heated to 70 - 80°C and filtered through a Whatman No. 42 filter paper into a 100 cm^3volumetric flask. The residue on the filter paper is washed three times with 15 cm^3 of hot 1 M nitric acid. The flask is cooled and made to 100 cm^3. A blank solution is prepared by passing the same volume of hot nitric acid through a Whatman No. 42 filter paper and making up to 100 cm^{-3}

The procedure is calibrated against a 5 ppm standard sodium solution. The sodium concentration of the sample is obtained as:

$$\text{Sodium (ppm)} = \frac{\text{Percentage deflection} \times 5 \times 100 \times F}{100 \times W}$$

where the percentage deflection is the difference between % sample and % blank deflections, W is weight of sample taken, and F is a calibration factor.

69.5.3 Emission Spectrography.

The ash is mixed with twice its weight of carbon powder containing 0.1% palladium. Two electrodes were filled with the ash/carbon mixture and each is burned as the cathode of an electric are under the conditions described below:

Electrodes:	sample - Ringsdorf RW v, 4.6 mm diameter with cup size 0.125 in o.d., 0.0625 in i.d. and depth 8 mm.
	counter Ringsdorf RW v, 4.6 mm diameter, plane ends.
Analytical gap	6 mm
Excitation unit	Hilger FF 402 rectifier unit, output 230 volts D.C., 15 amps maximum.
Arc current	8 amp
Preburn	None
Burn and exposure	3 min.
Spectrograph	Hilger large quartz, E 742, prism marked to admit light from cathode tip and 5 mm of arc column.
Wavelength range	nm 245 to 350.
Slit width, u	10.
Condensing lens	Hilger F 1025, 2 cm from slit.
Sector	None.
Filter	Hilger F 1534, nominal densitive, 0, 0.6 and 12, with step height 1.2 mm each.
Photographic emulsion	Kodak B 10 or equivalent.
Developer	Kodak D 19 b, or equivalent diluted to 1/3 strength, 20°C, 3 min.
Plate calibration	Step filter.
Line pair	Na 3303.0, Pd 2763.1.

Spectrograms are recorded on photographic plates and the log spectral line intensity ratio Na 3303.0/Pd 2763.1 evaluated. Emulsion calibrations of the two wavelengths are established using the microphotometer transmittance data from adjacent steps of the spectrograms.

Concentration calibration is carried out by compounding standard mixtures containing 10%, 3.16%, 1%, 0.32% and 0.1% sodium carbonate in magnesium sulphate. Spectral line intensity ratios corresponding to these sodium concentrations are obtained using the same spectrographic procedure as for the samples. An analytical curve is established relating log intensity ratio and log concentration for the standard and by interpolation of the sample log intensity ratios in this curve the sodium concentration in the ash is obtained.

The concentration of sodium in the polymer is obtained by multiplying the sodium concentration in the ash by the ratio of ash weight to sample weight.

69.6 TYPICAL RESULTS

The results in Table 7.37 show clearly that flame photometry following dope ashing at 500°C gives a quantitative recovery of sodium relative to results obtained by a non-destructive method of analysis vis neutron activation analysis. Direct ashing without an ashing aid at 500°C causes losses of 10% or more of the sodium whilst direct ashing at 800°C causes even greater losses.

69.7 DISCUSSION OF RESULTS

These methods are capable of determining down to 0.2 ppm of sodium in polyethylene and polypropylene with an accuracy of $\pm$ 5%.

METHOD 70 - DETERMINATION OF LITHIUM IN POLYALKENES AND COPOLYMERS. ASHING-FLAME PHOTOMETRIC METHOD.

70.1 SUMMARY

This flame photometric method determines lithium neutralizing aid in polyalkenes in amounts down to 0.2 ppm.

70.2 APPARATUS

Flame photometer.
Platinum dish, 100 ml.
Volumetric glassware.

70.3 REAGENTS

Lithium stock solution, 100 micrograms per cm^3.

Dissolve 0.9219g lithium sulphate monohydrate in 1 M nitric acid and make up to 1 cm^3 with 1 M nitric acid.

Nitric acid, 1 M solution.

Water, distilled or deionised.

Calibration solutions (a) 0 to 10 ppm Li, (b) 0 - 1 ppm Li.

Prepare standard solutions by diluting to 100 ml aliquots of the stock solution using 1 M nitric acid.

70.4 METHOD

The polymer is ashed in platinum and a dilute nitric acid solution of the ash analysed by flame photometry at 670.9 um.

Table 7.37 - The Effects of Modification of Ashing Procedure on the Flame Photometric Determination of Sodium (sodium ppm)

		By Flame Photometry		
By Neutron Activaton	By Emission Spectrography	Original (Ashed between 650°C and 800°C)	Dope Ash at 500°C	Direct Ash at 500°C
99, 96, 99	95	60, 75, 55	100	75
256, 247, 259	258, 259	160, 178, 271	225	208
343, 321, 339	339, 287	250, 312	282	265
213, 210, 212	218, 212	140, 196	210	191
194, 189, 192	209, 198	80, 158, 229	196	169
186, 191, 198	191, 191	95, 95, 173	193	173

70.5 EXPERIMENTAL PROCEDURE

70.5.1 Sample Preparation.

Weigh 10 g of the polymer sample into a platinum dish. Place the dish in the cold electric muffle which is then programmed as follows:

Time from start:	0 to 1 h	Heat to 200°C
	1 to 3 h	Heat to 200°C
	3 to 5 h	Heat to 450°C
	5 to 8 h	Hold at 450°C

Remove the dish from the muffle, when cool add 5 cm^3 1 M nitric acid and warm the dish on a hot plate to ensure complete dissolution of soluble salts. Transfer the solution to a 10 cm^3 volumetric flask. Rinse the dish with further 1 cm^3 portions of nitric acid and add each to the volumetric flask. When the flask and the contents are cool, make up the volume to 10 cm^3 and mix thoroughly.

Set the flame photometer wavelength to 670.9 nm and adjust the slit, zero and gain controls of the flame photometer until 0% and 100% scale readings are obtained for 1 N nitric acid solution and the 1 or 10 ppm Li standard solution. Obtain the scale readings for each standard solution in turn, followed by the sample at 670.9, 685.9 and 655.9 nm.

70.5.2 Calibration.

Subtract the mean of the 685.9 and 655.9 scale readings from the value obtained at 670.9 nm to give a background corrected scale reading for each solution. Plot a calibration graph of the background corrected scale readings for the standard solutions against lithium concentrations under the conditions given above, the standard solutions given above are equivalent to 0.3 and 10 ppm lithium in polymer. Interpolate the sample background corrected scale reading into the calibration to obtain the concentration of lithium in the polymer.

70.6 DISCUSSION OF RESULTS

This method is capable of determining down to 0.2 ppm of lithium in polyalkenes with an accuracy of ± 5%. Sodium in amounts up to 1 ppm does not interfere.

METHOD 71 - DETERMINATION OF TRACES OF NICKEL, COPPER, ZINC, IRON, MANGANESE, LEAD, CADMIUM AND CHROMIUM IN POLYMERS. ASHING - ATOMIC ABSORPTION SPECTROMETRY.

71.1 SUMMARY

This atomic absorption specroscopic procedure determines eight heavy metals in polymers in amounts down to 0.03 to 0.13 ppm.

71.2 APPARATUS

Platinum dishes, 100 cm^3.
Volumetric flasks, 10, 25 cm^3,
Atomic absorption spectrometer.

71.3 REAGENTS

Nitric acid, 1 molar prepared using Aristar nitric acid.

71.4 METHOD

The polymer is ashed in platinum. A nitric acid extract of the ash is examined by atomic absorption spectroscopy for concentrations of nickel, copper, zinc, iron, manganese, lead, cadmium and chromium at specific wavelengths. The method is calibrated against standard solutions of the heavy metals in nitric acid.

71.5 EXPERIMENTAL PROCEDURE

Accurately weigh 10 g of dry polymer into a platinum dish. Place the dish in a cold contamination-free muffle furnace and temperature programme as follows:

Time from start:	0 - 1 h	Heat to 200°C
	1 - 3 h	Hold at 200°C
	3 - 5 h	Heat to 450°C
	5 - 8 h	Hold at 450°C

Remove the dish from the furnace and allow to cool in a dessicator. When cool, add 5 cm^3 1 M nitric acid and warm the dish gently on a hot plate to ensure complete dissolution of the metallic salts. Transfer the solution quantitatively with pure water to a 25 cm^3 volumetric flask and make up to the mark (final solution 0.2 M with respect to nitric acid). Blanks and standard solutions should also be prepared in 0.2 M nitric acid.

Atomic absorption instrument operating conditions,
Instrument:
Single beam
Background correction using deuterium arc
Grating monochrometer.
Burner:
Single slot 10 cm long
Burner aligned along optical axis for each metal
Fuel gas - Acetylene 12 lb/sq in
Support gas - compressed air, oil-less. Compressor model SY 006.
Aspiration rate: 9.5 cm^3/min
Spoiler left in position for all elements.
Recorder single pen.

Renew the calibration standards every two weeks from 10 ug/cm^3 stock solutions, by dilution with 0.2 M nitric acid (Aristar). Renew the 10 ug/cm^3 stock solutions monthly.

71.6 TYPICAL RESULTS AND DISCUSSION

The detection limits for metals in polymers achievable by this procedure are given in Table 7.38.

METHOD 72 - DETERMINATION OF TRACES OF COPPER IN POLYALKENES. MATRIX ASHING. SPECTROPHOTOMETRIC PROCEDURE.

72.1 SUMMARY

This spectrophotometric procedure determines down to 0.1 ppm copper in polyalkenes.

72.2 APPARATUS

Silica crucibles, Vitrosil, cleaned periodically by soaking in concentrated nitric acid.

Miscellaneous volumetric glassware. Pipettes miscellaneous, graduated pipette 5 cm^3, volumetric flasks 50 cm^3, 25 cm^3. Graduated cylinder 10 ml. Separatory funnels 50 cm^3 (ungreased stopcock), cleaned by soaking in concentrated nitric acid followed by rinsing with distilled water before use.

Electric hotplate with Simmerstat control.

Absorption spectrophotometer.

Table 7.38 - Analytical Conditions. Metals in Polymers

Element	Wavelength nm	Band Pass	Operating Range (in Polymer)ppm	Detection Limit (in Polymer) ppm	Concentration of Standard Solution ug L^{-1}
Iron	284.3	0.3	5	0.57	500
Manganese	279.5	0.5	1.25	0.03	250
Chromium	357.9	0.5	2.5	0.03	500
Cadmium	228.8	1.0	0.5	0.015	50
Copper	324.7	1.0	1.25	0.045	250
Lead	217.0	1.0	5	0.15	1250
Nickel	232.0	0.15	2.5	0.07	500
Zinc	213.9	1.0	0.5	0.015	125

72.3 REAGENTS

Sodium diethyldithiocarbamate solution 1% aqueous.

EDTA - citrate solution. Dissolve 18.2 g of citric in distilled water and add the theoretical quantity of 0.880 S.G. ammonia (both B.D.H. lead-free reagent; for foodstuff analysis grade) to produce the triammonium salt. The final solution should be slightly alkaline to litmus, add 5 g of disodium EDTA "Analar" and dilute to 100 ml with distilled water.

Ammonium hydroxide (1:3). Mix one volume of S.G. 0.880 ammonia (B.D.H. lead-free reagents foodstuff analysis grade) with 3 volumes of distilled water.

Cresol red indicator, 0.04% ethanolic.

Carbon tetrachloride, redistilled, passed dithizone test.

Standard stock copper solution, dissolve 0.1 g copper in 15 ml 1 + 4 nitric acid, (use B.D.H. lead-free reagent for foodstuff analysis grade nitric acid), boil gently to expel nitrous fumes, cool and dilute to 100 cm^3 with distilled water.

Dilute copper solution for calibration (10 ug cm^3), dilute 5 cm^3 stock copper to 500 cm^3 as required. Use only on day of analysis.

Magnesium nitrate solution 7%. Dissolve 70 g of hydrated magnesium nitrate, $(Mg(No_3)_2 6H_2O)$ Analar in 1 litre 30:70 v/v absolute ethanol: distilled water.

Sulphuric : nitric acid mixture 1:3. Mix one volume of concentrated sulphuric acid with three volumes of concentration nitric acid (both B.D.H. lead-free reagents for foodstuff analysis grade).

72.4 METHOD

A known weight of the polymer is mixed with an aqueous alcoholic solution of magnesium nitrate. The mix is then ashed and carbon removal is completed by wet digestion with a mixture of nitric and sulphuric acids.

The acid digest is taken up in water and suitable complexing agents, which prevent interference by other metals are added. Copper is then extracted from this solution with carbon tetrachloride, as the coloured copper diethyl dithiocarbamate complex, and the colour is evaluated spectrophotometrically at 432 nm.

72.5 EXPERIMENTAL PROCEDURE

72.5.1 Ashing Procedure.

Weigh out 10 g sample into a silica crucible. Pipette in 10 ml magnesium nitrate solution. Stir the mixture with a thin glass rod to ensure thorough contact between the polymer and the liquid. The addition of a further 5 cm^3 of distilled water may be necessary at this stage. Heat strongly on a bunsen until most of the carbon is removed. Allow the crucible to cool. Run 2 cm^3 of 1:3 sulphuric-nitric acid mixture down the wall of the crucible using a thin glass rod to transfer any remaining carbon to the bottom of the crucible. Heat again until sulphuric acid fumes are produced, do not heat to dryness. Allow to cool and repeat the treatment with a further 2 ml of sulphuric - nitric acid mixture. Dilute the contents of the crucible with 10 cm^3 of cold distilled water and boil gently. Quantitatively transfer this solution, together with hot distilled water crucible washings, to a 50 cm^3 volumetric flask and make up to the 50 cm^3 mark with distilled water. Filter this solution if necessary.

Perform a reagent blank determination in exactly the same manner as described above, omitting only the addition of polymer. Make the blank solution also up to 50 cm^3.

72.5.2 Determination of Copper.

Transfer a suitable volume of the sample extract containing not more than 30 ug of copper and the same volume of the reagent blanks solution to two 50 cm^3 separatory funnels. Pipette 10 cm^3 of EDTA citrate solution and add 2 drops of cresol-red indicator into the sample and blank solutions and then add 1:3 ammonia until the solution becomes purple red in colour. Allow to cool and add 1 cm^3 of sodium diethyldithiocarbamate solution. Pipette in 10 cm^3 carbon tetrachloride and shake for 2 min. Allow the layers to separate, then filter the lower layer through dry filter paper into a 25 cm^3 volumetric flask. Extract the aqueous layer with further 5 cm^3 portions of carbon tetrachloride until no further colour is extracted. Make the combined filtrates up to the 25 cm^3 mark with carbon tetrachloride. Filter this solution through dry filter paper if cloudy.

Determine the optical density of the sample solution approximately 30 min after the addition of sodium diethyldithiocarbamate reagent under the following conditions.

Spectrophotometer conditions:	
Wavelength	432 nm
Cells	4 cm glass
Blank solution	This is the reagent blank which has been through the whole test procedure from the ashing stage.

72.5.3 Preparation of Calibration Graph.

Pipette 40 cm^3 distilled water into several clean 50 cm^3 separatory funnels and pipette in 0.5, 1.0, 2.0, 2.5, 3.0 and 3.5 cm^3 of freshly prepared dilute (10 ug per cm^3) standard copper solution. Include a further separatory funnel containing 40 ml distilled water only as a blank. Continue as described above from the addition of EDTA citrate reagent. Employ the extract obtained in the reagent blank determination in the comparison cell. Construct a graph relating these optical densities to the number of micrograms (ug) of copper present in the calibration solutions.

72.6 TYPICAL RESULTS

Calculations. Refer the optical density given by the carbon tetrachloride extract obtained from the sample to the copper calibration graph and read off the number of micrograms of copper present in 25 cm^3 of carbon tetrachloride solution. Calculate the copper content of the polymer sample from the following equation:

$$\text{Copper, ppm} = \frac{n \times 10^5 \times 50}{10^6 \times V} = \frac{5N}{V}$$

assuming of 10 g of polymer is ashed, where:

N = number of micrograms ug of copper present in 25 cm^3 of carbon tetrachloride solution.

V = volume (cm^3) of aqueous solution taken for extraction with carbon tetrachloride from the 50 cm^3 dissolved polymer ash solution.

The results in Table 7.39 show that distinctly higher copper determinations are obtained on polyolefins by the procedure involving the use of an ashing aid than are obtained without an ashing aid or by the use of a molten potassium bisulphate fusion technique to take up the polymer ash.

72.7 DISCUSSION OF RESULTS

This method is capable of determining down to 0.1 copper in polyalkenes with an accuracy of $\pm$ 5%. Bismuth is the only metal known to interfere in the determination of copper by the described procedure.

Table 7.39 - Determination of Copper in Polyolefins

Type of Polymer	Original Copper Content of Polymer Obtained by Proposed Method (10g sample) Described under Procedure	Known Synthetic Addition of Copper to Polymer	Total Expected Copper Content of Polymer (calculated)	Total Determined Copper Content of Polymer ppm					
				Obtained by Proposed Method Described under Procedure (10g sample)		Obtained by Direct Ashing without Ashing Aid aFusion b		Obtained by Potassium Bisulphate	
	ppm	ppm	ppm	ppm	% Rec.	ppm	% Rec.	ppm	% Rec.
Polyethylene	1.1	0.0	1.1	0.9, 1.0, 1.1	90	0.4	36	0.5	45
Polyethylene	1.1	1.5	2.6	2.5	96	0.6	22	1.0, 0.7, 0.5	28
Polypropylene	2.7	0.0	2.7	2.7, 2.6, 2.5	96	1.4, 0.9, 1.1	42	0.4	14
Polypropylene	2.7	1.5	4.2	4.2	100	1.4	33	2.3, 2.8, 2.0	56
Ethylene-propylene	4.8	0.0	4.8	4.8, 4.6, 4.5	96	2.0, 1.4, 1.6	35	0.6	13
copolymer	4.8	1.5	6.3	6.2	98	3.5	55	2.7	43

(a) 10g polymer ashed in silica at 600°C. Ash dissolved in 2 ml concentrated nitric-sulphuric acid mixture (3:1) and evaporated gently to dryness prior to dissolving in water.
(b) 10g polymer ashed at 600°C. Ash fused with 2.5g potassium bisulphate and residue dissolved in water.

METHOD 73 - DETERMINATION OF TRACES OF ARSENIC IN ACRYLIC FIBRES CONTAINING ANTIMONY TRIOXIDE FIRE RETARDENT AGENT. ACID DIGESTION ATOMIC ABSORPTION SPECTROMETRY.[46]

73.1 SUMMARY

This atomic absorption spectrometric method determines down to 0.02% arsenic trioxide fire retardent in acrylic fibres.

73.2 APPARATUS

Atomic absorption and flame-emission spectrophotometer, 10 cm slit burner, equipped with a arsenic measurement unit or equivalent and an arsenic hollow-cathode lamp. A schematic diagram of the apparatus used for the arsenic measurement is shown in Figure 7.16.

73.3 REAGENTS

All reagents used were of analytical reagent grade.

Arsenic III standard solution.

1000 ug l^{-1}. Dissolve 132.0 mg of arsenic (III) oxide (purity 99.9%) in 2 cm^3 of 4% m/v sodium hydroxide solution. Add dilute sulphuric acid until the solution becomes slightly acidic and then dilute to 100 cm^3 with distilled water.

Arsenic (V) standard solution, 1000 ug l^{-1}. Oxidize the above arsenic (III) standard solution with a sufficient amount of hydrogen peroxide and evaporate the excess by boiling. Cool the resulting solution and dilute it to the required volume with distilled water.

Nitric acid, concentrated, Analar.

Perchloric acid, concentrated, Analar.

Sulphuric acid, concentrated, Analar.

Benzene, Analar.

Zinc tablet.

Tablet comprising 1 g of zinc powder containing a binder.

Potassium iodide, 20% u.v aqueous made from Analar solid, prepared weekly.

Titanium (III) solution 10% m/V.

Dissolve 50.0 g of titanium (III) chloride in concentrated hydrochloric acid* and dilute to 500 cm^3 with concentrated hydrochloric acid. Store this solution in a refrigerator.

* Hydrochloric acid to have an arsenic content less than 0.5 ugl^{-1}.

73.4 METHOD

A concentrated nitric acid - perchloric acid - sulphuric acid digest of the polymer is analysed at 193.7 nm by atomic absorption spectrometry. The procedure is calibrated against standards of arsenic free acrylic fibre and pure arsenic trioxide.

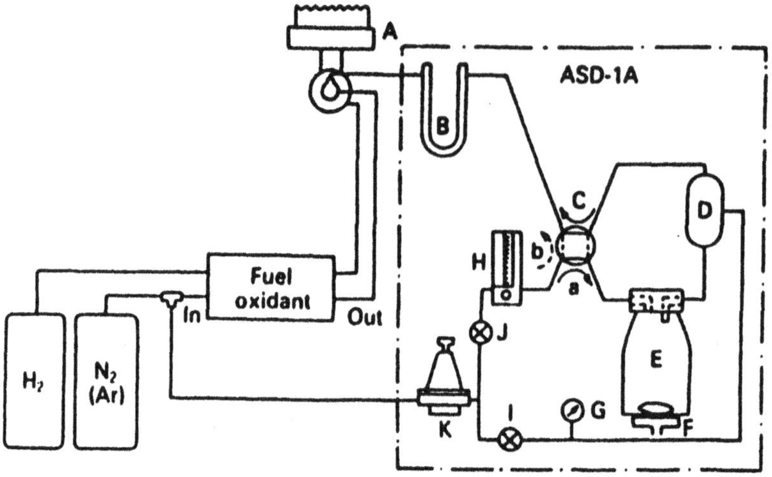

Figure 7.16. Schematic diagram of apparatus used for the arsenic measurement. A, slit burner; B, trap; C, change-over cock; D, buffer tank; E, arsine generator (glass reaction bottle); F, magnetic stirrer; G, pressure gauge; H, gas flow meter; I, back-sweep cock; J, stock-cock; and K, pressure controller. a. sweep; and b. by-pass. From Korenaga with permission.[46] Royal Society of Chemistry, London

73.5 EXPERIMENTAL PROCEDURE

73.5.1 Preparation of Sample Solution.

Weigh 1 g of sample (to contain less than 2 ug of arsenic) into a Kjeldahl decomposition flask. Add 3 cm^3 of concentrated nitric acid, 3 cm^3 of concentrated perchloric acid and 3 cm^3 of concentrated sulphuric acid. Heat the flask on an electric hot plate (about 1200 W) until the acrylic fibre sample is completely decomposed. If the sample cannot be decomposed completely, add a further 3 cm^3 of concentrated nitric acid and 3 cm^3 of concentrated perchloric acid to the flask and repeat the heating until the solution becomes clear, cool the resulting solution. After the digestion of the sample the amount of sulphuric acid remaining is about 3 cm^3.

Transfer the solution into a 50 cm^3 separating funnel and rinse the Kjeldahl decomposition flask with 7 cm^3 of concentrated hydrochloric acid. Add 5 cm^3 of 10% m/V titanium (III) chloride in concentrated hydrochloric acid solution then leave to stand for 30 min at 60°C in a waterbath (the funnel must occasionally be shaken and frequently de-gassed in order to avoid an explosion). Cool the solution to room temperature, add 10 cm^3 of benzene and shake the funnel for a few minutes. Discard the aqueous phase, add 10 cm^3 of distilled water and shake the funnel for a few min. In this way, arsenic (V) in the digested sample is reduced to arsenic (III), and only the arsenic (III) is extracted into benzene and then back-extracted quantitatively into the aqueous phase.

73.5.2 Atomic Absorption Spectrophotometry.

Transfer by pipette the aqueous solution containing arsenic III (not more than 2 ug of arsenic) obtained by the above procedure into a glass reaction bottle (E in Figure 7.16). Add 5 cm^3 of concentrated hydrochloric acid, 1 cm^3 of 20% m/V potassium iodide solution and 0.5 cm^3 of 20% m/V tin (II) chloride solution. Dilute the solution with distilled water to give a final volume of 25 cm^3 (final concentration of hydrochloric acid 2.4 M). Mix well and allow to stand for 20 min. Drop a zinc tablet containing about 1 g of zinc powder into the reaction bottle (E) and immediately connect the bottle to the collection tank (D). Store the generated arsine gas obtained by mixing the sample solutions with a magnetic stirrer (F) to ensure complete arsine generation in the 100 cm^3 collection tank (D) for about 1 min (until the pressure of arsine generation gas reaches 0.5 kg cm^{-2}). As soon as the pressure of arsine generation gas (H) reaches 0.5 kg cm^{-2} switch the stopcock (C) from the by-pass (b) and introduce the collected arsine into the flame with a stream of nitrogen carrier gas. Use the peak height on the recordings to determine the concentration of arsenic.

According to the analytical conditions given above, arsenic (III) obtained in the aqueous solution was determined at the 193.7 nm absorption line by using arsine generation and atomic absorption spectrophotometry (Figure 7.16). A calibration graph was prepared by using 0.25 - 2.0 ug of standard arsenic (III) solution throughout the procedure for the preparation of the sample solution and was then used for subsequent determinations of arsenic concentrations.

Arsenic (III) separated into the aqueous phase from the matrix antimony after the benzene extraction and back-extraction with distilled water was determined by using arsine generation atomic-absorption spectrophotometry as described under Atomic Absorption Spectrophotometry. A calibration graph was prepared by analysing a standard arsenic (V) solution and an acylic fibre containing no antimony oxide by the recommended procedure. Although the calibration graph obtained was not a straight line, amounts of arsenic in the range 0.25 - 2.0 ug could be determined with good reproducibility under the conditions given above.

73.6 TYPICAL RESULTS

The results for arsenic obtained with various acrylic fibre samples containing antimony oxide are given in Table 7.40. The relative standard deviations in these determinations were less than 7%.

73.7 DISCUSSION OF RESULTS

This method is capable of determining down to 0.02% arsenic in acrylic fibres with an accuracy of ± 5%. Antimony present in the sample does not interfere.

METHOD 74 - IDENTIFICATION OF PIGMENTS IN POLYMERS. THERMAL ULTRAVIOLET ABSORPTION SPECTROSCOPY.[47]

74.1 SUMMARY

This thermal UV absorption procedure is capable of identifying various coloured pigments in polymers.

74.2 APPARATUS

Graphite furnace atomic absorption spectrophotometer equipped with a flameless atomizer and deuterium lamp.

Argon purge gas.

Table 7.40 - Results for the Determination of Arsenic in Acrylic Fibres Containing Antimony Oxide

Sample no.	Supplier	Content of Sb_2O_5*%	No of Determinations	Arsenic Content ppm
1	A	5.1	4	50 ± 1
2	A	5.1	3	84 ± 4
3	A	4.5	2	94 ± 7
4	A	4.6	5	10.3 ± 0.5
5	A	5.0	5	4.1 ± 0.2
6	B	3.0 (Sb_2O_3)	3	45 ± 0
7	C	-	2	180 ± 11
8	C	-	4	103 ± 6
9	D	2.4 (Sb_2O_3)	2	8.5 ± 0.5
10	E	1.0 (Sb_2O_3)	4	3.2 ± 0.1
11	-	Sb_2O_5 50 mg + acrylic fibre (no Sb) 1g	2	0.47 ± 0.03
12	-	Sb_2O_5 10 mg + acrylic fibre (no Sb) 1g	1	0.08

From Korenaga[46]. Royal Society of Chemistry, London.

* Even when antimony (III) oxide was present in acrylic fibres (samples 6, 9 and 10), antimony (III) was easily and completely oxidised to antimony (V) by the wet digestion with a mixture of nitric, perchloric and sulphuric acids. Hence, arsenic in acrylic fibres containing antimony (III) oxide could be determined as well as that in acrylic fibre samples containing antimony (V) oxide by this method.

74.3 METHOD

Pellets of the polymer sample are introduced into the graphite furnace of an atomic absorption spectrometer and heated to temperatures between 300 and 900°C. The UV absorption bands occuring are examined in the atomic absorption spectrometer and the thermal ultraviolet profiles characteristic of the pigment obtained. Each pigment has a characteristic thermal UV profile at a particular temperature.

74.4 EXPERIMENTAL PROCEDURE

Pigments and organic compounds are investigated directly without dissolution. Powdered samples of the pigments are pressed in pellets about 0.5 mm thick at 10^7 Pa. Small pieces of 0.5 mm x 0.5 mm wide are introduced into the graphite tube through the central sampling hole. The pellet position must be checked carefully as its shift along the tube axis causes remarkable changes in the temperature (about 50 - 70°C) and absorbance (30%) of the thermal ultraviolet profiles. The tube is then inserted into the furnace and the thermal cycle for vapor phase measurements is performed.

The instrumental conditions of the atomic absorption spectrometer are as follows: 200 nm wavelength, 0.7 nm bandwidth, background mode and time constant depending on the selected thermal cycle. Three different thermal cycles are carried out (Table 7.41), the choice of the cycle is based on the type of pigment examined.

Thermal cycle 1 is used for rapid screening, ie. to investigate whether the pigments contain organic or inorganic compounds or both, thermal cycles 2 and 3 are employed respectively for identification or organic and inorganic pigments. During the heating from step 1 to step 2 (thermal cycles 1 and 2) or from step 2 to step 3 (thermal cycle 3) ultraviolet absorption is recorded at the temperature of the furnace to obtain TUV profiles of the vapours.

Purge gas (argon) flow is maintained at 300 cm^3/min during the whole thermal cycle, as its reduction gives rise to condensation of vapours and decomposition products in the cooler parts of the furnace with subsequent memory effect. This effect is noteworthy especially in the case of inorganic pigment identification. On these grounds the heating rate plays the main role in the resolution of TUV profiles.

74.5 TYPICAL RESULTS

A practical example of the identification of pigments is given in Figure 7.17. A 1:1 mixture of organic pigment yellow/(2-nitro-p-toluidine coupled with acetoacetanilide) and inorganic P.Y 34 (lead chromate) was vaporized following thermal cycle 1. The thermal ultraviolet profile shows clearly two absorption bands at about 500°C and 1250°C. The first band is attributable to the vapours which originate from the decomposition and pyrolysis of the organic pigment, the second band corresponds to the decomposition and vaporization of lead chromate at high temperature (mp 844°C). It is possible therefore to determine by a rapid run whether the pigment is a mixture or belongs to the organic or inorganic group.

74.6 DISCUSSION OF RESULTS

This procedure is useful for obtaining qualitative information regarding the types of organic and inorganic pigments in polymers.

METHOD 75 - DETERMINATION OF TRACES OF CHLORINE IN POLYALKENES AND POLYALKENE COPOLYMERS. SODIUM CARBONATE FUSION - TITRATION PROCEDURE.

75.1 SUMMARY

This sodium carbonate fusion - titration procedure is capable of determining chlorine residues in amounts down to 50 ppm in polyalkenes and polyalkene copolymers.

75.2 APPARATUS

Titration burette, 10 ml.
Platinum crucible.

75.3 REAGENTS

Silver nitrate, M/10.
Silver nitrate, M/100.

Table 7.41 - Experimental Conditions for TUV Profile Recording

	Thermal Cycle 1			Thermal Cycle 2				Thermal Cycle 3			
	1	2	3	1	2	3	4	1	2	3	4
Temperature °C	200	2000	2650	150	1000	1500	2650	150	700	2300	2650
Ramp Time,s	10	80	2	10	110	2	2	10	10	40	1
Hold Time,s	25	10	5	25	20	5	5	20	5	1	5
Recorder		*			*					*	
Time Constant,s		0.5			0.5					0.4	

From Tittareli et al[47]. American Chemical Society

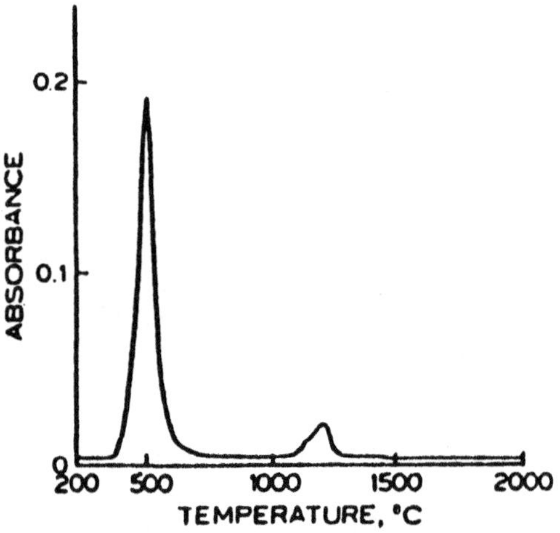

Figure 7.17. TUV profile of (a) 1.1 mixture of organic pigment yellow (2-nitro p-toluidine coupled with acetoacetanilide) and inorganic pigment PY 34 (lead chromate). Thermal cycle temp, °C 200 ramp times, S 10 hold temp, °C 25. From Tittareli et al with permission.[47] American Chemical Society.

Nitric acid, 30% aqueous, dilute 30 cm^3 concentrated nitric acid (M.A.R.) with 100 cm^3 distilled water.

Sodium carbonate, solid microanalytical reagent grade.

Xylene cyanol/methyl orange mixed indicator, alcoholic solution.

Acetone, A.R. grade.

Acid buffer solution: to approximately 200 cm^3 of distilled water, in a 500 cm^3 volumetric flask add 100 cm^3 glacial acetic acid and 6.5 ml concentrated nitric acid (sp gr. 1.42), dilute to the mark with distilled water.

Gelatine solution: add to 250 cm^3 of deionized water, 2.5 g gelatine and 5 g thymol blue. Heat slowly to the boil and stir until solution is complete. Add 0.5 g thymol as preservative and dilute to 500 cm^3 with distilled water. The solution is stable for up to 3 months at room temperature.

75.4 METHOD

In the method polyethylene is mixed with pure sodium carbonate and ashed in muffle furnace at 500°C. The residual ash is dissolved in aqueous nitric acid and then diluted with acetone. This solution is titrated potentiometrically with standard silver nitrate.

75.5 EXPERIMENTAL PROCEDURE

Accurately weigh 5 g of polymer into a platinum crucible and cover the polymer with a layer of 2 g sodium carbonate. Place in a cold muffle furnace and increase the temperature gradually to 500°C, maintaining this temperature for four h. Allow the crucibles to cool then dissolve the residue in 15 - 20 ml distilled water. Transfer this solution with crucible washings to a 100 cm^3 beaker. Adjust the final volume of water to 30 cm^3. Add 5 drops of screened methyl orange indicator solution and neutralise the mixture by dropwise addition of 30% nitric acid to the purple red coloured end point. Add a further 10 drops of 30% nitric acid solution to the beaker and then add 30 cm^3 of acetone. Titrate the resultant solution potentiometrically with silver nitrate solution (M/100) using an automatic titrator equipped with a silver measuring electrode and glass reference electrode. Carry out a blank determination exactly as described above, omitting only the sample addition.

75.6 TYPICAL RESULTS

Calculation:
ppm (w/w) chlorine in polymer

$$= \frac{(T_S - T_B) \times M \times 35.46 \times 10^3}{W}$$

where
T_B = titration of silver nitrate (mls) in blank determination
T_S = titration of silver nitrate (mls) in sample determiantion
M = molarity of silver nitrate solution
W = weight (g) of polymer sample.

Note: If a polyolefin sample does not contain any free residual alkali left in the manufacturing process then there exists a danger that some chlorine will be lost during the ignition process and low chlorine analyses will result. If this is suspected to be the case the polymer sample (50 g) should first be contacted with twice its volume of 2% W/V alcoholic potassium hydroxide solution and left in an oven at 70°C until the polymer is dry. The above method is then applied.

In Table 7.42 are compared chlorine contents obtained by x-ray fluorescence spectroscopy with those obtained by the above method involving fusion at 500°C with sodium carbonate. Analyses were carried out on samples which had been previously treated with alcoholic potassium hydroxide and samples which had not been so treated. It is seen in Table 7.42 that, when carried out on the same sample, this chemical method gives considerably more reproducible duplicate results than those obtained by the x-ray method (compare columns A and C). In the case of alcoholic potassium hydroxide treated samples the average of duplicate analyses carried out by the x-ray and the chemical method agree satisfactorily within ± 15% of each other. In the case of untreated samples the chemical method tends to give lower chlorine contents than the x-ray method.

75.7 DISCUSSION OF RESULTS

This method is capable of determining down to 50 ppm chlorine in polyalkenes and polyalkene copolymers with an accuracy of ± 5%.

Table 7.42 - Comparison of Chlorine Contents by X-Ray Fluorescence and Chemical Method

Polymer not Treated with Alcoholic Potassium Hydroxide Before Analysis, ppm Chlorine					Polymer Treated with Alcoholic Potassium Hydroxide Before ppm Chlorine		
Sample	X-Ray on Discs (avg in brackets) "A"	Chemical Method on Same Disc as used for X-Ray anal. "B"	Chemical Method on Powder (av. in brackets) "C"	Diff. (%) between Av. X-Ray & Chem. Anal. "D" = ("A" - "C") x 100 / "C"	X-Ray on Discs (av. in brackets) "E"	Chemical Method on Same Disc as used for X-Ray anal. "F"	Chemical Method on Powder (av. in brackets) "G"
1	480,540 (510)	456	502,499 (500)	+2	865,815 (840)	700	786,761 (773)
2	365,480 (422)	334	380,355 (367)	+15	535,570 (552)	606	636,651 (643)
3	435,445 (440)	338	349,353 (351)	+25	785,675 (730)	598	650,654 (652)
4	425,570 (497)	392	395,404 (400)	+24	625,675 (650)	600	637,684 (660)
5	390,530 (460)	371	416,399 (408)	+13	895,870 (882)	733	828,816 (822)

METHOD 76 - DETERMINATION OF CHLORINE IN CHLOROBUTYL AND OTHER CHLORINE CONTAINING POLYMERS. OXYGEN FLASK COMBUSTION - TURBIDIMETRY.[48]

76.1 SUMMARY

This oxygen flask combustion turbidimetric method is capable of determining total chlorine in amounts down to 5 ppm in chlorobutyl and other chlorine containing polymers.

76.2 APPARATUS

Conventional 1 and 2 litre Schoniger type oxygen combustion apparatus.

The absorbance measurements were obtained using a spectrophotometer with a 1 cm cell and a tungsten lamp as the energy source, or equivalent.

76.3 REAGENTS

Standard aqueous potassium chloride solution in demineralized water to contain 1000 ppm chloride. Dilutions made from this standard stock solution to those containing very low chloride levels.

Silver nitrate, 0.01 M aqueous.

Nitric acid, 0.01 M potassium nitrate 0.01 M aqueous.

76.4 METHOD

A weighed sample is combusted in an oxygen filled flask over dilute nitric acid. Silver nitrate solution is then added and the resulting silver chloride, ie. chlorine, estimated turbidimetrically at 420 nm using a grating spectrophotometer.

76.5 EXPERIMENTAL PROCEDURE

The preparation of the calibration curve involves the preparation of six standard chloride solutions to contain 0 to 4 ppm chloride. Acidified 0.01 M silver nitrate is added to each solution to form a silver chloride suspension. All the solutions are swirled briefly and allowed to stand at least 35 minutes with occasional shaking. The absorbance of each of these solutions is then measured at 420 nm in a 1 cm cell. A straight line curve passing through the origin is obtained covering chloride concentrations from 0 to 4 ppm. The ideal absorbance is approximately in the range of 0.03 to 0.2 absorbance units.

During the preparation of standards swirling each solution for a few seconds is sufficient. Prolonged agitation tends to agglomerate the silver chloride particles.

Generally, about 100 mg of a weighed sample are combusted in a 2 litre Schoniger flask containing 10 ml of 0.01 M nitric acid. After the combustion, the flask is allowed to stand at least 15 minutes with occasional shaking before 5 ml of 0.01 M silver nitrate solution are added.

If the polymeric sample is difficult to combust, the normal platinum sample basket may be wrapped with 52 mesh platinum gauze to prevent hot and partially burned sample from dripping out of the sample basket. For subsequent combustions of this sample, a smaller sample size and the use of twice the normal volume of nitric acid absorbent are recommended.

The measurement of the absorbance of the silver chloride suspensions should be done within 35 - 60 minutes of turbidity development of both the sample and standard chloride solutions. The absorbances of the standard solutions are measured first, followed by the sample solutions, then the standard solutions are measured again.

The calculation of results is made in two different ways. For samples of low halogen level and if the combustion results in a clear solution, no removal of a solution aliquot is necessary and the calculation is as follows:

$$\% \text{Cl} = \frac{\text{ppm chloride} \times 15 \times 100}{\text{sample weight (mg)} \times 1000}$$

The value for ppm chloride is obtained from the calibration curve.

If combustion of the sample is such that a clear solution is not obtained and an organic film results in the combustion flask, an 8 ml aliquot is usually removed and only 4 ml silver nitrate is added to the clear aliquot. Here again, the ratio of sample solution aliquot volume to that of silver nitrate is still 1:1 and the calculation for chloride is the same as above.

76.6 DISCUSSION OF RESULTS

This method is capable of determining down to 5 ppm total chlorine in chlorine containing polymers with an accuracy of ± 5%.

METHOD 77 - DETERMINATION OF UP TO 80% CHLORINE, BROMINE AND IODINE IN POLYMERS. OXYGEN FLASK COMBUSTION - TITRATION PROCEDURE.

77.1 SUMMARY

This oxygen flask combustion - titration procedure is capable of determining between 2 and 80% of total chlorine, bromine or iodine in polymers.

77.2 APPARATUS

Combustion flask - Pyrex, 500 ml capacity conical flask with B24 conical ground stopper.

Stopper B24 - with a fixed in platinum wire (30 mm long, 0.8 mm diameter) carrying a 15 x 20 mm piece of 40 mesh platinum gauze or carrying a 5/8" long x 1/4" diameter bucket fabricated in 40 mesh platinum gauze.

Safety jacket - for combustion flask to serve as a protection during the combustion. Conical shaped, detachable metal wire gauze fitting around the conical flask. (Figure 7.18).

Automatic titrimeter - with silver/glass electrode system. Wick lighter, small burner fed with sulphur and halogen free fuel.

Cellulose capsules - halogen free.

77.3 REAGENTS

Silver nitrate solution standard, 0.1 N.

Sodium hydroxide solution, 0.01 N approximately, dissolve 0.4 g of sodium hydroxide in distilled water and dilute to 1 litre.

Sodium metabisulphite solution, 0.05% w/v, (Micranalytical Reagent Grade, available from British Drug Houses Ltd., Poole, Dorset) dissolve 0.05 g of sodium metabisulphite in distilled water and dilute to 100 ml.

Barium nitrate, Analar, solid.

Oxygen, cylinder.

Nitric acid, Analar, concentrated.

Methyl alcohol, Analar.

Acetone, pure grade, redistilled.

Water, deionized halogen content less than 0.05 ppm.

Filter paper, any grade with a halogen content less than 50 ppm.

77.4 METHOD

The sample is burnt in oxygen in a sealed Schoniger flask and the resultant volatiles are absorbed in a suitable absorption solution. This solution is acidified and titrated potentiometrically with standard silver nitrate solution.

77.5 EXPERIMENTAL PROCEDURE

77.5.1 Combustion Technique

The method requires different absorption solutions for the various halogens as follows:

Chlorine	10 ml of 0.01 N sodium hydroxide solution.
Bromine and/or iodine	10 ml 0.05% w/v solution metabisulphite solution.
Chlorine and bromine and/or iodine	5 ml 0.01 N sodium hydroxide solution and 5 ml 0.05% w/v sodium metabisulphite solution.

Pipette the required absorption solution into the flask and fill with oxygen, stopper securely. Accurately weigh out 10 - 30 mg of sample into a cellulose capsule and place the capsule in the Schoniger basket with a strip of filter paper, to act as a fuse. Light the fuse and quickly insert the basket into the oxygen filled flask. During the combustion hold the basket in the flask firmly and at the same time lift the flask off the bench. Allow the mist formed in the flask to subside with periodic shaking over 15 to 30 minutes. Transfer the solution to a 250 ml beaker using 50 ml of distilled water and 120 mg of methanol. (If the halogen content of the sample is less than 5% then 120 ml acetone should be used instead of methanol).

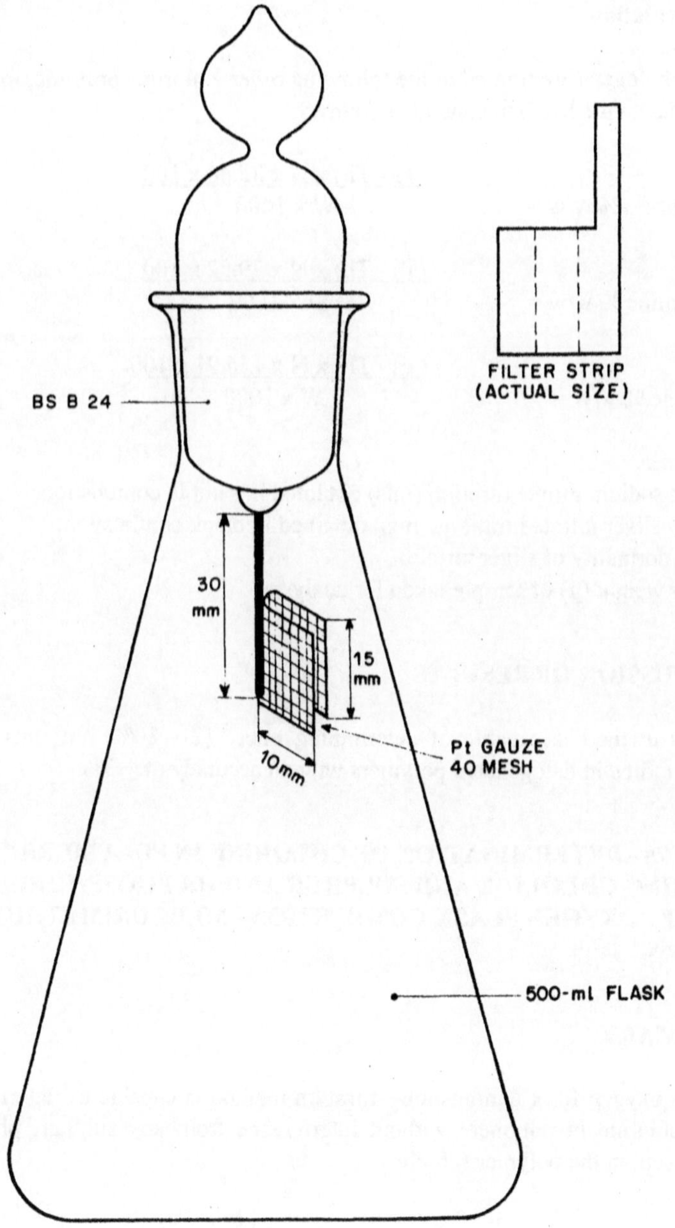

Figure 7.18. Stopper conical flask with platinum wire gauze to suspend the sample.

If any carbonaceous matter is evident in the solution then the determination should be abandoned and a further combustion carried out using a smaller sample weight.

Carry out a reagent blank combustion including the paper and all reagents, but omitting the sample.

Add 1 drop of concentrated nitric acid and, if a mixture of halogens is present, a few crystals of barium nitrate. Titrate the solution potentiometrically using 0.1 N silver nitrate solution.

77.5.2 Calculations.

The halogens are titrated in the following order: chlorine, bromine, iodine. Calculate the halogen content as follows:

$$\text{Chlorine \% w/w} = \frac{(Ts - Tb) \times N \times 35.46 \times 100}{W \times 1000}$$

$$\text{Bromine \% w/w} = \frac{(Ts - Tb) \times N \times 79.92 \times 100}{W \times 1000}$$

$$\text{Iodine \% w/w} = \frac{(Ts - Tb) \times N \times 126.91 \times 100}{W \times 1000}$$

where:
Ts = sodium nitrate titration (mls) obtained in sample combustion.
Tb = silver nitrate titrations (mls) obtained in blank combustion.
N = normality of silver nitrate.
W = weight (g) of sample taken for analysis.

77.6 DISCUSSION OF RESULTS

This method is capable of determining micro (2 - 80%) amounts of chlorine, bromine or iodine in halogenated polymers with an accuracy of $\pm$ 3%.

METHOD 78 - DETERMINATION OF CHLORINE IN POLYMERS CONTAINING CHLORIDE AND SULPHUR AND/OR PHOSPHORUS AND/OR FLUORINE. OXYGEN FLASK COMBUSTION - MERCURIMETRIC TITRATION.

78.1 SUMMARY

This oxygen flask combustion - titration method is capable of determining down to 5 ppm chlorine in polymers without interference from any sulphur, phosphorus or fluorine present in the polymer sample.

78.2 APPARATUS

Oxygen flask - 500 ml capacity, see Figure 7.18.

A conical flask provided with B24 ground glass stopper, sealed at the lower end to a glass tube 5 mm in external diameter. The lower end of this tube is sealed and flattened to form a slight flange. The length from the lower edge of the ground portion of the stopper to the flanged end of the tube is 70 mm. The sample holder is constructed from 80 mesh platinum gauze in the form of a cylinder 6 mm in diameter and about 10 mm long. It is closed at one end and fastened to the flanged tube by a length of 0.5 mm diameter wire, one end of which is bent over the edge of the open end of the cylinder and pinched so that the gauze is firmly gripped. The arrangement has significant advantages over the usual type of sample holder. The platinum gauze and wire can be renewed with ease and the thermal capacity of the assembly is relatively small.

Magnetic stirrer.

Content pipette, 2.0 ml capacity - conforming to B.S. 1428.

Microburette, 10 ml capacity - conforming to B.S. 1428.

78.3 REAGENTS

Sodium hydroxide, 0.1 N.

Nitric acid, 0.1 N.

Bromophenol blue indicator solution, dissolve 50 mg of bromophenol blue in 500 ml of ethanol.

Diphenylcarbazone indicator solution, dissolve 20 mg of diphenylcarbazone in 20 ml of ethanol. Keep the solution in the dark and renew it after 2 weeks.

Sulphur dioxide solution, saturate 50 ml of water with sulphur dioxide. Renew it after 2 to 3 days.

Hydrogen peroxide, 100 volume, micro-analytical reagent grade.

Barium nitrate solution, a saturated aqueous solution of analytical reagent grade barium nitrate.

Thorium nitrate solution, dissolve 15.0 g of analytical reagent grade thorium nitrate tetrahydrate in water and dilute the solution to 1000 ml.

Ethanol, absolute.

Mercuric nitrate, approximately 0.005 M, dissolve 3.5 g of mercuric nitrate in 570 ml of 0.01 N nitric acid and set the solution aside for at least 2 days. Filter off any precipitate and dilute the filtrate to 2 litres.

Sodium chloride, 0.025 M.

Standardisation of the mercuric nitrate solution. Measure out 2.0 ml of 0.025 M sodium chloride solution by means of a content pipette. Dilute to 15.0 ml with water and add 0.05 ml of bromophenol blue indicator solution. Add 0.1 N nitric acid until the yellow colour of the indicator appears and then add a further 0.5 ml of acid. After the addition of 100 ml of ethanol and 0.5 ml of diphenylcarbazone indicator solution, titrate the solution with the mercuric nitrate solution to the first appearance of a permanent voilet colour. The solution should be stirred magnetically throughout the titration period. The use of a content pipette for measuring the sodium chloride solution is recommended, since it has been found that the reproducibility of repeated titrations is better than that obtained with, for example, a 5.0 ml delivery pipette, in conjunction with 0.01 M sodium chloride solution. The titration carried out as described must be corrected for the indicator blank value, obtained by the titration of 15.0 ml of water. This blank value is usually less than 0.05 ml of mercuric nitrate solution but much higher values have been obtained occasionally and these have been found to be due to the presence of chloride in the ethanol. In such instances, a suitable stock of ethanol may be mixed with a predetermined volume of the mercuric nitrate solution, sufficient to reduce the blank value to less than 0.05 ml.

78.4 METHOD

The polymer sample is combusted in an oxygen filled flask over water, after neutralization to the bromophenol blue end-point and the addition of thorium nitrate, diphenyl carbazide and ethanol the solution is titrated to the violet end-point with standard mercuric intrate solution:

$$2H\,Cl + Hg\,(NO_3)_2 = HgCl_2 + 2HNO_3 \qquad 7.15$$

The chlorine content of the polymer can then be calculated from the consumption of standard mercuric nitrate solution. Modifications to the method are described for overcoming interferences by any sulphur, phosphorus or fluorine present in the polymer.

78.5 EXPERIMENTAL PROCEDURE

The procedures described below are restricted to the determination of chlorine in organic compounds containing phosphorus or fluorine, or both, in the presence or absence of sulphur. For compounds containing little if any hydrogen, a preferred alternative procedure is given. The determination of iodine or bromine in the presence of phosphorus or fluorine is not included since the determination of these halogens is in this instance best completed by other well known volumetric methods, such as Leipert's method for iodine and oxidation to bromate for bromine.

78.5.1 Compounds Containing Phosphorus or Fluorine, or Both.

Weigh out an amount of sample corresponding to 1.0 to 2.0 mg of chlorine and decompose it in an oxygen flask containing 5.0 ml of water. Wash the stopper and sample holder with 10.0 ml of water, add 0.1 ml of bromophenol blue indicator solution and neutralise the solution with 0.1 N sodium hydroxide to the blue colour of the indicator. Add 0.75 ml of thorium nitrate solution, 0.5 ml of diphenylcarbazone indicator solution and 100 ml of ethanol and titrate the stirred solution with standard mercuric nitrate solution to the appearance of a permanent violet colour.

78.5.2 Compounds Containing Sulphur and Either Phosphorus or Fluorine or all Three Together:

Proceed as before up to the washing of the stopper, etc. Boil the solution for 60 seconds, holding the flask directly over a small bunsen burner flame and swirling the solution continuously. Add 0.5 ml of barium nitrate solution, 0.1 ml of bromophenol blue indicator solution and sufficient 0.1 N sodium hydroxide to produce the blue colour of the indicator. After the addition of thorium nitrate, complete the determination as before.

78.5.3 Compounds Containing Phosphorus or Fluorine or Both, and Little Hydrogen:

Decompose the weighed sample in an oxygen flask containing 5.0 ml of water and 0.25 ml of saturated solution of sulphur dioxide. Wash the stopper etc. with 10.0 ml of water and complete the determination as described for the analysis of compounds containing sulphur. In this procedure the volume of barium nitrate should be increased if 0.5 ml is found to be insufficient for complete precipitation of the sulphate.

Blank determinations on the reagents, etc. should be carried out in conjunction with each of the procedures described above. These blank values should not exceed about 0.2 ml of 0.005 M mercuric nitrate.

78.6 DISCUSSION OF RESULTS

The method has a precision of approximately 0.1 ppm at the 22% chlorine level.

METHOD 79 - DETERMINATION OF FLUORINE IN FLUORINATED POLYMERS. OXYGEN FLASK COMBUSTION - SPECTROPHOTOMETRIC PROCEDURE.[49]

79.1 SUMMARY

This oxygen flask combustion spectrophotometric procedure is capable of determining fluorine in fluorinated polymers such as PTFE in amounts down to 0.5%.

79.2 APPARATUS

The combustions apparatus consists of silica or boronfree glass 500 ml Erlenmeyer flask constructed of suitable glass, (see Method 76). Into the stopper is fused one end of a length of platinum wire, 1 mm in diameter, to the free end of which is attached a piece of 36 mesh platinum gauze, 1.5 cm x 2 cm to act as a sample holder. The flask shall be essentially free from boron and aluminium. Low results are obtained using borosilicate glass flasks.

Optical densities were measured in 4 cm cells with a Unicam SP600 visual range spectrophotometer.

79.3 REAGENTS

Alizarin complexan, 0.0005 M, transfer 0.385 g of alizarin complexan to a 2 litre calibrated flask by means of 20 ml of recently prepared 0.5 N sodium hydroxide and set aside for 5 minutes with occasional swirling, to ensure complete solution. Dilute to about 1500 ml with water, add 0.2 g of hydrated sodium acetate and adjust the pH to about 5 by careful addition of 1 N hydrochloric acid. Dilute to the mark and filter into a brown glass bottle. This solution is stable for at least 4 months.

Cerous nitrate, 0.0005 M, standardise approximately 0.02 M cerous nitrate by titration against standard ethylenediaminetetra-acetic acid solution at pH 6 with xylenol orange as indicator. To a suitable volume of this solution (about 50 ml) add 0.2 ml of concentrated nitric acid, 0.1 g of hydroxylamine hydrochloride and sufficient water to produce 2 litres and filter.

Acetate buffer solution, pH 4.6, dissolve 150 g of hydrated sodium acetate in about 600 ml of water, add 75 ml of glacial acetic acid, dilute to 1 litre with water and filter.

Standard fluoride solution, 5 ug per ml, dissolve about 22 mg (accurately weighed) of dried analytical reagent grade sodium fluoride in water and adjust the volume to 2 litres. Store in a polythene container.

79.4 METHOD

The method for the determination of fluorine is based on decomposition of the sample by oxygen flask combustion followed by spectrophotometric determination of the fluoride produced by a spectrophotometric procedure involving the reaction with the cerium III complex of alizarin complexan (1,2-dihydroxy-anthraquinone 2-ylmethyl-amine N, N-diacetic acid). The blue colour of the fluoride containing complex (absorption maximum, 565 nm) is completely distinguishable from either the yellow of the free dye (maximum absorption, 423 nm) or the red of its cerium III chelate (maximum absorption, 495 nm).

79.5 EXPERIMENTAL PROCEDURE

79.5.1 Preparation of Calibration Graph.

In each of a series of 100 ml calibrated flasks place 50 ml of distilled water, an accurately measured volume between 2 and 8 ml of standard fluoride solution, 10 ml of alizarin complexan solution and 3 ml of acetate buffer solution. Mix each solution thoroughly, add 10 ml of 0.0005 M cerous nitrate, dilute to the mark with distilled water and set aside, protected from direct light for 1 hour. At the same time, prepare a blank solution in a similar fashion by omitting the standard fluoride solution. Measure the optical densities of the test solutions against the blank in 4 cm cells at 610 nm and plot a graph of optical density against amount of fluoride present.

79.5.2 Procedure.

Accurately weigh an appropriate amount of the sample (5 to 25 mg) on a strip of filter paper approximately 3 cm x 4 cm (Whatman No. 1 grade is suitable) that has been folded into three along its length. Enclose the sample in the paper, insert a narrow strip of filter paper to act as a fuse and fix it in the platinum gauze sample holder. Place 20 ml of water in the combustion flask, fill the flask with oxygen, ignite the fuse and immediately insert the stopper. Carefully tilt the flask and when combustion is complete, set it aside the 10 minutes, with intermittent shaking. Quantitatively transfer the liquid to a 250 ml calibrated flask, dilute to the mark and treat an aliquot expected to contain about 25 mg of fluoride by the procedure described for colour development under Preparation of Calibration Graph. At the same time prepare a standard colour from 5 ml of standard fluoride solution to serve as a check on the calibration graph.

Liquid samples can be satisfactorily decomposed by burning in a small gelatin or preferably, methylcellulose capsule containing approximately 30 mg of powdered cellulose.

For solutions derived from the combustion of sulphur containing compounds, boil gently for about 10 seconds with 1 ml of 100 volume hydrogen peroxide, neutralise to phenolphthalein with 1 N sodium hydroxide and then add 1 ml in excess; boil to destroy excess of peroxide, cool and adjust the pH to about 4 with 1 N hydrochloric acid.

79.6 TYPICAL RESULTS

Using this method Johnson and Leonard[49] obtained from polytetrafluoroethylene a content of 75.8% fluorine using a silica or boron-free glass combustion flask against a theoretical value of 76%. Using a boroscilicate glass combustion flask they obtained a low fluorine recovery of 72.1%.

79.7 DISCUSSION OF RESULTS

This procedure is capable of determining down to 0.5% total fluorine in fluorinated polymers with an accuracy of $\pm$ 5%.

METHOD 80 - DETERMINATION OF SULPHUR IN POLYMERS. OXYGEN FLASK COMBUSTION - DIRECT TITRATION PROCEDURE.

80.1 SUMMARY

This oxygen flask combustion - titration method is capable of determining total sulphur in amounts down to 50 ppm in sulphur containing polymers.

80.2 APPARATUS

Conical flask - Pyrex or Jala glass, 500 ml capacity, provided with conical ground joint BS-B 24, (see Method 76).

Stopper, BS-B 24, with fused in platinum wire (30 mm long, 0.8 mm diameter) carrying a 15 x 20 mm piece of 40 mesh platinum gauze.

Safety jacket, to serve as a protection during the combustion. Detachable metal wire gauze jacket fitting around the conical flask.

Syringe - 0.05 ml capacity, with 0.001 ml divisions.

Lighter - any suitable small flame fed with sulphur-free fuel, eg. alcohol.

Micro burette assembly, with 10 ml burette and Pyrex supply bottle, two required.

Ultramicro burette, with 0.001 ml divisions and provided with a glass capillary delivery tube.

Photoelectric colorimeter equipped with two 100 ml optical cells and two 520 nm interference filters in rim, or equivalent.

Electric stirrer, of suitable dimensions to fit on and close the colorimeter compartment, provided with small glass propeller-shaped stirrer.

Filter paper for sample wrapping. Any grade with sulphur content less than 100 ppm eg. Whatman No. 41 or 42. Cut out paper, sized and shaped according to Figure 7.18. Fold them along the dotted line to the U shape and store in a closed bottle to protect from any sulphur present in the atmosphere. As the paper contains a small amount of sulphur it is necessary to ensure that the same weight (30 to 40 mg) is used in the sample and the blank combustions.

80.3 REAGENTS

Barium perchlorate standard solution 0.0005 M, dissolve 0.17 g barium perchlorate in 200 cm^3 of deionized water and make up to 1 litre with redistilled isopropanol. Adjust to pH 3.5 by additions of 10% perchloric acid solution. Standardize in an optical cell against sulphuric acid using 1 cm^3 of isopropanol. Add 1 drop (approx.

0.05 cm^3) of perchloric acid solution 3% and 200 ± 1 ul of 0.07% w/v Thorin indicator solution and proceed as described under Procedure. Run a blank titration in the same way using exactly the same amounts of reagents but omitting the sulfuric acid and applying 20 ± 0.5 cm^3 of deionized water instead of 10 cm^3. Obtain the net normality of the barium perchlorate solution.

Barium perchlorate solution standard 0.005 M.

Dissolve 1.7 g barium perchlorate in 200 cm^3 of deionized water and make up to 1 litre with redistilled isopropanol. Adjust to pH 3.5 by additions of 10% perchloric acid. Standardize in an optical cell against sulfuric acid using 10 cm^3 of standard 0.005 M sulphuric acid solution. Dilute with 10 ± 0.5 cm^3 of deionised water and add 80 ± 2 cm^3 of isopropanol. Add 1 drop (approx 0.05 cm^3) of perchloric acid solution 3% and 200 ± 1 ul (1 ul = 0.001 cm^3) of 0.02% w/v Thorin indicator solution and proceed as described under Procedure. Run a blank titration in the same way using exactly the same amounts of reagents but omitting the sulfuric acid and applying 20 ± 0.5 ml of deionized water instead of 10 cm^3. Obtain the net normality of the barium perchlorate solution.

Hydrogen peroxide: 100 volume Microanalytical reagent grade.

Isopropyl alcohol, redistilled.

Oxygen, free from sulphur compounds.

Perchloric acid, 0.01 M microanalytical reagent grade in glacial acetic acid.

Sulphuric acid, 0.005 M prepared by accurate dilution of 0.05 M acid.

Sulphuric acid, 0.00025 M, prepared by accurate dilution of 0.005 M acid. Prepare the 0.0005 N reagent at least once daily as required.

Thiorin indicator solution: 0.02% aqueous Thorin (sodium salt of 2(2-hydro-3,6-disulpho-1-napthylazo) benzene arsonic acid).

Water, deionised, sulphate content less than 0.05 ppm.

80.4 METHOD

The sample is wrapped in a piece of filter paper and burnt in a closed conical flask filled with oxygen at atmospheric pressure. The sulphur dioxide produced in the reaction reacts with dilute hydrogen peroxide solution contained in the reaction flask to produce an equivalent amount of sulphuric acid which is estimated by visual titration with 0.005 M or 0.0005 M barium perchlorate using Thorin indicator.

80.5 EXPERIMENTAL PROCEDURE

80.5.1 Sample Combustion.

The weight of sample taken should be sufficient to produce a titration of about 4 cm^3 of 0.005 or 0.0005 M barium perchlorate. The maximum permissible sample weight for complete combustion is 30 mg. Weigh a blank piece of filter paper of similar size and check that its weight is within ± 5% of the weight of the paper used in the sample analysis. Wrap the sample up in the following way. First cover the sample with the raised edges of the paper strip and then roll up the body of the strip towards the narrow strip at the end, which will serve as a fuse. Now clamp the packet in the platinum gauze or bucket on the flask stopper, keeping the fuse free from and in line with the platinum suspension wire. Repeat this operation with the blank piece of filter paper. Pipette 4.0 ± 0.5 ml deionised water and 0.15 cm^3 hydrogen peroxide solution into a clean combustion flask. Place the flask in the safety jacket. Replace the air in the conical flask with a rapid stream of oxygen. Ignite the end of the fuse of the sample packet and immediately insert the sample into the combustion flask. Keep the flask firmly closed and keep it upside down for a minute. Shake the flask for one min and allow it to stand for 15 min. When all the mist has disappeared, wet the rim of the flask with isopropyl alcohol, carefully open the flask and transfer the flask contents quantitatively to a 100 cm^3 beaker by means of 65 cm^3 isopropyl alcohol and 12 cm^3 water. To the beaker add 2 cm^3 0.02% Thorin indicator, 3 drops of 0.05 M perchloric acid and 1 cm^3 of 0.00025 M sulphuric acid (by pipette). Carry out a blank combustion including the paper and all the reagents but omitting the sample.

Titrate the sample and blank solutions with 0.005 M or 0.0005 M barium perchlorate solution using a 5 cm^3 syringe.

80.5.2 Calculations.

Calculate the sulphur content of the sample as follows:

$$\text{Sulphur (ppm)} = \frac{(V - B) \times M \times 16 \times 10^3}{W}$$

where:

V = volume (cm^3) of barium perchlorate solution consumed in actual determination

B = volume (cm^3) of barium perchlorate solution consumed in blank determination

W = weight of sample in g

M = molarity of barium perchlorate.

80.6 DISCUSSION OF RESULTS

The procedure is capable of determining sulphur in amounts from 500 ppm upwards in polyolefins and in other polymers. The repeatability of the method is $\pm$ 40% of the determined sulphur content at the 500 ppm sulphur level, improving to $\pm$ 2% at the 1% level. Chlorine and nitrogen concentrations in the sample may exceed the sulphur concentration several times over without causing interferences. Fluorine does not interfere unless present in concentrations exceeding 30% of the sulphur content. Phosphorus interferes even when present in moderate amounts. Metallic constituents also interfere when present in moderate amounts.

METHOD 81 - DETERMINATION OF MACRO-AMOUNTS OF SULPHUR IN POLYMERS. SODIUM PEROXIDE FUSION - TITRATION PROCEDURE.[50]

81.1 SUMMARY

This sodium peroxide fusion - titration procedure is capable of determining total sulphur in polymers in amounts down to 500 ppm.

81.2 APPARATUS

Nickel "fluorine bomb" capacity 2 ml - obtainable from Charles W. Cook and Sons Ltd., 97 Walshall Road, Perry Barr, Birmingham.

Conical flask with ground glass stopper, capacity 150 or 200 ml.

Magnetic stirrer.

Microburette, capacity 10 ml - conforming to BS. 1428.

81.3 REAGENTS

Sodium peroxide, MAR grade.

Ethanol, absolute.

Barium perchlorate, 0.01 N aqueous, adjust to about pH 3.0 by adding perchloric acid.

Thorin indicator solution, dissolve 25 mg of thorin in 5 ml of distilled water.

Methylene blue indicator solution, dissolve 15 mg of methylene blue in 50 ml of water.

Cation exchange resin, Amberlite IR-120 (H).

Pre-treatment of the resin. By the column method wash about 400 g of the analytical grade resin with about 700 ml of 3.0 N hydrochloric acid and then with 4 or 5 litres of water. Transfer the resin to a 1 litre flask, remove most of the water by decantation and shake the resin thoroughly for several minutes with three succesive 100 ml portions of ethanol. Remove the residual ethanol by repeated washing with water and shake the resin vigorously with 400 ml of 0.5 N sodium hydroxide for about 10 minutes. Decant the slightly turbid solution, wash the resin with water until all suspended fine particles have been removed, and filter on a Buchner funnel. Press the resin between filter papers and store it in a glass stoppered bottle sealed with adhesive tape.

81.4 METHOD

The polymer sample is fuzed with sodium peroxide in a sealed bomb. The fusion product is dissolved in water and sodium ions removed on a cation exchange column. The sodium-free eluate is titrated with standard barium perchlorate solution to the thorin - methylene blue end-point to estimate the amount of sulphate present.

81.5 EXPERIMENTAL PROCEDURE

81.5.1 Fusion of the Sample with Sodium Peroxide:

Place in the dry bomb 0.5 g of powdered sodium peroxide, a suitable weighed amount of the sample (5 to 15 mg) and a further 0.5 g of sodium peroxide. Close the bomb and mix the contents by rotation. Heat the bomb in a muffle furnace for 3 minutes at 650°C. Cool the bomb, remove the lid and extract the fusion product by placing the bomb in a small beaker containing 10 to 15 ml of water and warming until effervescence ceases. Remove the bomb, rinse it with water and then rinse the lid. Transfer the combined solutions quantitatively to a 50 ml calibrated flask and dilute to the mark.

81.5.2 Removal of Sodium from Fusion Product and Titration

Place about 30 g of the cation exchange resin in a conical flask. Add about 25 ml of ethanol, shake the stoppered flask for about 1 minute, remove the ethanol by decantation and repeat the operation with a further 20 ml of ethanol. Transfer 25 ml of the fusion product extract to the resin, shake for about 5 minutes and decant the solution into a suitable conical titration flask containing a magnetic stirrer bar. Rinse the resin with four successive 25 ml portions of ethanol, add all washings to the contents of the titration flask. Add 0.1 ml each of thorin and methylene blue indicator solutions and titrate with 0.01 N barium perchlorate to a pink end-point colour persisting for about 20 seconds. Carry out a blank determination in the same manner, omitting only the sample. The blank value should lie between 0.1 and 0.2 ml.

Precautions:

When the sodium peroxide fusion is carried out as described, only a small amount of insoluble matter should be found in the aqueous extract of the fusion product, but this material can interfere with the detection of the end-point of the titration. For the best results the extract should be set aside overnight, so that a clear portion can be withdrawn for titration.

End-points in the barium perchlorate titration may be unsatisfactory if the ethanol washed resin is set aside for more than about 2 hours before use. If such a delay is unavoidable, the resin should be washed once more with about 20 ml of ethanol immediately before use.

With some batches of resin it has been observed that after pre-treatment in the prescribed manner, unsatisfactory end-points are obtained in the titration of the resin treated sample solution. If this defect is found with any portion of a given batch of resin, it is recommended that all subsequent portions should be vigorously shaken with about 25 ml of 0.5 N sodium hydroxide for about 10 minutes and then washed with water and ethanol in succession before use.

81.6 DISCUSSION OF RESULTS

This procedure is capable of determining down to 500 ppm of total sulphur in sulphur containing polymers with an accuracy of $\pm$ 5%.

Chlorine, fluorine and nitrogen in amounts up to 2 mg in the sample are without serious effect on the determination of sulphur. The effect of larger amounts of fluorine can be suppressed by the addition of boric acid.

METHOD 82 - DETERMINATION OF SULPHUR IN POLYMERS. OXYGEN FLASK COMBUSTION - PHOTOMETRIC TITRATION PROCEDURE.

82.1 SUMMARY

This oxygen flask combustion photometric titration procedure is capable of determining total sulphur in polymers in amounts down to 50 ppm.

82.2 APPARATUS

A general view of the titration arrangement is given in Figure 7.19.

Conical flask - pyrex or Jena glass, 500 cm^3 capacity, provided with conical ground joint B 24, (see Method 76)

Stopper, B 24 with fused in platinum wire (30 mm long, 0.8 mm diameter) carrying a 15 x 20 mm piece of 40 mesh platinum gauze (see Figure 7.18 and Method 76).

Safety jacket, to serve as a protection during the combustion. Detachable metal wire gauze jacket fitting around the conical flask.

Syringe - 0.05 cm^3 capacity, with 0.001 cm^3 divisions.

Lighter - any suitable small flame fed with sulfur-free fuel, eg. alcohol.

Microburette assembly, with 10 cm^3 burette and Pyrex supply bottle, two required.

Ultramicro burette, with 0.001 cm^3 divisions and provided with a glass capillary delivery tube.

Photoelectric colorimeter equipped with two 100 cm^3 optical cells and two 520 nm interference filters in rim.

Electric stirrer of suitable dimensions to fit on and close the colorimeter compartment, provided with small glass propeller-shaped stirrer (Figure 7.19).

82.3 REAGENTS

Barium perchlorate solution, standard 0.005 M, prepare as described in Method 80.

Barium perchlorate solution standard 0.0005 M, prepare as described in Method 80.

Filter paper, ash-free. Schleicher and Schull No. 5892 is recommended. Store in a closed bottle.

Hydrogen peroxide, 30% AR.

Isopropanol, cp, denatured.

Oxygen, compressed, free of sulphur compounds.

Perchloric acid, 10% AR.

Perchloric acid, 3%, AR.

Sulphuric acid solution, standard, 0.01 N prepare by accurate dilution of standardized 0.1 N sulphuric acid solution.

Sulphuric acid solution, 0.0005 N, prepare by accurate dilution of 0.01 N sulphuric acid solution.

Thorin indicator solution, 0.2% aqueous solution of thorin (sodium salt of 2(2-hydroxy-3, 6-disulfo-1-napthylazo) benzene-arsonic acid). The thorin quality marketed by Merck, Darmstadt, West Germany, is recommended.

Water, deionized.

82.4 METHOD

The sample, contained in a piece of filter paper and suitably suspended, is rapidly and completely burnt in a closed conical flask filled with oxygen at atmospheric pressure. The products of combustion are allowed to enter a hydrogen peroxide solution. The amount of sulfuric acid formed is determined by titration with barium perchlorate, using

Thorin as the indicator. The equivalence point is obtained photoelectrically by comparing the optical density of the solution with that of a standard reference solution.

82.5 EXPERIMENTAL PROCEDURE

82.5.1 Sample Combustion

Weigh a filter strip to the nearest 0.1 mg. Place approximately 30 mg of sample onto the middle of the filter strip and weigh again. Wrap the sample up in the following way: first cover the sample with the raised edges of the strip and then roll up the body of the strip towards the end, which will serve as a fuse. Now clamp the packet in the platinum gauze of the stopper, keeping the fuse free and in line with the platinum suspension wire.

Introduce 4 $\pm$ 0.4 ml of deionized water and 3 drops of hydrogen peroxide solution 30% into the conical flask. Install the safety jacket.

Replace the air in the conical flask by oxygen by introducing a rapid stream of oxygen at a point near the bottom for 30 seconds. (Place a G 3 porosity glass filter in the outlet end of the oxygen line to prevent contamination of the absorption liquid).

Place the flame of the lighter close to the top of the conical flask, ignite the end of the fuse of the sample packet and immediately insert the stopper.

Keep the flask firmly closed and keep it upside down to prevent the flame from touching the walls or bottom of the flask. When the combustion slows down bring the flask in an inclined position to promote complete combustion also of any dropping particles. Remove the safety jacket, shake the bottle for one minute and allow it to stand for 15 minutes.

82.5.2 Spectrophotometric Evaluation

In an optical cell prepare a colour reference solution by mixing 100 ml of deionized water and 200 $\pm$ 1 ul of thorin indicator solution (measured from the ultramicro burette). Measure the optical density using the deflection method. This value should be 0.17 $\pm$ 0.01, adjust if necessary by adding a known amount of either 0.0005 N sodium hydrogen carbonate solution or 0.0025 N sulphuric acid solution.

Prepare a double amount of the colour reference solution described in the previous paragraph in a beaker. Introduce half of it into each optical cell and check the optical density of the solution. With both cells in the compartments and the photometer adjusted for highest sensitivity bring the pointer at position 50 of the linear scale marked "abs". Check whether the cells are matched by interchanging them. During the actual determination keep the right-hand cell in its place and titrate in the left-hand cell.

Prepare and measure a fresh colour reference solution at least once a day and whenever a new batch of deionized water or thorin indication solution is used. Switch on the photometer at least two hours before use.

Introduce 1 ml of 0.00025 N sulphuric acid solution by means of a pipette and 15 $\pm$ 0.5 ml of deionized water into the optical cell. Wet the rim between flask and stopper with some isopropanol and open the flask to draw the liquid in. Use 80 $\pm$ 2 ml of isopropanol to wash the stopper and the platinum gauze and to transfer the solution under test quantitatively to the cell. Add 1 drop (approx 0.05 ml) of 3% perchloric acid solution and 200 $\pm$ 1 ul of thorin indicator solution.

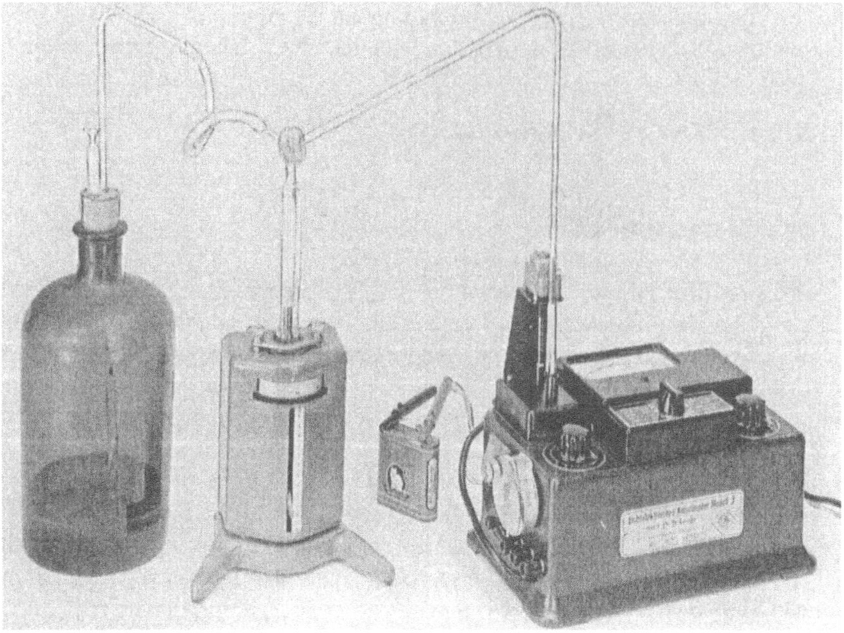

Figure 7.19. Titration assembly for the photometric sulfate titration.

Place the optical cell in the photometer and install the electric stirrer. Immerse the burette tip into the cell solution and stir. Titrate at a slow rate with barium perchlorate solution, until the pointer reaches 50 again. Use the 0.005 M barium perchlorate solution for sulphur contents exceeding 0.15% and the 0.0005 M barium perchlorate solution for the lower sulphur concentrations.

Especially in the lower sulphur range, a minimum titration time of two minutes is recommended. Run a blank determination including all reagents used (also filter paper) but omitting the sample.

82.5.3 Calculation.

Calculate the sulphur content by means of the following equation:

$$\text{Sulphur } \%\,w = \frac{(V - B) \times N \times 16 \times 10^3}{W}$$

where:

V = volume of barium perchlorate solution consumed in the actual determination, millilitres.

B = volume of barium perchlorate solution consumed in the blank determination, millilitres.

N = normality of the barium perchlorate solution.

W = weight of sample milligrams.

82.6 DISCUSSION OF RESULTS

The following data should be used for judging the acceptability of results (95% probability). Duplicate results by the same operator should not differ by more than the following amounts;

Sulphur content	Repeatability
Above 0.5%	0.05
Below 0.5%	2% of amount present but not better than 0.005

Chlorine and nitrogen concentrations in the sample may exceed the sulfur concentrations several times over without causing interference. Fluorine does not interfere unless present in concentrations exceeding 30 per cent of the sulfur content. Phosphorus and metallic constituents interfere even when present in moderate amounts.

METHOD 83 - DETERMINATION OF DOWN TO 50 PPM ORGANIC NITROGEN IN POLYOLEFINS. KJELDAHL DIGESTION - BORIC ACID TITRATION METHOD.

83.1 SUMMARY

This kjeldahl digestion - titration procedure determines organic nitrogen in amounts down to 50 ppm in polyalkenes.

83.2 APPARATUS

Kjeldahl digestion flasks: 500 cm^3. Kjeldahl flask heater, 120 watts. Mount flasks in heater at an angle of 30 degrees to the horizontal.

Distillation apparatus.

Microburette, capacity 10 cm^3.

83.3 REAGENTS

Boric acid, 4% w/v.

Mercuric oxide/cupric sulphate catalyst: grind together 10 g mercuric oxide and 5 g cupric sulphate (anhydrous).

Selenium powder.

Iron III chloride: 40% w/v aqueous, to 100 cm^3 of this solution add 5 drops of concentrated sulphuric acid.

Sulphuric acid, 98% use nitrogen-free grade. Keep well stoppered to ensure minimum contamination of this reagent by nitrogen containing impurities in the laboratory atmosphere.

Potassium sulphate - AR.

Sodium hydroxide, 40% w/v aqueous, made from AR reagent.

Hydrochloric acid, 0.02 M.

Tashiro indicator, 0.125% w/v methyl red and 0.0083% w/v methylene blue in ethyl alcohol.

83.4 METHOD

A sample of polymer (up to 5 g) is digested with a mixture of sulphuric acid, mercuric oxide, copper sulphate and selenium catalysts to convert nitrogen containing organic compounds to ammonium sulphate. Digestion is completed by adding sufficient potassium sulphate to the digestion mixture to raise its boiling point to 380°C followed by heating to completely remove any carbonaceous matter present.

The solution is then transferred to a distillation apparatus and excess sodium hydroxide added to convert ammonium sulphate to free ammonia. The ammonia is distilled off and collected in boric acid solution and the ammonia content of the steam distillate is determined by titration with standard hydrochloric acid solution.

83.5 EXPERIMENTAL PROCEDURE

83.5.1 Sample Digestion.

Weigh out a known amount of the polymer (maximum 5 g) and carefully transfer into a 500 cm^3 Kjeldahl digestion flask. Into a second (blank) flask transfer the same weight of nitrogen-free polymer. To each flask add 0.2 g mercuric oxide/cupric sulphate mixture, 10 mg selenium, some glass beads and 15 cm^3 concentrated sulphuric acid. In addition to each flask add a further 10 cm^3 concentrated sulphuric acid per g of polymer present in the sample flask. Place the digestion flasks on the electric heaters and raise to the boiling point as rapidly as possible with occasional swirling. Continue heating until the sample solution becomes yellow/brown in colour (this usually takes 2 to 4 h). If necessary add further similar quantities of concentrated sulphuric acid to the sample and blank flasks to maintain a minimum volume of 20 cm^3 of acid. Leave the flasks to cool and note the weight of the flask contents. To the sample and blank flask add 0.72 g potassium sulphate per g of flask contents. Again place the flasks in the heating mantle and boil for a further 90 min to complete the digestion process. Allow the flasks to cool and to each add approximately five volumes of deionized water per volume of acid present (care). Add 2 drops iron III chloride reagent.

Fill the steam generator with water and add to it a few pellets of sodium hydroxide. Connect the generator to the steam distillation apparatus and steam out the apparatus for 20 min (run steam distillate to waste). Transfer the sulphuric acid digestion mixture into the 1L sample flask and pass steam through the sample of a further twenty min (removes any volatile acids present in the sulphuric acid digest). Into a 250 cm^3 conical flask (B 24 socket) pipette 25 cm^3 4% boric acid solution and add 100 ml deionized water and 5 drops of Tashiro indicator. If necessary add a few drops of M/10 sodium hydroxide solution to produce a green colour and then back titrate each flask with M/50 hydrochloric acid to a turquoise blue colour. Transfer 30 cm^3 of this solution into each of the two 250 cm^3 clean stoppered conical flasks. Retain one of the flasks (flask A) as a comparison standard for the final ammonia titration. At the end of the steam purging, connect the second 250 cm^3 titration flask (flask B) to the freshly purged steam distillation apparatus. Continue passing steam through the apparatus and fill the separatory funnel on the steam distillation apparatus with 40% sodium hydroxide. Run this solution into the hot acid mixture until a slight excess of alkali is present (indicated by formation of brown precipitate of cupric hydroxide). Run in a further 5 to 10 cm^3 excess of sodium hydroxide. Continue the steam distillation until about 130 cm^3 of liquid is present in conical flask B. Remove this flask from the distillation apparatus and leave until about 130 cm^3 of liquid is present in the flask.

83.5.2 Sample Titration.

Titrate the contents of the sample conical flask B with 0.02 M hydrochloric acid until it has the same shade (turquoise or bluish grey) as the contents of the comparison flask A. Note the volume of hydrochloric acid used in the titration.

83.5.3 Calculation.

Calculate the nitrogen content of the polymer as follows:

$$\text{Nitrogen (ppm)} = \frac{(VI - V2) \times M \times 14 \times 10^6}{1000 \times W}$$

where:

VI = titration (cm^3) of standard hydrochloric acid in sample determination.

M = molarity of hydrochloric acid.

V2 = titration (cm^3) of standard hydrochloric acid in reagent blank determination.

W = weight (g) of sample taken for analysis.

83.6 DISCUSSION OF RESULTS

This procedure is capable of determining down to 50 ppm of organic nitrogen in polymers with an accuracy of ± 10%.

METHOD 84 - DETERMINATION OF 1 TO 90% ORGANIC NITROGEN IN POLYMERS. KJELDAHL DIGESTION - BORIC ACID TITRATION METHOD.

84.1 SUMMARY

This Kjeldahl digestion - titration procedure is capable of determining between 1 and 90% of organic nitrogen in polymers.

84.2 APPARATUS

a) Kjeldahl digestion apparatus consisting:
Kjeldahl digestion flasks - 100 ml Pyrex glass.
Electric heating "Bun ray" electric bunsens suitable (minimum loading 375 watts) or Kjeldahl digestion flask heaters with flask supports (100 ml flask size).

b) Ammonia recovery apparatus consisting of (see Figure 7.20). Condenser (B24 cone and socket), Leibig 12". Round bottomed steam generation flask, 1000 ml central B24 and 2 x B19 side. Three way stopcock, vertical arm connected to B24 cone. Separatory funnel 250 ml with B19 cone.

Recovery apparatus proper:

Consisting of steam stripping vessel (interior vessel large enough to hold 150 ml solution). Separatory funnel 50 ml with B14 cone and short delivery stem. Steam trap leading to vertical condenser and guard tube containing activated silica gel which is connected by a suitable adaptor (note 2). Activated silica gel. Self indicating, mesh 6-20, available from British Drug Houses Ltd., Poole. To regenerate gel, heat for 6 hours at 150° C in a vacuum oven.

c) Apparatus for determination of ammonia in distillate, miscellaneous glassware: Burette, 50 ml pipettes.

84.3 REAGENTS

Sulphuric acid, preferably nitrogen-free quality supplied by British Drug Houses Ltd., Poole.

Glucose, micro analytical reagent (MAR) grade, British Drug Houses Ltd., Poole.

Kjeldahl digestion catalysts, the following supplied by British Drug Houses Ltd., Poole.

Tablets each consisting of 1 g sodium sulphate and the equivalent of 0.05 g selenium (as selenium dioxide).

Tablets each consisting of 1 g sodium sulphate and equivalent of 0.05 g of mercury (as mercuric oxide).

Cupric sulphate, 25% w, dissolve 25 g $CuSo_44H_20$ Analar in water and make up to 100 ml (0.2 ml of this solution contains 0.05 g $CuSo_45H_20$.

Mercuric oxide, pure grade. .

Potassium sulphate, Analar.

Sodium hydroxide, (use when mercury - free catalysts are used for Kjeldahl digestion). Pour 1600 ml distilled water into a 4 litre beaker and add 1500 g of sodium hydroxide Analar. When dissolved allow to cool and transfer to a winchester.

Sodium hydroxide - sodium thiosulphate (use when mercury containing catalysts are used for Kjeldahl digestion). Pour 1400 ml distilled water into a 4 litre beaker and add 1500 g sodium hydroxide Analar. When dissolved allow to cool. In a further beaker dissolve 150 g sodium thisulphate $5H_20$ Analar in 200 ml distilled water. Mixt the two solutions and transfer to a winchester. Remove nitrogenous impurities from either reagent as follows: Add 5 g zinc powder and pass a slow stream of nitrogen for 48 hours to sweep our ammonia. Filter the cold solution through a 40 cm Whatman No. 1 filter paper to remove sodium carbonate. Store the reagent in polyethylene bottles. Carry out these operations in a fume cupboard to avoid contamination of the solutions by any ammonia impurity present in the laboratory atmosphere.

Boric acid, 4% w, dissolve 40 g boric acid (H_3BO_3) Analar in 1000 ml distilled water contained in a 2 litre round bottomed flask. Boil for 20 minutes to expel carbon dioxide. Connect a carbon dioxide absorbing guard tube and cool. Transfer to a 1 litre volumetric flask. Make up to 1 litre with cold boiled out distilled water and mix.

Methyl red, bromo-cresol green indicator: Mix 5 volumes of 0.2% bromo-cresol green with 1 volume of 0.2% methyl red, both in 95% ethanol.

Hydrochloric acid, 0.05 N.

Hydrochloric acid, 0.02 N prepared daily by dilution of the 0.05 N hydrochloric acid stock.

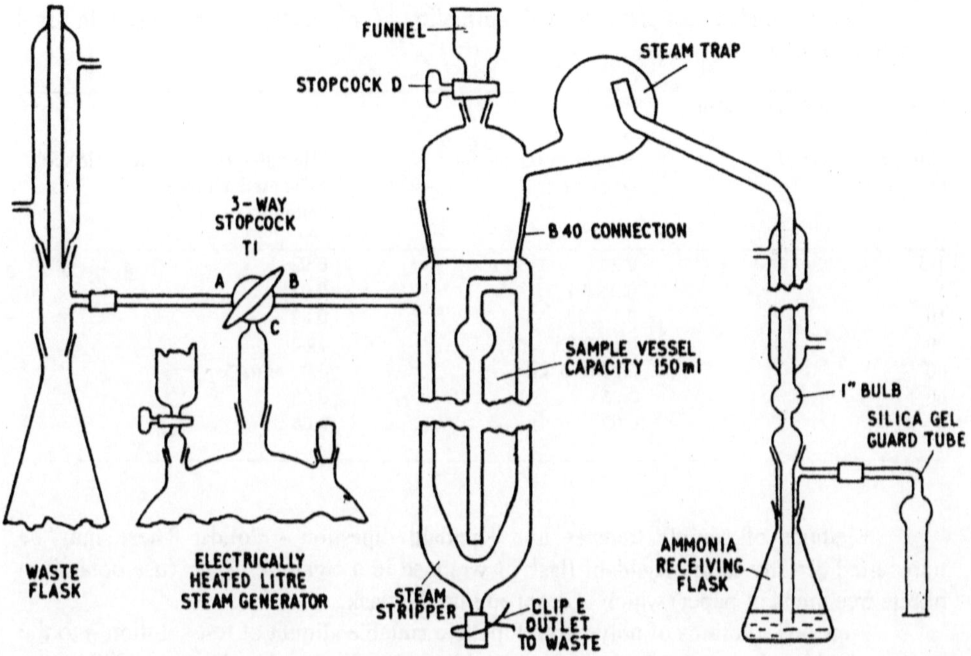

Figure 7.20. Ammonia recovery steam stripper assembly.

84.4 METHOD

The procedure involves Kjeldahl digestion followed by collection of generated ammonia in boric acid.

A suitable weight of sample is digested with concentrated sulphuric acid, sodium or potassium sulphate and digestion catalysts using an electrical heater (note 1). A reagent blank determination is also carried out. The digested sample, after dilution, is transferred to a steam distillation apparatus. The addition of sodium hydroxide liberates ammonia equivalent to the nitrogen content of the sample. Liberated ammonia is collected in boric acid solution and determined by titration with standard hydrochloric acid.

84.5 EXPERIMENTAL PROCEDURE

84.5.1 Sample Digestion

The weights of sample required for the determination of nitrogen in compounds containing between 1% and 90% of nitrogen are shown below. Also shown is the normality of the hydrochloric acid required for the titration of ammonia at the end of the determination.

Recommended sample size and normality of hydrochloric acid used in final ammonia titration.

Weighing with 4-place balance		
Nitrogen content of sample %	Weight of sample for digestion (g)	Strength (N) of hydrochloric acid to be used in titration of ammonia, normality
1-3	0.2	0.02
5	0.2 - 0.1	0.02
10	0.1	0.02
20	0.1 - 0.05	0.02
40	0.03 or 0.05	0.02 or 0.05
60	0.05	0.05
80-90	0.03	0.05

Methods of sample transfer into Kjeldahl digestion - Solids: These may be transferred directly to the Kjeldahl flask or wrapped in a cigarette paper, (use potassium nitrate free brand of paper) which is dropped into the flask.

Aqueous solutions of polymers: Pipette a suitable aliquot of test solution into the flask and add 6 drops of 0.5 N sulphuric acid (include in blank). Heat at 50°C under vacuum to reduce the volume to approximately 1 ml. The sample is now ready for analysis.

In the Kjeldahl flask containing the sample introduce a small piece of clean porous pot. Depending on the weight of sample to be digested, add the recommended quantities of a suitable catalyst mixture and of concentrated sulphuric acid as indicated in Table 7.43 also 0.05 g glucose (note 3).

To the flask used for the reagent blank determination introduce the same amount of catalyst mixture, sulphuric acid and glucose. Suitable catalysts and length of sulphuric acid digestion time for the digestion of various types of nitrogen compounds and polymers are shown in Table 7.44.

Mount the digestion flask containing sample, sulphuric acid, catalysts and glucose on an electrical heater (note 2) and insert a small funnel in the neck of the flasks. Digest gently at first and then increase the digestion temperature steadily, at five or ten minute intervals over the recommended digestion period (Table 7.44). After the first 45 minutes digestion the heating rate should be such that the acid is condensing about half way up the neck of the digestion flask (maximum digestion temperature 380°C). Ensure that the carbon char becomes completely oxidized during digestion. Continue the digestion 20 minutes at least, at maximum temperature after the solution has clarified. When digestion is complete, switch off the heaters. Leave to cool at room temperature. At least half of the original sulphuric acid addition should remain in the digestion flasks at this stage (note 5). Add distilled water to the contents of each digestion flask.

84.5.2 Ammonia Recovery by Steam Distillation.

Set up the distillation apparatus shown in Figure 7.20. To the 1 litre steam generation flask add several pellets of sodium hydroxide and some boiling chips and three quarters fill with distilled water. Turn stopcock T1 to connect A to C. Boil the contents of the generator for 5 minutes (note l) to sweep out dissolved ammonia into the waste flask. Then turn stopcock Tl to connect B to C thereby passing steam through the steam stripper (stopcock D and clip at E closed). Again divert the steam supply to the waste flask.

Transfer the contents of a digestion flask with two or three 5 ml distilled water washings together with 0.2 ml 25% cupric sulphate solution via the funnel on the steam stripper to the interior sample vessel and close the stopcock D. Open the outlet E to waste and turn stopcock T1 to reconnect B and C. Pass steam until it exits at E. Meantime pour an excess of 40% sodium hydroxide into the (closed) funnel at D. Use 10 ml, 20 ml or 40 ml sodium hydroxide respectively, depending on whether 2 ml, 5 ml or 10 ml of concentrated sulphuric acid was used for acid digestion. If the digestion catalyst used contains mercury or mercuric oxide use instead, the same volumes of sodium hydroxide - sodium thiosulphate solution (note 6). Use the same amount of alkali in both sample and blank determinations.

Connect to the vertical condenser of the steam stripper a 250 ml flask. Connect a guard tube containing freshly activated silica gel in position above this receiving flask (note 2). When steam is emitting from the steam stripper outlet E close the clip thereby allowing steam to pass through the sample solution. Steam distil until about 50 ml distillate has been collected (note 7) and disconnect the flask from the apparatus.Into a clean 250 ml ammonia receiving flask pipette 25 ml 4% boric acid solution and add 4 drops of bromo-cresol green - methyl red mixed indicator. If this solution is blue-green in colour (due to the presence of a trace of dissolved ammonia), then add single drops of 0.01 N hydrochloric acid from a burette until the solution becomes neutral grey in colour. Connect this flask to the vertical condenser of the steam stripper, (silica gel guard tube still in position).

When steam is emitting from the steam stripper outlet E, close the clip thereby allowing steam to pass through the sample solution. Open stopcock D and slowly admit the alkaline reagent in several portions. Leave 1 ml of alkali in the funnel to act as a seal. It is necessary to ensure that an excess of alkali is added at this stage. Alkalinity is indicated by the formation of, first, a soluble deep blue cupramine salt then a precipitate of brown copper oxide.

Steam distil until the volume of liquid in the receiver is approximately 100 ml, ie. for about 20 minutes. Remove the receiver flask from the apparatus. Clean the steam stripper ready for the next analysis (note 8).

Titrate the contents of the receiver flask with either 0.05 N or 0.02 N hydrochloric depending on the amount of sample digested and its nitrogen content.

The colour change at the end-point is from green to neutral grey. The addition of a further drop of hydrochloric acid beyond this end-point should produce a pale pink colour.

84.5.3 Calculation.

The nitrogen content of the sample is given by:

$$\%\,N\,(\%) = \frac{(Ta-Tb) \times f \times 14}{W \times 10}$$

where:

Ta = titration (ml) of hydrochloric acid, sample.
Tb = titration (ml) of hydrochloric acid, blank.
f = normality of hydrochloric acid.
W = weight (g) of sample taken for analysis.

Table 7.43 - Catalyst and Sulphuric Acid Additions

Weight of Sample Digested (g)	Catalyst and Alkali Sulphate Mixture		Volume of Concentrated Sulphuric Acid Used for Digestion, ml
0.1 - 0.3*	A**	0.4g Sodium Sulphate 0.02 g Selenium (as Dioxide) 0.02 g $CuSo_45H_20$	2
	or		
	B	0.4 g Sodium Sulphate 0.02 g Selenium (as Dioxide) 0.02 g Mercury (as Oxide)	2
	or		
	C	1.3 g Potassium Sulphate 0.04 g Mercuric Oxide	2
0.03 - 0.08*	A**	1 g Sodium Sulphate 0.05 g Selenium (as Dioxide) 0.05 g $CuSO_45H_20$	5
	or		
	B	1 g Sodium Sulphate 0.05 g Selenium (as Dioxide) 0.05 g Mercury (as Oxide)	5
	or		
	C	3 g Potassium Sulphate 0.1 g Mercuric Oxide	5
0.08 - 0.2	A**	2 g Sodium Sulphate 0.1 g Selenium (as Dioxide) 0.1 g $CuS0_45H_20$	10
	or		
	B	2 g Sodium Sulphate 0.01 g Selenium (as Dioxide) 0.01 g Mercury (as Oxide)	10

* Use a micro or semi-micro balance when weighing out 0.01 - 0.03 g samples

** Recommended as a good general digestion catalyst. The sodium sulphate-selenium catalyst is available in tablet form from B.D.H. Ltd.

Note 1. Electrical heating during the Kjeldahl digestion and ammonia recovery is preferable to gas heating. The chance of contamination of the sample by nitrogenous impurities in coal gas is thereby avoided.

Note 2. Pick up of ammonia impurity from laboratory atmosphere. The apparatus used for steam distillation from alkali to recover ammonia from the acid digest is fitted with an activated silica gel guard tube. This prevents contamination of the distillate with any ammonia impurity that is present in the laboratory atmosphere.

Note 3. Addition of glucose in Kjeldahl digestion. Normally no organic matter would be present in the blank digestion flask. If the nitrogen impurity in the reagents is in both oxidized and reduced forms, therefore, it is necessary to add some organic material to the contents of the blank flask in order to ensure that nitrogen impurity is

Table 7.44 - Selection of Suitable Catalysts and Digestion Times

Type of Compound Digested	Catalyst and Alkali Metal Sulphate Mixture Recommended* (see Table 7.43)	Total Digestion Time with Sulphuric Acid (hrs)
Amines, amino compounds amino acids, amides and their simple derivatives	A, B or C	1.5 - 2
Nitriles and their simple derivatives also: styrene-acrylonitrile copolymers and polyacrylonitrile	A	3
Refractory nitrogen compounds	C	4
Nitro, nitroso and azo compounds hydrozones, oximes and some heterocyclic nitrogen compounds	See Note 4	4
Many azo compounds, volatile nitro compounds, diazo ketones and certain semicarbazones	No successful Kjeldahl digestion procedure known	-

completely reduced. The addition of 0.05 g pure glucose to the blank (and sample) is sufficient for this purpose.

Note 4. Reduction of nitrogen compounds prior to Kjeldahl digestion. Nitro, nitroso and azo compounds also, hydrazones, oximes and some heterocyclic compounds are not quantitatively digested by the catalysts A, B or C shown in Table 7.43. If compounds of these types are being analysed either add 0.5 g pure sucrose (or glucose) to the mixture of catalyst A or B, sulphuric acid and a suitable weight of sample or carry out a preliminary reduction of the sample with hydriodic acid and red phosphorus as described below.

Transfer a suitable weight of sample (catalysts and sulphuric acid absent) and 5 ml of pure hydriodic acid (Analar about 55%) to a clean Kjeldahl flask with B19 quickfit socket. Warm gently and introduce 50 mg pure red phosphorus and some porous pot. Reflux for 45 minutes and then dilute with 5 ml water. Add 5 ml concentrated sulphuric acid and mix. Boil the mixture rapidly (without condenser) to remove hydriodic acid and iodine vapour. If iodine is not completely removed add a further 5 ml of water and boil again until the mixture fumes. Carry out an identical reagent blank omitting only the sample additions. Add a suitable quantity of Kjeldahl digestion catalysts A or B (Table 7.43), concentrated sulphuric acid and 0.05 g glucose (note 3) to the reduced sample and blank solutions.

Suitable quantities of catalyst and acid for the digestion of various sample weights are shown in Table 7.43. Continue by the normal Kjeldahl digestion as described in section b) under procedure.

Various other methods have been proposed for the reduction of nitro groups prior to digestion. These include the use of thiosalicylic acid, salicylic acid - sodium thiosulphate, sodium hydrosulphite-ethanolic hydrochloric acid and zinc dust - pyrogallic acid.

To obtain satisfactory results in the digestion of pyridine it is necessary to leave the sample in contact with catalysts and sulphuric acid for 8 hours at room temperature. Then digest at a low temperature for 1 hour and at a higher temperature for 4 hours. Alternatively add a crystal of iodine to the sample, sulphuric acid and catalysts and digest in the normal manner.

No satisfactory general method for the digestion of nitrogen compounds with an N-N linkage is known.

Note 5. Loss of nitrogen during digestion. Some sulphuric acid is usually lost during digestion. The boiling temperature of the digestion mixture then increases due to the increase in the alkali metal sulphate concentration in the digestion mixture. If the temperature of the mixture becomes too high, due to loss of acid, then a partial loss of nitrogen also occurs. No loss of nitrogen occurs, however, provided the volume of sulphuric acid left at the end of the digestion is at least half of the amount originally added.

Note 6. Modification of ammonia recovery stage when mercury containing catalysts are used for digestion. Low ammonia recoveries result during steam distillation from sodium hydroxide when mercury containing catalysts are used in the Kjeldahl digestion. This is due to the partial formation of a mercuric-amine which is not completely decomposed by sodium hydroxide. The addition of a mixture of sodium hydroxide and sodium thiosulphate completely degrades this complex and quantitative ammonia recoveries result.

Note 7. Volatile acids in digestion mixture. It has been found that traces of steam volatile acids remain in the acid digestion mixture after digestion of certain organic materials. These are removed by passing steam through the diluted digestion mixture prior to sodium hydroxide addition. This acidic distillate is rejected.

Note 8. Cleaning of steam distillation apparatus. Clean the steam stripper ready for the next determination as follows. With clip E closed, turn stopcock T1 to divert the steam supply to the waste flask (ie. connect A to C). The sample solution now syphons from the sample vessel into the outer vessel and is disposed of by opening clip E. Now fill the sample vessel with water via the funnel (clip E again closed) and pass steam through this water until it boils. Syphon the water from the sample vessel as before. Repeat this cleaning operation until the interior of the vessel is perfectly clean.

84.6 DISCUSSION OF RESULTS

The method is capable of determining nitrogen in polymers which contain between 1% and 90% of this element. The accuracy of results obtained is of the order of $\pm$ 1.0% of the true nitrogen content.

The Kjeldahl digestion procedures described quantitatively decomposes amines, amino compounds, amino acids, amides, nitriles and their simple derivatives and also many refractory nitrogen compounds. Quantitative decomposition of nitrogen containing polymers, eg. styrene-acrylonitrile copolymers and polyacrylonitrile, is achieved.

Reduction of nitrogen compounds prior to Kjeldahl digestion-Nitro, nitroso and azo compound also, hydrazones, oximes and some heterocyclic compounds are not quantitatively digested by the catalysts A, B or C shown in Table 7.43. If compounds of these types are being analyzed either add 0.5 g pure sucrose (or glucose) to the mixtue of catalyst A or B, sulphuric acid and a suitable weight of sample or carry out a preliminary reduction of the sample with hydriodic acid and red phosphorus.

Various other methods have been proposed for the reduction of nitro groups prior to digestion. These include the use of thiosalicylic acid, salicylic acid - sodium thiosulphate, sodium hydrosulphite, ethanolic hydrochloric acid and zinc dust and pyrogallic acid.

To obtain satisfactory results in the digestion of pyridine it is necessary to leave the sample in contact with catalysts and sulphuric acid for 8 h at room temperature. Then digest at a low temperature for 1 h and at a higher temperature for 4 h. Alternatively add a crystal of iodine to the sample, sulphuric acid and catalysts and digest in the normal manner.

No satisfactory general method for the digestion of nitrogen compounds with an N-N linkage is known.

Normally no organic matter would be present in the blank digestion flask. If the nitrogen impurity in the reagents is in both oxidized and reduced forms, therefore, it is necessary to add some organic material to the contents of the blank flask in order to ensure that nitrogen impurity is completely reduced. The addition of 0.05 g pure glucose to the blank (and sample) is sufficient for this purpose.

METHOD 85 - DETERMINATION OF BETWEEN 0.002% AND 75% ORGANIC NITROGEN IN POLYMERS. KJELDAHL DIGESTION - SPECTROMETRIC INDOPHENOL BLUE METHOD.

85.1 SUMMARY

This Kjeldahl digestion - spectrophotometric method is capable of determining organic nitrogen in polymers in amounts between 0.002 and 75%.

85.2 APPARATUS

Kjeldahl digestion apparatus consisting of (see Figure 7.21).

Kjeldahl digestion flasks 100 cm^3 Pyrex glass, B19 sockets.

Kjeldahl flask glass bulbs, with inlet and outlet and B19 cone.

Kjeldahl digestion flask heaters, with flask supports (100 ml flask size).

Nitrogen purifying train (see Figure 7.21), for purifying nitrogen supply to digestion flasks.

Two glass tubes, (approximately 18 in x 2 in) of approximately 1 cm^3 capacity with No. 1 sintered glass discs and gas inlet at base and gas outlet at top. Third tube (with gas inlet and outlet) packed with glass wool to remove spray. Glass connections by ball and socket joints.

Ammonia recovery apparatus consisting of:

(Figure 7.20) miscellaneous items.

Condenser, Liebig 12 in (B24 cone and socket) round bottomed steam generator flask B24, central and 2 x B19 side, 1 dm^3. Three-way stopcock, with vertical arm connected to B24 cone. Separatory funnel with B19 cone, 250 cm^3, Conical flasks B24, 250 cm^3.

Recovery apparatus proper:

Consisting of steam stripping vessel (interior vessel large enough to hold 150 cm^3 solution). Separatory funnel 50 cm^3 with B14 cone and short delivery stem, steam trap leading to vertical condenser. Guard tube containing activated silica gel connected by suitable adaptors to lower end of condenser. (Silica gel to be regenerated daily, note 3).

Activated silica gel:

Self indicating, mesh 6 - 20, regenerate gel: heat for 6 h at 150°C in a vacuum oven.

Apparatus for concentration of ammonia steam distillate consisting of:

Concentration apparatus B24 to B24 (70 degree bend) distillation flask head connected to slightly (horizontally) inclined 12 inch Liebig condenser with B24 cone and socket. Connected to outlet end of Liebig condenser (ie. to B24 cone) an 8 in vertical delivery tube (B24 socket) with 1 in diameter bulb 4 in from the open end. Open end of delivery tube dipping in dilute sulphuric acid solution.

Silica gel guard tubes, 8 in x 1 in with B24 cone at base, packed with freshly activated silica gel.

Apparatus for preparation of ammonia free distilled water.

Round bottomed flask (2 dm^3) in electric mantle: connected via adaptor and steam trap to slightly (horizontally) inclined 14 in Liebig condenser (B24 cone and socket).

Delivery end of condenser (B24 cone) connected by suitable B24 socket to B24 cone) adaptor to 2 dm^3 separatory funnel receiver (with activated silica gel guard tube connected by PVC tubing on side arm of the adaptor). The apparatus is totally enclosed - the only exit to atmosphere being via the silica gel guard tube (silica gel tube regenerated daily).

Colorimetric determination of ammonia.

Stoppered cylinders graduated, 50 cm^3. Pipette 5 cm^3 bulb, 1 cm^3 graduated.

Spectrophotometer - 4 cm glass cells.

85.3 REAGENTS

Ammonia free water. Use water which has been redistilled from 100 cm^3 5% sulphuric acid and fresh boiling chips in an apparatus which is sealed from atmosphere by a guard tube containing freshly regenerated silica gel.

Sulphuric acid, (93%). Nitrogen free quality.

Dextrose, Analar.

Kjeldahl digestion catalysts, the following supplied by British Drug Houses Ltd., Poole.

Tablets each consisting of 1 g sodium sulphate and the equivalent of 0.05 g selenium as selenium dioxide.

Tablets each consisting of 1 g sodium sulphate and the equivalent of 0.05 g mercury as mercuric oxide.

Tablets each consisting of 1 g sodium sulphate and 0.05 g hydrated cupric sulphate.

Tablets each consisting of 1 g sodium sulphate.

The following reagents may also be required for Kjeldahl digestion:

Copper II sulphate (25% w/v), prepared from Cu SO4 $5H_20$, aqueous.

Potassium sulphate (nitrogen free) Purify Analar potassium sulphate by recrystallization in the following way:

Introduce into a 4 cm^3 beaker 3 cm^3 of distilled water, 2 g sodium hydroxide and fresh boiling chips, heat to incipient boiling and add 675 g of Analar potassium sulphate. Adjust to pH 10 or higher by sodium hydroxide addition. Boil for 20 mins to expel ammonia. Filter the boiling solution on a buchner funnel through (previously hot water washed) Whatman No. 42 filter paper. Cool the filtrate to 5°C and vacuum filter off the small uniform crystals. Dry the crystals in an air oven (yield 450 g). Allow to cool in a desiccator and store in well stoppered glass bottle.

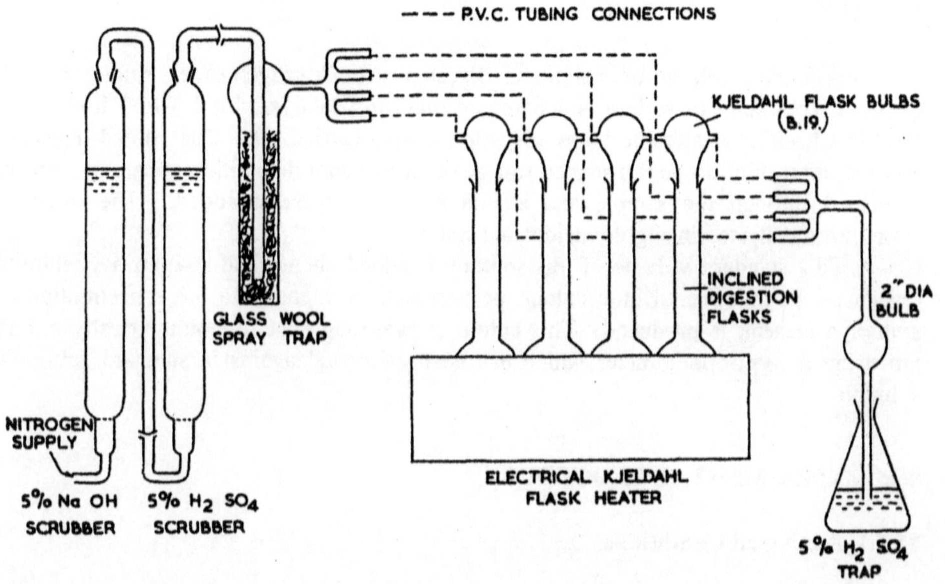

Figure 7.21. Kjeldahl digestion apparatus.

Sodium hydroxide (use when mercury free catalysts are used for Kjeldahl digestion). Pour 1600 cm^3 distilled water into a 4 cm^3 beaker and add 1500 g sodium hydroxide Analar. When dissolved allow to cool and transfer to a winchester.

Sodium hydroxide - sodium thiosulphate (use when mercury containing catalyst are used for Kjeldahl digestion), (see Method 84).

Sulphuric acid 0.25 M. Prepare by dilution of nitrogen free concentrated sulphuric acid.

Phenol, 8% in ammonia-free water. Store in a brown glass bottle.

Sodium hypochlorite (6 w/v available chlorine). Made by diluting sodium hypochlorite solutions: (10 - 14% available chlorine) to 6.0 $\pm$ 0.1% w/v available chlorine content with ammonia free water.

Standard nitrogen solution (for calibration of the indophenol blue method).

Stock solution, dry ammonia sulphate Analar at 100 to 110°C for 1 h and weigh out exactly 0.118 g. Transfer to a 250 cm^3 volumetric flask and make up to volume with ammonia free water. This solution contains 100 microgram nitrogen per cm^3.

Working solution: for calibration purposes take 25 cm^3 of the solution and dilute to 250 cm^3 with ammonia free water. The solution contains 10 microgram nitrogen per ml. Prepare this solution daily by dilution of the stock solution.

85.4 METHOD

A known weight of sample is digested with "nitrogen-free" sulphuric acid, dextrose, potassium or sodium sulphate and digestion catalysts for 1.5 to 2 h using an electrical heater. A suitable blank digestion is also carried out. The cooled digestion mixture, after dilution, is transferred to a semimicro steam distillation apparatus. Steam is passed through the sample and excess sodium hydroxide added. The liberated ammonia is collected in slightly acidulated water.

To a standard volume of this solution is added phenol and sodium hypochlorite reagents. An indophenol blue colour, proportional in intensity to the concentration of ammonia present, is produced. This colour is evaluated spectrophotometrically at 625 nm using a spectrophotometer which has been calibrated against a standard ammonia solution.

85.5 EXPERIMENTAL PROCEDURE

85.5.1 Analytical Conditions.

Choose an appropriate weight of sample as indicated below:

Approximate nitrogen content of sample		Amount of sample required for analysis		Volume of H_2SO_4 needed in Kjeldahl	Dilution to be applied to 50 ml of steam distilate
%w/w	ppm	g	mg	digestion	cm^3
0.0025	25	0.6	600	10	nil
0.005 - 0.05	50 - 500	0.3	300	5	nil
0.075	750	0.15	150	5	nil
0,1	1000	0.1	100	5	nil
0.15	1500	0.075	75	5	nil
0.2	2000	0.05	50	5	nil
0.2 - 2	-	0.05	50	5	nil or x 10
1.0 - 2.5	-	0.01-0.005	10	5	x 10
75	-	0.005		5	x 50

85.5.2 Setting up Nitrogen Purifying Train (Figure 7.21).

Charge the first bubbler with 250 to 300 cm^3 5% sodium hydroxide solution and the second bubbler with 250 to 300 cm^3 5% sulphuric acid solution (replace these solutions daily). Connect four digestion flask bulbs (see Figure 7.21) in parallel, at the outlet end of the nitrogen purifying train by means of PVC tubing. Purge the system by passing nitrogen for 1 hour before commencing sample digestion.

85.5.3 Sample Digestion.

Into two 100 cm^3 digestion flasks containing a suitable weight of sample and two blank flasks containing no sample, (ie. perform sample and blank determinations in duplicate), introduce 0.05 g dextrose and one small piece of clean porous pot. Keep the flasks stoppered between reagent additions to prevent contamination with any ammonia in the laboratory atmosphere. From this point treat the blank and sample determinations in an identical manner throughout the whole determination. Now add alkali metal sulphate, digestion catalysts and acid to sample and blank digestion flasks, as indicated below:

Amounts of reagents for the digestion of up to 0.3 g sample.

Digestion reagents, g	Concentrated sulphuric acid, cm^3
Sodium sulphate, 1 g (selenium, 0.05 g copper II sulphate $5H_20$), 0.05 g	5 ± 0.2
Sodium sulphate, 1 g mercury added as metal or mercury II oxide, 0.05 - 0.1 g	5 ± 0.2
Sodium sulphate, 1 g: copper II sulphate $5H_20$, 0.05, - 0.1 g	5 ± 0.2
Sodium sulphate, 1 g: mercury 0.05 g copper II sulphate $5H_20$, 0.05 g	5 ± 0.2

Digest the samples gently and at an even rate for 20 min. Now increase the digestion temperature steadily at five min intervals over a period of one h so that eventually the acid is condensing about half way up the neck of the digestion flask. Ensure that the carbon char is completely oxidized (maximum digestion temperature 380°C), Continue digestion for 15 to 20 min at maximum temperature after the solution has clarified.

Maintain a gentle purge of pure nitrogen through the bulbs during the acid digestion and also, during the cooling period which follows digestion. If the nitrogen compound is known to be refractory then extend the total acid digestion period to 4 h. Switch off the heaters and allow the flasks to cool to room temperature with the pure nitrogen stream still flowing. At least half of the original sulphuric acid additions should remain in the digestion flasks at this stage. Carefully add 5 cm^3 of ammonia free distilled water to each flask, remove the flasks from the digestion rack and stopper tightly.

85.5.4 Ammonia Recovery by Steam Distillation.

To the 1 litre steam generator (Figure 7.20) add some pellets of sodium hydroxide and boiling chips. Three quarters fill the flask with distilled water. Turn stopcock T1 to connect A to C. Boil the contents of the generator for 5 to 10 min on an electric hotplate, to sweep dissolved ammonia impurity into the waste flask. Then turn stopcock T1 to connect B to C thereby passing steam through the steam stripper for 5 min to clean the stripper (stopcock at D and clip at E closed). Again divert the steam supply to the waste flask.

Quickly transfer the contents of a digestion flask with two or three 5 cm^3 ammonia free water washings together with 0.2 ml 25% copper II sulphate solution via the funnel on the steam stripper to the interior of the sample vessel. Close stopcock D immediately. Open outlet E to waste and turn stopcock T1 to reconnect B and C. Pass steam until it exists at E. Meantime pour an excess of 40% sodium hydroxide into the (closed) funnel at D. Use 20 cm^3 or 40 cm^3 respectively, depending on whether 5 cm^3 or 10 cm^3 of concentrated sulphuric acid was used for acid digestion. If, however, the digestion catalyst used contains mercury or mercury II oxide use the same volumes of sodium hydroxide - sodium thiosulphate instead. Use the same amount of alkali in both sample and blank determinations.

Connect to the vertical condenser of the steam stripper a clean 250 cm^3 flask. Connect a guard tube containing freshly activated silica gel in position above this receiving flask.

When steam is emitting from the steam stripper outlet E close the clip thereby allowing steam to pass through the sample solution. Steam distill until about 50 cm^3 distillate has been collected. Disconnect the flask from the apparatus and replace immediately by a clean 250 cm^3 ammonia receiving flask containing 50 cm^3 ammonia-free water and 6 drops 0.25 M nitrogen free sulphuric acid (silica gel guard tube still in position). Open stopcock D and slowly admit the alkaline reagent in several portions. Leave 1 cm^3 of alkali in the funnel to act as a seal.

It is necessary to ensure that an excess of alkali is added at this stage. Alkalinity is indicated by the formation of first, a soluble deep blue cupramine salt then a precipitate of brown copper oxide (formed by reaction between free ammonia and sodium hydroxide, respectively, with the previously added copper II sulphate).

Steam distil for about 20 min ie. until about 100 cm^3 to 120 cm^3 of liquid is in the receiver. Remove the B24 neck ammonia receiving flask from the apparatus and tightly connect an 8 in vertical guard tube (24 cone) containing activated silica gel to prevent contamination of the distillate by any ammonia impurity in the laboratory atmosphere.

Remove the silica gel guard tube from the ammonia receiver flask and stand on an electric hotplate. Connect the flask immediately to a horizontally inclined Liebig condenser fitted with a delivery tube immersed in dilute sulphuric acid solution.

Boil until the contents of the distillation flask reduce to 20 to 25 cm^3 Immediately disconnect the flask and connect a B24 silica gel guard tube and leave to cool. If up to 150 microgram of nitrogen is present in the concentrated sample solution transfer this solution and the blank solution (with 2 x 5 cm^3 ammonia free water washings) to two clean 50 cm^3 graduated cylinders and make up to 40 cm^3 with ammonia free water. If more than 100 to 150 micrograms of nitrogen is present in the sample solution transfer this solution and the blank solution to two 50 cm^3 volumetric flasks and make up to the mark with ammonia free water. Then transfer to a 50 cm^3 graduated cylinder a volume of the sample solution containing 100 to 150 micrograms nitrogen. Transfer the same volume of blank solution from the 50 cm^3 volumetric flask to a further 50 cm^3 graduated cylinder. Make the volume of the sample and blank solutions in the graduated cylinders up to 40 cm^3 with ammonia free water.

85.5.5 Spectrophotometric Evaluation.

Pipette into the 50 cm^3 graduated cylinders containing the blank and the sample solutions 5 cm^3 8% phenol solution and mix. Then pipette 5 cm^3 sodium hypochlorite solution into each cylinder and again mix. Immerse the two cylinders up to the neck (stoppers loosened) into a 2 dm^3 beaker of water which has been brought previously to the boil. Remove the cylinders from the water bath after exactly 7 min and cool to 20 $\pm$ 1°C by standing in a beaker of water.

Determine the optical density of the indophenol blue colour produced between 15 and 30 min after addition of the phenol and sodium hypochlorite reagents.

Spectrophotometer conditions:
Instrument: Visible spectrophotometer.
Wavelength: 625 nm

Cells: 4 cm glass

Blank solution: Use the reagent blank referred to in the text, ie. a test solution prepared identically to the sample test solution omitting only the addition of sample at the Kjeldahl digestion stage in the determination. To determine the magnitude of the blank optical density, (as a check the reproducibility obtained in duplicate determinations) measure the optical density of the blank solution at 625 nm relative to distilled water in the comparison cell.

85.5.6 Preparation of Calibration Graph.

Accurately pipette 2, 5, 10, 15 and 20 cm^3 of a freshly prepared standard nitrogen solution (containing 10 microgram nitrogen per cm^3) into five clean 50 cm^3 graduated cylinders, ie. between 20 to 200 micrograms nitrogen. Include a further 50 cm^3 graduated (blank) cylinder containing no standard nitrogen solution addition.

Make the contents of the cylinders up to 40 cm^3 with ammonia free distilled water. Continue as described under spectrophotometric evaluation.

Measure the optical densities at 625 nm of the standard nitrogen solutions in 4 cm glass employing the reagent blank solution (ie. containing no addition of standard nitrogen solution), in the comparison cell. Construct a weekly calibration graph relating these optical densities to the number of micrograms of nitrogen present in the nitrogen calibration solutions.

85.5.7 Calculations.

Refer the optical density (reagent blank solution in comparison cell), given by the sample solution to the nitrogen calibration graph and read off the number of micrograms of nitrogen present. Calculate the nitrogen content of the sample from the following equations:

$$\text{Nitrogen, \%} = \frac{N \times 10^{-6} \times 100}{W}$$

$$\text{Nitrogen, ppm} = \frac{N}{W}$$

where:

N = number of micrograms of nitrogen present in 50 cm^3 final test solutions used for colour development.

W = weight of sample (g) which contains N microgram nitrogen.

N.B. any dilution of the test solution prior to the development of colour with phenol and sodium hypochlorite must be allowed for when calculating W.

85.6 DISCUSSION OF RESULTS

The overall reproducibility of this nitrogen determination method is ± 5% in the nitrogen content range 0.002 to 75%. The sodium sulphate - selenium - copper sulphate mixture is particularly suitable for the quantitative digestion of all compounds in which the nitrogen is present in an easily decomposed form (eg. as amine, amino compounds, amino acids, amide or nitrile and their simple derivatives). Forms of nitrogen for which accurate results are not usually obtained include those with -N-N (eg. diazo) and N-O (eg. nitro) linkages and some resistant heterocyclic structures.

METHOD 86 - DETERMINATION OF 1 - 50% ORGANIC NITROGEN IN POLYMERS. MICRO DUMAS COMBUSTION PROCEDURE.

86.1 SUMMARY

This Micro Dumas combustion procedure is capable of determining organic nitrogen in amounts down to 1% in a wide range of polymers.

86.2 APPARATUS

Dumas nitrogen apparatus: as shown diagrammatically in Figure 7.22.

Combustion tube, quartz, 11 to 12 mm o.d., 610 mm long, equipped with an 18/9 socket joint at one end. The tube is shown in Figure 7.23.

Auxiliary combustion tube, quartz, 10 mm o.d. and 230 mm long, equipped with a 12/5 socket joint at each end, as shown in Figure 7.24.

Carbon dioxide absorption tower: an ungraduated glass tube, approximately 15 mm o.d. and 310 mm long with a 7 mm section of l mm capillary tubing at the upper end (Figure 7.22). A 100 cm^3 levelling bulb is attached to a side arm near the bottom of the tower by means of a length of rubber tubing. A stopcock is provided at the upper end of the tower and a ring is etched at about the centre of the capillary section.

Weight azotometer, as shown in Figure 7.25.

Electrolytic oxygen generator. Direct current (0.08 amps and 817 volts) is supplied to the electrodes by a suitable rectifier, the positive lead being attached to the arm having the oxygen reservoir.

Combustion boats, platinum, micro.

Capillary sample bulbs. Draw 6 mm Pyrex tubing to form a capillary. Heat the tubing with a small oxygen gas flame, 2 to 6 mm from the base of the capillary and draw out to form another capillary, thus forming a bulb with capillaries at each end. The size of the bulb is determined by the amount of sample to be taken. The bore of the capillary depends upon the volatility of the sample: use sample bulbs with fine capillaries for volatile liquids.

Filling device for capillary sampling bulbs. To one end of a glass T-tube, attach a short piece of rubber tubing and close the open end of the rubber tubing with a small cork with 0.5 mm hole through its centre. Attach the opposite end of the T-tube to the vacuum line. The other arm of the T allows finger control of the vacuum.

Weighing bottle, glass, micro.

Quartz plug, approximately 7 mm in diameter and 50 mm long. Make a lip at one end of the plug loop to facilitate removal from the combustion tube.

Hook, stainless steel rod, approximately 2 mm in diameter and 270 mm long, with a small hook formed at one end.

Platinum tray, approximately 5 mm wide, 4 mm deep and 60 mm long.

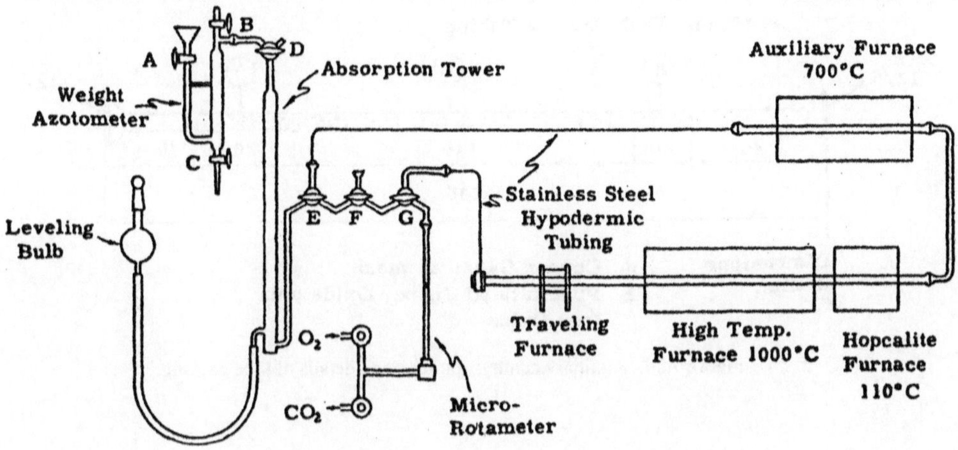

Figure 7.22 Gas flow system for micro-dumas apparatus.

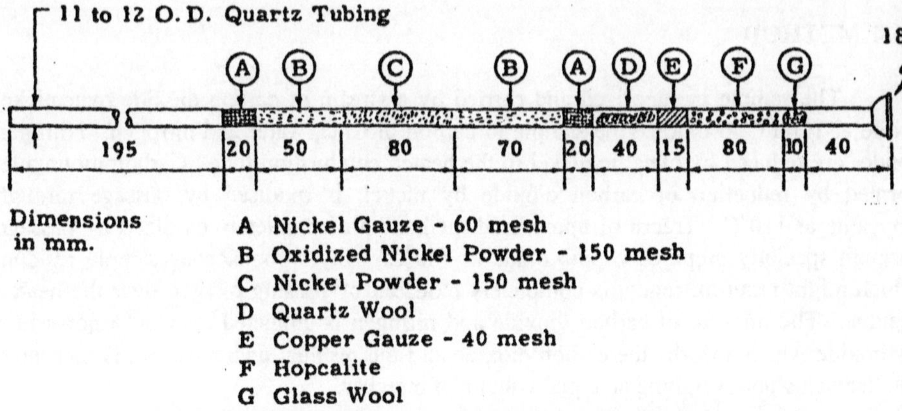

Figure 7.23. High temperature combustion tube and details of tube packing.

86.3 REAGENTS

Nickel gauze, 60 mesh.
Nickel powder, 150 mesh.
Quartz wool.
Hopcalite. Obtainable from Mine Safety Appliances Co., Pittsburgh, Pennyslvania.
Glass wool, pyrex.
Copper oxide, precipitated, containing 1% iron oxide.
Copper gauze, 40 mesh.
Potassium hydroxide solution, 30% w/v.
Mercury.
Potassium hydroxide solution, 2.5% w/v.

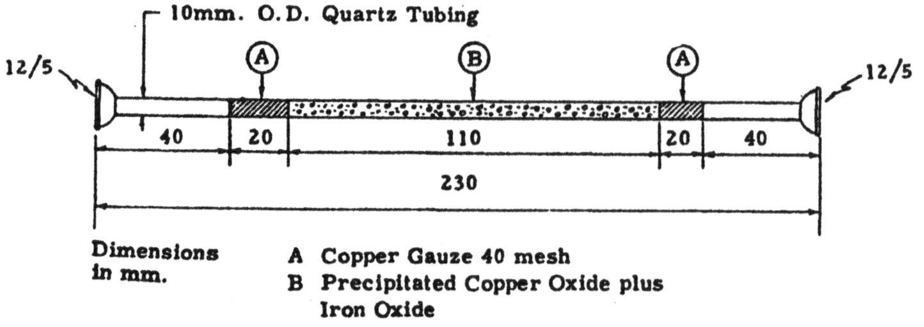

Figure 7.24. Auxiliary combustion tube and details of tube packing.

Carbon dioxide gas, containing less than 0.02% by volume of nitrogen or inert gases.

Hydrogen, commercial.

Oxygen, commercial.

86.4 METHOD

The sample is vaporized and carried by a stream of carbon dioxide over nickel oxide at 1000°C to oxidize the sample to carbon dioxide, water and nitrogen. Nitrogen oxides are reduced by metallic nickel in the heated combustion tube. Carbon monoxide, formed by reduction of carbon dioxide by nickel, is oxidised by passage through hopcalite at 110°C. Traces of unoxidized methane are completely oxidized by passage through specially prepared copper oxide at 700°C. Any carbonaceous sample residue, which might retain nitrogen, is completely oxidized by passing oxygen over the heated residue. The mixture of carbon dioxide and nitrogen is collected over 30% potassium hydroxide which absorbs the carbon dioxide and the residual gas, nitrogen, is measured by displacing and weighing an equal volume of mercury.

86.5 EXPERIMENTAL PROCEDURE

86.5.1 Apparatus Preparation.

Electrolytic oxygen generator. Place 500 cm^3 of 2.5% potassium hydroxide solution in a liter flask and boil for 5 min. Pour the 2.5% potassium hydroxide into the generator. By means of a glass tube inserted above the liquid level in the oxygen reservoir, flush the air from the reservoir with oxygen to expedite the necessary flushing with generated oxygen from the electrodes. Seal all connections with sealing wax. Add sufficient mercury to the mercury by-pass line to give a 5-10 cm head above the sintered disk. Connect the electrodes to the rectifier.

Combustion tubes. Pack the 610 mm tube with nickel powder and nickel gauze as shown in Figure 7.22. Hold the tube in a horizontal position

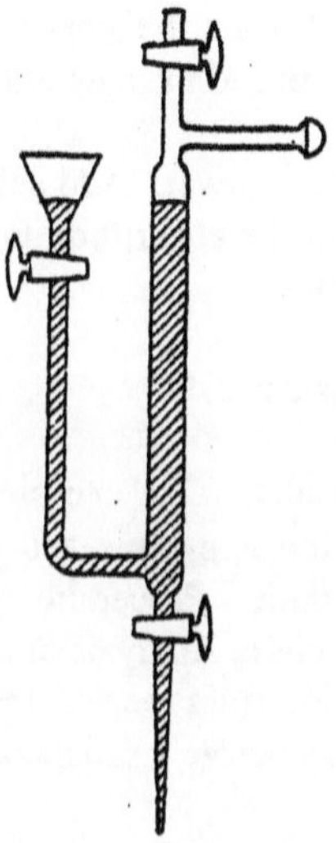

Figure 7.25. Weight azotometer (approximately 1/2 scale).

and tap to form a channel along the top of the nickel powder. Insert the tube into the furnaces as shown in Figure 7.22 and attach the inlet assembly. Adjust the temperature of the main furnace to 800°C and that of the hopcalite to 110°C. Pass carbon dioxide through stopcocks G, F, E and D to remove air from the gas manifold, then turn stopcock G so that carbon dioxide flows through the tube. By means of tubber tubing, attach stopcock F to a source of commercial hydrogen. Pass a small amount of hydrogen through F, E and D to remove air from the connecting tubing. Turn off the carbon dioxide flow and pass hydrogen, at a rate of 10 cm^3 per minute, into the tube through stopcocks F and G until manifold with carbon dioxide. Without removing the tube from the furnace, pack the exit end of the tube with hopcalite as shown in Figure 7.23. Pack the auxiliary tube with special copper oxide as described by Figure 7.24. Insert the auxilliary tube into its furnace and connect one end to the tube containing the nickel, using a section of glass tubing equipped with ball joints at each end. Connect the other end to stopcock E by means of hypodermic tubing fitted with metal ball joints. Seal the joints with sealing wax. Set the temperature of the auxiliary tube at approximately 300°C. Disconnect stopcock F from the hydrogen source and sweep hydrogen from the rubber tubing by introducing carbon dioxide through stopcocks G and F. Sweep hydrogen from the absorption tower by passing carbon dioxide through stopcocks G, F and E. Attach the rubber tubing to a source of commercial oxygen. Remove the cap from the front end of the tube and turn stopcocks F and E so that oxygen can be introduced into the combustion tube through the auxiliary furnace. Adjust the flow of oxygen to 10 cm^3 per min and oxidize the nickel for a distance of 90 mm. Recap the front end of the tube and turn stopcocks F, G and E so that oxygen flows through the combustion tube and into the carbon dioxide absorption tower: leave stopcock D open to the atmosphere. Oxidize about a 70 mm section of the nickel, then, slowly (to prevent surface oxidation of the nickel) sweep oxygen from the tube with carbon dioxide. Adjust the temperature of the main furnace to 1000°C and that of the auxiliary furnace to 700°C. When not in use, keep the system filled with carbon dioxide and at operating temperatures. Disconnect the tubing from F.

Weight azotometer. Lubricate the stopcocks of the weight azotometer and fill the unit with mercury to within 10 mm of the side arm.

Carbon dioxide absorption tower. Add sufficient 30% potassium hydroxide to the levelling bulb to completely fill the absorption tower. Cover the bottom of the absorber with mercury to a height of 10 mm.

86.5.2 Conditioning of Apparatus.

To obtain accurate results, it is necessary to condition the high temperature combustion tube. Assemble the apparatus and perform carbon dioxide and operation blanks as described below under Procedure. Determine the nitrogen content of acetanilide as described below under Analysis of Samples. Repeat the test until a value is obtained which shows a deviation from the true result of 0.1% nitrogen or less. Usually 4 - 5 determinations must be made before a satisfactory value is obtained.

86.5.3 Procedure. Blanks.

Before beginning the first analysis of the day, purge the carbon dioxide and oxygen lines by venting the gasses to the atmosphere through stopcock F and then perform carbon dioxide and operational blanks as described below.

Carbon dioxide blank. Open stopcock D to the atmosphere and raise the levelling bulb until the potassium hydroxide solution rises to the mark etched on the capillary portion of the absorption tower: mark the position of the levelling bulb so that it can be returned to this position. Close stopcock D and lower the levelling bulb. Pass 60 cm^3 of carbon dioxide, at a rate of 10 cm^3 per min, through stopcock G, the combustion tubes, stopcock E and into the absorption tower. Open stopcock D and return the level of potassium hydroxide solution to the mark. Close stopcock D and lower the levelling bulb. Again, pass 60 cm^3 of carbon dioxide into the absorption tower at a rate of 10 cm^3 per min. Close stopcocks A, B and E and raise the levelling bulb to the previously fixed upper position. Turn stopcock D to connect the absorption tower to the weight azotometer and, by means of stopcock C, drain mercury into a weighed container, until the level of the potassium hydroxide solution returns to the mark. Weigh the mercury. The amount of mercury displaced should be less than 0.1 gram.

Operational blank. Open stopcock D to the atmosphere, bring the level of the potassium hydroxide solution to the mark. Close stopcock D and open E to the absorption tower: lower the levelling bulb. Pass carbon dioxide through the apparatus and into the absorption tower for 17 m at a rate of 3 ml per m. Change the gas flow to oxygen and pass oxygen through the tube for 6 m at a rate of 5 cm^3 per m. Follow with carbon dioxide at a rate of 5 cm^3 per m for 4 m. Increase the rate of flow of carbon dioxide to 10 ml per m and maintain this flow for 5 m. Open stopcocks A and B to the atmosphere for a few s and determine the amount of residual gas by displacing mercury as directed under carbon dioxide blank. Weigh the mercury; less than 0.2 g of mercury should be obtained.

86.5.4 Analysis of Samples.

Bring the level of potassium hydroxide solution to the mark, close stopcock D and lower the levelling bulb. Accurately weigh sufficient sample to yield 0.5 to 2.0 mg of nitrogen, but do not take more than 25 mg. Weigh solids and nonvolatile liquids in platinum boats. Weigh hygroscopic samples in boats enclosed in a glass piggy. Weigh volatile samples in a previously weighed glass capillary with a bulb in the middle. To fill, insert one end of the sample bulb into the hole in the cork stopper of the special filling device and completely fill the sample tube with sample by applying vacuum. Seal one end of the sample tube. Weigh the capillary plus sample and quickly dip the open end into melted paraffin. Place the capillary on a platinum tray before putting it in the combustion tube.

Place a quantity of solid carbon dioxide over that section of the combustion tube which contains the sample. Open the front end of the combustion tube and reverse purge by passing carbon dioxide through stopcocks G, F and E and into the exit end of the auxiliary tube. Adjust the rate of flow to 40 cm^3 per m. Insert the sample and, with the hook, push the sample to within 4 - 6 cm of the high temperature furnace.

Insert the quartz plug so that it touches the boat, replace the cap on the front end of the combustion tube: screw the cap on firmly. Quickly turn stopcocks G and E so that carbon dioxide flows through the combustion tubes and into the absorption tower. Reduce the flow of carbon dioxide to 10 ml per m and maintain this flow for 6 m. Open stopcock D to the atmosphere and allow the potassium hydroxide solution to flow into the levelling bulb. Bring the level of potassium hydroxide to the mark on the capillary as before: close stopcock D and perform a carbon dioxide blank as described under Analysis of Sample, above. Repeat if not normal. Continued high blanks indicate that the sample is vaporizing or the system is leaking. Open stopcock D and return the level of potassium hydroxide solution to the mark. Close stopcock D, lower the levelling bulb, turn on the carbon dioxide and adjust the flow to 3 cm^3 per m. Remove the solid carbon dioxide from around the combustion tube and turn on the heater of the travelling furnace. Move the furnace to within 1 - 2 cm of the sample and allow it to warm up for about 2 m. Start the furnace travelling by increasing the drive motor voltage with the variable autotransformer located on the front panel of the unit. Adjust the rate of travel so that the sample is burned slowly and at a uniform rate, as indicated by the rate of formation and size of the gas bubbles in the absorption tower. Conduct the combustion at such a rate that the travelling furnace reaches the high temperature furnace in less than 17 m. When a total of 17 m has elapsed from the time the travel of the furnace was started, return the travelling furnace to a position directly over the sample. Turn off the carbon dioxide, turn on the oxygen and adjust its flow to 5 cm^3 per m. After two m has elapsed, start the furnace travelling at such a rate that it will reach the high temperature furnace in 4 m. Turn off the oxygen and pass carbon dioxide through the tube for 4 m at a rate of 5 cm^3 per m; then for 5 m at a rate of 10 cm^3per m. Turn off the carbon dioxide, the heater of the travelling furnace, and close stopcock E. Momentarily open stopcocks A and B to the atmosphere, place the levelling bulb in its upper position and open stopcock D to the weight azotometer. Drain mercury into a weighed container through stopcock C until the level of potassium hydroxide solution returns to the mark on the capillary. Weigh the mercury to the nearest 0.005 g.

86.5.5 Calculations.

Calculate the nitrogen content of the sample by means of the following equation:

$$\text{Nitrogen \% w} = \frac{(G - g)\,(f)\,100}{S}$$

where:
G = grams of mercury from sample.
g = grams of mercury from operational blank.
f = factor for converting grams of mercury to milligrams of nitrogen, the factor is obtained from the nomograph (Figure 7.26).
S = weight of sample in milligrams.

86.6 DISCUSSION OF RESULTS

This method gives results which do not differ from the true nitrogen content of the sample by more than 0.2%. Replicate analysis agree within 0.1%.

METHOD 87 - DETERMINATION OF 0.01 - 2% PHOSPHORUS IN POLYMERS. OXYGEN FLASK COMBUSTION - SPECTROPHOTOMETRIC METHOD.

87.1 SUMMARY

This oxygen flask combustion - spectrophotometric method determines 0.01 to 2% total phosphorus in phosphorus containing polymers.

87.2 APPARATUS

500 ml conical combustion flask fitted with a reducing adaptor and joint carrying approximately 1" square platinum gauze, (see Method 76).

Unicam SP 500 spectrophotometric with 1 and 4 cm glass cells and tungsten lamp.

Methyl cellulose capsules - No. 4 small, Arthur H Thomas Co. Catalogue No. 6471-0.

Sodium carbonate scoop - fabricated from glass tubing - 4 mm in outside diameter and marked to contain 50 $\pm$ 10 mg of anhydrous sodium carbonate powder.

Filter paper fuses approximately 1.5 x 1/8" in size made from ashless filter paper.

100 ml volumetric flasks: 150 ml conical flasks: pipettes, etc.

87.3 REAGENTS

Demineralised water used throughout.

Ammonium molybdate solution, dissolve 40 g reagent grade ammonium molybdate $(NH_4)_6Mo_7O_{24}\cdot 4H_2O)$ in a cooled mixture of 450 ml concentrated sulphuric acid and l litre water. Dilute to 2 litres with water.

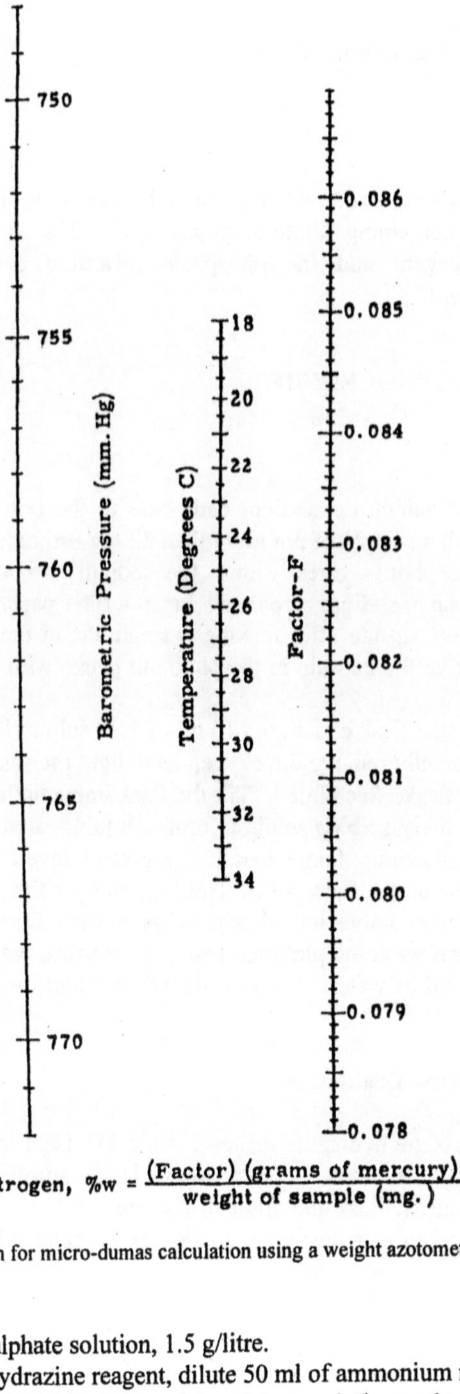

$$\text{Nitrogen, \%w} = \frac{(\text{Factor})\ (\text{grams of mercury})\ (100)}{\text{weight of sample (mg.)}}$$

Figure 7.26. Nomograph for micro-dumas calculation using a weight azotometer.

Hydrazine sulphate solution, 1.5 g/litre.

Molybdate-hydrazine reagent, dilute 50 ml of ammonium molybdate solution with 130 ml of water, add 20 ml of hydrazine sulphate solution and mix well. Use 50 ml for each determination and prepare no earlier than 1 hour before use as this mixture is unstable.

Stock standard phosphorus solution, dissolve 4.393 g of dried Analar potassium dihydrogen phosphate (KH_2PO_4) in 150 ml of 1:10 sulphuric acid and dilute to 1 litre with water (1.0 mg/ml phosphorus). From this prepare solutions containing 0.01 mg P/ml and 0.001 mg P/ml.

Oxygen.
Sodium carbonate, anhydrous, Analar.

87.4 METHOD

The ground polymers mixed with sodium carbonate is decomposed by igniting in an oxygen filled flask containing dilute sulphuric acid. The solution is treated with molybdate-hydrazine reagent and the phosphate measured colorimetrically as its heteropoly-blue equivalent.

87.5 EXPERIMENTAL PROCEDURE

87.5.1 Sample Digestion.

Place 1 scoop of anhydrous sodium carbonate in the bottom half of a methyl cellulose capsule. Weigh a sample of not more than 32 mg estimated to contain between 0.001 and 0.06 mg phosphorus, directly onto the sodium carbonate bed. Cover the sample with another scoop of sodium carbonate. Insert a filter paper fuse between the top and bottom halves of the capsule after making a small slit in the top half to give the clearance required. Secure the capsule in the platinum gauze with the wick towards the stopper.

Pass oxygen into the flask containing 10 ml of 1:10 sulphuric acid for 30 seconds. Holding the flask horizontally remove the oxygen lead, light the sample fuse and quickly insert the stopper in the flask. See Note 1. Tilt the flask immediatley at an angle of 135° from the vertical so that the absorbing solution forms a liquid seal at the neck of the flask. During the period of maximum flame height completely invert the flask to prevent impingement of the flame on the flask walls. Hold the stopper firmly in place during the combustion (Note 1). After combustion allow the flask to stand for 10 to 15 minutes.

Open the flask and wash the platinum gauze and surfaces of the reducing adaptor with a maximum of 35 ml of water. Remove the adaptor and lower the gauze into the solution.

87.5.2 Spectrophotometric Evaluation.

Add 50 ml molybdate hydrazine reagent, (Note 2). Heat the solution rapidly to boiling and boil for 2 - 3 minutes. Cool to room temperature in an ice-water bath, transfer to 1000 ml volumetric flask and dilute to volume.

Measure the absorbance against water in appropriate cells at 830 nm.

Carry out a blank determination.

87.5.3 Calibration Curve.

Prepare separate calibrations from standard phosphorus solutions for 1 cm and 4 cm cells containing 0-0.06 mg P/100 ml and 0-0.015 mg P/100 ml respectively as follows:

Measure appropriate amounts of the phosphorus standards into 150 ml conical flasks and dilute to approximately 30 ml with water. Add 10 ml of 1:10 sulphuric acid and 50 ml of molybdate/hydrazine reagent and develop the colour as before.

Measure the absorbance against water in appropriate cells at 830 mu.

Carry out a blank determination.

Prepare graphs of corrected absorbance against weight of phosphorus (mg per 100 ml). Linear calibrations are obtained.

Note 1. Gloves and goggles should be worn and the ignition carried out behind a protective glass screen.

Note 2. It is essential that the solution is diluted before the molybdate hydrazine reagent is added.

87.6 DISCUSSION OF RESULTS

This method is capable of determining total phosphorus in polymers in amounts down to 0.01% and up to 2% with an accuracy of $\pm$ 5%.

METHOD 88 - DETERMINATION OF 2 - 13% PHOSPHORUS IN POLYMERS. OXYGEN FLASK COMBUSTION - SPECTROPHOTOMETRIC METHOD.

88.1 SUMMARY

This oxygen flask combustion - spectrophotometric procedure determines total phosphorus in polymers in the concentration range 2 to 13% phosphorus.

88.2 APPARATUS

500 ml conical flask fitted with a reducing adaptor and joint carrying approximately 1" square of platinum gauze, (see Method 76).

Methyl cellulose capsules - No. 4 small, Arthur H Thomas Co. Catalogue No. 6471-0.

Sodium carbonate scoop, fabricated from glass tubing 4 mm in outside diameter and marked to contain 50 $\pm$ 10 mg of anhydrous sodium carbonate powder.

Filter paper fuses, approximately 1.5" x 1/8" in size made from ashless filter paper.

Unicam SP 500 spectrophotometer with l and 4 cm cells and tungsten lamp.

100 volumetric flasks, 150 ml conical flasks, pipettes, etc.

88.3 REAGENTS

Oxygen.

Sodium hydroxide, 0.5 N.

Saturated bromine water.

Percolated water, prepared by percolating distilled water through a mixed resin bed containing Amberlite IR 120 (H) and Amberlite IRA 400 (OH).

Ammonium molybdate, 5%v. Dissolve 50 g ammonium molybdate in a litre of warm distilled water.

Ammonium vanadate, 0.25% w. Dissolve 2.5 g of ammonium vandate in 500 ml hot water. Cool and add 20 ml of concentrated nitric acid. Cool and dilute to 1 litre.

Aqueous sulphuric acid, 25% w.

88.4 METHOD

The ground polymer, mixed with sodium carbonate, is decomposed by igniting in an oxygen-filled flask containing saturated bromine water and sodium hydroxide. The solution is acidified, boiled and made up to 100 ml. The phosphorus is determined colorimetrically as molydovanadophosphoric acid.

88.5 EXPERIMENTAL PROCEDURE

88.5.1 Sample Digestion.

Pass oxygen into a flask containing 5 ml of 0.5 N sodium hydroxide and 4 ml saturated bromine water for 30 seconds.

Place 1 scoop of anhydrous sodium carbonate in the bottom half of a methyl cellulose capsule supported in a bored cork. Weigh a sample of not more than 32 mg, estimated to contain between 0.05 and 2.5 mg phosphorus, directly onto the sodium carbonate bed. Cover the sample with another scoop of sodium carbonate. Insert a filter paper fuse between the top and bottom halves of the capsule after making a small slit in the top half to give the clearance required. Secure the capsule in the platinum gauze with the wick towards the stopper.

Holding the flask horizontally remove the oxygen lead, light the sample fuse and quickly insert the stopper in the flask. Tilt the flask immediately at an angle of 135° from the vertical, so that the absorbing solution forms a liquid seal at the neck of the flask. During the period of maximum flame height, completely invert the flask to prevent impingement of the flame on the flask walls. Set aside until the mist has cleared (10 to 15 minutes is usually sufficient).

Open the flask and wash down the platinum gauze and surfaces of the reducing adaptor with demineralised water. Remove the adaptor and lower the gauze into the solution.

88.5.2 Spectrophotometric Evaluation.

Add to the solution 6 ml of 25% v. sulphuric acid and boil the solution for 15 minutes. Cool and transfer the solution to a 100 ml graduated flask. Add 10 ml of ammonium vanadate, followed by 10 ml of ammonium molybdate and make up to 100 ml. Allow 30 minutes for colour development.

Using either 1 or 4 cells, read off the optical density at 460 nm.

Carry out a blank determination.

88.5.3 Preparation of Calibration Curve.

Prepare separate calibration curves from standard phosphorus solutions for 1 cm and 4 cm cells containing 0 - 0.25 mg P/100 cm^{-3} and 0 - 0.06 mg P/100 cm^3 respectively, as follows:

Measure appropriate amounts of phosphorus standards into 100 cm^3 graduated flasks. Add 10 cm^3 of ammonium molybdate and 10 cm^3 of ammonium vanadate and make up to 100 cm^3. Allow 30 minutes for colour development.

Using either 1 cm or 4 cm cells read off the optical density at 460 nm. Carry out a reagent blank.

Prepare graphs of corrected absorbance against weight of phosphorus (mg per 100 cm^3).

88.6 DISCUSSION OF RESULTS

This method is capable of determining phosphorus in polymers in amounts down to 2% and up to 13% with an accuracy of $\pm$ 5%.

METHOD 89 - MICRO DETERMINATION OF PHOSPHORUS IN POLYMERS. ACID DIGESTION - SPECTROPHOTOMETRIC METHOD.

89.1 SUMMARY

This acid digestion - spectrophotometric method determines phosphorus in amounts down to 0.5% in polymers.

89.2 APPARATUS

Hot plate capable of maintaining a temperature of 250 - 300°C.
Fume chamber connected to water vacuum pump.
Pyrex flasks, 100 ml.
Spectrophotometer with matched cells (1 cm).
Small glass sample cups, with ground glass lids.

89.3 REAGENTS

Ammonium molybdate, 10% w aqueous.
Ammonium vanadate, 1% w/v in 2 $N.HNO_3$.
Perchloric acid, 70% A.R.
Sulphuric acid, A.R.
Standard phosphate solution, 0.1 mg P/cm^3.

89.4 METHOD

The sample is decomposed and oxidized in sulphuric acid-perchloric acid. After cooling, water, ammonium vanadate and ammonium molybdate are added to form the yellow phosphovanadomolybdate complex, which is measured colorimetrically at 430 nm.

89.5 EXPERIMENTAL PROCEDURE

89.5.1 Sample Digestion.

Weigh out a suitable amount of polymer (ranging from 25 - 50 mg at the 0.1 to 0.5% phosphorus level to 3 - 5 mg at the 20% phosphorus level).

Place the weighed sample in the flask. Add 3 cm^3 of concentrated sulphuric acid, then 0.5 cm^3 perchloric acid and digest on the hotplate at 250 - 300°C. After about 15 s the perchloric acid decomposes with formation of dense white fumes and the solution becomes yellow. Heating is continued for a further two m.

Remove the flask from the hotplate, allow to cool and then cautiously add a few drops of water and mix. Dilute to about 50 cm^3 with distilled water.

89.5.2 Spectrometric Evaluation.

Add 5 cm^3 of ammonium vanadate solution, followed by 10 cm^3 of ammonium molybdate solution, make up to 100 cm^3 and mix. Allow 15 m for colour development and read off the optical density at 430 nm. Carry out a blank determination.

89.5.3 Preparation of Calibration Curve.

Transfer by use of pipettes or a burette 0, 2, 5, 8, 10, 12, 15 and 20 cm^3 aliquots of standard phosphate solution into separate 100 cm^3 volume reaction flasks. Add 3 cm^3 of concentrated sulphuric acid and 0.5 ml of 70% perchloric acid.

Add a small glass bead to each flask and boil off the water by heating on the hotplate. Continue heating until the dense cloud of decomposing perchloric acid appears and then for a further two minutes.

Cool, add water cautiously to a volume of 50 - 60 cm^3. Add 5 cm^3 ammonium vanadate solution followed by 10 cm^3 of ammonium molybdate, make up to 100 cm^3 and leave for at least 15 m.

Read off the optical densities of the solutions using 1 cm cells at 430 nm. Plot milligrams of phosphorus/cm^3 against optical density. The average factor (F) may be calculated from the expression:

$$F = \frac{\text{mg P per cm}^3}{\text{O.D. reading}}$$

89.5.3 Calculation of Results

$$\text{Phosphorus, \% w} = \frac{(\text{O.D.}_1 - \text{O.D.}_2) \times F \times 100}{W}$$

where:

$O.D._1$ = optical density of test solution.

$O.D._2$ = optical density of blank.

F = factor obtained by standardisation against known concentrations of standard phosphate solution.

W = sample weight in mg.

89.6 DISCUSSION OF RESULTS

This method is capable of determining total phosphorus in polymers in amounts down to 0.5% and up to 20% with an accuracy of ± 5%.

CHAPTER 7.4

COMPOSITIONAL ANALYSIS

Methods 90 - 97

METHOD 90 - DETERMINATION OF NATURAL RUBBER, STYRENE-BUTADIENE RUBBER AND ETHYLENE-PROPYLENE DIENE TERPOLYMER IN COMPOUNDED RUBBER STOCKS. PYROLYSIS - GAS CHROMATOGRAPHY.[51]

90.1 SUMMARY

This pyrolysis - gas chromatography procedure enables the determination of monomer units such as butadiene, isoprene and propylene to be determined in natural and synthetic rubbers in amounts down to 0.5%.

90.2 APPARATUS

Pyrolysis unit. Loenco Division Infotronics Corp, Altadene, Calif. this Unit, Model 260 uses a 3" x 1/8" quartz U-tube as the pyrolysis chamber. It is heated by sliding a thermostatically controlled heated furnace over it for a fixed time. A sliding gas valve arrangement in the unit allows uninterrupted carrier gas flow through the gas chromatographic column while the sample chamber is being reloaded or purged with helium.

Stripping and backflushing valve. A Carle Micro Volume Eight Port Gas Chromatography Valve No. 2012 (Carle Instruments, Fullerton, Calif.) is used to avoid the introduction of high boiling pyrolysis products onto the chromatographic column. It is connected to the columns and the gas chromatograph. The pyrolysis products are allowed to go onto the main column for the first 15 minutes only, (Figure 7.27).

Gas Chromatograph. The pyrolysis products are analysed with an F & M 1609 gas chromatograph equipped with a flame ionization detector. The chromatographic peaks are recorded at electrometer range of 100 and a chart speed of 1/2 inch per minute. The area measurements are made by the triangulation method (peak height x width at half peak height). The main column was aluminium tubing 40 ft x 3/16 inch o.d. 0.032 in wall thickness. It is packed with 10% tricresylphosphate on 60-80 mesh Chromosorb P, the stripper and restrictor columns are 10 ft lengths of the 2/16 inch aluminium tubing packed as the main column. The columns are operated at 35°C. The injection port and the detector block are maintained at 175°C. The helium, hydrogen and air pressure are 30 psi (flowrato at 4.0), 15 psi (flowrato at 7.50) and 20 psi (flowrato 9.0) respectively. Relative retention data for the pyrolysis products are shown in Table 7.45.

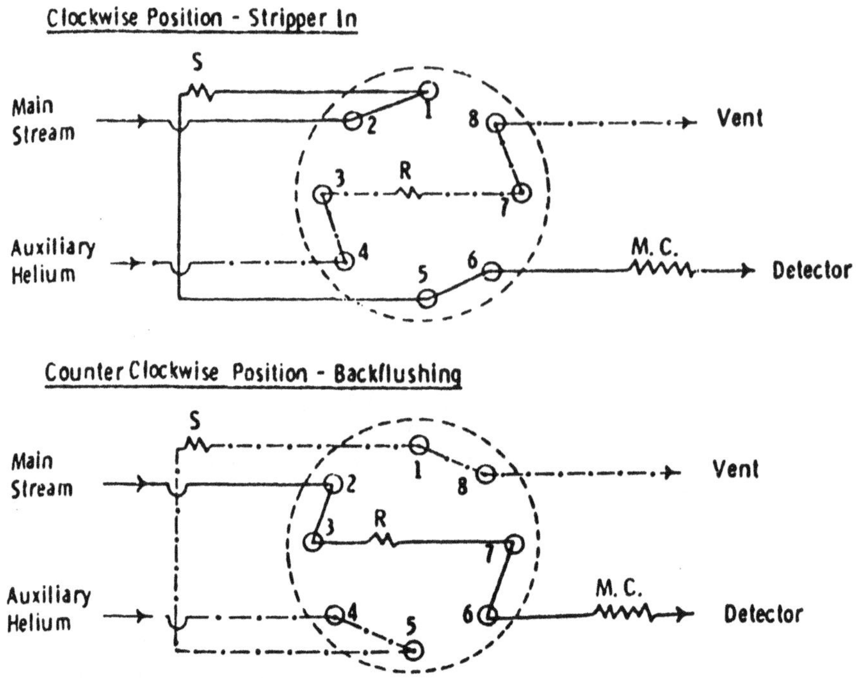

Figure 7.27. Stripping and backflushing valve connections.
M.C. Main column, 40 ft x 3/16 in. 10% tricresylphosphate on 60-80 mesh Chromosorb P.
S. Stripping column 10 ft x 3/16 in. 10% tricresylphosphate on 60-80 mesh Chromosorb P.
R. Restrictor column 10 ft x 3/16 in. 10% tricresylphosphae on 60-80 mesh Chromosorb P.
From Krishen with permission.[51] Americal Chemical Society.

Table 7.45 - Peak Identification and Relative Retention Data. (tricresylphosphate column at 35°C)

Peak no. in Figure	Compound	Relative Retention Nonane = 1.0000
1	Methane	0.0009
2	Ethane + Ethane	0.0082
3	Propane	0.0268
4	Propene	0.0347
5	2-Methylpropane	0.0849
6	Propadiene + Butane	0.0948
7	1-Butene + 2-Methylpropene	0.1048
8	Trans-2-Butene	0.1409
9	Cis-2-Butene	0.1649
10	1,3-Butadiene	0.1769
11	2-Methyl-1-Butene	0.2074
12	1-Pentene	0.3038
13	2-Methyl-1-Butene	0.2074
14	Trans-2-Pentene	0.3848
15	Cis-2-Pentene	0.3993
16	2-Methyl-2-Butene	0.4476
17	Isoprene	0.5397

From Krishen[51]. American Chemical Society

90.3 REAGENTS

Tricresylphosphate stationary phase.
Chromosorb P, 60-80 mesh.
Methanol, Analar.

90.4 METHOD

A methanol extract of a weighed portion of the polymer is pyrolysed at 700°C and the pyrolysis products swept on to a gas chromatograph and then determined. For example if a styrene-butadiene copolymer is being examined then determination of 1,3 butadiene in the pyrolysis products enables the bound butadiene content of the copolymer to be determined. The procedure is calibrated against copolymers of known composition, or, alternatively against injections of the pure monomers.

90.5 EXPERIMENTAL PROCEDURE

The cured rubber samples are subdivided into small pieces. A 0.1 to 1 gram sample of these pieces is extracted with methanol in Underwriters' extraction apparatus for 8-12 hours. The solvent is completely removed from the rubber by placing the sample in a vacuum oven at 60°C for 30 minutes. A 500 to 800 ug sample of the dried rubber is weighed on the microbalance and carefully placed inside the quartz pyrolysis tube. Small borosilicate glass wool plugs are put on each end of the U-tube. This tube is then installed in the pyrolyser unit, purged with a helium flow of 30 ml per minute for 2 minutes, and then heated for 30 seconds at 700°C by positioning the furnace over the tube. At the end of 30 second, the furnace is move away; the pyrolysis tube is connected in series with the column carrier flow using the appropriate valve on the pyrolysis unit; and the start of the chromatogram marked on the recorder chart. At the end of about 60 seconds, the carrier gas flow through the pyrolysis tube is terminated by sliding up the valve on the pyrolysis unit. The tube is removed from the pyrolysis unit, glass wool plugs are removed and the tube cleaned by heating it thoroughly in a burner flame to burn off all combustible material. Another weighed sample is then placed in the tube and it is installed in the pyrolysis unit as before ready for the next run. After 15 minutes, the flow of the carrier gas through the stripper column is terminated and the restrictor column placed in series with the main column by turning the Carle valve knob to the counter clockwise position. The stripper is back flushed while the chromatogram is being recorded. At the end of the chromatographic run, the Carle valve knob is turned back to the clockwise position and the apparatus is ready to accept the next run.

90.6 TYPICAL RESULTS

Peak identification and retention data are given in Table 7.45 for a range of pyrolysis products.

Krishen[51] quantitatively analysed the gaseous pyrolysis products from natural rubber, styrene-butadiene rubber and ethylene-propylene diamine terpolymer rubber by gas chromatography. He showed that the 2-methyl-2-butene peak was linear with the natural rubber content of the sample. Styrene-butadiene rubber was determined from the peak area of the 1,3-butadiene peak. The ethylene-propylene-terpolymer content was deduced from the 1-pentane peak area of the pyrolysis products.

90.7 DISCUSSION OF RESULTS

This method is capable of determining down to 0.5% of monomer units such as bound butadiene and isoprene in synthetic and natural rubbers with an accuracy of $\pm$ 5%.

METHOD 91 - DETERMINATION OF THE BOUND ETHYLENE IN ETHYLENE-PROPYLENE COPOLYMERS (>95% PROPYLENE). INFRARED SPECTROSCOPY.[52]

91.1 SUMMARY

This infrared spectroscopic procedure determines bound ethylene in propylene rich ethylene copolymers in amounts down to 0.1%.

91.2 APPARATUS

Fourier Transform infrared system at four wavenumber resolution in double beam operation. Standard double precision computer software were used to present data properly scale expanded in absorbance form.

91.3 METHOD

Determination of infrared spectra of moulded films of the polymer at 732 cm^{-1} enables the propylene unit content of the polymer to be determined. The absorbance is measured of the 732 cm^{-1} infrared band, attributed to $gamma_r$ $(CH_2)_3$ is characteristic of an ethylene unit isolated between two head to tail propylene units. The method is calibrated against known copolymers of similar constitution to the copolymers being analysed.

91.4 EXPERIMENTAL PROCEDURE

Infrared spectra are obtained on moulded films of a nominal 1 mm thickness. Approximately 0.52 g of sample is placed between 31.8 mm diameter aluminium foil disks on a 1 mm brass spacer of 19.0 mm i.d. and 31.8 mm o.d. resting in a Buehler 20-2112 specimen mould. The temperature of the assembled mould is maintained at 175°C by a standard thermocouple temperature controller. After the sample is pressed to 6000 psig in a Buehler 1315 AB specimen mould and cooled to 35°C under pressure, the film is removed and mounted in a holder. Five measurements with a micrometer are averaged to provide a film thickness for the ethylene calculations. Since small variations in thickness introduce substantial errors in final calculations, the film is discarded if the readings varied by more than 0.05 mm and another sample is moulded.

Infrared spectra are obtained on a Digilab Model 15 Fourier Transform system at four wavenumber resolution in double beam operation. An adequate signal-to-noise ratio resulted from 1000 scans. Standard double precision Digilab/Data General computer software are used to present data properly scale expanded in absorbance form. The absorbance of the 732^{-1} micron infrared band, attributed to gamma $_r$ $(CH_2)_3$ is recorded. The equation relating the infrared absorbance to the ethylene content is:

$$Y = 2.465\,X + 0.451$$

where Y is the infrared absorbance at 732 wavenumbers divided by the film thickness in centimetres and X is the wt% ethylene. The standard errors are 0.110 and the intercept 0.051 for the slope.

The infrared absorbance at $732cm^{-1}$ is sufficiently sensitive to the ethylene incorporation to determine the wt% ethylene within 0.1-0.2% at the 95% confidence level. From the 13-C NMR data, it can be concluded that the propylene units occur in predominantly isotactic, heat to tail sequences and that the ethylene units are incorporated as isolated units only. Thus, this structural prequisite is a requirement for application of this method because it has not been tested on copolymers containing propylene configurational irregularities or ethylene sequences two units and longer.

91.5 TYPICAL RESULTS

The infrared absorbance at $732cm^{-1}$ is sufficiently sensitive to the ethylene incorporation to determine the wt% ethylene to within 0.1 - 0.2 % at the 95% confidence level.

91.6 DISCUSSION OF RESULTS

This method enables down to 0.1% propylene to be determined in ethylene rich (> 95% propylene) ethylene-propylene copolymers with an accuracy of $\pm$ 5%.

METHOD 92 - DETERMINATION OF ETHYLENE IN ETHYLENE-PROPYLENE COPOLYMERS (>95% PROPYLENE). NMR SPECTROSCOPY.[52]

92.1 SUMMARY

This NMR method is capable of determining down to 0.1% propylene units in propylene rich (>95% propylene) ethylene-propylene copolymers.

92.2 APPARATUS

NMR Spectrometer Varian, XL0100 15 NMR equipped with Varian FT-100 pulsed NMR Fourier transform system and disc accessory or equivalent.

92.3 REAGENTS

1,2,4 trichlorobenzene.
Perdeuterobenzene.

92.4 METHOD

A 1,2,4 trichlorobenzene - perdeuterobenzene solution of the polymer is examined on a NMR spectrometer to evaluate methine resonances. The method is calibrated against known copolymers of similar constitution to the copolymers being analysed.

92.5 EXPERIMENTAL PROCEDURE

NMR spectra are measured from polymer samples dissolved in a mixture of 1,2,4-trichlorobenzene and perdeuterobenzene at concentrations between 10 - 15 wt %. Sufficient perdeuterobenzene is added to maintain a lock signal on the NMR spectrometer at a temperature of 125°C. The 13-C NMR spectra is obtained utilizing the following conditions for the pulse sequences:

pulse angle	90°	transients accumulated	2000 - 5000
pulse delay	8.5 s	spectral width	5000 Hz
acquisition time	1.5 s	no signal enhancement used	

A pulse interval of 10 s is sufficient to satisfy the spin-lattice relaxation time, T of the methyl group, which is the slowest relaxing nucleus with a T_1 of 2 s at 125°C. The number of transients accumulated depended upon the ethylene content and in each case, accumulations are allowed to continue until a satisfactory signal-to-noise ratio was achieved. Each 13C NMR spectrum is recorded with proton noise-decoupling to remove unwanted 13C - 1H scalar couplings. No corrections are made for differential nuclear Overhauser effects since these have been shown to be constant for the major resonances in low ethylene - propylene copolymers.[53]

92.6 TYPICAL RESULTS

The best results are obtained using methine resonances 4 and 5 (Table 7.46) to determine copolymer composition. This is because the methine carbon resonance is the least sensitive towards configurational differences and also the least affected by overlap from neighbouring resonances. Similar results are obtained whether one uses peak heights or peak areas.

92.7 DISCUSSION OF RESULTS

From the NMR data it can be concluded that the propylene units occur in predominantly isotactic head to tail sequences and that the ethylene units are incorporated as isolated units only. The NMR method can be used to provide reference standards for the less time consuming infrared method (Method 91). Provided the infrared method is calibrated in this way excellent agreement is obtained between the infrared and NMR methods for copolymers containing >95% propylene.

METHOD 93 - DETERMINATION OF BOUND STYRENE IN STYRENATED ALKYL RESINS. INFRARED SPECTROSCOPY.[54]

93.1 SUMMARY

This infrared spectroscopic method is capable of determining bound styrene units in styrenated alkyl resins in amounts down to 1%.

93.2 APPARATUS

Infrared spectrometer equipped with 0.062 cm path length liquid cell with potassium bromide windows.

93.3 REAGENTS

Tetrahydrofuran.
Acetone, AR.

93.4 METHOD

A tetrahydrofuran or acetone solution of the sample is run on an infrared spectrometer and the true absorbance of the 700 cm^{-1} absorption calculated. The method is calibrated against styrenated alkyl resins of known composition.

93.5 EXPERIMENTAL PROCEDURE

To determine styrene units, solutions of the polymers are prepared in volumetric flasks by dissolving known weights of the dried polymers in tetrahydrofuran or acetone. The infrared cell is rinsed twice with the sample solution before use. A liquid cell, having potassium bromide windows and a path length of 0.062 cm was used. The optimum region for performing quantitative infrared analysis is between 0.3 and 0.6 absorbance unit. Typical polymer concentrations were 3 to 8 wt% within the optimum region. Before scanning the spectrum, the filled infrared cell is placed in the instrument for 15 minutes to allow the sample to reach temperature equilibrium. The region of the spectrum between 15.38 and 13.33 micron is scanned at 17 cm^{-1}/min and the absorbance of the 13.33 micron band is determined using the "basic line technique".

93.5.1 Determination of Absorbtivity Value.

Polystyrene solutions are prepared between 10 and 60 g/l, and their infrared spectra determined 15.38 to 13.33 microns. The true absorbance of the 14.28 micron band is then calculated using the following equation:

$$A = Ar - Ao$$

where
A = true absorbance.

Ar = the total recorded absorbance of the absorption maxima.
Ao = the general beckground absorption at the absorption maxima.

The true absorbance of the 14.28 micron styrene band is then plotted against sample concentration. Using Beer's law an absorptivity value is readily calculated. The absorptivity values determined in acetone and tetrahydrofuran are given in Table 7.47.

93.5.2 Determination of Styrene Content.

By knowing the absorbance, cell path length and absorptivity value, the concentration (g/l) of styrene is a solution is calculated using the equation:

$$c = A/ab$$

where
c = styrene concentration (g/l) in the solution.
A = true absorbance of 14.28 micron band.
b = cell path length (cm).
a - absorptivity (1/g cm)

The concentration of styrene in the polymer is calculated using the equation:

$$\% \text{ styrene} = \frac{C}{W} (100$$

where
C = styrene concentration (g/l) in solution.
W = sample concentration in solution (g/l).

93.6 TYPICAL RESULTS

The absorptivity values determined in acetone and tetrahydrofuran are given in Table 7.47.

93.7 DISCUSSION

The accuracy of this method on typical acid modified styrenated acrylic polymers is shown in Table 7.48. In the 5 to 70% styrene range, determined values are better than 95% relative.

METHOD 94 - DETERMINATION OF STYRENE AND METHACRYLATE UNITS IN STYRENE METHYL-METHACRYLATE AND STYRENE-*N*-BUTYL METHACRYLATE COPOLYMERS. NMR SPECTROSCOPY.[55]

94.1 SUMMARY

This NMR method determines styrene and methacrylate units in styrene-methyl methacrylate and styrene-n-butyl methacrylate copolymers in amounts down to 0.5%.

Table 7.46 - Observed and Reference ^{13}C NMR Chemical Shifts in ppm for Ethylene-Propylene Copolymers and Reference Polypropylenes as Measured with Respect to an Internal TMS Standard

Resonance Line	Carbon	3/97 E/P[a]	E/P	97/3[a] E/P	Sequence Assign-ment	Reference Crystalline Polypropylene	Amorphous Polypropylene
1	alpha alphaCH_2	46.4	46.3		PPPP	46.5	47.0-47.5 r 46.5 m
2	alpha alphaCH_2	46.0	45.8		PPE		
3	alpha gammaCH_2	37.8	37.8		PPEP		
4	CH	30.9	30.7		PPE		
5	CH	28.8	28.7		PPP	28.5	28.8 mmm 28.6 mmmr 28.5 rmmr 28.4 mr+rr
6	beta betaCH_2	24.5	24.4		PPEPP		
7	CH_3	21.8	21.6		PPPPP	21.8	21.3-21.8 mm 20.6-21.0 mr 19.9-20.3 rr
8	CH_3	21.6	21.4		PPPE		
9	CH_3	20.9	20.7		PPPEP		
	CH_3		19.8	19.8	EPE		
	CH		33.1	33.1	EPE		
	alphaCH_2			37.4	EPE		
	betaCH_2		EPE	27.3			
	$(CH_2)_n$		29.8	29.8	EEE		

From Paxton and Randall[52]. American Chemical Sociey
[a] As measured

Table 7.47 - Absolute Absorptivity Values Obtained for Styrene

Component	Absorptivity Value 1/cm g	Solvent
Styrene	1.903	Acetone
Styrene	1.911	Tetrahydrofuran

From Anderson et al.[54] American Chemical Society

94.2 APPARATUS

Proton NMR spectrometer, Joel C-60H Mhz spectrometer used in original work or equivalent.

94.3 REAGENTS

Deuterated chloroform.
Trimethylsilane.

Table 7.48 - Infrared Analysis for Bound Styrene in Styrenated Acrylic Polymers

Sample	Styrene	
	Calculated %	Observed %
1	5.0	5.0, 4.9
2	7.5	7.1, 6.9
3	10.0	19.6, 19.5
4	24.3	24.4, 24.3
5	24.3	25.6, 24.6
6	25.0	25.1, 25.8, 24.2[a]
7	35.0	36.3, 35.6
8	35.0	34.7, 35.2
9	40.0	28.9, 40.1
10	60.7	59.1, 60.3, 59.7[a]
11	70.0	69.3, 70.3, 69.5[a]

From Anderson.[54] American Chemical Society

[a] Percent styrene values, determined in tetrahydrofuran. All the remaining values were determined using acetone as the solvent.

94.4 METHOD

A deuterated chloroform solution of the polymer containing trimethylsilane internal standard is run on the proton NMR spectrometer. Integration of peak areas at delta 8 and 1-4 delta respectively gives a measure of signal strengths from aliphatic protons (methacrylate units) and aromatic protons (styrene). These areas can be used as a basis for calculating the methacrylate and styrene unit contents of the polymer.

94.5 EXPERIMENTAL PROCEDURE

The proton NMR spectra are obtained using a Jeol C-60H MHz spectrometer. Sweep width of 50 Hz and sweep time of 2.5 min are employed. For integration, sweep time is increased to 1.25 min. Deuterated chloroform is used as a solvent and trimethylsilane is used as an internal standard. The spectra are run at ambient temperature using 10-15 wt% solutions. Small variations in room temperature do not affect the precision of spectra as demonstrated by 1% standard deviation. Solutions of 10-15 wt% of copolymer are used since this gives the best spectra. If the number average molecular weight (Mn) of a copolymer is high, a lower wt% gives better results and vice versa. The sample size is optimized by varying the sample concentration to obtain the best possible spectra. Too high a concentration dampens the signals. One can easily differentiate the aromatic protons from the aliphatic protons since the former absorb at 8 delta whereas the latter absorb between 1-4 delta. Areas under these peaks are integrated. The integrated areas are related to either the aliphatic or aromatic protons of the copolymer under study. These areas can then be converted to weight per cent of the monomers using simultaneous equations.

94.6 DISCUSSION OF RESULTS

This procedure enables down to 0.5% styrene and methacrylate units to be determined in styrene methyl methacrylate and styrene-n-butyl methacrylate copolymers with an accuracy of ± 2%.

METHOD 95 - DETERMINATION OF STYRENE AND METHACRYLATE UNITS IN STYRENE-METHYL METACRYLATE AND STYRENE *N*-BUTYL METHACRYLATE COPOLYMERS. PYROLYSIS - GAS CHROMATOGRAPHY.[55]

95.1 SUMMARY

This pyrolysis gas chromatography method determines styrene and methacrylate units in styrene-methyl methacrylate and n-butyl methacrylate copolymers in amounts down to 0.5%.

95.2 APPARATUS

The pyrolyser. Chemical Data System Pyroprobe Model 100 are used as the pyrolyser. Quartz sample tubes, 30 mm long and l mm i.d. are placed in the platinum coil heating probe. Pyrolysis conditions are varied until the most reproducible pyrograms were obtained. The optimum operation conditions are interface temperature, 120°C; ramp, 20°C; interval, 20 s; final temperature 600°C. The ramp indicates rise in temperature °C/m. Interval indicates the period for which the final temperature is held.

The gas chromatograph. Hewlett Packard 5750 gas chromatograph with 12 ft x 1/8 inch stainless steel column packed with 80/100 mesh Chromosorb P (Applied Science) coated with 10% W-982 are used. Operating conditions: helium flow 25 ml/min; injection port temperature, 150°C; detector temperature, 250°C; flame ionization detector range 10 with 4 x attenuation; oven temperature programmed from 70-270°C at 8°C/min.

95.3 REAGENTS

Methylene dichloride.

95.4 METHOD

A methylene dichloride solution of the polymer is pyrolysed and the integrated areas of the styrene and acrylic monomers obtained from a gas chromatographic scan. The ratio of these two monomer peak areas proportional to the styrene and methacrylate unit contents of the original polymer is compared with that obtained for standard copolymers of known composition, and , hence the composition of the sample obtained.

95.5 EXPERIMENTAL PROCEDURE

The copolymer material (100 mg) is weighed in a 10 ml volumetric flask and dissolved in 10 ml methylene chloride. Two microlitres of this solution are placed at the centre of a quartz sample tube and allowed to dry. The sample tube is placed inside the platinum coil of the pyrolysis probe which is then inserted into the heated interface. Before initiating the run, 10 min are allowed for any entrapped air, residual solvents or monomers to escape.

For calculation of percent styrene and methacrylate in the sample, the ratio of integrated areas of acrylic monomer to that of the styrene monomer is obtained from the gas chromatographic scan. This ratio is then compared to the one obtained from a standard copolymer of known composition which is prepared by 100% conversion of the starting monomers.

95.6 DISCUSSION OF RESULTS

The analysis of comonomer composition does not appear to be significantly affected by the extent of randomness observed for these materials. The difference between NMR and gas chromatography pyrolysis results are in the range of 0.4% and 0-4.8% for styrene-n-butyl methacrylate and styrene-methyl methacrylate, respectively. The difference in carbon analysis and pyrolysis gas chromatography results is also in the same range for both the copolymers. Standard deviation for pyrolysis gas chromatography ranges from 1.2-2.1%. Precision for NMR analyses is better than 1%.

Pyrolysis of methyl methacrylate-ethylene dimethacrylate copolymer gives only once major peak (methyl methacrylate) using a hot filament detector. The composition of methyl methacrylate-ethylene dimethacrylate copolymer can, however, be determined by pyrolysing a weighed sample and using the ratio of sample weight to area of the methyl methacrylate peak for obtaining a standard analysis curve. Under favourable conditons and with careful control of pyrolysis column, and detector variable, constituents can be determined within $\pm$ 0.5%.

The differences in the pyrograms of block and random copolymers allows estimation of the comonomer distribution. Random copolymers of ethylene with methyl acrylate or methyl methacrylate yield on pyrolysis a lower ratio of methanol/methyl acrylate or methyl methacrylate, respectively, than block polymers of the same composition. Differential thermal analysis measurements give a first-order transition for block polymers only, and by measuring the area under the transition an indication of the minimum chain length between acrylate units can be obtained.

METHOD 96 - DETERMINATION OF COMPOSITIONAL ANALYSIS OF METHYLMETHACRYLATE- METHACRYLIC ACID COPOLYMERS. FOURIER TRANSFORM 13_C NMR SPECTROSCOPY.[56]

96.1 SUMMARY

This Fourier transform 13C NMR method determines the compositional analysis of methylmethacrylate-methacrylic acid copolymers.

96.2 APPARATUS

13C NMR spectrometer, various CFR-20 pluse Fourier transform spectrometer used in original work, or equivalent.

96.3 REAGENTS

Pyridine: pyridine d_5, 1:1 v/v.

96.4 METHOD

A pyridine-pyridine d_5 solution of the polymer is run of a 13C NMR spectrometer and acidic (methacrylic acid) and ester (methyl methacrylate) group resonances evaluated. The procedure is calibrated against standards of known composition.

96.5 EXPERIMENTAL PROCEDURE

Samples are prepared by dissolving or swelling ca. 0.3 g of copolymer in 2 g solvent - both components being weighed directly into a 10 mm NMR sample tube. A 50/50 mixture of pyridine and pyridine d_5 is used as the solvent and provided the deuterium internal lock for the spectrometer. Spectra are obtained on a Varian CFR-20 pulse Fourier transform spectrometer operating at 18.7 kg (20.0 Mhz 13C frequency). The pulse power delivered to the single coil 10 mm o.d. probe is sufficient to rotate the 13C magnetization by 90° in 15 us. Probe temperature under these conditions was 38°C. Spectrometer parameters for the determination of spectra are: ca 65 degree (10 us) pulse, 4 Khz spectral width, 1 second data acquisition time and 3 seconds delay between repetitive pulses (ie. a total experiment recycle time of 4 seconds). Typically 12,000-15,000 free induction decays (FID) are accumulated (ie. an "overnight" run mode was employed) for each quantitative determination. The accumulated FID is digitally filtered with a time constant that produces a 1.7 Hz line broadening. The 8 K data table is then Fourier transformed to yield a 4 Khz spectrum defined by 4K real points (ie. digital resolution ca 1 Hz).

Spin lattice relaxation times (T 1) for the carbonyl carbides are estimated from 180-5-90 pulse sequences and found to be in the range 0.8-1.1 seconds for ungassed solutions. It should be noted that the solutions were not degassed and that the presence of dissolved oxygen may influence the observed relaxation rate. If samples were degassed, the carbonyl relaxation could be longer in which case a longer delay between repetitive pulses would be required. The imposition of a 3 second pulse delay and employment of a 65° pulse ensured the recovery of the carbonyl magnetization during the recycle time, thus eliminating possible errors in quantitation resulting from differential relaxation times. Spectra are obtained under fully proton decoupled conditions using a gated-decoupling scheme in which the proton decoupler was off during the 3 seconds delay time and gated on at the start of the 1.0 seconds acquisition period. In this manner, problems associated with differential nuclear Overhauser enhancements (NOE) for different carbonyl carbons are avoided.

A method of checking that the spectrometer conditions are sufficient to yield quantitative data is to compare the total integral of the carbonyl region with those obtained for alpha-CH_3>C< CH2 since all integrals should be the same under quantitative

quantitative conditions. In all cases the integral values for all four resonance regions are within 2-3% with the deviations being random from sample to sample - ie. in one case the carbonyl might be the greatest of the four in magnitude and in another case the lowest.

A 13C NMR spectrum of a polymethylmethacrylate - methacrylic acid copolymer demonstrates the acidic group and the ester group resonances.

In Figure 7.28 is shown the fully proton decoupled 13C spectrum obtained in pyridine at 38°C of a (MMA/MAA) copolymer of ca. 33% acid composition. While the spectrum resolves into alpha CH3, CH2, OCH_3 >C<, and C = O regions, the structural similarity of the comonomers results in resonance overlap (at 20 Mhz) between ester and acid carbons in all regions but the carbonyl. The complete structure in the carbonyl region arises from the sensitivity of the carbons to the microstructural features of the copolymer chain, ie. sequence and tacticity effects. To assure these effects do not lead to overlap of ester and acid carbonyl resonance, chemical shifts of both homopolymers and homosteric copolymers were examined with the result that the carbonyl resonance region does separate into distinct acid and ester regions in pyridine.

96.6 DISCUSSION OF RESULTS

In order to test the reliability of compositional results obtained by the 13_C NMR, analyses of several copolymers were carried out by both 13_C NMR and titration. Titrations were carried out in 50% aqueous ethanol medium to the phenol phthalein end-point using 0.15 N aqueous potassium hydroxide as the titrant. As evidenced in Table 7.49 there is an excellent correlation between the compositional analysis by NMR and titration (correlation coefficient = 0.998) with most of the comparative analyses differing by 2% or less.

METHOD 97 - DETERMINATION OF HEXAFLUOROPROPYLENE AND VINYLIDENE FLUORIDE UNITS IN POLYHEXA FLUOROPROPYLENE - VINYLIDENE FLUORIDE CO POLYMERS. PYROLYSIS - GAS CHROMATOGRAPHY.[57]

97.1 SUMMARY

This pyrolysis - gas chromatography method determines hexafluoropropylene and vinylidene fluoride units in polyhexafluoropropylene-vinylidene fluoride copolymers in amounts down to 0.5%.

97.2 APPARATUS

Curie point pyrolyser (Figure 7.30).

Gas chromatograph, (Hewlett Packard HP5750 or equivalent, Figure 7.29, equipped with gas sampling valve). Magnetic induction coil located near to injection port. The solenoid gas valve is operated by a switch on the control panel. The HP-5750 gas chromatograph is equipped with a gas sampling valve and the solenoid valve is connected in place of the sample loop. With the gas sample valve open, the solenoid valve can be used to divert the carrier gas flow either directly through the injection port or through the pyrolysis chamber.

Table 7.49 - Compositional Analysis of Poly(methyl methacrylate-co-methacrtylic acid) by Titration and ^{13}C NMR

Test Sample Number	Acid Content	
	Base Titration %	^{13}C NMR %
1	14	13
2	15	15
3	21	22
4	23	24
5	23	26
6	24	27
7	24	24
8	25	27
9	24.5	23.5
10	25	24
11	25.5	25.5
12	29	28
13	33	33
14	36	33
15	48	48
16	48	49
17	45	49
18	50	49
19	97	100

From Johnson et al[56]. American Chemical Society

Gas chromatography conditions. Gaseous pyrolyzate separated on a 10 foot x 1.8 inch o.d. stainless steel column packed with beta,beta'-oxydipropionitrile Porasil C Durapak (Waters Associates), 80 to 100 mesh. The carrier gas is argon flowing at the rate of 15 cm 3 min^{-1} The temperature of the column oven is initially 90°C; 3 min after pyrolysis, the temperature of the column oven is raised to 120°C and maintained there. The flame ionization detector is optimized for fluorocarbons. The complete elution of all fragments takes 30 min.

Hypodermic needles. Huler point 24 gauge with syringe fitting removed and replaced by 3 mm od quartz tubes sealed with epoxy resin suitable for use in vacuum systems.

97.3 REAGENTS

Acetone, reagent grade.

97.4 METHOD

An acetone extract of the polymer is injected into a pyrolysis unit and the pyrolysis products swept into a gas chromatograph. The copolymer composition is calculated from the amount of vinylidine chloride and trifluoromethane produced during pyrolysis. The procedure is calibrated against standard polymers whose composition has been determined by 19_F NMR analysis.

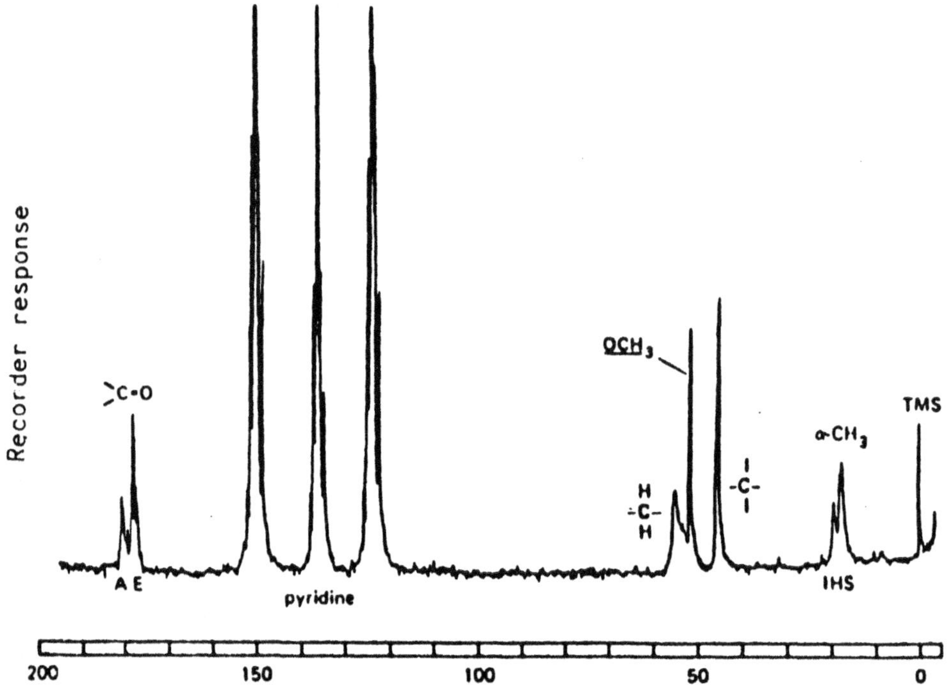

Figure 7.28. The 20Mhz 13C NMR spectrum of P(MMA-co-MAA)-33% acid dissolved in (50/50) v/v pyridine pyridine-d_5. The I,H,S notation refers to isotactic, heterotactic and syndiotactic stereochemical triads: A and E refer respectively to acid and ester. From Johnson et al with permission.[56] American Chemical Society

7.16

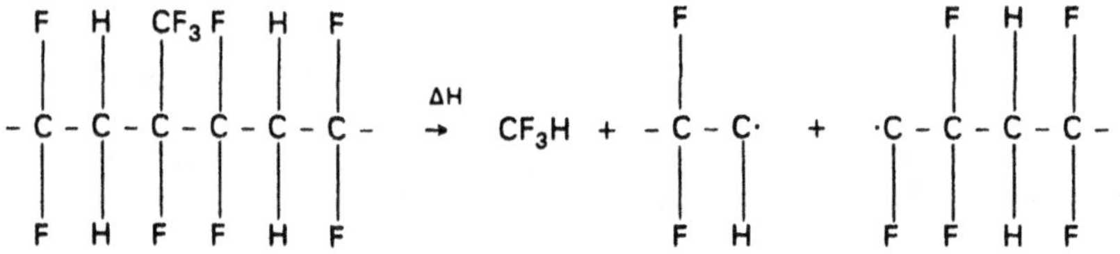

97.5 EXPERIMENTAL PROCEDURE

97.5.1 Sample Preparation.

A 10% solution of each hexafluoropropylene-vinylidene fluoride copolymer is prepared by dissolving 10 g of polymer in 100 ml of reagent grade acetone. A vial 58 mm high is filled with the sample solution. A pyrolysis wire, previously cleaned in a bunsen burner until red hot, then cooled, is dipped into the vial. The dipped wire is then dried in a vacuum oven at 90°C for 30 min. Use of lower drying temperatures or shorter times results in an acetone peak in the pyrogram. This is due to mechanically trapped

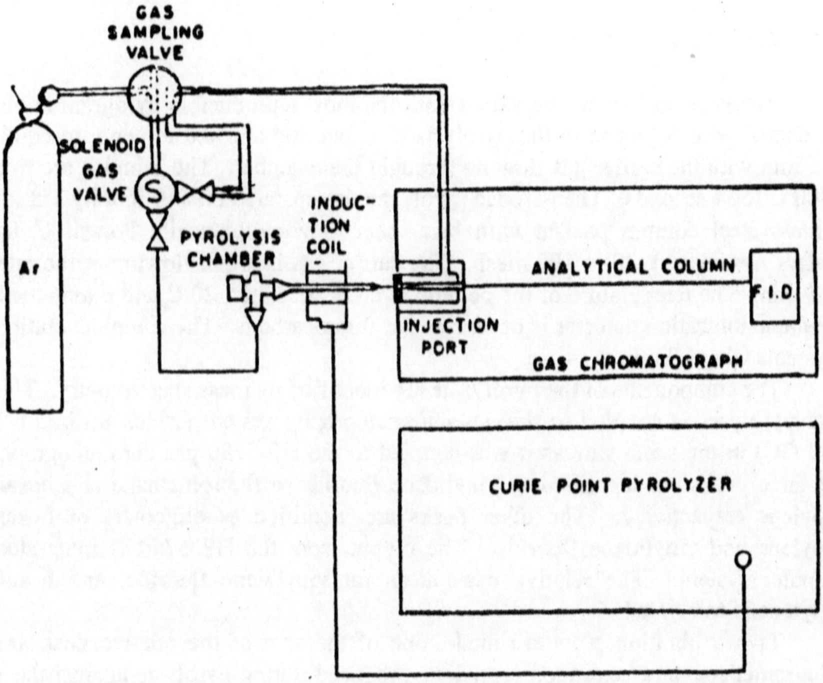

Figure 7.29. Schematic of the coupling of the Curie point pyrolyser to the gas chromatograph. The induction coil was positioned as close as possible to the injection port. The solenoid valve and the gas sampling valve were mounted directly on the gas chromatograph. From Blackwell with permission.[57] American Chemical Society.

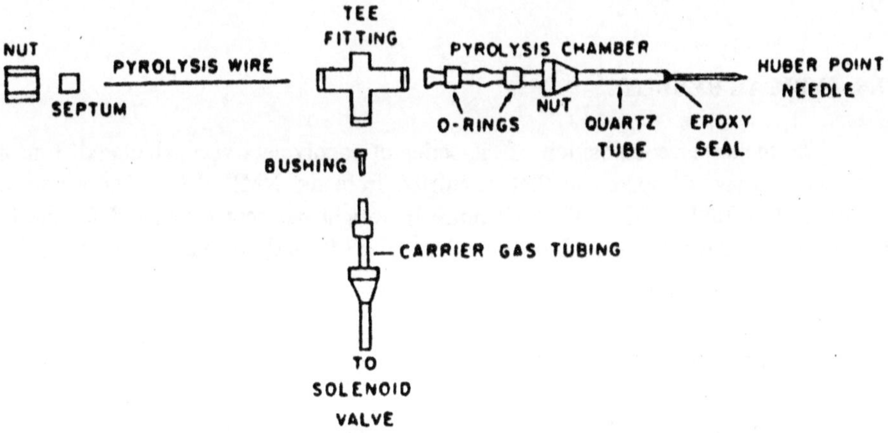

Figure 7.30. Schematic of the pyrolysis chamber. From Blackwell with permission.[57] American Chemical Society

acetone being liberated. This procedure will provide a pyrolysis wire coated with a thin film of uniform thickness and reproducible length.

97.5.2 Procedure.

Pyrolysis conditions are varied until the most reproducible pyrogram is obtained. The coated wire is placed in the pyrolysis chamber and allowed to come to equilibrium for 5 min with the carrier gas flowing throught the chamber. The samples are pyrolysed at 800°C for 4 seconds. The gaseous pyrolyzate is separated on a 10 foot by 1/8 inch o.d. stainless steel column packed with beta, beta' oxydipropionitrile Porasil C Durapak (Waters Associates), 80 to 100 mesh. The carrier gas is argon flowing at the rate of 15 cm 3 min. The temperature of the column oven is raised to 120°C and maintained there. The flame ionization detector is optimized for fluorocarbons. The complete elution of all fragments takes 30 min.

The components of the pyrolyzate are identified by mass spectrometry. The Curie point pyrolyser is coupled to the gas chromatography system (which utilizes a Varian 1800 GC) in the same way as it was coupled to the HP-5750 gas chromatograph. The first three peaks are identified as vinylidene fluoride, trifluoromethane and hexafluoropropylene respectively. The other peaks are identified as oligomers or hexafluoropropylene and vinylidene fluoride. The output from the HP 5750 is integrated by a computer system. The relative peak areas for vinylidene fluoride, and hexafluoropropylene is calculated.

Two calibration plots are made, one of the sum of the relative peak areas for trifluoromethane and hexafluoropropylene produced during pyrolysis against the weight per cent of hexafluoropropylene calculated from 19F NMR data and the other of the relative peak area for vinylidene fluoride produced during pyrolysis against the weight per cent vinylidene fluoride calculated from 19F NMR data. A least square fit was calculated for each calibration plot and the slope and intercept were determined.

The two calibration plots demonstrate a linear relationship between the pyrolysis products and the NMR data. The slopes of the two calibration plots are close to 1, the slope for weight per cent hexafluoropropylene is 0.99 and that for vinylidene fluoride is 0.97.

97.6 TYPICAL RESULTS

The monomer composition of the series of copolymers was calculated from the pyrolysis data and compared to that calculated from the NMR data. These data are summarized in Table 7.50. The difference in weight per cent calculated for the two techniques, averaged for hexafluoropropylene ± 6.0% and for vinylidene fluoride ± 0.85%.

97.7 DISCUSSION OF RESULTS

The components of the pyrolyzate can be identified by mass spectroscopy. The Curie point pyrolyzer is coupled to the gas chromatograph system (which utilizes a Varian 1800 GC) in the same way as it was coupled to the HP-5750 gas chromatograph (see Figure 7.29). The first three peaks are identified as vinylidene fluoride, trifluoromethane and hexafluoropropylene respectively. The other peaks are identified as oligomers of hexafluoropropylene and vinylidene fluoride.

The reproducibility of the pyrolysis step, achieved by the Curie point pyrolyser permitted the monomer composition to be determined with a reproducibility of ± 1%.

Table 7.50 - Comparison of the Monomer Composition of HFP/VF_2 Copolymer as Calculated by Pyrolysis Data and NMR Data

Weight % HFP			Weight % VF_2			Std Dev For Curie Point Pyrolyser
^{19}F NMR	Pyrolysis		^{19}F NMR	Pyrolysis		
28.6	28.8	0.2	71.4	70.9	0.5	1.04
29.0	28.8	0.2	71.0	71.0	0	0.34
34.79	36.04	1.25	65.21	62.63	2.58	0.52
36.46	36.73	0.27	63.96	62.91	1.05	0.82
37.24	37.33	0.11	62.76	62.61	0.15	0.39
39.04	39.07	0.03	60.96	60.88	0.08	0.67
39.04	38.35	0.69	60.96	61.55	0.59	0.63
39.82	38.46	1.36	60.18	61.89	1.71	0.30
40.10	39.45	0.65	59.90	61.49	0.57	0.10
42.4	43.2	0.8	57.6	56.7	0.9	0.39
44.8	46.2	0.2	55.2	53.8	1.4	0.48
44.8	45.4	1.4	55.2	54.3	0.9	0.59
47.3	47.0	0.3	52.7	52.8	0.1	0.92
49.95	49.45	0.50	50.05	50.55	0.50	0.30

CHAPTER 7.5

ADDITIVE MIXTURES

Methods 98 - 108

METHOD 98 - DETERMINATION OF MIXTURES OF ADDITIVES IN POLYMERS. THIN-LAYER CHROMATOGRAPHY - INFRARED AND ULTRAVIOLET SPECTROSCOPY.

98.1 SUMMARY

This thin-layer chromatography - spectroscopic procedure is capable of identifying and determining a wide range of additives in polymers in amounts down to a few hundred ppm. Several additives can be evaluated during the same analysis.

98.2 APPARATUS

Thin layer plates. Plates (20 cm x 20 cm and 20 cm x 5 cm) coated with 250 u thickness of Merck silica gel G 254 or GF 254 (ultraviolet fluorescent). These plates can either be prepared in the laboratory using an adjustable spreader or prepared plates can be obtained commercially.

Alumina type D5F (Fluka), fluorescent MN cellulose powder 300 F254 (Mackery Nagel and Co), fluorescent silica gel HF 254 + 366 (Merck) fluorescent.

The glass used to prepare plates in the laboratory should be cleaned with chromic acid, then with deionised water avoiding detergents and dried in an air oven prior to coating with silica gel in a clean laboratory atmosphere; conditioning in an air oven for 30 min at 120°C in a vertical position should be followed by storage in a desiccant box until required for use. The plates should preferably be used within 1 - 2 h of preparation so that their activity does not change appreciably and to reduce the possibility of contamination of the silica gel layer by volatile impurities in the laboratory atmosphere which might interfere in the subsequent evaluation of the plates.

Volumetric flasks, 0.5 and 1 cm^3.

Automatic sample applicator, for applying polymer extracts to thin-layer plates capable of applying 35 cm band of extract to plates.

Ultraviolet lamp, shortwave (254 nm) and longwave (366 nm).

Electrically heated copper wire.

Stylus.

Extraction thimble, sintered glass, see Figure 7.31.

Elution column, see Figure 7.32.

Infrared spectrometer, with beam condenser and ultramicro cavity cells.

Ultraviolet spectrometer with rectagular microcells (Figure 7.33)

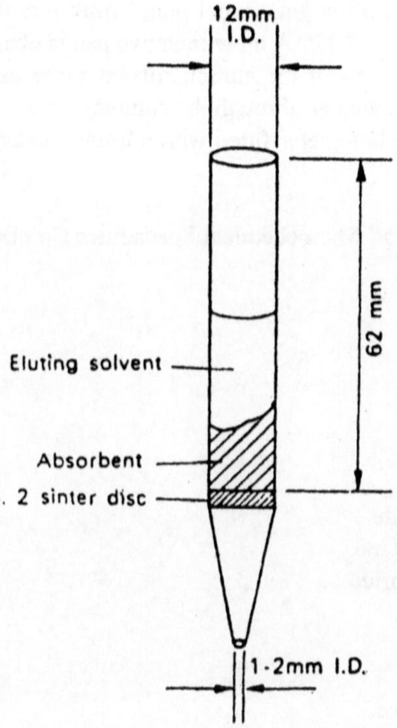

Figure 7.31. Filtration apparatus for extracting separated additives from adsorbent isolated from thin-layer chromatography plates.

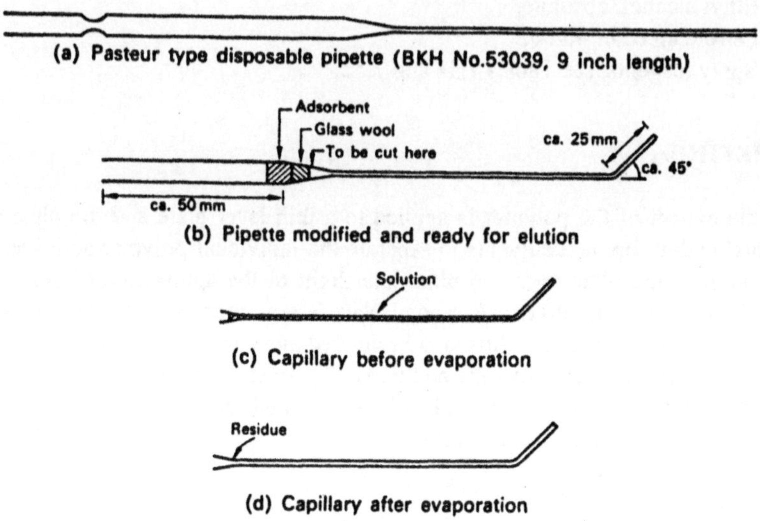

Figure 7.32. Steps in preparation and use of elution column used for GLC fraction collecting; (a) pasteur-type disposable pipette (CKH No. 53039, 9 in length); (b) pipette modified and ready for elution; (c) capillary before evaporation; (d) capillary after evaporation.

Elution columns, disposable, made from Pasteur-type pipettes (230 mm size) as shown in Figure 7.32. Pack a glass wool plug firmly into the larger half of the tapered section as shown in Figure 7.32. A more retentive pad is obtained if the glass wool pad is dampened with a few drops of the elution solvent to be used just before use. Excess liquid is removed by drawing air through the column.

Infrared spectrophotometer fitted with a beam condenser and "ultramicro" cavity cells.[69]

Centrifuge.

Micromanipulator, Microchemical Specialities Co., Berkeley, Calif., No. 5020, or equivalent.

Hair dryer

98.3 REAGENTS

Acetone.
n-hexane.
Carbon disulphide.
Tetrachloroethylene.
Carbon tetrachloride.
Toluene.
Benzene.
Chloroform.
Methyl cellosolve.
Dioxan.
Ethylene dichloride.
Ethyl acetate.
Nitromethane.
Acetone.
Methyl alcohol.
Ethyl alcohol, absolute.
Petroleum ether 40/60.
Spray reagents, see Tables 7.51 and 7.52.

98.4 METHOD

A suitable extract of the polymer is applied to a thin layer plate and the plate migrated with suitable development solvents to separate the individual polymer additives present. Examination of the plate under an ultraviolet light or the application of spray reagents reveals the separated additives. A second plate is now prepared and the bands of plate which contain the separated additives are removed and each band is separately extracted with a suitable solvent to extract the additives. These extracts are then concentrated and dissolved in suitable spectroscopic solvents for infrared and/or ultra-violet spectroscopy to identify and/or determine each additive.

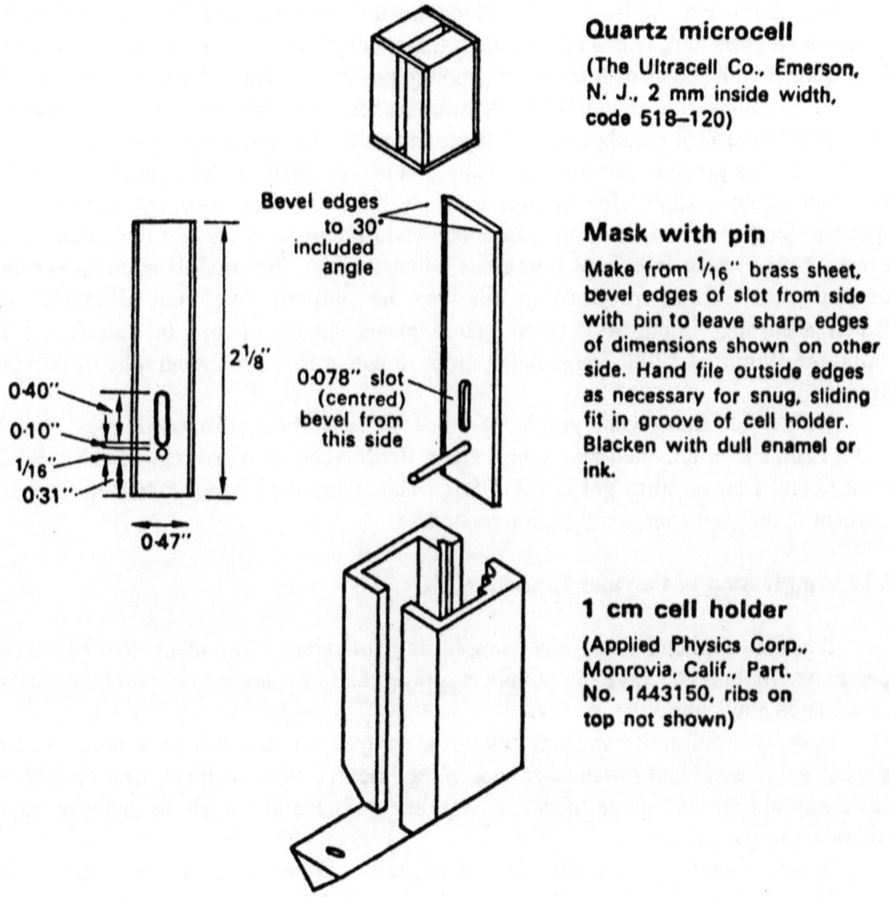

Figure 7.33. Microcell, mask and cell holder used in ultraviolet spectroscopy.

98.5 EXPERIMENTAL PROCEDURE

98.5.1 Premigration of Plate.

Thin-layer chromatographic grades of silica gel usually contain traces of inorganic impurities. If, during the development of the chromatogram these impurities migrate to regions of the plate which coincide with the Rf values of separated additives then the impurities will interfere with the interpretation of the plate following application of spray reagents. A plate premigration procedure for overcoming such interferences is discussed in Note 1.

The following procedure is suitable for the premigration of adsorbent impurities from thin-layer plates prior to the use of the plates for separations of polymer additives.

In the case of silica gel, at least, methyl alcohol is the recommended solvent. Methanol is poured into a glass tank with a ground glass lid to a depth of 1 - 2 cm and the walls of the tank are lined with sheets of filter paper dipping into the solvent. The tank is left for 30 - 60 min in a draught-free area until the interior has become saturated with solvent vapour. The coated 20 cm x 20 cm plate is then supported vertically in the methanol layer and left until solvent has ascended to the top of the plate. The plate is then removed from the tank, conditioned at 120°C (for 30 min) and the top 5 - 10 cm of adsorbent containing the impurities scraped off with a sharp instrument and thrown away leaving the remainder of the adsorbent coating free from impurities. Plates should be used as soon as possible after activation. Plates prepared in this way are virtually free from ultraviolet and infrared absorbing impurities. Premigrated plates should always be reactivated by heating for 30 min at 120°C immediately prior to use in the chromatography of polymer extracts.

Ultraviolet fluorescent grades of silica gel absorbent, discussed later, usually contain manganese activated zinc silicate as a fluoroescening agent, eg. Merck GF 254 Merck G and Camag silica gel G (SG-05F). This compound is not extracted from the adsorbent in the methanol premigration procedure.

98.5.2 Application of Polymer Extract to Plate.

Regular application of the sample is particularly important for successful separations. In this procedure the sample is sprayed on to the layer as a band by means of an automated applicator.

In order to minimize washing-out of the sprayed solution at high sample volumes the most volatile solvent possible is chosen, eg. diethyl ether or methylene dichloride. Where appropriate the plate may be warmed beforehand so as to achieve rapid evaporation of the solvent.

A convenient trial sample size of solvent solution of a polymer extract for application as a band on a 250 u thick 20 cm x 20 cm plate is 1 cm^3 of a 1% solution, ie. 10 mg.

98.5.3 Selection of Chromatographic Elution Solvent.

Chromatograph an unknown mixture with solvents of different polarities to obtain an idea of the types of compounds on the plate, see Note 2.

In order to avoid 'missing' any sample components it is highly advisable to apply the sample location techniques described in the last section to chromatograms of the sample obtained using several plate materials such as silica gels and alumina, and with each adsorbent to use as wide a variety as possible of different types of development solvents. These steps reduce to a minimum mistakes brought about by supposedly pure separated compounds consisting, in fact, of two or more unresolved compounds and enable a suitable development solvent to be selected.

98.5.4 Detection of Separated Compounds on the Plate.

Detection techniques should be carried out immediately after the chromatogram has been developed, in order to reduce to a minimum any opportunity for volatile sample constituents to be lost by evaporation from the plate. Detection of the separated compounds on the plate is achieved by examination under ultraviolet light which locates

Table 7.51 - General Spray Reagents for Location of Compounds on 20 cm x 20 cm Silica Gel Coated Thin-Layer Chromatography Plates

A. Reagents Applied to FG 254 Plate Without Subsequent Heating*		
1	Potassium permanganate(0.1 N) in aqueous sodium carbonate	(5% w/v)
2	Potassium permanganate	(2% w/v) in aqueous sulphuric acid (6% v/v)
3	Potassium permanganate	0.1% w/v in sulphuric acid (96%)
4	Antimony pentachloride	(2% w/v) in carbon tetrachloride
5	Phosphomolybdic acid	(3% w/v) in ethanol, then expose plate to amonia vapour
B. Reagents Applied to G254 Plate with Subsequent Heating*		
1	Sulphuric acid aqueous (20% w/v)	5-15 minutes at 120°C, then 5 minutes at 150°C
2	As (2) under A	5-15 minutes at 120°C, then 5 minutes at 150°C
3	Phosphoric acid (10%) methanolic	5-15 minutes at 120°C, then 5 minutes at 150°C
4	Perchloric acid (2%) methanolic	5-15 minutes at 120°C, then 5 minutes at 150°C
5	As (4) under A	5-15 minutes at 120°C, then 5 minutes at 150°C
6	Phosphomolybdic acid (20% w/v in methanol or methyl cellosolve, then expose plate to ammonia vapour	5-15 minutes at 120°C

* Spray 20 cm x 2 cm wide sections of plate with each reagent using an aluminium or glass mask with suitable aperture

some, but not all, types of compounds and by spraying with a range of general or specific spray reagents. (Tables 7.51 and 7.52 and Note 3).

98.5.5 Removal of Separated Compounds from Plate.

The next stage is to run a chromatogram on fresh 20 cm x 20 cm plates using a suitably sized sample (say 1 cm^3 of a 1% solution) and mark off with a sharp stylus the bands corresponding to the known positions of the separated compounds. No detection reagents are applied to these plates although they may be examined under the ultraviolet light to precisely locate any components which show up under these conditions. The simplest method of removing the zones containing the separated compounds (after allowing solvent to evaporate from the plate) is to hold the plate vertically its side resting on a sheet of paper and to scrape off the desired zone with a spatula. For substances which are not sensitive to oxidation, the zones may be sucked from the layer directly into an extraction thimble by using a small 'vacuum cleaner'.[72] Each separated adsorbent band can then be bottled off in 5 ml polyethene stoppered tubes and retained for further examination.

Each portion of adsorbent is transferred from the storage tube to a small sintered glass extraction thimble (Figure 7.31) and the organic compounds leached out with a suitable solvent such as anhydrous absolute ethanol, diethyl ether or methylene dichloride (see Note 4).

Table 7.52 - Chemical Analysis of Additives in Plastics: Specific Spray Reagents for Locations of Compounds by Thin-Layer Chromatography

Additive Type	Spray Reagent	Reference
Phenolic antioxidants	1. 2,6-dichloro-benzoquinone chlorimine (1-2% in ethanol followed 15 mins later by 2% borax in 40% aqueous ethanol)	75 - 77
	2. Alpha, alpha'-diphenyl picryl hydrazyl (0.1% in 95% aqueous ethanol)	77
	3. Palladium chloride (mix 150 ml palladium chloride with 100 ml of 2 N hydrochloric acid)	77
	4. Diazotized p-nitroaniline (mix 5 ml 0.5% p-nitroaniline in 2 N hydrochloric acid with 0.5 ml 5% sodium nitrite until colourless and 15 minutes later add 15 ml 20% sodium acetate)	74
Amine antioxidants	Diazotized p-nitroaniline (mix 5 ml 0.5% p-nitroaniline in 2 N hydrochloric acid with 0.5 ml 5% sodium nitrite, until colourless and 15 minutes later add 15 ml 20% sodium acetate)	74
Dialkyl thiodipropionates	Potassium platinoiodide (mix 5 ml 5% platinum tetrachloride in 1 N hydrochloric acid with 45 ml 10% potassium iodide and 100 ml water)	77
Phthalate ester plasticizer	Resorcinol. (Spray with 20% aqueous resorcinol in 2% aqueous zinc chloride and heat to 150°C. Then spray with 4 N sulphuric acid and heat for 20 minutes at 120°C. Spray with 40% potassium hydroxide to produce orange spots). Phthalic acid and phthalates also react	78, 79
Acids and bases	Bromocresol green (0.5% in 50% aqueous ethanol) Bromocresol purple (0.5% in 50% aqueous ethanol) Bromophenol blue, Methyl red (0.5% in 50% aqueous ethanol)	80, 81 80, 82, 83
Carboxylic acids, Aliphatic (primary, secondary, tertiary) amines, long chain quaternary salts and amine oxides	Sodium dichlorophenolindophenol (1% ethanolic). Cobalt thiocyanate 10 g $Co(NO_3)_2 6H_2 0$ and 10 g ammonium thiocyanate made up to 100 ml. Produces blue clour	84
Alkanolamines	Ninhydrin, heat plate for 5 minutes at 110°C, spray with 0.2% ninhydrin in acetone and heat 5 minutes at 110°C to produce colours. Further colours then produced upon spraying plate with 0.2% alizarin in acetone	79
Alkyl phenols	Phenols coupled as p-nitrophenol azo dyes applied to plate of silica gel impregnated with alkali. Separated azo dyes located as yellow/red colours upon exposure of plate to ammonia vapour	
Carbonyl compounds	Carbonyl compounds in sample converted to 2,4-dinitrophenylhydrazones, applied to thin-layer plate and plate developed. Separated 2,4-DNPH compounds located as yellow or brown colours upon spraying plate with 2% sodium hydroxide in 90% ethanol	85
Organic peroxides	Hydriodic acid (spray plate with a reagent comprising 40 ml glacial acetic and 0.2 g zinc dust added to 10 ml of 4% aq. potassium iodide, then spray with fresh 1% starch solution). Peroxides (and certain other types of oxidizing agents) revealed by liberation of free iodine. Alternatively use 2,6-dibromobenzoquinone chlorimine	86

* Spray 20 cm x 2 cm wide sections of the plate with various reagents using an aluminium or glass mask with suitable aperture

98.5.6 Preparation of Infrared Spectrum of Separated Additives.

Evaporate the low boiling solvent from the solution obtained in the previous section and dissolve the residue in a suitable spectroscopic solvent (chloroform, carbon tetrachloride or carbon disulphide). Transfer this solution into an ultramicro cavity cell.

The infrared spectrum of the solvent solution can then be obtained by normal infrared spectroscopic techniques.

Alternatively, especially if the compound is insoluble in the usual spectroscopic solvent, the solid can be dispersed in well-ground solid dry potassium bromide using a dental mixing machine and the mixture pressed into a 1 mm thick 5 mm diameter disc. This disc can then be mounted in a cardboard or plastic holder and used to prepare a spectrum.

98.5.7 Preparation of Ultraviolet Spectra of Separated Additives.

Measure the absorbance of the solution at the desired wavelength or record its spectrum using small volume microcells. Either air or the elution solvent may be used as the reference. Blank determinations which were carried through the entire procedure, including chromatography, should always be made and measured or recorded. It is important that the sections of adsorbent eluted for the blanks come from chromatograms prepared and handled in the same manner as the sample chromatogram and that they represent the same areas and locations of adsorbent. This is necessary because impurities, if present, are not distributed uniformly on the chromatogram. Removal of impurities from the region of the adsorbent layer to be used for the sample separation by washing the layer prior to migration of the sample using the methanol premigration technique described earlier is a convenient and effective method of reducing blank absorption.

Standards. When quantitative determinations are desired using this technique, it is necessary to known or to determine the absorptivities of the particular compounds. This is done by preparing and measuring solutions of known concentration of the pure compounds. It is also advisable to chromatograph known amounts of the pure compounds to verify the applicability of the technique to the particular compounds. This is recommended because unexpected errors can occur if compounds have enough volatility to escape from the adsorbent or are unstable and change during the chromatography and drying.

Note 1. Premigration of plate. Thin-layer chromatographic grades of silica gel ususally contain traces of inorganic impurities. If, during development of a chromatogram, these impurities migrate to regions of the plate which coincide with the Rf values of separated additives, then the impurities will interfere in the interpretation of the plate, following spraying with aggressive detection reagents, such as concentrated sulphuric acid and antimony pentachloride, as the organic impurities in the adsorbent may also react with these reagents and show upon the sprayed chromatograms. Also the impurities absorb strongly in the ultraviolet region, especially below 250 nm. The adsorbent impurities do not have an appreciable absorption in the infrared region. As organic and inorganic impurities are extracted from the silica gel with organic solvents such as diethyl ether, ethanol, acetone, benzene, chloroform and many others, they could occur as contaminants in some of the fractions of separated additives isolated from the plate by solvent extraction of the gel and consequently interfere in the interpretation of the spectra of the additives, particularly in the case of additives which absorb below 250 in the ultraviolet. For this reason an identical blank chromatogram (only sample absent) should always be run in parallel with the sample chromatogram in order to check whether such interference effects exist.

Ultraviolet absorbing impurities in adsorbants are influenced by the nature of the migration solvent. Depending on the polarity of the migration solvent used the impurities migrate to a greater or lesser extent up the plate towards the solvent front with the result that the lower part of the chromatogram nearer the baseline is cleared of impurities and the impurities become concentrated in the upper section of the plate nearer the solvent front. Subsequently, when a section of the absorbent is removed and eluted with a further solvent to recover a separated compound the amount of impurities contaminating the compound will depend on the location of the compound on the chromatogram (Rf values) and contamination may range from negligible to substantial.

In circumstances where slow moving compounds are being separated, the impurities may move away from the polymer additives towards the solvent front and thus not interfere in the subsequent examination of the separated compounds. This behaviour leads to a convenient method for moving the impurities beyond the section of the chromatogram to be used for the separation of polymer additives by migrating the chromatogram with appropriate washing solvents before applying and migrating the sample mixture. If this premigration washing covers a longer distance on the plate than is to be used in the sample migration then it is possible to move interfering impurities out of the way.

Note 2. Selection of chromatographic solvent. An unknown mixture should be first chromatographed on 20 cm x 5 cm plates with solvent of different polarities to obtain an idea of the polarities of compounds present in the sample and to reduce the possibility of missing any of the sample components. Solvents of low polarity, such a n-hexane, tetrachloroethylene and carbon tetrachloride, cause polar sample constituents to migrate more readily. Solvents of intermediate polarity such as toluene, benzene, chloroform and methyl cellusolve have a greater elutive effect on polar sample components, whereas highly polar solvents such as dioxan, methylene dichloride, ethyl acetate, nitromethane, acetone, lower alcohols and water elute polar sample constituents towards the solvent front, ie. Rf values near unity. Mixtures of 40/60 petroleum spirit and up to 10% (v/v) ethyl acetate are very useful general solvents for the separation of unknown mixtures.

Note 3. Detection of separated compounds on the plate. Alumina and cellulose powders are both available in fluorescent forms, respectively, alumina Fluka type D5F and MN cellulose powder 300 F254 (Macherey Nagel & Co.). Exposure of plates to 254 nm radiation permits visualisation of substances on the plate which absorb above 230 nm as a dark area on a blue fluorescent background. Fluorescent indicators which are activated by long-wave length ultraviolet light may also be incorporated in the adsorbent. The sodium salts of hydroxy-pyrene-sulphonic acids are particularly suited for this purpose. The separated polymer constituents show on the chromatogram, partly as dark zones, partly as zones which fluoresce brightly. The Merck silica gel HF 254+366 contains two fluorescence indicators, long and short ultraviolet light sensitive. To incite fluorescence, apparatus is needed which will allow a choice of short or long wave ultraviolet radiation. The non-fluorescent silica gel, Merck G254, reveals the presence of polymer constituents which have an intrinsic fluorescence under short wave (254 nm) or long wave (366 nm) ultraviolet light.

A further general test for organic compounds on the plate involves holding an electrically heated 25 cm long copper wire, set at red heat, at a few millilitres above the plate along the length in which the chromatogram has been developed. After a few seconds exposure, many types of organic compounds reveal themselves by charring or otherwise discolouring.

Note 4. Removal of separated compounds from the plate. The extraction solvent must: (l) be a good solvent for the additive, (2) be sufficiently polar to desorb the additive from the absorbent (successive desorption with different solvent may be necessary at this stage), (3) have a low boiling point to facilitate subsequent removal of solvent and reduce to a minimum evaporation losses of any volatile sample constituents, and/or, (4) not interfere in the subsequent spectroscopic examination of extracts.

Provided the desorption solvent is sufficiently powerful and polar, it should recover between 50 and 100% of the additive present in the silica gel fraction and provide sufficient material for examination by ultraviolet or infrared spectroscopy.

During the solvent elution of compounds from the isolated bands of adsorbent, the elution solvent effectively displaces the components from the adsorbent so that all of the component is contained in the first liquid emerging from the sintered glass thimble. Loss of the component by its failure to elute is not usually serious, provided that the eluant selected is one which would elute the particular compounds to the solvent front (ie. R f near unity) on a thin-layer plate. In fact, the behaviour of the component on the plate with plate elution solvents is a good guide to the selection of an appropriate solvent for desorbing the same component from the isolated gel in the extraction thimble. If the elution solvent used is less effective than this, then losses can be expected of the compound at the adsorbent extraction stage.

In extractions from silica gel, care should be taken to ensure that, according to the solvent used, components of the adsorbent are not coextracted with polymer additives. Silica gel and cellulose powder adsorbents containing gypsum as a binder cannot be extracted with water or with aqueous organic solvent mixtures because calcium sulphate is quantitatively extracted at the same time. A significant extraction residue is obtained even using chloroform, benzene or acetone as extractants. A significant inorganic extraction residue is also obtained from gypsum-free silica gel to which silicon dioxide of particle size less than 40 nm is added as a binder; this extract decreases with decreasing polarity of the extraction solvent. The extract solution should, therefore, be filtered through a hardened filter at the base of the extraction thimble (Figure 7.31) and/or through short columns of Celite. Alternatively, the solvent extract can be evaporated to dryness and the silica contaminated residue contacted with a good solvent for the polymer additive and a bad solvent for the silica.

98.6 TYPICAL RESULTS

The method was applied to a polymer suspected to contain one or more of the following additives; (1) Topanol CA (4,4′,4″ - butylene -1,1 3 tris (trimethyl-6-*tert* butyl phenol). (2) DLTDP (dilauryl thiodipropionate) and (3) Ionox 330 (1,3,5 tri methyl - 2,4,6-tri (3,5-di-*tert*-butyl)-4-hydroxy benzyl) benzene.

Additives were removed from 30 g of each of the polymers by Soxhlet extraction overnight with diethyl ether. The extract residue was made up to 10 cm^3 with chloroform and portions of these solutions and of solutions of the trace additives mentioned above were spotted on to silica gel plates. Viewing the developed plates under 254 nm ultraviolet light located Ionox 330 and Topanol CA but not DLTDP. All three additives were however located with aggressive spray reagents and with a reagent for phenols (2,6 dibromo benzoquinone-4-chlorimine) which gives a strong colour with hindered phenols and also a yellow colour with DLTDP. Polymers 1 and 2 both contained DLTDP and Ionox 330, polymer 2 also contained Topanol CA, and polymer 3 contained DLTDP and Topanol CA. Both of the aggressive spray reagents revealed the presence of low molecular weight polymer at the solvent front. These reagents also revealed the presence in Topanol CA of impurity which appears at the solvent front. Obviously, this impurity is of low polarity and is not phenolic as evidenced by the fact that it does not produce a colour with the phenol reagent. It is probably a hydrocarbon.

The method was also applied to the identification of an ultraviolet light stabiliser and an antioxidant in a sample of a polyolefin. Again the polymer was extracted with diethyl ether to isolate total additives and a portion of a chloroform solution of the extract run on a silica gel coated plate in parallel with various known light stabilisers and antioxidants commonly used in polymer formulations.

The polymer contained two additives appearing at Rf 0.6 and 0.85 which coincided in Rf value and in the colour obtained with 2,6 dibromobenzoquinone -4-chlorimine with known specimens of UV 531 (2-hydroxy-4-*n*-octoxy benzophenone) and Ionol CP (2,6-di-*tert* butyl-*p*-cresol). Spraying the plate with 2,6 dibromo benzoquinone-4-butyl chlorimine also revealed an additional orange coloured spot at Rf 0.8 (degraded Ionol) which did not coincide with any of the known additives tried. All three compounds showed up under ultraviolet light.

In Table 7.53 are summarized the results obtained in some experiments carried out to determine the recovery from thin-layer plates of compounds adsorbing at shorter and longer ultraviolet wavelengths, respectively, di-n-butyl phthalate (222 nm) and Ionox 330 (277 nm). These compounds were carried through the whole series of operations involving application of sample to a silica gel plate, solvent development, separation of adsorbent from plate and finally, solvent extraction of the compound from the adsorbent. Ionox 330 has a low Rf value with 4:1 cyclohexane:benzene development solvent hence, during solvent development, ultraviolet absorbing adsorbent impurities are swept well away from this compound to the solvent front. Premigration of the plate with methyl alcohol was unnecessary and not used, therefore, prior to application of Ionox 330 to the plate. Premigration of the plate with methanol before sample application was however, carried out in the case of di-n-butyl phthalate. This was because this compound has a fairly high R f value with the 9:1 iso-octane:ethyl acetate development solvent used, with consequent possible contamination of the di-*n*-butyl phthalate band with ultraviolet absorbing adsorbent impurities near the solvent front.

Table 7.53 - Reproducibility of Recovery from Polypropylene of Ionox 330 and di-n-butyl phthalate from Thin-Layer Plate (Silica Gel G254)

Compound	Absorbance Maximum	No. of Plates Prepared	Sample Size (ul)	Sample Concentration (% w/v)	Development Solvent	Plate Drying Time (h)	Section of adsobent removed from Plate Rf	Solvent used to Desorb Compound from Adsorbent	Adsorbtivity of Standard Solution of Compound (litres g cm)	Recovery of Compound	Standard Deviation
Ionox 330	277u	10	1	1	4:1 cyclohexane:benzene	18	0.03 - 0.14	Methanol (extract made up to 1 ml)	8.0	100	1.2
Di-n-butyl-p-phthalate	222*u*	5	1	5	9:1 iso-octane: ethyl acetate	18	-	Methanol (extract made up to 1 ml)	29.0	99.6	1.2

METHOD 99 - DETERMINATION OF ADDITIVES IN POLYETHYLENE AND PVC. HIGH PERFORMANCE LIQUID CHROMATOGRAPHY.[58]

99.1 SUMMARY

This hplc procedure is capable of determining a wide range of types of additives in polyethylene and PVC in amounts down to 0.01%.

99.2 APPARATUS AND EXPERIMENTAL PROCEDURE

99.2.1 Apparatus

The high pressure liquid chromatogrpahic system is shown in Figure 7.34. The unit employed in this particular method was a Whitey Micro-regulating high pressure feed pump (Whitney Research Tool Co. Oakland, Calif) equipped with a ll mm plunger, capable of a maximum flow rate of 7.3 cm^3 per h and output pressure of 5,000 psi. The solvent reservoir is placed a few feet above the pump since a slightly positive inlet pressure was required for operation. The degassed solvent is slowly stirred by means of a magnetic stirrer and heated externally slightly above room temperature to keep the solvent degassed. A Whitey Union Bonnet Bar stock valve is placed on the high pressure side of the pump to aid in priming the pump or to shut off the solvent flow when required. A Circle Seal (Circle Seal Co., Anaheim, Calif.) adjustable relief valve set at 5,000 psi is placed on the high pressure side of the pump for safety in case a blockage occurred upstream. A Hellicoid Pressure Gauge 0 - 5,000 psi with a snubber and cleaned for oxygen is used to monitor the output pressure. Downstream from the pressure gauge, the entire system is connected with 0.04 in i.d. 0.063 in o.d. stainless steel tubing to ensure a minimal dead-volume in the system which is particularly desirable when changing solvent or solvent programming. Up to 2,500 psi output pressure on ALC-100 LC pulse damper (Water Associates, Framingham, Mass.) are used.

The system is operated at ambient temperatures. The system is shielded from drafts and the components beyond the pulse damper are insulated with glass wool which was wrapped with several layers of aluminium foil. This procedure eliminated short term, localized temperature fluctuations and appeared to be satisfactory as far as system stability was concerned. No base-line drift occurs under these conditions.

To ensure saturation of the carrier liquid when doing liquid-liquid chromatography, a pre-saturator column is placed before the chromatographic column. It consists of a 8 in x 3/8 in o.d. stainless steel tube packed with 62 - 100 micron Porasil A. (Waters Asssociates) loaded with 5% by weight of the same liquid phase used in the chromatographic column.

When necessary, the liquid flow can be split between the analytical column and the reference column. A valve placed before the reference column permits the flow rate through that column to be completely shut-off or varied depending on the requirements of the system. This is particularly useful when using the refractive index monitor at higher flow rates. For column flow rates up to 2 cm^3 min^{-1} good stability could be obtained by careful balance of flow. The reference column was either filled with the same liquid-liquid support employed in the analytical column or merely filled with uncoated glass beads. With the ultraviolet detector the reference column is not used. Column head pressures are read on 1 - 3,000 psi or 0 - 5,000 psi standard gauges.

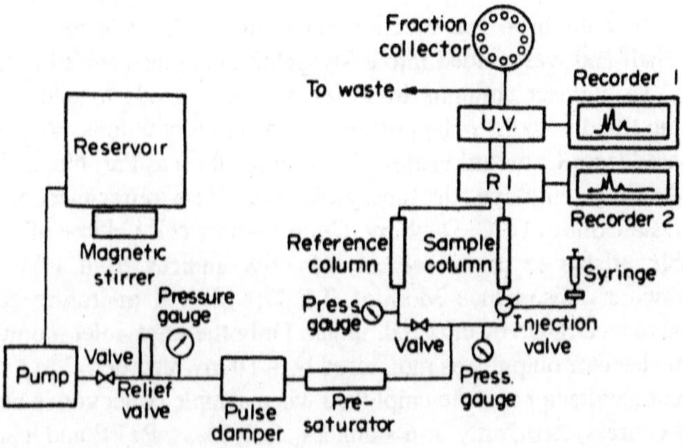

Figure 7.34. Schematic of high pressure liquid chromatographic system. From Majors with permission.[58] Preston Publications Inc., New York

A six-port rotary liquid sampling valve is used for high pressure injections in the continuous mode of operation. The basic design of the valve is that of Scott et al[73] but with the volume reduced by careful machining to approximately a hundredth of that in the original specifications. The ports are constructed with 0.010 in ie. 0.063 in o.d. stainless steel tubing and the slots in the plastic cylindrical sleeve which rotates to permit injection of the filled sample loop are drilled with a 0.010 in drill to the smallest possible depth. Because of the extremely close tolerances of the internal connections great care is needed to avoid the introduction of particulate matter which may accidently enter the stream and clog the tiny holes. Immediately before the eluent-in port, a high pressure stainless steel filter with a 5 micron frit is placed. Likewise, the sample loop is filled at ambient pressure by connection to a five cm^3 Luer-lok syringe equipped with a Millipore Filter holder. A five micron PTFE Millipore membrane is used as the filter material. These precautions permitted the valve to be used for hundreds of injections with a wide variety of sample without any clogging of the ports or internal sleeve.

The sample loop is connected to the valve with zero-dead volume Swagelok stainless steel capillary unions. The size of the loop can be varied depending on the injection size desired. For the minimum sample size, a three inch piece of 0.01 in i.d. tubing is used.

Since a six-port valve rotates to divert the eluent stream through the filled sample loop, there is a small internal volume (ie. the volume of the valve without the sample loop) which must necessarily be injected. This internal volume can be determined by filling the sample loop with a solution of N,N-dimethyl-p-phenylazoaniline dye dissolved in benzene and injecting the contents into a volumetric flask. The solution is diluted to volume with benzene and the dye concentration determined spectrophotometrically. The sample loop is calibrated externally in a similar manner. The volume of the small sample loop (4 ul) is subtracted from total volume injected to give the true sample injected volume (2 - 10 ul).

The columns used by Majors[58] were 2.1 mm i.d. x 0.125 in o.d. precision bore stainless steel. The columns were normally one metre in length and were rinsed thoroughly with dilute hydrochloric acid, water and acetone in that order and dried, prior to filling. A special stainless steel modified Swagelok union was constructed for connection of the column to the injection valve and to the detector. A capillary union was sawed in half and was welded into a Swagelok union also sawed in half as shown in Figure 7.35. To connect columns in series one only needs to add a short length of capillary tubing fitted with the male portion of the capillary unions. A very similar union can be purchased from Crawford Fitting Co. (Solon, Ohio) as Part No. 200-6-1-GC-316.

The detectors employed in the system are: (1) a refracto-monitor model 1103 (Laboratory Data Control LDC, Danbury, Conn.) with a cell volume of 3 microlitres and interchangeable prisms to cover solvent refractive indices from 1.31 - 1.55 and (2) Monitor Ultraviolet Absorbance Monitor (LDC) with a 8 microlitre cell volume and minimum absorbance range of 0.02 o.d. units. Only the ultraviolet monitor was used by Majors.[58] The detector output was monitored by a 10 mv recorder.

A fraction collector may be employed when sample collection is necessary.

In the entire system, only 3=6 stainless steel, glass, PTFE and a small amount of polyacetal in the injection valve contacted the solvent.

Chromatographic supports. (20 - 30 microns). E.I. du Pont de Nemours and Co., Wilmington, Delaware.

Corasil I and II (particle range 37 - 50 microns), Waters Associates, Framingham, Mass. (solid glass bead with either single (Type I) or a double (Type II) porous silica layer).

Durapak/OPN (36 - 75 microns), Waters Associates.

99.2.2 Column Packing Procedure.

Two small discs of Millipore PTFE filter (0.125 in in diameter) are placed inside an end fitting and the fitting placed on the column. Regular 1/8 in Swagelok stainless steel nuts and ferrules are used to attach the column to the end fitting. A short extension of 1/8 o.d. tubing is added directly to the top of the column by the use of a Swagelok union which was bored out internally to a 0.125 in diameter. A small funnel is used to add small, incremental amounts of packing and the bottom and sides gently trapped during the addition. Once the entire length has been filled, the column is tapped for an additional ten minutes. Finally the column with the small extension is connected to the outlet of the pump and the chromatographic solvent allowed to flow slowly through the column in an upward direction to sweep out air bubbles. As the liquid emerged from the column, the pressure is gradually increased to approximately a thousand pounds above the pressure at which the column would be used. The liquid is permitted to flow at this pressure for about 5 min. This final treatment insured that the column is packed to its maximum density. The extension was removed and the upper end fitting with a small disc of Millipore filter carefully pressed into the column end of it was attached to the column. Then the column is affixed to the chromatographic system.

99.2.3 Coating of the Support.

For liquid-liquid chromatography, the liquid phase is dissolved in a sufficient amount of dichloromethane to give Z-factor of 1.4. The Z-factor is defined as the ratio of the volume of solvent to the weight of support. The solution is added to the dry support in a round bottom flask. The volatile solvent is slowly removed by evaporation in a stream of dry nitrogen while slowly rotating the flask.

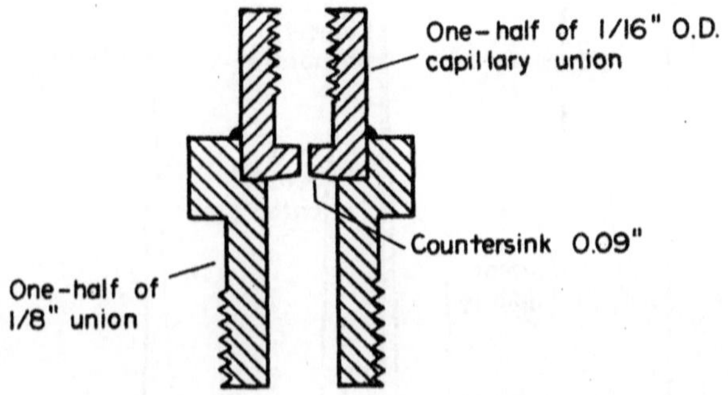

Figure 7.35. High pressure column chromatography. Low dead-volume column end fitting (not to scale). From Majors with permission.[58] Preston Publications Inc., New York.

99.3 TYPICAL RESULTS

To determine Irganox 1076, CAO-40, and Santonax R antioxidants in polyethylene the additives were extracted from the polymer by grinding them in a freezer mill then extracting with diethyl ether in a Soxhlet extractor for two days. The ether was then evaporated off and the residues dissolved in 25 cm^3 of 1% by volume of isopropanol in hexane. Using the chromatographic system of a 1% isopropanol in hexane solvent and Corasil II adsorbent (surface area $14 \pm 2\ m^2g^{-1}$) and the conditions outlined below the amount of hindered phenolic antioxidant in the ether extract of the polymer was determined. Column 1000 mm x 2.1 mm, i.d.; packing: 37-50 u Corasil II (activated at 110°C) carrier 1% (v/v) iso-propanol in hexane; flow rate - 0.95 ml/min.

Calibration curves over the concentration range of interest were obtained for the antioxidants known to be present. Figure 7.36 shows the separation achieved of three phthalate plasticizers used in poly(vinyl chloride). The separation of all three is obtained in less than 8 minutes on Zipax with HETP values less than a millimeter as summarized in Table 7.54. At the same flow rate the separation times on Corasil are considerably longer and the plate heights greater although not excessive.

99.4 DISCUSSION OF RESULTS

The method is capable of determining a wide range of antioxidants and other polymer additives in polyethylene and PVC extracts in amounts down to 0.01% with an accuracy of ± 5%.

METHOD 100 - DETERMINATION OF 0.01 - 0.1% ADDITIVES IN POLYPROPYLENE. CHEMICAL IONIZATION MASS SPECTROMETRY.[59]

100.1 SUMMARY

This chemical ionization mass spectrometric technique determines down to 0.01% additives in polypropylene without the necessity of a prior solvent extraction of the additives from the polymer.

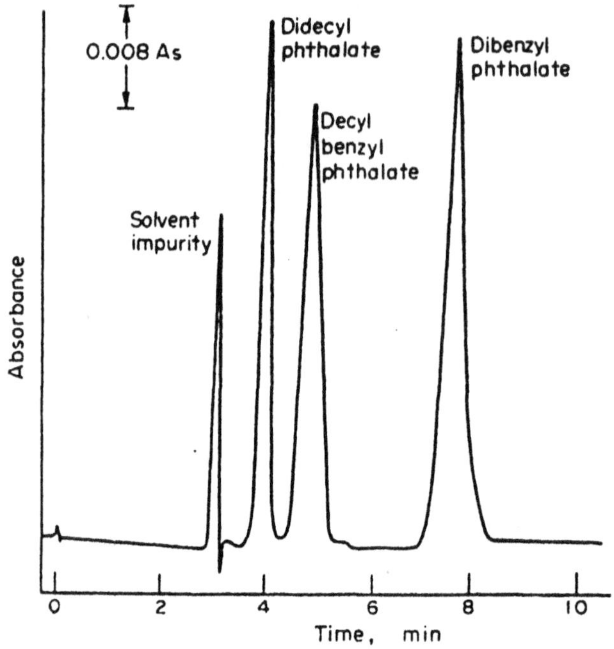

Figure 7.36. Separation of phthalate plasticisers using Zipex support column 1000 mm x 2.1 mm i.d., Zipax support, carrier-iso-octane; flow rate 0.50 ml/min sample - 10.6 ul of a mixture of 0.40 ul/ml each of didecyl phthalate and decyl benzyl phthalate and 0.35 mg/ml of dibenzyl phthalate in heptane.
From Majors with permission.[58] Preston Publications Inc., New York.

100.2 APPARATUS

Mass spectrometer, Dupont 492B used in original work or equivalent modified for chemical ionization operation.

Computer with data system, Hewlett Packard 210-MX used in original work or equivalent.

Capacitance Manometer, MKS Instruments Burlington or an instrument used in original work or equivalent.

100.3 REAGENTS

1.1% ammonia in methane mixture.

100.4 METHOD

A small pellet of the sample is placed in the heatable glass probe and spectra produced evenly using the computer and data system. The identity and concentration of antioxidants present in the polymer is obtained from the profiles for the additives being determined.

Table 7.54 - Evaluation of Solid Core Supports for Phthalate Plasticizers

	Zipax		Corasil 1		Deactivated Corasil	
Solute	V_R,ml	HETP,mm	V_R,ml	HETP,mm	V_R,ml	HETP,mm
Didecyl Phthalate	2.0	0.38	3.11.9	3.1	1.8	
Decyl Benzyl Phthalate	2.4	0.93	5.13.3	4.7	2.7	
Dibenzyl Phthalate	3.8	0.51	11.6	1.9	8.4	2.5

From Majors.[58] Preston Publications Inc, New York

100.5 EXPERIMENTAL PROCEDURE

A small slice of a polypropylene pellet (1 - 2 mg) is placed in the well of a heatable glass probe.[54] The probe is placed into the source of a mass spectrometer and heated from 30° C to 350° C at 20 or 30° C/min. Spectra are obtained every 6 s with the computer and data system. Spectra of the pure additives are also obtained by introducing the samples withthe heatable glass probe. The ion accelerating voltage is approximately 1750 V. The source repeller voltage is kept at 0 V to maximize the ionic residence times. A mixture of 1.1% ammonia in methane is used as the reagent gas at 0.5 torr, with the source temperature kept at 225° C for the pure additives and 240°C for the polymer samples. The source pressure is measured with a capacitance manometer through a hollow glass probe connected directly to the ion source.

100.6 TYPICAL RESULTS

Using this technique Rudewicz and Munson[59] determined Ionox 330 and Irganox 168 antioxidants and UV 531 ultraviolet absorber in polypropylene. Figure 7.37 shows two ion profiles when a 2.0 mg sample of polypropylene containing 0.15 wt % Irganox 168 (m/z 647 (M + H +) and 0.05 wt % Ionox 330 (M/z 792 (M + NH_4)+ is heated at 30° C/min. Both additives begin to vaporize rapidly at temperatures near the melting point of polypropylene, about 176°C. The two additives, however, can clearly be distinguished by their different masses and their very different heating profiles. The lower molecular weight additive gives a relatively sharp peak with a maximum at approximately 210°C and the higher molecular weight additive gives a broader peak that maximizes near 300°C.

It is possible to identify these additives from a small set of likely compounds from the masses of their characteristic (M + H)+ or (M + NH_4)+ ions in the ammonia chemical ionization spectra. However, greater confidence in the identification would be provided by spectra obtained with a more reactive chemical ionization reagent gas, like methane, which gives significantly more fragmentation than does ammonia. The ammonia chemical ionization spectrum of Ionox 330 contains (M + H)+ as the most abundant ion and abundant fragment ions at m/z 569, 319, 203 and 163. This compound may be readily identified in a simple mixture, by noting that these ions have the same temperature profiles. Similarly, the methane chemical ionization spectrum of Irganox 168 contains a reasonably abundant (M + H)+ ion, the base peak at m/z 441 and another

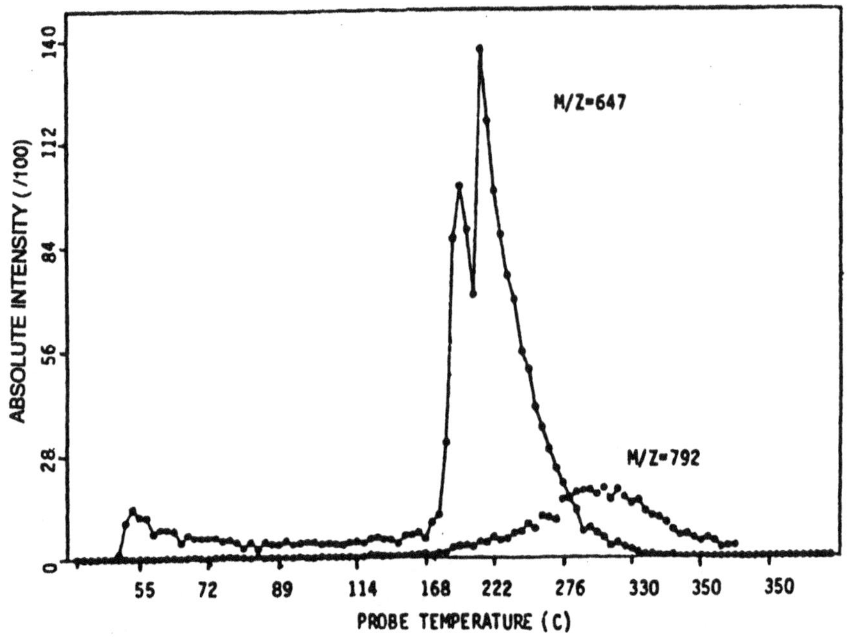

Figure 7.37. Heating profiles for $(M + H)^+$ for Irganox 168, m/z 647, and $(M + NH_4)^+$ for Ionox 330, m/z 792, from polypropylene. Probe heating rate was 30°C/min. From Rudewicz and Munson with permission.[59] American Chemical Society.

fragment ion at m/z 147. Similar temperature profiles for these ions indicate the presence of this additive. One could probably identify the additives from a limited set of likely possibilities by comparing temperature (or time) plots of characteristic ions, even when several additives are present.

100.7 DISCUSSION OF RESULTS

This procedure enables additives to be determined in polypropylene in amounts down to 0.01% with an accuracy of $\pm$ 5%.

METHOD 101 - DETERMINATION OF ADDITIVES IN POLYPROPYLENE. LIQUID CHROMATOGRAPHY - MASS SPECTROMETRY.[61]

101.1 SUMMARY

This liquid chromatography - mass spectrometric system enables additives to be determined in polypropylene in amounts down to 0.01%.

101.2 APPARATUS

Quadrupole mass spectrometer, Finnigan MAT Model 4165 or equivalent equipped with moving belt LC/MS interface.

High performance liquid chromatography dual pump gradient system with ultraviolet detector (variable wavelength) and 0.5 ml flow through cell.

Absorbance detection was done at 280 nm. Chemical separations were achieved on a 1/8 o.d. x 2.1 mm i.d. x 25 cm long column packed with 5 um diameter ODS particles (Alltech Associates Inc., Deerfield, I.L.O.). Sample injections were made with a Valco injection valve equipped with a 10 ul loop. A precolumn filter was used to remove particulate material from the injection sample.

101.3 REAGENTS

Standard stock solutions made up in acetonitrile with enough tetrahydrofuran (THF) added to dissolve the compounds.

Acetonitrile, HPLC grade.

Tetrahydrofuran, freshly distilled.

101.4 METHOD

Applicable to any acetonitrile extract of the polymer. Separation of additives were achieved using an acetonitrile - tetrahydrofuran gradient elution system and the separated additives were detected using an ultraviolet detector. This method was calibrated against standard solutions of the additives put through the analytical procedure.

101.5 EXPERIMENTAL PROCEDURE

101.5.1 Chromatography

Approximately 1 g of plastic shavings is extracted overnight with 5 cm^3 of acetonitrile at ambient temperature in a sealed amber vial with constant stirring. The extract solutions are filtered prior to analysis. The gradient elution scheme shown in Table 7.55 is used for the liquid chromatographic separations. The gradient controller is set for a flow rate of 0.2 cm^3/min^{-1}. At this flow rate, it takes approximately 10 min for the mobile phase in the solvent mixing chamber to reach the column inlet. Therefore, injections were not made until the gradient controller was 7.0 min into the gradient elution programme. This reduces the amount of computer time and disk space required to record an LC/MS chromatogram.

101.5.2 Mass Spectrometry.

Mass spectrometric detection was achieved with a quadrupole mass spectrometer which was equipped with a moving belt LC/MS interface. Methane chemical ionization is used. The ion source is pressurized to 0.3 torr with methane reagent gas, which is ionized with 70 eV electrons. Solutes are desorbed from the polyimide LC/MS interface belt at 230°C. The ion source temperature is 120°C. Chemical ionization spectra from m/z 200 to 1200 are recorded repetitively at 3 s/scan. Electron ionization spectra are recorded from m/z 50 to 800 at 3 s/scan.

The LC/MS moving belt interface was modified so that the column effluent was deposited on the belt in a fine spray. With this system, the LC/MS interface is capable of efficient operation with the mobile phases used at flow rates up to 0.2 cm^3/min^{-1}. There is no significant loss of resolution or efficiency in going from the absorbance detector to

Table 7.55 - Gradient Elution Scheme

Time, min.	% Solvent A[a]	% Solvent B[b]
0	100	0
10	60	40
20	0	100
30	0	100
32	100	0

From Vargo and Olsen[61]. American Chemical Society
[a] Solvent A 75% acetonitrile and 25% water. [b] Solvent B 50% THF and 50% acetonitrile

the mass spectrometer when the two detectors were used in series. When the absorbance detector and mass spectrometer are operated in series, a 20 cm length of 0.01 in i.d. stainless steel tubing is used to connect the absorbance detector outlet to the LC/MS interface.

101.6 TYPICAL RESULTS

The procedure was used to characterise additives in polypropylene samples of unknown composition. Butylated hydroxy toluene and Irganox 1076 were identified in the extract of an automative component moulding. In addition, several other additives were identified (palmitic acid, dioctylphthalate, stearic acid and octadecanol).

101.7 DISCUSSION OF RESULTS

The procedure enables additives to be identified and determined in polypropylene in amounts down to 2 mg (0.01% in polymer) with an accuracy of $\pm$ 5% to 10%.

METHOD 102 - IDENTIFICATION AND DETERMINATION OF ADDITIVES IN POLYMERS. PHOTOLYTIC DEGRADATION-GAS CHROMATOGRAPHIC METHOD.[62]

102.1 SUMMARY

This photolytic degradation technique enables additives to be identified and determined in polymers in amounts down to 0.01%.

102.2 APPARATUS

Photolytic degradation apparatus, see Figure 7.38.
Capillary crusher, model SI-4 solid samples, (Hewlett Packard).
Potassium bromide press.
Gas chromatograph, with flame ionization detector.
Flame photometer, providing light at 360.0 nm.
Hydraulic press, 30 ton.
Gas chromatographic column, 10% w/w SE-30 on S0/200 chromo-ports.

102.3 METHOD

Films of the polymer are irradiated with light at 360 nm. The photolysis products are then examined by gas chromatography to identify them. The structure of the polymer additive is elucidated from the nature of the photolysis products.

102.4 EXPERIMENTAL PROCEDURE

102.4.1 Sample Preparation.

Pure polymer samples are prepared for photolysis with the apparatus shown in Figure 7.38. A sample of up to 50 mg is placed between two plane glass microscope slides which are then inserted between the two large aluminium blocks. The temperature of the blocks is maintained slightly above the glass transition temperature of the polymer (usually ca 150°C) by controlling the voltage to the 65 watt cartridge heaters. The 3/8 inch machine screw is tightened and the sample flattened into a thin film rapidly (about 5 s) to preclude thermal degradation of the polymer. The resulting film is placed between two glass plates with one edge slightly protruding and sliced into strips ca. 0.8 mm wide with a scalpel. These strips are weighed and transferred to 1.0 mm i.d. quartz capillaries, previously sealed at one end. The capillary is evacuated for 5 min to remove air and is then sealed.

To prepare samples of known composition for calibration purposes, weighed amounts of additives are mixed with pure polymer samples prior to casting into thin films with the apparatus shown in Figure 7.39. A stainless steel cylinder kept at a temperature slightly above the glass transition temperature of the polymer is placed under the top stop of 30 ton hydraulic infrared laboratory KBr pellet press. The film is placed on the cylinder and compressed until it forms a disk, the diameter of which equals that of the cylinder. The pressure is released, the cylinder withdrawn, and the sample is further mixed by folding and compressing several times with a small spatula, after which it is recompressed with the hydraulic press. Small portions of the disk thus formed are cast into thin films between glass slides as above. The film is attached to the burner mounting of a Heath flame photometer module, the position of which can be varied by a micrometer adjustment system. Heath light source, photomultiplier tube and monochromator modules are used to isolate light at 360.0 nm. The light beam from the exit slit is focused to dimensions of ca. 0.2 mm wide and 1.0 mm high. Absorbance measurements are made at 1.8 mm intervals across the full 18 mm diameter of the disk.

102.4.2 Sample Irradiation.

To provide maximum ultraviolet intensity for photolysis in the solid phase, an irradiation apparatus using an Engelhard Hanovia Type 612C medium-pressure mercury source was used. Photolysis period of 5 minutes were used in quantitative determinations of polymer additives. The irradiation cavity is of an elliptocylindrical shape in which the source and sample are positioned at the foci. The cavity was formed by placing a cylinder of spring brass, 13 cm high and 0.22 mm wall thickness within four upright posts arranged in a rectangular pattern of 8 x 14 cm. The brass cylinder assumes the shape of an ellipse having major and minor axes of 18 cm and 14 cm respectively. A lining of highly reflecting aluminium foil concentrates a large percentage of the emitted radiation on the sample.

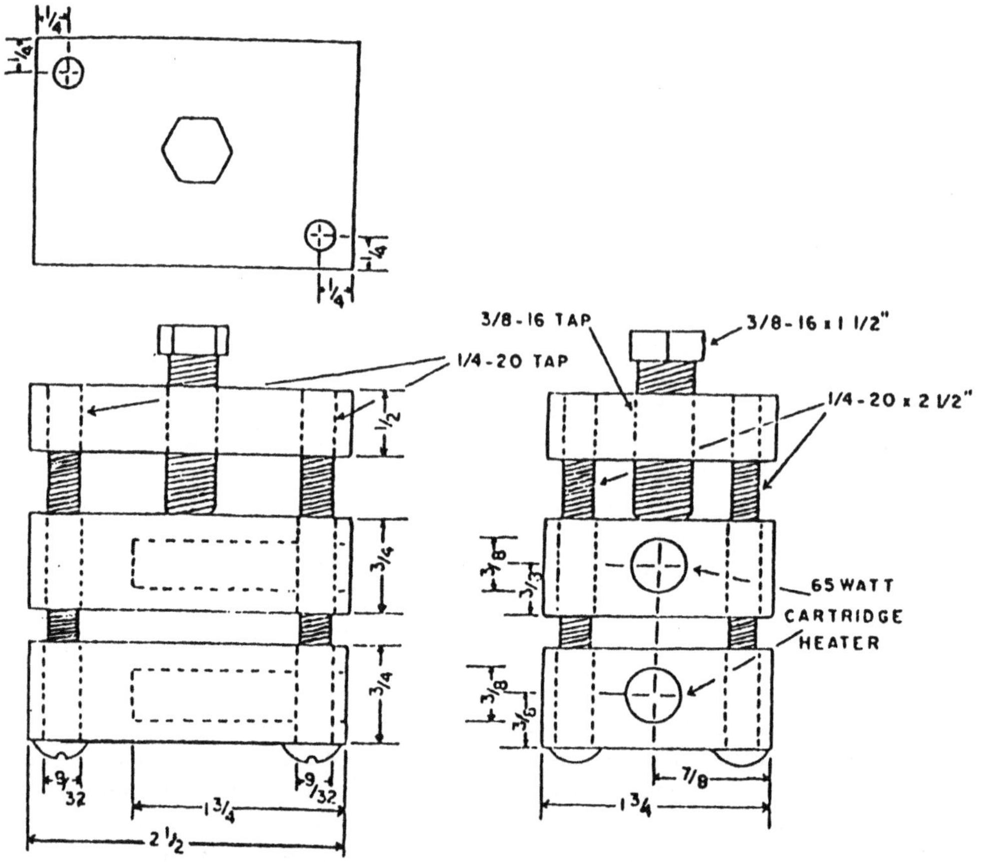

Figure 7.38. Heated compressing unit for rapid preparation of polymer films prior to photolytic degradation and GC analysis.
Compression time, 5 sec. Temperature maintained above polymer glass transition temperature (usually 150°C). From Juvet et al with permission.[62] American Chemical Society.

The water jacketed sample compartment consists of an inner 2 mm i.d. quartz tube surrounded by a 25 mm i.d. quartz tube with sidearms for cooling water circulation. Directional irregularities in light intensity are minimized by rotating the sample with a 1 rpm synchronous motor.

102.4.3 Product Separation and Identification.

After irradiation, the samples are introduced onto the chromatographic column with a capillary crusher to accommodate 1.0 mm i.d. quartz tubing.

A gas chromatograph with flame ionization detector was used to obtain chromatographic data using a 6 foot 1/8 inch o.d. stainless steel column packed with 10% w/w SE-30 on 80/300 mesh Chromport S. The injection port temperature is maintained at about 150°C. A temperature programming rate of ca $6^{o}/min^{-1}$ and a helium carrier gas flow rate of ca 30 cm^3 min^{-1} are employed. Products can also be separated, either isothermally or under temperature programmed conditions, on a 6 foot 1/4 inch o.d. stainless steel column containing 15% w/w Carbowax 20M on silane-treated 60/80 mesh Chromosorb P. Retention data are reported as Kovats retention indices. Peak area measurements are made with a planimeter and are reproducible to $\pm$ 0.1 cm^2.

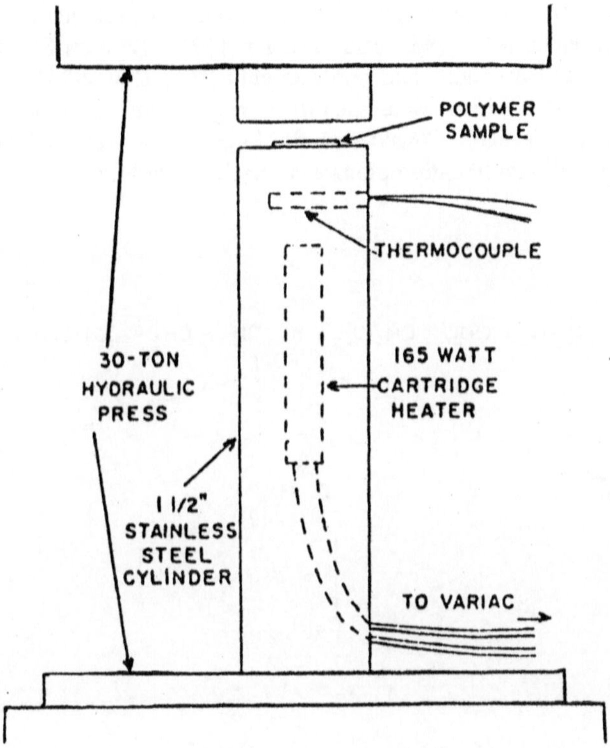

Figure 7.39. Heated compressing unit for use with IR-laboratory KBr pellet press for preparation of polymer mixtures of known composition. From Juvet et al with permission.[62] American Chemical Society

To aid in the determination of the retention indices of the photolysis products an iron constantin thermocouple device is used which automatically indicates column oven temperature on the recorder chart during temperature programming. With proper calibration, the column oven temperature is indicated on the chart of a 1 mV recorder and can be read directly in degrees centigrade. A mixture of n-hydrocarbons from *n*-pentane to *n*-pentacosane is chromatographed daily from which a calibration curve of column oven temperature vs Kovats retention index is prepared. The retention indices of eluted compounds are obtained by graphical comparison of the retention temperature with the calibration curve.

102.5 TYPICAL RESULTS

Juvet et al[62] found that with low concentrations of additives, the photodegradation pattern of a polymer is not affected and new products appear from photolysis of the additives. As an example Table 7.56 lists the retention indices on SE-30 for the additional photolysis products from two similar sample, 1.0% dilaurylthiodipropionate in polyethylene and 1.0% distearylthiodipropionate in polyethylene. No difficulty is found in differentiating between these two additives by photolysis gas chromatography although they are of closely related structure; indeed the similarity in structure is readily apparent from the constant difference of 600 I units in the homologous peaks of these two additives. The product from dilaurylthiodipropionate eluting at I = 1690 was identified from retention data and mass spectrometry as dodecyl propionate. The overlapping doublets which elute at I = 995 - 1000 and I = 1194 - 1200 on SE-30 are resolved in Carbowax 20M and the individual products elute at I values of 1000, 1045, 1200 and 1242. These four products were identified, respectively, as n-decane, n-decene, n-dodecane and n-dodecene. Considering the kinetic data presented in this figure, the decomposition of dilaurylthiodipropionate is very likely to follow the mechanism:

7.17

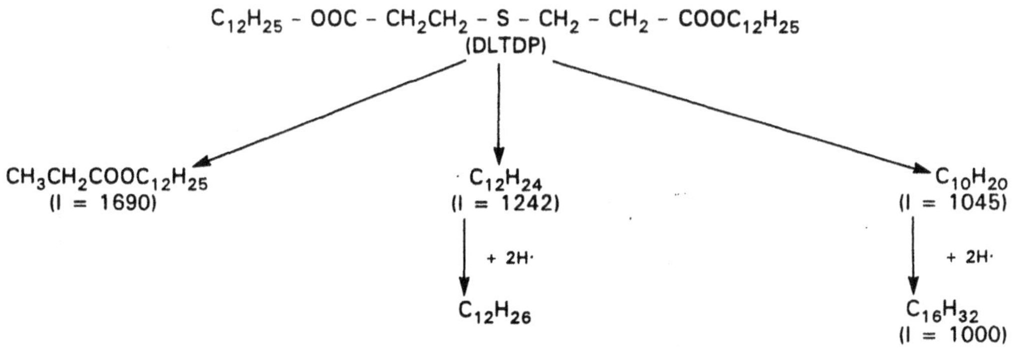

Addition of 1.0% dilaurylthiodipropionate to poly(methyl methacrylate) has no effect on the basic polymer photolysis pattern, but the formation of C 10 and C 12 hydrocarbons relative to the peak at I = 1690 reduced to ca. 0.01 and 0.02 respectively. In this matrix the concentration of the product eluting at I = 1690 increases continuously throughout a two-hour irradiation further strengthening the argument that the hydrocarbon fragments obtained in the polyethylene matrix are secondary photolysis products which show a decrease in quantum yield in the methyl methacrylate matrix toward production of the hydrocarbon fragments.

102.5.1 Quantitative Determination of Additives.

Quantitative measurements are carried out by irradiating the sample for a carefully controlled time period using a light source of constant intensity. In the analysis of the antioxidant, dilaurylthiodipropionate in polyethylene an irradiation time of 5.00 $\pm$ 0.02 min was chosen and a calibration curve constructed by plotting the peak area of the product peak, dodecyl propionate at I = 1690 vs the per cent LTDP in the sample. The average relative mean deviation of the 11 points used in constructing the graph is 2.7%. Since the product peak tails to a small extent under the chromatographic conditions used, the error in peak area measurement is increased somewhat and is thought to be the major source of error in this determination.

As a typical example of the determination of plasticizers, dinonyl phthalate at concentrations up to ca. 9% was studied in a poly (methyl methacrylate) matrix. The rate of formation of several photolysis products from samples containing 4.30% dinonyl phthalate is shown in Figure 7.40. The peaks at I < 500, I = 756 and I = 812 are irradiation products of the polymer and the product eluting at I = 658 is from dinonyl phthalate. The nonlinear calibration curve produced using irradiation times of 15.0 $\pm$ 0.1

Table 7.56 - Photolysis Products of Antioxidants and Polyethylene

Additive	Concentration %	Additional Products, I Values on SE-30	Relative Peak Area for 30 Minutes Irradiation
Dilaurylthio-dipropionate	1.0	995-1000 ± 4[a]	0.22
		1100 ± 6	0.09
		1194- 1200 ± 4[a]	1.00
		1690-1200 ± 5	0.21
Distearylthio-dipropionate	1.0	1592-1602 ± 4[a]	0.21
		1700 ± 6	0.34
		1792-1805 ± 6[a]	1.00
		2298 ± 8	0.50

From Juvet[62]. American Chemical Society
[a] Overlapping doublet peaks

min was prepared. It is interesting to note that dioctyl phthalate in poly (vinyl chloride) shows linear dependence on concentration as high as 19 weight per cent for the decomposition product peak appearing at I = 796 (probably n-octane) on SE-30 liquid phase when an irradiation time of 5.00 ± 0.02 minutes is used. In the case of dinonyl phthalate, variation in plasticizer concentration affects the relative amounts of the polymer photolysis products at I = 756 and I = 812. Since the relative product yield from the basic polymer is altered it is apparent that the matrix in which photolysis of the additive occurs in changing with plasticizer concentration.

102.6 DISCUSSION OF RESULTS

This technique enables additives to be identified in polymers and to be determined in amounts down to 0.01% with an accuracy of ± 5% to ± 10%.

METHOD 103 - IDENTIFICATION OF ADDITIVES IN POLYMERS. GAS CHROMATOGRAPHY - INFRARED SPECTROSCOPY.[63]

103.1 SUMMARY

This gas chromatography infrared spectroscopic procedure enables a wide range of additives to be determined in polymers in amounts down to 0.01%.

103.2 APPARATUS

The fraction-collecting apparatus is used in conjunction with gas chromatographic columns operated under reduced pressure and with katharometer detection.

Apparatus involving a column thermostatically controlled within the range 0°C to 130°C and a katharometer at room temperature. This apparatus is used for separating mixtures when the highest boiling constituent boils below 160°C at atmospheric pressure. If the polymer constituent boils between 160 and 250°C then the column should be thermostatically controlled in the range 100 - 200°C.

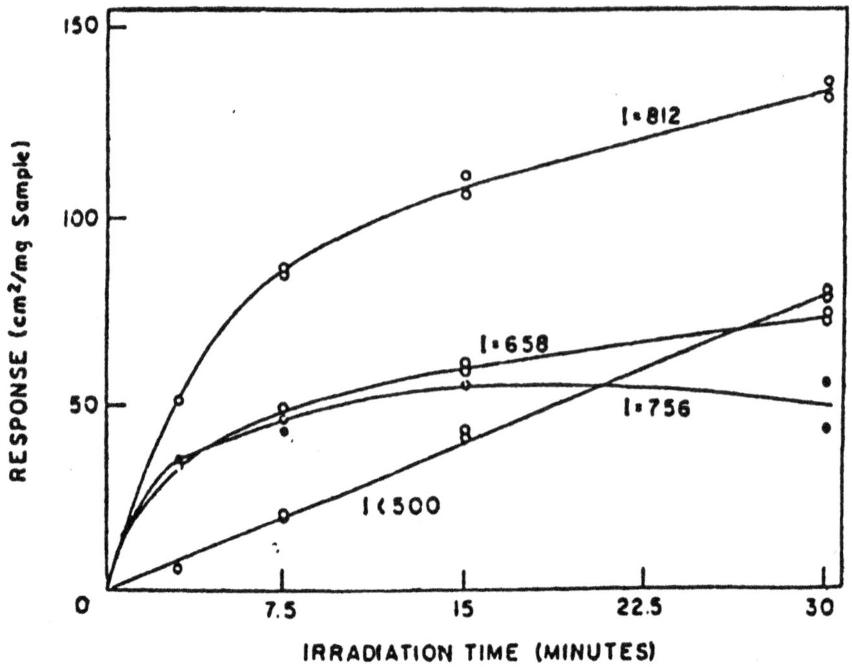

Figure 7.40. Radiation kinetics for 4.30% dinonyl phthalate plasticizer in poly(methylmethacrylate). Radiation products peak at I = 658 from dinonyl phthalate; others from the polymer matrix. Retention indicies measured on a Carbowax 20M column at 85°C. From Juvet et al with permission.[62] American Chemical Society

Infrared spectrometer with 9 cm^3 cavity cells, length 10 cm, height 1.5 cm width 0.6 cm. (Figure 7.43).

The trap shown in Figures 7.41 and 7.42 has been successfully used for the condensation of substances boiling in the range 100° to + 160°C. The packing used is small Dixon rings held in place by a small loose fitting piece of glass rod. The efficiency of this type of trap varies between 85 and 95% depending on the chemical class of substance. The trap is connected to the exit of the katharometer by means of the minimum length of stainless steel tubing (internal diameter 1 mm) brazed into a metal cup. This in turn is cemented to a glass socket with Araldite cement. The exit ends of the four traps are connected via rubber tubing and separate stopcocks to a vacuum manifold. The traps are partly immersed in liquid nitrogen contained in dewer flasks.

Having condensed the desired component in a trap it has to be decided whether to record the spectrum in the vapour state (method 1 below) or the liquid phase (method 2 below).

103.3 METHOD

The sample is heated in a trap device and the volatiles produced transferred to an infrared gas cell. Examination of the infrared spectrum provides information on the identity of volatile additives present in polymers which boil between 0 and 250°C.

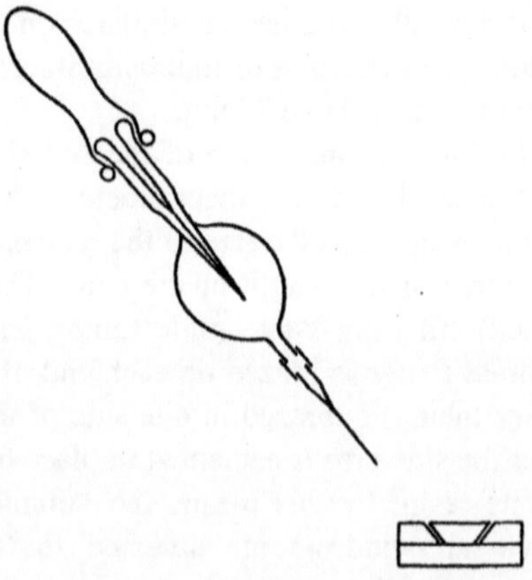

Figure 7.41. Pipette for transferring liquid to micro infrared cell.
From Anderson with permission.[63] Royal Society of Chemistry, London.

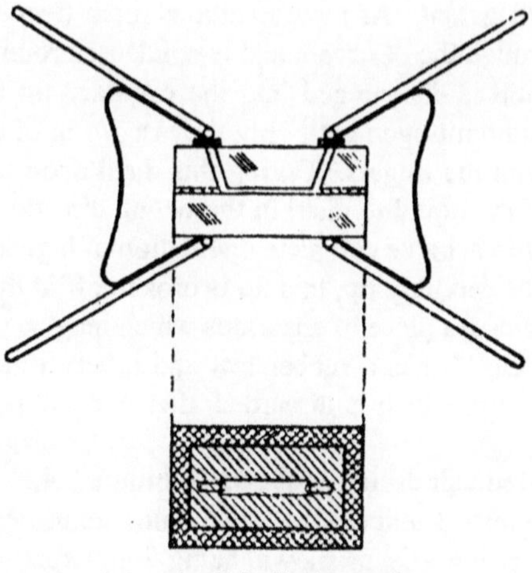

Figure 7.42. Micro infrared cell for liquid samples.
From Anderson with permission.[63] Royal Society of Chemistry, London

103.4 EXPERIMENTAL PROCEDURE

103.4.1 Method 1 - Liquids of Boiling Point up to 60°C.

Figure 7.43 shows how the low-boiling component is transferred to a 9 cm^3 infrared gas cell. No heat is applied to the trap.

In order to attain optimum light transmission, such a cell is inserted at a point where the area of the spectrometer beam is least; this will be where a focus of the beam is formed in the centre of the cell. At a few centimetres on either side of the focal position the beam is rectangular. Condensation of liquids boiling up to 100°C may occur on the walls of a 9 cm^3 cell at pressures above 80 nm.

The image to be found in the centre of the cell should normally have the length and width of the entrance slit of the spectrometer. Knowing the f number of the instrument, it is possible to calculate the size of the rectangular aperture required to admit the radiation cone in terms of the length of the cell. The body consists of rectangular brass tube (1 mm thick) cut from wave-guide tubing and has the internal dimensions stated above. A flat brass flange is brazed on each end of the tube, and a short side arm (stainless steel capillary tubing) is brazed in one side of the body. A glass capillary tap that will just slide over the side-arm is cemented in place by filling the space between the two tubes with Araldite resin; by this means the volume of the side arm is rendered negligible. Two rock-salt windows are attached to the flanges with Picene wax. Although this cell is designed for an f 11 system, it will give acceptable performance, although with some loss of energy, on an f 7 spectrometer.

103.4.2 Method 2 -Liquids Boiling above 60°C.

If the gas chromatographic evidence indicates that the substance boils above 60°C the fraction condensed in the trap is treated as shown in Figure 7.44. The trap and contents are held in liquid nitrogen while the apparatus is completely evacuated. After testing the system for leaks, the source or vacuum is cut off. The flask containing the liquid nitrogen is removed from the trap and placed over the side-arm so that the bulb is just immersed in the refrigerant. As the trap attains room temperature, the small amount of liquid in it distils through the desiccant and is condensed round the sides of the bulb of the side-arm. The liquid is encouraged into the capillary tip by lowering the vacuum flask containing the liquid nitrogen until only the last 1/4 in of the tip is immersed. The bulb is then warmed with the fingers. During this distillation the trap is not warmed as this may cause some of the liquid to distil in the wrong direction. The apparatus may be left for 1 or 2 h in order to achieve complete distillation of higher boiling liquids. Having obtained the liquid in the capillary tip, this tip is broken off at the constriction just below the bulb. Figure 7.41 shows a piece of apparatus which may be used to transfer the liquid to the micro infrared cell. It has a rubber teat and a very fine internal capillary. The device has a detachable tip, which is discarded after use and permits close control over the transfer operation.

The micro infrared cell design is shown in Figure 7.42. The two rock-salt plates are about 20 mm x 10 mm x 3 mm in size. Two holes are drilled in one plate with a No. 86 drill; these holes taper inwards, as shown, being 7 mm apart on the inside and 12 mm apart on the outside of the plate. A gold-foil washer (25 u thick) is cut, the inner space being 1.5 mm wide and approximately 7 mm in length (just sufficient to clear the edges of the filling holes); the width of this washer is about 2 mm. The cell is first stuck together with gold amalgam. The washer is immersed in mercury for a short period and withdrawn after a layer of amalgam has formed on its surface. The cell is then carefully assembled in a small clamp and left under pressure for 24 h for the amalgam to set. The space between the plates outside the washer is then filled with Araldite 700 resin and the surplus of resin is allowed to bridge the edges of the plates on all four sides.

The cell is filled by inserting a capillary needle containing the liquid into one filling hole. Provided that the thickness of the cell is less than the diameter of the needle, the liquid will run in to fill the cell area. It is essential that no free space be left in the cell behind the filling holes; such space will not immediately fill by capillary action and air

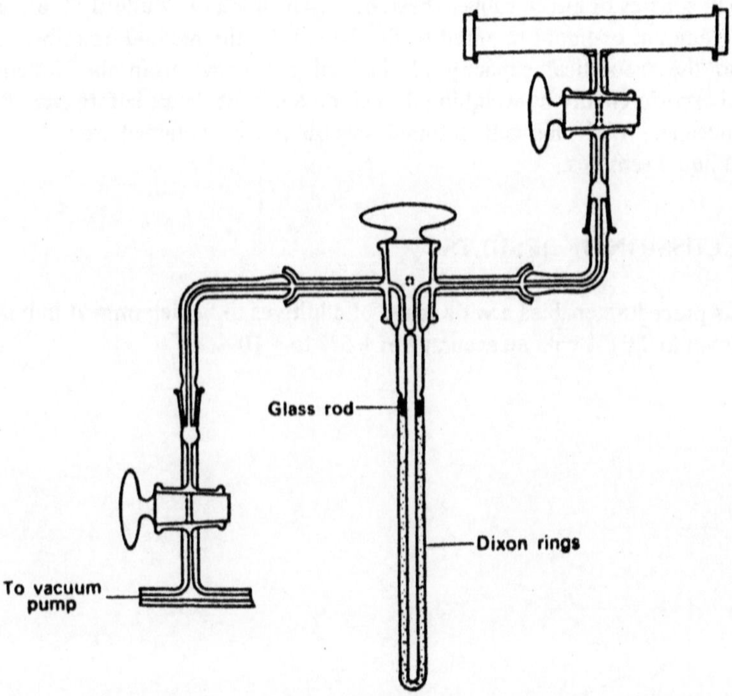

Figure 7.43. Apparatus for transfer of volatile components from trap to infrared gas cell. From Anderson with permission.[63] Royal Society of Chemistry, London.

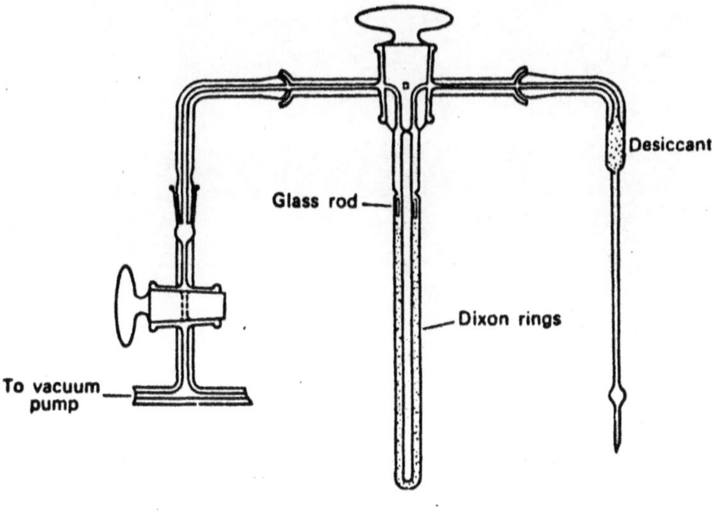

Figure 7.44. Apparatus for transfer of liquid components from trap to tip of side arm. From Anderson with permission.[63] Royal Society of Chemistry, London

bubbles trapped here may subsequently run into the useful part of the cell and are most difficult to remove. When the cell is filled, the outer ends of the filling tubes are covered by two small squares of nitrile-rubber sheet, held in position by "butterfly" spring clips.

The amount of liquid required to fill the cell by the method described above (as opposed to the theoretical capacity of the cell calculated from the dimensions) is determined experimentally by weighing the cell on a microbalance before and after filling with *n*-dodecane. With this cell, infrared spectra can be obtained from 0.5 to 1.0 ul portions of liquid samples.

103.5 DISCUSSION OF RESULTS

This procedure enables a wide range of additives to be determined in polymers in amounts down to 0.01% with an accuracy of $\pm$ 5% to $\pm$ 10%.

CHAPTER 7.6

PRELIMINARY QUALITATIVE IDENTIFICATION

Methods 104 - 108

METHOD 104 - QUALITATIVE DETECTION OF ELEMENTS IN POLYMERS. OXYGEN FLASK COMBUSTION.[64]

104.1 SUMMARY

This qualitative oxygen flask combustion method enables nitrogen, fluorine, chlorine, bromine, iodine, sulphur and phosphorus to be identified in polymers in amounts down to 0.01%.

104.2 APPARATUS

Combustion unit - the electrically fired combustion unit was used in this work, although there is no reason why other forms of oxygen flask should not be used, with appropriate calibration.

Filter paper - Whatman No. 541 filter paper. When not in use, store filter paper in a sealed container out of contact with the laboratory atmosphere.

Cotton wool - B.P.C. super quality.

104.3 REAGENTS

Sodium hydroxide, N.
Nitrogen.
Resorcinol - AnalR
Acetic acid, glacial.
Ammonium ferrous sulphate - AnalR.

Fluorine.

Reagents. Buffered alizarin and complexan solution - weigh 40 mg of 3-amino-ethylalizarin-NN′-diacetic acid (Hopkins and Williams Ltd.) into a beaker and add 2 drops of N sodium hydroxide and approximately 20 ml of distilled water. Warm the solution to dissolve the reagent, cool and dilute to 208 ml. Weigh into another beaker 4.4 g of sodium acetate ($CH_3COONa3H_20$) and dissolve in water. Add 4.2 ml of glacial acetic acid and dilute to 42 ml. Pour this sodium acetate solution into the alizarin complexan solution and mix to give the final buffered alizarin complexan solution.

Cerous nitrate, 0.0005 M - dissolve 54.3 mg of cerous nitrate ($Ce(NO_3)_36H_20$) in water and dilute to 250 ml.

Chlorine.

Reagents. Ammonium ferric sulphate solution - dissolve 12 g of AnalaR ammonium ferric sulphate in water and add 40 ml of AnalaR nitric acid. Dilute to 100 ml and filter.

Mercuric thiocyanate solution - dissolve 0.4 g of recrystalized mercuric thiocyanate in 100 ml of absolute ethanol.

Bromine.

Reagents. Fluorescein solution - dissolve 0.1 g of fluorescein in 25 ml of 0.1 N sodium hydroxide and dilute to 100 ml with water.

Sodium acetate buffer solution - mix 100 ml of N sodium acetate with 15 ml of N acetic acid.

Chloramine-T solution - dissolve 12 g of chloramine-T in 100 ml of distilled water.

Sodium thiosulphate solution - prepare a 0.5% w/v solution of sodium thiosulphate in 5% w/v sodium hydroxide solution.

Hydrochloric acid, N.

Iodine.

Reagents. Starch solution - dissolve 0.2 g of soluble starch in 100 ml of distilled water.

Sulphuric acid, dilute - prepare an approximately 10% w/v solution of concentrated sulphuric acid in distilled water.

Qualititative Determination of Sulphur - Barium Chloride Method.

Reagents. Hydrochloric acid, N.

Precipitating reagent solution A - dissolve 0.2 g of peptone in 50 ml of 1 per cent barium chloride ($BaCl_22H_20$) solution. Buffer to a pH of 5.0 with 0.02 N hydrochloric acid, add 20 g of sodium chloride (Analar) and dilute to 100 ml. Heat on a water bath for 10 minutes, and add a few drops of chloroform. Filter if necessary.

Precipitating reagent solution B - dissolve 0.04 g of gum ghatti in 200 ml of distilled water by warming slightly. When solution is complete, add 2.0 g of barium chloride ($BaCl_22H_20$). Filter if necessary.

Store solutions A and B separately and prepare the final reagent just before use by diluting 10 ml of solution A to 100 ml with solution B.

Hydrogen peroxide, 100 volume - AnalaR.

Qualitative Determination of Sulphur -4-Amino -4'Chlorodiphenyl Hydrochloride Method

Reagents. 4-amino-4' chlorodiphenyl hydrochloride - peptone solution (solution A). Dissolve 0.025 g of peptone (Hopkin and Williams Ltd.) and 0.125 g of 4-amino-4'chlorodiphenyl hydrochloride (L. Light and Co. Ltd.) as completely as possible by warming with 50 ml of 0.05 N hydrochloric acid. Cool, and filter through a double Whatman No. 40 filter paper.

4-amino-4'chlorodiphenyl hydrochloride - gum ghatti solution (solution B). Dissolve 0.20 g of gum ghatti (Hopkin and Williams Ltd., finely ground) and 1.00 g of 4-amino-4'chlorodiphenyl hydrochloride as completely as possible by warming to about 70°C with 400 ml of 0.05 N hydrochloric acid. Cool and filter through a double Whatman No. 40 filter paper.

Final precipitation reagent - dilute 10 ml of solution A to 100 ml with solution B. This solution should be freshly prepared.

Hydrochloric acid, N.

Qualitative Determination of Hydrolysable Phosphorus.

Reagents. Ammonium molybdate solution - dissolve 10 g of Analar ammonium molybdate ($NH_4MO_7O_{24}4H_2O$) in about 70 ml of water and dilute to 100 ml. Add this solution with stirring to a cooled mixture of 150 ml of sulphuric acid and 150 ml of water.

Asorbic acid - B.D.H. laboratory reagent grade.

104.4 METHOD

Approximately 20 mg of polymer is combusted in an oxygen filled flask over dilute sodium hydroxide solution. Specific colorimetric tests for nitrogen, sulphur and phosphorus and the four halogens are applied to reveal the presence or absence of these elements.

104.5 EXPERIMENTAL PROCEDURE

104.5.1 Combustion Procedure.

Weigh out approximately 20 mg of sample and transfer it to the centre of a small piece of the Whatman No. 541 filter paper weighing approximately 0.1g. Fold the filter paper so that the sample is completely enclosed and before making the final fold, insert a small wick of cotton wool weighing about 6 mg.

Carry out the combustion using 5 ml of N sodium hydroxide in the bottom of the flask as absorption solution. Set the flask aside for 15 minutes to allow the gases formed in the combustion to be absorbed and then wash the contents of the flask quantitatively into a 25 ml measuring cylinder with distilled water. Dilute the solution to 25 ml and mix well. This constitutes the test solution, aliquots of which are taken for the detection of the individual elements by the colorimetric methods detailed below. For comparison purposes in the tests, prepare a blank test solution by carrying out the combustion procedure on the filter paper and cotton wool only.

104.5.2 Determination of Nitrogen.

Weigh 0.1 g of resorcinol into a clean dry 50 ml beaker and dissolve in 0.5 ml of glacial acetic acid. Add 5 ml of the test solution and, after mixing, add 0.1 g of ammonium ferrous sulphate. Carry out the same test on the blank test solution. The development of a green colour in the sample test solution, compared with a pale yellow in the blank, indicates the presence of nitrogen in the sample.

If a semi-quantitative estimation of the nitrogen content of the sample is required, set both sample and blank solutions aside for 20 minutes. Add 10 ml of distilled water to each, mix and measure the optical density of the sample solution against the blank at 690 nm in a 4 cm cell. The blank in this test is low, giving an optical density of 0.07 measured against water at 690 nm in the 4 cm cells.

104.5.3 Determination of Fluorine.

Transfer 20 ml of distilled water and 2.4 ml of buffered alizarin complexan solution to a 50 ml beaker. Add 1 ml of test solution and mix swirling the solution. Finally, add 2 ml of cerous nitrate solution and mix again. Treat the blank solution in a similar manner. When fluorine is present in the sample a mauve colour will be developed in the test solution compared with the pink coloured blank solution. If a semi-quantitative estimation of fluorine is required, set the solutions aside for 10 minutes and measure the optical density of the test solution against the blank solution at 600 nm in 1 cm cells. Sulphur, chlorine, phosphorus and nitrogen do not interfere in this procedure.

104.5.4 Determination of Chlorine.

Transfer 5 ml of the test solution to a 50 ml beaker and add 1 ml of ammonium ferric sulphate solution. Mix the solution and add 1.5 ml of mercuric thiocyanate solution. Mix and dilute to 10 ml. Treat the blank solution in a similar manner. When chlorine is present in the sample an orange colour will be developed in the test solution compared with yellow coloured blank solution. If a semi-quantitative estimation of chlorine is required, set the solutions aside for 10 minutes and measure the optical density of the test solution against the blank solution at 460 nm in 2 cm cells.

104.5.5 Determination of Bromine.

Transfer 5 ml of the test solution to a 50 ml beaker. Add 1 ml of N hydrochloric acid and then 0.5 ml of sodium acetate buffer solution and 1 drop of fluorescein solution. Mix thoroughly and then add 1 drop of chloramine T solution. Mix by swirling and set aside for 30 seconds, then stop the reaction by adding 2 drops of alkaline thiosulphate reducing agent. Treat the blank solution in a similar manner. When bromine is present in the sample, a rose-pink colour will be developed in the test solution compared with the yellow-green blank solution, note: This test is also given by iodine.

104.5.6 Determination of Iodine.

Transfer 5 ml of test solution to a 50 ml beaker and add a few drops of starch solution. Mix the solution and then acidify with dilute sulphuric acid. Treat the blank in a similar manner. When iodine is present in the sample, the characteristic blue colour of starch iodide will be developed in the test solution compared with the colourless blank solution.

104.5.7 Determination of Sulphur (Barium Chloride Method).

Transfer 5 ml of the test solution to a 6" x 1" test tube and add 2 drops of 100 volume hydrogen peroxide, add then 1.2 ml of N hydrochloric acid. Mix well and add 2.0 ml of precipitating reagent with continued shaking. A distinct turbidity will be produced in the mixed solution if hydrogen peroxide decomposable sulphur is present in the sample; the blank test under the same conditions will be perfectly clear. If a semi-quantitative estimation of the sulphur content is required, add 5 ml of distilled water to both blank and test solutions, mix, and set aside for 30 minutes. Mix the solutions and measure the optical density of the test solution in a 4 cm cell at 700 nm with the blank solution in the comparison cell.

104.5.8 Determination of Sulphur (4 Amino-4'Chlorodiphenyl Hydrochloride Method).

Transfer 5 ml of the test solution to a 6″ x 1″ test tube and add 1-2 ml of N hydrochloric acid. Mix well and add 10.0 ml of the prepared final precipitation reagent. A distinct turbidity will be produced in the mixed solution if hydrolysable sulphur is present in the sample. A blank test under the same conditions will be perfectly clear. If a semi-quantitative estimation of the sulphur content is required, set the solutions aside for 30 minutes, then mix and measure the optical density of the test solution against the blank in 2 cm cells at 700 nm.

104.5.9 Determination of Phosphorus.

Transfer 2 ml of the test solution to a 100 ml beaker. Add 40 ml of distilled water and 4 ml of ammonium molybdate solution. Mix thoroughly, then add 0.1 g of ascorbic acid and boil the solution for 1 minute. Cool in running water for 10 minutes and dilute to 50 ml with distilled water. Treat the blank solution in a similar manner. When hydrolysable phosphorus is present in the sample, a blue colour will be developed in the test solution as compared with a pale yellow in the blank. If a semi quantitative estimation of the phosphorus is required, measure the optical density of the test solution against the blank solution at 820 nm in 2 cm cells.

104.6 DISCUSSION OF RESULTS

Nitrogen, fluorine, chlorine ,bromine and iodine, sulphur and phosphorus are quantitatively detected in polymers in the solution resulting from a single oxygen combustion. Semi quantitative data can be obtained on the same solutions for nitrogen, fluorine, chlorine, sulphur and phosphorus.

METHOD 105 - IDENTIFICATION OF POLYMER FILMS. THERMAL METHODS.

105.1 SUMMARY

Simple burning tests are described for the identification of certain types of polymers. These tests are of limited value, but taken together with results obtained in other tests might provide information leading to an identification of the polymer.

105.2 APPARATUS

Heating test.
Bunsen burner.
Pair of tweezers.
Copper wire test.
Copper wire: 1 mm.

105.3 METHOD

The polymer sample is held in a flame or heated in a test tube or on a copper wire and observations made (colour of flame, smoke, odour etc) to provide information on the identity of the polymer.

105.4 EXPERIMENTAL PROCEDURE

Heating test. Hold a 1 in x 3/8 in specimen of polymer in tweezers over bunsen burner flame.

What to look for. Colour of flame, visible changes in film sample, smoke formations.

Interpreting observations. If film shrinks a great deal, this usually means material was stretched or oriented in production process (see Table 7.57).

What to look for. Green colour in flame indicates presence of halogens, eg. chlorine or fluorine.

Interpreting observations. Caution should be used in interpreting results because test sample may have been varnished or laminated with halogen containing products. In such cases, more detailed analysis may be necessary to clarify inconclusive results.

To analyse chemical nature of fumes, insert a strip of wet litmus paper in the upper part of test tube. Alkaline reaction will turn paper blue, acid reaction will turn paper red. If acid reaction is very strong, insert a strip of Congo paper. If it turns blue the acid is mineral (see Table 7.57).

Copper wire test. Insert one end of copper wire into a cork. Holding the cork, heat the wire in flame until only a very slight yellow colour is seen in the flame. Now melt the film sample onto the wire while it is still hot. Hold the wire in outer part of fully burning flame.

105.5 TYPICAL RESULTS

Characteristic results obtained by these tests are shown in Table 7.57.

Table 7.57 - Identification of Polymer Films

Film Type	Physical Characteristics	1. Burning Test - Watch Both Flame and Film	2. Copper-Wire Test - Look for Flame and Film	3. Heating Tests - Check Odour and Acid-Alkalinity	4. Solubility Test - Dissolves and May Swell in
EXTENSIBLE					
Fluorocarbon	Waxy Feel	No flame produced, sample deforms slowly	Flame turns green	Fumes may be toxic - do not inhale	Virtually inert
Polyolefins, (low and high density PE,PP)[a]	Wax-like feel, good extensibility	Burns freely, smokeless, bluish flame, melts, drips like wax[c]	None	Burnt paraffin wax aroma	Tetralin at 275 F
Polyvinyl Chloride (flexible)	Softness range is that of rubber, extensibility proportionate to plasticizer used	Burns with yellowish sooty flame, melts freely to form pearl like drops	Flame turns green	Sharp acrid fumes turn Congo paper blue	Cyclohexane, tetrahydrofuran and pyridine, solution turns red in presence of alcoholic potassium hydroxide solution
Rubber Hydrochloride	Clear and tough, relatively good extensibility[c]	Will not burn, edges tend to singe, dark smoke given off, green-blue seam in flame	Flame turns green	Rubber odour, fumes turn Congo paper blue	Chloroform, solution shows no change in presence of pyridine and alcoholic potassium hydroxide solution
SEMI EXTENSIBLE					
Cellulose Acetate	Poor extensibility	Melts and drips	None	Acetic odour	Acetone

Table 7.57 (continued)

Cellophane[b]	Transparent, paper-like feel when dry, crumples easily, poor extensibility	Burns like paper, with light flame	None	Burn paper aroma	Schweitzer's reagent (cupro-ammonia)
Cellulose Acetate Butyrate	Horn-like feel	Melts and drips, drippings may burn	None	Rancid butter	Ketones, lactates, Me, Et propyl chlorides
Cellulose Nitrate	Similar to cellulose	Burns rapidly with an intense, white flame	None	Odour of camphor	Acetone at high nitrogen content
Polyamide (Nylon)	Opaque to clear, hard feel, poor extensibility	Burns with yellowish, sooty flame, melts to form pearl-like drops	None	Sour aroma, like burnt horn or hair, fumes turn litmus paper red	Formic acid
Polyester Film, (Polyethylene terephthalate type)	Clear, very tough, non-tearable, poor extensibility	Ignites, but extinguishes on removal of flame, shrinks without singeing of edges, melts to form pear-shaped drops	None	Faintly sweet	Chlorophenol
Polyvinyl Chloride (rigid)	Hard, greasy feel, white fracturing will show up in creasing line	Self extinguishing	Flame turns green	Sharp, acrid olour, fumes turn Congo paper blue	Cyclohexane, tetrahydrofuran and pyridine, solution turns red in presence of alcoholic potassium hydroxide solution
Polyvinylidene Chloride	Either milky or clear film, softer feel than PVC	Self extinguishing	Flame turns green	Sharp, acrid odour, fumes turn Congo paper blue	Cyclohexane, solution black in presence of pyridine and alcoholic potassium hydroxide solution

Table 7.57 (continued)

NONEXTENSIBLE					
Oriented Polystyrene	Clear, nonextensible, when crumpled, gives metallic sound, sharp folds displays mother of pearl effect	Burns with luminous, soot-producing flame, smoke is dense	None	Repugnantly sweet, like artificial illuminating gas	Benzene, hydrocarbons

[a] Oriented polyolefin films will shrink substantially on initial heating

[b] Absence of condensation when breathing on film indicates film is unvarnished. To test for commonly applied nitrocellulose varnish, apply several drops of a mixture of one part diphenyl and 100 parts concentrated sulfuric acid to film surface. Appearance of blue colour indicates presence of varnish. Polymer varnishes are readily identified by the Beilstein test for halogens

[c] Extensible > 200% elongation, Semi-extensible 25 - 200% elongation, Non-extensible < 25% elongation

METHOD 106 - PYROLYSIS - GAS CHROMATOGRAPHY OF POLYMERS. CURIE POINT FILAMENT PYROLYSER TECHNIQUE.[66]

106.1 SUMMARY

This Curie point filament pyrolysis technique enables unknown polymers to be identified.

106.2 APPARATUS AND CONDITIONS

Pyrolyser - Pye Unicam Curie point, attached directly to the column, pyrolysis temperature 610°C maintained for 10 seconds.

Gas chromatograph - dual columns, twin flame-ionisation detectors.

Columns - standard Pye 5 feet long x 4 mm i.d. glass (silanised).

Carrier gas - nitrogen with a flow rate of approximately 60 ml min.

Column packing - Porapak Q, 50 to 80 mesh or 80 to 100 mesh.

Temperature programme - from 100 to 200°C at 8°C min-1 (the temperature was maintained at 200°C for up to 25 minutes and the programmer started on completion of 10 seconds pyrolysis).

The columns are silanised before being packed by passing a 5 per cent solution of dichlorodimethylsilane in toluene through the columns and then drying at 100°C. Packing is facilitated by applying suction from the detector end of the column, accompanied by gentle tapping. The columns are aged by heating at 250°C for 48 hours, with a nitrogen flow of 60 ml min-1. The hydrogen and air flows are adjusted to give approximately the maximum sensitivity.

106.3 METHOD

In this Curie point filament pyrolysis technique the polymer is pyrolysed at a controlled temperature and the pyrolysis products passed into a gas chromatograph. Relative retention data and peak height ratio data obtained when compared with tabulations of data obtained from known polymers provide a means of identifying the unknown polymer.

106.4 EXPERIMENTAL PROCEDURE

The flow rate required for pyrolysis is determined by injecting through the pyrolyser head 1 ul of headspace gas from a retention time standard comprising methanol -n-propanol (50 + 50) with the column temperature at 100°C and immediately programming at 8°C min-1. The flow rate of nitrogen is adjusted to give retention times of 2.6 and 9.1 minutes for methanol and n-propanol respectively.

After fixing the flow rate of carrier gas, the upper temperature limit is adjusted to give a retention time of 4.2 minutes for benzene or 4.4 minutes for cyclohexane when these compounds are injected at the upper temperature limit. These two compounds have retention times of 13.4 minutes (benzene) and 14.0 minutes (cyclohexane) when injected at 100°C and the oven is programmed as described.

The pyrolysis wire is prepared in the following way. A 5 to 10 mm length at one end of the wire is flattened until the flattened portion is approximately 1 mm wide. A "hook" is then made by doubling over about 2 mm of the flattened end. At this stage, the hook end plus a few centimetres of the wire adjacent to it are heated to red heat in a bunsen flame so as to remove contaminants. The sample is placed in the elbow of the hook, care being taken not to touch or otherwise contaminate that end of the wire. The sample is then pressed into close contact with the wire surface by crimping the hook.

106.5 TYPICAL RESULTS

In developing a systematic scheme of polymer identification, pyrolysis conditions must be such that all polymers degrade rapidly. However, at temperatures above 1000°C the pyrograms will also be less suitable for identification, since secondary reactions become predominant, leading to increasing amounts of simple molecules such as carbon dioxide, acetylene, ethylene or benzene which gives less characteristic patterns than those observed for monomers or primary degradation products. Pyrolysis temperatures between 500 and 800°C (optimum 610°C) for 10 s are recommended. Although pyrograms containing from 30 to 50 peaks (for polyolefins) have been reported (Voigt[67-69]) the relative retention times and peak height ratios of three to five major peaks are usually sufficient for identification. Thus a single set of column operating conditions is desirable that will give characteristic pyrograms for degradation products of all types of polymers.

METHOD 107 - PYROLYSIS - GAS CHROMATOGRAPHY OF POLYMERS. PLATINUM RIBBON OR COIL PROCEDURES (CHEMICAL DATA SYSTEMS PYROPROBE 190).

107.1 SUMMARY

This platinum ribbon or coil pyrolysis - gas chromatographic procedure enables unknown polymers to be identified from their pyrolysis patterns.

107.2 APPARATUS

Pyrolysis apparatus, Chemical Data Systems Pyroprobe 190 or equivalent.

107.3 METHOD

A solvent solution of the polymer is applied to the pyrolysis filament and pyrolysed at a controlled temperature. Comparison of relative retention time and peak height ratio data for the unknown with that for known polymers enables the unknown to be identified.

107.4 EXPERIMENTAL PROCEDURE

107.4.1 The Chemical Data System, Pyroprobe 190 Platinum Ribbon Coil Pyrolyser.

The two types of pyrolyser probe available are based on a platinum ribbon or coil. If a sample can be dissolved in a suitable solvent, it is best applied to the ribbon as a solution. This should be prepared at a concentration of about 10mg/1 ml (approximately 1%). About 1-5 ul (representing 10-50 ug of sample) is applied to the centre of the ribbon using a micro syringe (this may take several applications). The solution should be spread as evenly as possible so that the entire sample will experience the same heating profile during pyrolysis. It is essential to avoid the extreme ends of the ribbon as the points of attachment to the probe mounting provide a large heat-sink and do not reach the set pyrolysis temperature.

Before inserting the probe into the injection adaptor, the final temperature should be set to a low value (such as 100°C) with an interval of 5 to 10 seconds, so that the residual solvent is flashed off without pyrolysing the sample. For most samples the solvent can be flashed off in the atmosphere - but for extremely sensitive materials, the probe should be located in the injector adaptor prior to solvent flash. After the solvent has been removed and the detector response has returned to base-line, the sample can be pyrolysed.

Insoluble but meltable samples should be applied by placing fine particles of material onto the ribbon and gradually raising the temperature with the interval set a 1 to 5 seconds until the particles adhere to the ribbon.

Samples that are insoluble and that will not melt should be pyrolysed on a quartz tube located inside the coil probe. The quartz tube must be inseted into the platinum coil probe very carefully to avoid distortion and damage.

107.4.2 Application of Sample to the Probe.

Quartz tubes must be inserted into the platinum coil probe very carefully to avoid distortion and damage. This is best achieved by passing the coil probe in the probe stand and holding the tube by one end with thumb and index finger, gently inserting it into the coil, rotating it as it enters. When the tube is through the coil part it can then be gently pushed all the way in so that the end rests on the base of the coil. After insertion, a careful check of the element is advisable to ensure that there is not shorting to any parts of the probe assembly and that the coil is even and not distorted.

Samples can then be inserted directly into the quartz tube and, for best results, should be between 10 and 50 ug and placed in the centre of the tube. As far as possible, the sample size should be kept constant from analysis to analysis, but for optinum repeatability of pyrolysis patterns, better results will be achieved by placing the sample onto a small amount of quartz wool which is itself located in the centre of the tube. Sample solutions can also be analysed in this way by injecting 1-3 ul from a microsyringe of a 1% solution of sample onto the quartz wool. The solvent should be evaporated to dryness by flashing the coil at 100°C with interval set a 1 second. For most samples, the solvent can be flashed off into the atmosphere but for extremely sensitive material the probe should be located in the injector adaptor prior to solvent flash. After the solvent has been removed and the signal from the gas chromatograph has returned to base-line, the sample can be pyrolysed.

The coil probe generally requires a final temperature of about 150°C higher than that required by the ribbon probe.

107.4.3 Cleaning the Probes.

Between each sample the probes should be cleaned in the atmosphere by setting an interval of 2-5 seconds and a final temperature of 1000°C and pressing the RUN button.

107.4.4 Selection of Pyrolysis Temperature.

An investigation to determine the lowest temperature at which pyrolysis will occur for a particular sample can be made with repeated runs on the same sample at increasing pyrolysis temperatures without removing the probe from the adaptor.

Alternatively, arbitrary pyrolysis conditions may be selected and the sample pyrolysed. The amount of residual organic material remaining can be determined by repyrolysing the sample at 980°C for 1 second for the ribbon probe and 10 seconds for the coil probe. A large residual peak under these conditions indicates that only a small part of the sample was initially pyrolysed. A higher final pyrolysis temperature should then be used.

METHOD 108 - PYROLYSIS - GAS CROMATOGRAPHY OF POLYMERS. LASER PYROLYSIS TECHNIQUE.[70,71]

108.1 SUMMARY

This laser pyrolysis gas chromatographic technique is used to identify unknown polymers from the pattern of the breakdown products of their pyrolysis products.

108.2 APPARATUS

A Gen-a-lite Model 3R laser (General) (Laser Corporation, Natick, Mass 01760) and a Focuscope (General) Laser Corporation, focussing device. The laser uses a 3.25″ ruby rod with a maximum energy output of 2 joules with a pulse langth of 600 micro seconds. The instruments are mounted on an optical rail so that the focused beam entered a box containing a sample holder and a peak-diverting prism. The prism is mounted so that the beam can be moved in two directions and can thus be aimed at any desired portion of the sample. The sample holder is a borosilicate glass tube 6 mm o.d. by 25 mm in length. It is clamped between two Teflon (Du Pont) gaskets so that carrier gas can sweep through the tube in to the chromatograph. The carrier gas enters the chromatograph through a 1/16 inch o.d. tube which pierces the injection port septum and extends into the injection port.

Figure 7.45 is a schematic drawing of the apparatus showing the light paths.

The chromatograph used had a pair of flame ionization detectors. The flame ionization detectors and the associated duel electrometer were from a Varian Aerograph (Varian Aerograph, Walnut Creek, Calif. 94598) instrument. The output signal from the electrometer is connected to an Infotronics Digital Readout Systems Model CRs-104

(Infotronics, Inc., Houston, Texas 77042) and the output from this system went into a 10 mV Honeywell (Honeywell, Inc., Philadelphia, Pa. 19144) strip chart recorder.

Column A. A 223 cm (7 ft) 0.25 cm i.d. (3/16 inch o.d.) stainless steel column of 70-80 mesh Anakrom ABS coated with 10% UCW-98. This column is operated at room temperature for one minute, then heated to 70°C the end of the second minute; its temperature is increased 20°C/min after a maximum temperature of 210°C is reached.

This temperature is maintained until all of the components are apparently eluted. Helium is used as a carrier gas at a flow rate of 40-50 ml/min. Hydrogen and air flows are optimized and maintained at those values.

Column B. 476 cm (15 ft) 0.25 cm i.d. (1/8 inch o.d.) stainless stee column of 70-80 mesh Anakrom ABS coated with 10% UCW-98. It is operated at room temperature for 2 minutes, at 70°C for another minute, then programmed at 10°C/min to the end of the analysis or to a temperature of 260°C which ever came first. Column flow rate is 22 ml/minute. Detector flows are the same as with the other column. Energy output of the laser rod is measured with a Quantromix Model 504 energy meter equipped with a Model 500 energy receiver.

108.3 METHOD

The pyrolysis products produced when the polymer is exposed to laser light are swept into a gas chromatograph. The pattern of the decomposition products is compared with that obtained with known polymers thus facilitating the identification of the unknown polymer.

108.4 EXPERIMENTAL PROCEDURE

Three different types of sample preparation were used. All samples were run as solids.

1. None. Small pieces of the sample were run as-is.
2. Coated. Sample pieces were coated with graphite by rubbing them with a 6B drawing pencil. Gold coating was accomplished in a vacuum evaporator.
3. Added carbon. Weighed amounts of sample and finely ground permium coke are melted and intimately mixed in a porcelain dish. In some cases the sample was ground and mixed with coke before melting and further mixing.

The polymer is laser pyrolysed using the conditions described under Apparatus.

Folmer[70,71] studied the effects of different operating conditions and methods of sample preparation on fragmentation patterns. Clear or translucent samples give reproducible results if mixed with carbon. The concentration of carbon is critical; it has a great effect on the fragmentation pattern. Some polymers were run whose pyrolysis chromatograms show great difference in pattern. Others had patterns which were quite similar. To better compare these similar patterns and the patterns arising from different operating conditions, a statistical method was devised. An attempt was made to correlate these comparisons of patterns with some of the known characteristics of the polymers.

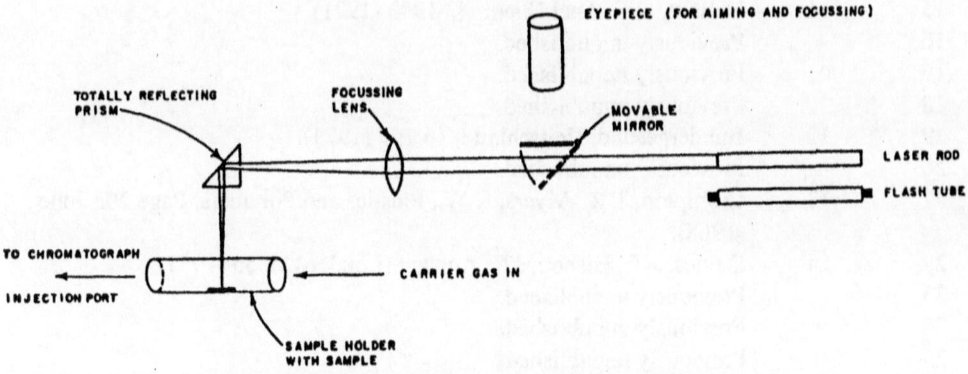

Figure 7.45 - Schematic drawing of laser pyrolysis apparatus. From Folmer et al with permission.[70,71] American Chemical Society

Method	Ref	
1	1	Jeffs, A.R. Analyst (London) **94** 249 (1969).
2		Previously unpublished.
3		Metcalf, K. Tomlinson, R. Plastics (London) **25** 319 (1960).
4	3	Hilton, C.L. Anal. Chem. **32** 383 (1960).
5		Previously unpublished.
6		Previously unpublished.
7	4	Soncek, J. Jelinkova, E. Analyst (London) **107** 623 (1982)
8	5	Ruddle, L.H. Wilson, J.R. Analyst (London) **94** 105 (1969)
9		Previously unpublished.
10	6	Dobies, R.S. J Chromatography **40** 110 (1969)
11	7	Simpson, D. Currell, B.R. Analyst (London) **96** 515 (1971).
12	8	Schabon, J.F. Fenska, L.E. Anal Chem. **52** 1411 (1980).
13	9	Crompton, T.R., Appl. Polymer Science, **6** 538 (1962).
14	10	Hilton, C.L. Rubber Age **84** 263 (1958).
15	11	Kellum, G.E. Anal Chem. **43** 1843 (1971).
16		Previously unpublished.
17		Previously unpublished.
18		Previously unpublished.
19	12	Bundergesund, Heitschlatte, **16** 155 (1973).
20		Previously unpublished.
21	13	Crompton, T.R., Myers, C.W., Plastics and Polymers, Page 205 June (1968).
22	14	Davies, J.T., Bishop, J.R. Analyst (London) **96** 55 (1971).
23		Previously unpublished.
24		Previously unpublished.
25		Previously unpublished.
26		Previously unpublished.
27		Previously unpublished.
28		Previously unpublished.
29		Previously unpublished.
30		Previously unpublished.
31		See reference 3 above.
32		Previously unpublished.
33		Previously unpublished.
34		Previously unpublished.
35		Previously unpublished.
36		Previously unpublished.
37		Previously unpublished.
38		Previously unpublished.
39,40	15	Udris, J Analyst (London) **96** 130 (1971).
41	16	Milligen, M.B., Anal. Chem. **46** 746 (1974).
42	17	Lattimer, R.P. Harris, R.E. Rhoe, C.D. Schulten, H.R. Anal. Chem. **58** 3188 (1986).
43	18	Allen, B.J. Elsea, G.M. Keller, K.P. Kinder, H.D. Anal. Chem. **49** 741 (1977).
44	19	Hoffman, E.R., Anal. Chem. **48** 445 (1976).

45	20	Brown, L. Analyst (London) **104** 1165 (1979).
46	21	Fritz, D.E. Sahil, A. Keller, H.P. Kovat, E. Anal. Chem. **51** 7 (1979)
47	22	Protivova, J. Pospisil S.J. Chromatography **88** 99 (1974).
47	23	Yoshikova, T. Kimura, K.J. Appl. Polymer Science, **15** 2513 (1971).
47	24	Colburn, A.P. Pigford, R.L. Perry, H. (Editor) Chemical Engineers Hand Book. McGraw Hill, New York, 34th edition. P538 (1950).
47	25	Edwards, G.D. Polymer Sci. Part 21 105 (1967).
47	26	Henrickson, J.G. Moore, J.C. J Polymer Sci Part A - 1,**4** 167 (1966).
48		Previously unpublished.
49	27	Frankoski, S.P. Siggia, S. Anal. Chem. **44** 507 (1972).
50	28	Betso, S.R. McLean, J.D. Anal. Chem. **48** 766 (1976).
51	29	Skelly, N.E. Husser, E.R. Anal. Chem. **50** 1959 (1978).
51	30	Husser, E.R. Stehl, R.H. Price, D.R. Delap, R.A. Anal Chem **49** 154 (1977).
52	31	Guthrie, J.L. McKinney, R.W. Anal. Chem. **49** 1676 (1977).
53	32	Crompton, T.R., Myers, L.W. European Polymer Journal **4** 355 (1968).
54		Previously unpublished.
55	33	Crompton, T.R. Buckley, D. Analyst (London) **90** 76 (1965).
55	34	Mino, G. J. Polymer Science, **22** 369 (1956).
56	35	Roy, S.S. Analyst (London) **102** 302 (1977).
57	36	Steichen, R.J. Anal. Chem. **48** 1398 (1976).
58	37	Kaduji, I.I. Rees, J.H. Analyst (London) **99** 435 (1974).
58	38	Bark, L.S. Grime, J.K. Analyst (London) **97** 911 (1972).
59		Previously unpublished.
60	39	Tsuji, K. Kouishi, K. Analyst (London) **99** 54 (1974).
61	40	Sharp, J.L. Paterson, G. Analyst (London) **105** 517 (1980).
62	41	Anderson, D.G. Isakson, K.E. Snow, D.L. Tessari, D.J. Vandeberg, J.T., Anal. Chem. **43** 894 (1971).
63	42	Campbell, D.R. Anal. Chem. **47** 1477 (1975).
64	43	Anderson, D.G. Isakson, K.E. Vandeberg, J.T. Jao, M.Y.T. Tessari, D.J. Afremow, L.C. Anal. Chem. **47** 1008 (1975).
65		Previously unpublished.
66	44	Smith, A.J. Anal. Chem. **36** 944 (1964).
67	44	Smith, A.J. Anal. Chem. **36** 944 (1964).
68	45	Battiste, D.R. Butler, J.P. Cross, J.B. McDaniel, M.P. Anal. Chem. **53** 2232 (1981).
69		Previously unpublished.
70		Previously unpublished.
71		Previously unpublished.
72		Previously unpublished.
73	46	Korenaga, T. Analyst (London) **106** 40 (1981).
74	47	Tittareli, P. Zerlia, T. Colli, A, Ferrari, G. Anal. Chem. **55** 220 (1983).
75		Previously unpublished.
76	48	Falcon, J.Z. Love, J.L. Gacta, L.J. Altenau, A.G.G. Anal. Chem. **47** 171 (1975).
77		Previously unpublished.
78		Previously unpublished.
79	49	Johnson, C.A. Leonard, M.A. Analyst (London) **111** 101 (1961).

80		Previously unpublished.
81	50	Colson, A.S. Analyst (London) **113** 791 (1963).
82		Previously unpublished.
83		Previously unpublished.
84		Previously unpublished.
85		Previously unpublished.
86		Previously unpublished.
87		Previously unpublished.
88		Previously unpublished.
89		Previously unpublished.
90	51	Krishen, A. Anal. Chem. **44** 494 (1972).
91	52	Paxton, J.R. Randall, J.C. Anal. Chem. **50** 1777 (1978).
92	52	Paxton, J.R. Randall, J.C. Anal. Chem. **50** 1777 (1978).
92	53	Saunders, J.M. Komoroski, K. Macromolecules, **10** 1214 (1977).
93	54	Anderson, D.G. Isakson, K.E. Snow D.L. Tessari, D J. Vandeberg, J.T. Anal. Chem. **43** 894 (1971).
94	55	Evans, D.L. Weaver, J.L. Mukherji, A.K. Beatty, C.L. Anal. Chem. **50** 857 (1978).
95	55	Evans, D.L. Weaver, J.L. Mukherji, A.K. Beatty, C.L. Anal. Chem. **50** 857 (1978).
96	56	Johnson, D.E. Lyeria, J.R. Horikawa, T.T. Pederson, L.A. Anal. Chem. **49** 77 (1977).
97	57	Blackwell, J.T. Anal. Chem. **48** 1883 (1976).
98		Previously unpublished.
99	58	Majors, R.E. J Chromat. Sci. **8** 338 (1970).
100	59	Rudewicz, P. Munson, B. Anal Chem. **58** 358 (1986).
100	60	Spreen R.C., University of Delaware, Newark, D.E. (1983).
101	61	Vargo, J.D. Olsen, K.L. Anal. Chem. **57** 672 (1985).
102	62	Juvet, R.S. Smith L.S. Li, K.P. Anal. Chem. **44** 49 (1972).
103	63	Anderson, D.M.W. Analyst (London) **84** 50 (1959).
104	64	Haslam, J. Squirrel, D.C.M. Analyst (London) **86** 239 (1961).
105	65	Bird, V. Plastics Technology **40** September (1963).
106	66	May, R.W. Pearson, E.T. Porter, J. Scothern, M.D. Analyst (London) **98** 364 (1973).
106	67	Voigt, J Kunststoffe **54** 2 (1964).
106	68	Voigt, J Kunststoffe **51** 18 (1961).
106	69	Voigt, J Kunststoffe **51** 314 (1961).
107		Previously unpublished.
108	70	Folmer, O.F. Anal. Chem. **43** 1057 (1971).
108	71	Folmer, O.F. Azarraga, LV. J Chromatography Sci. **7** 665 (1969).
98	72	Shakl, E. Laboratory Practice June 1964.
99	73	Scott, C.D. Johnson, W.T. Walker, V.E. Anal. Biochem **32** 182 (1969).
98	74	Van der Heide R.F. Z Lehensum. Untersuch Forsch. **124** 198 (1962).
98	75	Van der Neut, J.H. Maagdenberg, A.C. Plastics **31** 66 Jan (1966).
98	76	Frew, N.M. Johnson, C.G. Bromund, R.M. (Chapenter, B.A., Sevenants, M.R. (editors) In "Supercritical Fluid Extraction and Chromatography" ACS Symposium Series No. 366. American Chemical Society pp 208 - 228 (1988).
98	77	Waggon, H. Jehle, D. Die Nahrung **4** 495 (1965).

98	78	Braun D. Chimica Ind. **19** 77 (1965).
98	79	Wandel, M. Tengler, H. Kunststoffe, **55** 11 (1965).
98	80	Healpaap, H. Chem. Org. Tech. **35** 488 (1963).
98	81	Braun, D. Greenen, I.T. J. Chromatography **7** 56 (1962).
98	82	Knappe, E. Peteri, D.Z. Analyst Chem. **188** 184 (1961).
98	83	Lynes, A.J. Chromatography **15** 108 (1964).
98	84	Lane, E.S. J. Chromatography **18** 426 (1965).
98	85	Crump, J.B., Anal. Chem. **36** 2447 (1964).
98	86	Dallas, H.S.J. J. Chromatography **17** 267 (1965).

INDEX

Tris (dinonyl phenyl) phosphite, det of in polyalkenes, 198, 199

GPSR Compliance

The European Union's (EU) General Product Safety Regulation (GPSR) is a set of rules that requires consumer products to be safe and our obligations to ensure this.

If you have any concerns about our products, you can contact us on ProductSafety@springernature.com

In case Publisher is established outside the EU, the EU authorized representative is:

Springer Nature Customer Service Center GmbH
Europaplatz 3
69115 Heidelberg, Germany

Batch number: 10762185

Printed by Printforce, the Netherlands